Linear Algebra

LINEAR ALGEBRA

Surjeet Singh

VIKAS® PUBLISHING HOUSE PVT LTD

VIKAS® PUBLISHING HOUSE PRIVATE LIMITED
E-28, Sector-8, Noida-201301 (UP), India
Phone: +91-120-4078900 • Fax: +91-120-4078999
Registered Office: A-27, 2nd Floor, Mohan Co-operative Industrial Estate, New Delhi-110044
E-mail: helpline@vikaspublishing.com • Website: www.vikaspublishing.com

SALES AND BRANCH OFFICES OF VIKAS® AND S CHAND AND COMPANY LIMITED

Chennai	:	Ph: 044-2363 2120, chennai@schandpublishing.com
Guwahati	:	Ph: 0361-2738 811, 2735 640, guwahati@schandpublishing.com
Hyderabad	:	Ph: 040-4018 6018, hyderabad@schandpublishing.com
Jalandhar	:	Ph: 0181-4645 630, jalandhar@schandpublishing.com
Kolkata	:	Ph: 033-2335 7458, 2335 3914, kolkata@schandpublishing.com
Lucknow	:	Ph: 0522-4003 633, lucknow@schandpublishing.com
Mumbai	:	Ph: 022-2500 0297, mumbai@schandpublishing.com
Patna	:	Ph: 0612-2260 011, patna@schandpublishing.com

Linear Algebra
ISBN: 978-81259-0482-3

Product Code: V6LAL61MATH10ENAA97O

First Published 1997
Reprinted several times between 1997 and 2020
Reprint 2023

PRINTED IN INDIA
By Vikas Publishing House Private Limited, Plot 20/4, Site-IV, Industrial Area Sahibabad, Ghaziabad-201010 and Published by Vikas Publishing House Private Limited, E-28, Sector-8, Noida-201301

Preface

In the last few decades, there have been extensive developments in linear algebra. It has widespread applications in basic sciences, engineering, social sciences and many other areas of studies. Because of its widespread applications, linear algebra is becoming an integral part of training in mathematics for students in various disciplines in basic sciences, etc. This book has been written to provide introduction to this fast growing subject. Almost all textbooks on introductory linear algebra deal with the subject over the real numbers and do not contain detailed proofs of results they intend to teach. On the other hand, advanced textbooks on linear algebra, not only dismiss these topics with brief discussions, but also switch over to discussions of advanced topics in the subject over general fields. Thus, an average student is left with a gap in his training in linear algebra, with a likelihood of its being never filled in future.

The present book is intended to be a bridge between introductory and advanced textbooks on linear algebra. The present day students are made aware of the various algebraic structures such as groups, rings and fields, in their early training in mathematics. Because of this reason, this book is not confined to linear algebra over real numbers, but also covers general fields. However, in view of the needs of students in a beginning course, the real numbers are emphasised by giving almost all examples over such numbers.

The first chapter deals with matrices, vector space F^n and Gauss-Jordan method of solving a system of linear equations. These are the most basic topics in linear algebra, that a student must learn. The discussion is kept at a leisurely pace, and is supplemented with a large number of examples. The second chapter gives introduction to various concepts on vector spaces, the basis of a vector space, and the rank of a matrix. The third chapter deals with linear transformations and their role in studying matrices. The determinant of a square matrix is discussed in the fourth chapter. Symmetric groups are used in discussing various proofs of the results on determinants. Most of the concepts, discussed in these chapters, the author believes, should be taught to any student who wishes to know linear algebra. Every effort has been made to give detailed proofs and illustrations of various concepts discussed in these chapters. Depending on the level of the student, an instructor can be selective in giving proofs of the results. However, the student will find this book, a ready source, to know the detailed proofs of the results he/she was taught without proofs.

After the fourth chapter, the book deals with advanced topics in the subject. The theory of rings and modules has greatly influenced the theory of linear algebra. With this in mind, the first two sections of the fifth chapter are devoted to rings of polynomials over fields, and modules over these rings. The rest of the fifth chapter is devoted to various canonical forms of linear transformations and matrices. Most of these are deduced from the cyclic decomposition theorem. The sixth chapter deals with the concept of an inner product space, central in studying mathematical analysis. Among the many concepts and results, the spectral theorem and the sesquilinear form are discussed. The seventh chapter is devoted to bilinear forms, of interest to algebraist as well as analysisists.

The author gratefully acknowledges his gratitude to all those, whose text books on linear algebra, have been used by him in learning the subject and in writing this textbook. He is grateful to his colleagues who have helped him in writing this book, by making criticisms of the various drafts of the book. These include Dr. Hassan Al-Zaid, Dr. Ismat Beg, Dr. Bela Brindza and Dr. M. Deaconescu.

The author would like to put on record, the painstaking efforts of Mr. Muhammad Sahid Rafique, in accomplishing the task of typing this book. The author conveys his sincerest thanks to him for all his efforts, and also thanks all the others who have played a role, by giving helpful suggestions.

Surjeet Singh

List of Symbols

Symbol	Meaning	Page
Z (or Z)	set of integers	1
Q	set of rational numbers	1
R	Set of real numbers	1
C	set of complex numbers	1
$\mathbf{X}^*$	Set of nonzero members of X	1
Z_n	$\{0,1,\ldots,n-1\}$	2
$+_n$	Additional modulo n	2
$\times_n$	multiplication modulo n	2
$R[X]$	ring of polynomials over a ring R	2
N	set of natural numbers	3
Z_e	set of even integers	3
Z_o	set of odd integers	3
a^{-1}	inverse of a	3
S^n	set of ordered n-tuples over S	4
ϕ	empty set	3
$A \subset B$	subset but $A \neq B$	3
$A - B$	complement of B in A	3
A^T	transpose of a matrix A	16
A^{-1}	inverse of an invertible matrix	23
V_F	vector space over a field F	60
$\langle S \rangle$	span of a set S of vectors	67
$R[[X]]$	ring of formal power series over a ring R	72
$\deg(f(x))$	degree of a polynomial $f(X)$	73
$P_n(F)$	the vector space of polynomial over a field F, of degree $\leq n$	73
$\dim_F V$	dimension of a vector space V_F	84
$V = W_1 \oplus W_2 \oplus \ldots \oplus W_n$	direct sum of vector spaces	92
$[\alpha]_B$	co-ordinate vector of a vector α	112
ker (T)	kernal of a linear transformation T	124
$\nu(T)$	nullity of a linear transformation T	124
$\rho(T)$	rank of a linear transformation T	126

Contents

Chapter 1

Algebra and Matrices

1.1 Groups, Rings and Fields

The basic concepts of algebra, viz., groups, rings and fields, are at present, taught as an integral part of undergraduate curriculum of mathematics. These concepts are of fundamental importance for the understanding of linear algebra. The reader is presumed to be familiar with them; he may consult any of the books listed at the end of this chapter. However, for the purpose of quick reference, we give some definitions and examples.

Definition 1. Groups. A group is a set G, together with a binary operation*, satisfying the following axioms:

I. Associativity : $(a*b)*c = a*(b*c)$ for all $a, b, c \in G$.

II. Existence of identity : There exists an $e \in G$, called an identity element, such that $a*e = a = e*a$ for all $a \in G$.

III. Existence of inverses : For each $a \in G$, there exists $a' \in G$, called an inverse of a, such that $a*a' = e = a'*a$

We shall, normally, identify a group with its underlying set G, and for any $a, b \in G$, the product $a*b$ will be simply written as ab. Sometime to emphasize the binary operation *, the group will be denoted by the pair $\langle G, *\rangle$. If in a group $\langle G, *\rangle$, the binary operation * is commutative, in the sense that $a*b = b*a$ for all $a, b \in G$, the group is called an *abelian group*. Throughout **Z** (or Z), **Q**, **R** and **C** will denote the sets of integers, rational numbers, real and complex numbers, respectively. Further, among numbers, +, ., – will denote the usual operations of addition, multiplication and subtraction, respectively. Familiar examples of abelian groups are $\langle \mathbf{Z}, +\rangle$, $\langle \mathbf{Q}, +\rangle$, $\langle \mathbf{R}, +\rangle$. For any set X of numbers, X^* denotes the set of non-zero members of X. Then, $\langle \mathbf{Q}^*, \cdot\rangle$ and $\langle \mathbf{R}^*, \cdot\rangle$ are abelian groups.

Definition 2. Rings. A set R, together with two binary operations, addition (+) and multiplication (.), is called a ring, if it satisfies the following axioms :

I. $\langle R, +\rangle$ is an abelian group.

II. Associativity of multiplication :

$$(a.b).c = a.(b.c) \qquad \text{for all } a,\ b,\ c \in R.$$

III. Laws of distribution :

For any $a,\ b,\ c \in R$ $\qquad a.(b+c) = a.b + a.c$

$$(b+c).a = b.a + c.a$$

If in a ring R, the multiplication operation is commutative, then, R is called a *commutative ring*. In a ring R, the identity element, under addition, is denoted by 0, and we call it the zero element of R. For any $a \in R$, its inverse, under addition, is called the negative of a, and is denoted by $-a$. The rings $\langle \mathbf{Z},\ +,\ \cdot \rangle$, $\langle \mathbf{Q},\ +,\ \cdot \rangle$ and $\langle \mathbf{R},\ +,\ \cdot \rangle$ are commutative.

If in a ring R, there exists an element e such that $e.x = x = x.e$ for all $x \in R$, then, e is called the *identity* (or *unity*) element of R. If $R \neq \{0\}$, the identity e in R, if it exists, is non-zero. A commutative ring F, with an identity element $e \neq 0$, is called a *field* if for each $a \neq 0$ in F, there exists $a' \neq 0$ in F, called the inverse of a, such that $a.a' = e$. Thus, **Q**, **R** and **C** are fields under usual addition and multiplication operations. However, **Z** is not a field under these operations. In analogy with unity 1 in rings of numbers, unity element, in a ring R, is usually denoted by 1. Throughout the book, the unity element 1 of a ring is taken to be non-zero.

Example 2.a : Let n be any positive integer and $Z_n = \{0,\ 1,\ 2\ ,...,\ n-1\}$. For example, $Z_3 = \{0,\ 1,\ 2\}$. For any $a,\ b \in Z_n$, the least non-negative remainders, obtained or dividing $a + b$ and ab by n, are called the sum and product modulo n of a and b, respectively; they are denoted by $a +_n b$ and $a \times_n b$, respectively. For example, in Z_3, $2 +_3 1 = 0$, $2 \times_3 2 = 1$. Then, $\langle Z_n,\ +_n,\ \times_n \rangle$ is a commutative ring with identity. It is a field if n is a prime number. This gives infinitely many finite rings and fields.

Example 2.b Polynomial Rings : Let R be any ring with identity 1. Analogous to polynomials with coefficients as numbers, we have polynomials

$$f(X) = a_0 + a_1X + a_2X^2 + ... + a_nX^n \quad \text{with coefficients } a_i \in R.$$

Let $R[X]$ denote the collection of all such polynomials.

Given another, $\qquad g(X) = b_0 + b_1X + a_2X^2 + ... b_mX^m \in R[X]$,

define $\qquad f(X) + g(X) = c_0 + c_1X + ... + c_kX^k + ...,$

where $\qquad c_i = a_i + b_i \qquad$ for $i \geq 0$

Further define $\qquad f(X)g(X) = d_0 + d_1X + d_2X^2 + ...$

where
$$d_0 = a_0b_0,$$
$$d_1 = a_0b_1 + a_1b_0,$$
$$d_2 = a_0b_2 + a_1b_1 + a_2b_0,$$

so on.

These two operations make $R[X]$ a ring, called the *ring of polynomials* over R.

We end this section by introducing a few more notations to be used throughout. ϕ denotes an *empty set*. For any two sets A and B, $A \subset B$ denotes that A is a subject of B but $A \neq B$, *i.e.*, A is a proper subject of B. The Symbol $A–B$ denotes the complement of B in A, i.e., $A–B = \{x : x \in A \text{ and } x \notin B\}$. The symbols N, Z_e, Z_0 denote the sets of natural numbers (including zero), even integers and odd integers, respectively. In a group G, the inverse of an element a is denoted by a^{-1}. However, if the binary operation in G is denoted by +, the inverse of $a \in G$, is denoted by $-a$, and is called the negative of a.

Exercise 1.1

1. Which of the following sets are groups under usual addition operation in numbers?

(*i*) $\mathbf{Z}_e$ (*ii*) $\mathbf{Z}_0$ (*iii*) $\left\{\frac{a}{b} : a, b \in \mathbf{Z}, b \text{ odd}\right\}$

2. Which of the following sets of numbers are groups under multiplication operation?

(*i*) $\mathbf{Z}^*$ (*ii*) $\mathbf{R}^*$ (*iii*) $\left\{\frac{a}{b} : a, b \text{ odd integers}\right\}$

(*iv*) $\left\{a + b\sqrt{2} : a, b \in \mathbf{Q} \text{ with } a \neq 0 \text{ or } b \neq 0\right\}$

3. A non-empty subset H of a group G is called a *subgroup* of G, if H itself is a group under the binary operation in G. Answer the following

(i) Is $\mathbf{Z}_e$ a subgroup a $\langle \mathbf{Z}, + \rangle$?

(*ii*) Is $\mathbf{Q}^*$ a subgroup of $\langle \mathbf{R}^*, \cdot \rangle$?

(*iii*) Is $\mathbf{Q}^*$ a subgroup of $\langle \mathbf{Q}, + \rangle$?

(*iv*) Is $\mathbf{Z}_3$ a subgroup of $\langle \mathbf{Z}_6, +_6 \rangle$?

4. Verify that each of the following is a ring under usual addition and multiplication operations.

(*i*) $Z[i] = \left\{a + bi : a, b \in Z, i = \sqrt{-1}\right\}$

(*ii*) $\mathbf{Q}\left[\sqrt{2}\right] = \left\{a + b\sqrt{2} : a, b \in Q\right\}$

(*iii*) $F = \left\{a + b\sqrt{2} + c\sqrt{3} + d\sqrt{6} : a, b, c, d \in \mathbf{Q}\right\}$

(*iv*) $K = \left\{a + b\omega + c\omega^2 : a, b, c, \in \mathbf{Q}\right\}$,

where $\omega = -\frac{1}{2} + \frac{\sqrt{-3}}{2}$. Prove further that the rings at (*ii*), (*iii*) and (*iv*) are fields.

5. Prove that any field is an integral domain, but the converse is not true (Definition : An *integral domain* is a commutative ring R with unity $1 \neq 0$ such that for any two non-zero elements a, b in R, $ab \neq 0$).

6. A subset of a ring R is called an *ideal* of R, if A is a subgroup of R under addition and $ra \in A$, $ar \in A$ for every $a \in A$, $r \in R$. Prove the following

(*i*) Z_e = set of all even integers is an ideal of the ring Z of integers.

(*ii*) In a field F, $\{0\}$ and F are the only ideals.

(*iii*) $J = \{a+bi : a \in Z,\ b \in Z_e\}$ is not an ideal of the ring $Z[i]$, $i = \sqrt{-1}$.

7. Let A and B be two ideals of a commutative ring R. Prove that $A \cap B$ is an ideal of R. Also, prove that $A+B = \{a+b : a \in A,\ b \in B\}$ is an ideal of R, and $A \cup B \subseteq A+B$.

8. Let A be any ideal of a ring R. For any $x \in R$, let $x+A = \{x+a : a \in A\}$. Prove the following :

(*i*) For $x, y \in R$, $x+A = y+A$ iff $x-y \in A$

(*ii*) $\frac{R}{A} = \{x+A : x \in R\}$ is a ring in which

$$(x+A)+(y+A) = (x+y)+A$$

and

$$(x+A)(y+A) = xy+A.$$

1.2 Matrices

Let S be any ring (in particular, S could be $\mathbf{R}$, the set of real numbers) and let n be a positive integer. Then, S^n denotes the set of all ordered n-tuples $(a_1, a_2, \ldots, a_n)$ with $a_i \in S$. Thus, $(1, 3) \in \mathbf{R}^2$, but $(1, i) \notin \mathbf{R}^2$. However, $(1, i) \in \mathrm{C}^2$ consider any

$$\alpha = (a_1, a_2, \ldots, a_n)$$

$$\beta = (b_1, b_2, \ldots, b_n)$$

in S^n and any $a \in S$.

We define

$$\alpha+\beta = (a_1+b_1,\ a_2+b_2, \ldots,\ a_n+b_n)$$

$$a\alpha = (aa_1,\ aa_2, \ldots,\ aa_n).$$

Clearly $\alpha+\beta \in S^n$, $a\alpha \in S^n$. So, we get binary operation + in S^n and a product $a\alpha$ in S^n. For any $\alpha = (a_1, a_2, \ldots, a_n)$, a_i is called the ith *component* of α, for $1 \le i \le n$. Thus, in (1, 0, 3), the second component is 0. The following theorem gives some basic properties of these two operations in S^n.

Theorem 1. Let S be any ring with identity 1 and n be any positive integer. Then :

I. S^n is an abelian group under addition

II. For any $a \in S$, $\alpha \in S^n$, $a\alpha \in S^n$

III. For any α, $\beta \in S^n$ and a, $b \in S$, the following hold :

(*i*) $a(\alpha+\beta) = a\alpha + a\beta$

(*ii*) $(a+b)\alpha = a\alpha + b\alpha$

(*iii*) $(ab)\alpha = a(b\alpha)$

(*iv*) $1\alpha = \alpha$

Proof. Let $\alpha = (a_1, a_2, \ldots, a_n)$

$\beta = (b_1, b_2, \ldots, b_n),$

$\gamma = (c_1, c_2, \ldots, c_n) \in S^n$

and $a, b \in S.$

I. S^n is an abelian group under addition.

(*i*) *Commutativety* :

$$\alpha + \beta = (a_1 + b_1, a_2 + b_2, \ldots, a_n + b_n),$$

$$\beta + \alpha = (b_1 + a_1, b_2 + a_2, \ldots, b_n + a_n).$$

As $a_i, b_i \in S, a_i + b_i = b_i + a_i.$

Consequently, $\alpha + \beta = \beta + \alpha.$

(*ii*) *Associativity* : In $(\alpha + \beta) + \gamma$ the *i*-th component is

$$(a_i + b_i) + c_i = a_i + (b_i + c_i),$$

the ith component of $\alpha + (\beta + \gamma)$.

Consequently, $(\alpha + \beta) + \gamma = \alpha + (\beta + \gamma).$

(*iii*) *Existence of identity* :

$$\mathbf{0} = (0, 0, \ldots, 0) \in S^n$$

and $\alpha + \mathbf{0} = \alpha.$

(*iv*) *Existence of inverses* :

For each $\alpha = (a_1, a_2, \ldots, a_n) \in S^n,$
its inverse under addition is

$$-\alpha = (-a_1, -a_2, \ldots, -a_n),$$

as $\alpha + (-\alpha) = \mathbf{0}.$

II. (*i*)
$$\begin{aligned} a(\alpha + \beta) &= a(a_1 + b_1, a_2 + b_2, \ldots, a_n + b_n) \\ &= \{a(a_1 + b_1), a(a_2 + b_2), \ldots, a(a_n + b_n)\} \\ &= (aa_1 + ab_1, aa_2 + ab_2, \ldots, aa_n + ab_n) \\ &= a\alpha + a\beta. \end{aligned}$$

(*ii*) Similarly, one gets $(a + b)\alpha = a\alpha + b\alpha$

(*iii*)
$$(ab)\alpha = \{(ab)a_1, (ab)a_2, \ldots, (ab)a_n\}.$$

However, in S $(ab)a_i = a(ba_i).$

Consequently, $(ab)\alpha = a(b\alpha).$

(*iv*)
$$\begin{aligned} 1\alpha &= (1a_1, 1a_2, \ldots, 1a_n) \\ &= (a_1, a_2, \ldots, a_n) \\ &= \alpha. \end{aligned}$$

The most interesting case of S^n is when $S = F$, a field, more so when $S = \mathbf{R}$ or $\mathbf{C}$. This F^n is an example of a vector space, the concept to be discussed in later chapters.

Definition 2. Let G be a non-empty set. For any two positive integers m and n, an $m \times n$*-matrix* over G is an array A of mn-elements of G (not necessarily distinct) arranged in m rows and n columns :

$$A = \begin{bmatrix} a_{11} & a_{12} & a_{13} & \cdots & a_{1n} \\ a_{21} & a_{22} & a_{23} & \cdots & a_{2n} \\ \vdots & & & & \vdots \\ a_{m1} & a_{m2} & a_{m3} & \cdots & a_{mn} \end{bmatrix}.$$

For $1 \le i \le m$, the ith *row* of A is the ordered n-tuple $(a_{i1}, a_{i2}, \ldots, a_{in})$.

For $1 \le i \le n$, the jth *column* of A is the ordered m-tuple

$$\begin{bmatrix} a_{1j} \\ a_{2j} \\ \vdots \\ a_{mj} \end{bmatrix},$$

written vertically down-ward.

The element a_{ij}, that is at the intersection of the ith row and j^{th} column of A, is called $(i, j)^{\text{th}}$ entry (or element) of A.

Instead of writing the matrix A in full, we shall use abbreviated notation $[a_{ij}]_{m \times n}$ (or simply $[a_{ij}]$). Thus, $B = [b_{ij}]_{2\times 3}$ means the 2×3-matrix whose $(i, j)^{\text{th}}$ entry is b_{ij}.

So,
$$B = \begin{bmatrix} b_{11} & b_{12} & b_{13} \\ b_{21} & b_{22} & b_{23} \end{bmatrix}.$$

A matrix in which the number of rows equals the number of columns is called a *square matrix*.

Any matrix over $\mathbf{R}$ is called a real matrix and any matrix over $\mathbf{C}$ is called a complex matrix. We are mainly interested in studying matrices and other concepts over a field F. All the examples, that will be given, will be over $\mathbf{R}$, unless, otherwise, specified.

Example 2.a. $A = \begin{bmatrix} 1 & 0 & -1 & 3 \\ 2 & 1 & 5 & 7 \\ 0 & 3 & 5 & 1 \end{bmatrix}$ is a 3×4-matrix over Z.

Its third row is (0 3 5 1) and its second column is $\begin{bmatrix} 0 \\ 1 \\ 3 \end{bmatrix}$.

The entry at the intersection of the third row and , the second column is 3, i.e., (3, 2)th entry is 3. Further, (2, 3)th entry is 5. If we write this $A = \left[a_{ij}\right]_{3\times 4}$, we can say $a_{11} = 1$, $a_{34} = 1$.

A is not a square matrix, $B = \begin{bmatrix} 1 & -1 \\ 1 & 2 \end{bmatrix}$

is a square matrix over Z.

However, $Z \subseteq \mathbf{R}$, we can say that A and B are real matrices.

Definition 3. Let $A = \left[a_{ij}\right]$ be a (square) $n \times n$-matrix over a ring S with unity. The n-tuple $(a_{11}, a_{22}, \ldots, a_{nn})$ is called the *main diagonal* (or simply, the diagonal) of A. Each entry a_{ii} is said to be on the main diagonal of A. If all the entries of A, off the main diagonal, are zeros, then, A is called a *diagonal matrix.* A diagonal matrix, whose all entries on the main diagonal are equal, is called a *scalar matrix.* A scalar matrix over S, whose every entry on the diagonal is 1, is called an *identity* matrix.

Example 3.a The diagonal of $A = \begin{bmatrix} 0 & 1 & -1 \\ 0 & 3 & 4 \\ 4 & 1 & 1 \end{bmatrix}$ is (0 3 1). A is not a diagonal matrix.

$\begin{bmatrix} 0 & 0 & 0 \\ 0 & 3 & 0 \\ 0 & 0 & 1 \end{bmatrix}$ is a diagonal matrix, but not a scalar matrix, $\begin{bmatrix} 2 & 0 & 0 \\ 0 & 2 & 0 \\ 0 & 0 & 2 \end{bmatrix}$ is a scalar matrix.

Finally $\begin{bmatrix} 1 & 0 \\ 0 & 1 \end{bmatrix}$ is a 2×2-identity matrix.

Hence-forth, the matrices, etc., will be over a ring S with identity 1; in particular, S could be **R**.

Definition 4. Let $A = \left[a_{ij}\right]_{n\times n}$ be a matrix over a ring S. If $a_{ij} = 0$ whenever $i > j$, then, A is called an *upper triangular matrix*.

In other words, a square matrix, in which all its, entries below the diagonal of A are zeroes, is called an upper triangular matrix.. Similarly, a square matrix, whose all entries above the diagonal are zeroes, is called a *lower triangular matrix.* Clearly, a diagonal matrix is upper and lower triangular. Thus, $\begin{bmatrix} 1 & 0 & 3 \\ 0 & 2 & 5 \\ 0 & 0 & -1 \end{bmatrix}$ is an upper triangular matrix, $\begin{bmatrix} -1 & 0 \\ 2 & 1 \end{bmatrix}$ is a lower triangular matrix.

Definition 5. Addition Operation. Let S be any ring with identity 1. For any positive integers m and n, let $S^{m\times n}$ denote the set of all $m \times n$-matrices over S. Let $A = \left[a_{ij}\right]$

and $B=\left[b_{ij}\right]\in S^{m\times n}$. Then, the *sum* $A + B$ is $m\times n$-matrix $C=\left[c_{ij}\right]$ such that $c_{ij}=a_{ij}+b_{ij}$, i.e., in $A+B$, (i, j)th entry is the sum of the (i, j)th entry of A and the (i, j)th entry of B, taken in S. For example, if

$$A=\begin{bmatrix}-1 & 2\\ 0 & 3\end{bmatrix},$$

and

$$B=\begin{bmatrix}2 & -1\\ 1 & 3\end{bmatrix}$$

are in $\mathbf{R}^{2\times 2}$, then, $$A+B=\begin{bmatrix}-1+2 & 2+(-1)\\ 0+1 & 3+3\end{bmatrix}=\begin{bmatrix}1 & 1\\ 1 & 6\end{bmatrix}.$$

For $$A_1=\begin{bmatrix}1 & 0 & 1\\ 1 & 1 & 0\end{bmatrix},$$

and $$B_1=\begin{bmatrix}0 & 1 & 1\\ 0 & 0 & 1\end{bmatrix}$$ in $\mathbf{Z}_2^{2\times 3}$,

$$A_1+B_1=\begin{bmatrix}1 & 1 & 0\\ 1 & 1 & 1\end{bmatrix},$$ as $1+1=0$ in $\mathbf{Z}_2$.

Theorem 6. For any ring S, $S^{m\times n}$ is an abelian group under matrix addition.

Proof. Let $A=\left[a_{ij}\right]$, $B=\left[b_{ij}\right]$ and $C=\left[c_{ij}\right]$ be any three members of $S^{m\times n}$. Observe that $(S, +)$ is an abelian group.

1. *Closure* : $A+B=\left[a_{ij}+b_{ij}\right]$, being an $m\times n$ matrix over S, is in $S^{m\times n}$.
2. *Commutativity* : The (i, j)th entries of $A+B$ and $B+A$ are

$$a_{ij}+b_{ij}$$

and $$b_{ij}+a_{ij},$$

respectively.

However in S, $$a_{ij}+b_{ij}=b_{ij}+a_{ij}.$$

This gives $$A+B=B+A.$$

3. *Associativity* : The (i, j)th entry of $(A+B)+C$ is

$$(a_{ij}+b_{ij})+c_{ij},$$

and the (i, j)th entry of $A+(B+C)$ is

$$a_{ij}+(b_{ij}+c_{ij}).$$

In S, $$(a_{ij}+b_{ij})+c_{ij}=a_{ij}+(b_{ij}+c_{ij}).$$

Consequently, $$(A+B)+C=A+(B+C).$$

4. *Existence of identity* : Consider the $m \times n$ matrix $0_{m\times n} = [\omega_{ij}]$ such that every $\omega_{ij} = 0$.

Then,
$$\begin{aligned} A + 0_{m\times n} &= [a_{ij} + \omega_{ij}] \\ &= [a_{ij} + 0] \\ &= A. \end{aligned}$$

Thus, $0_{m\times n}$ is the identity under addition in $S^{m\times n}$.

5. *Existence of negatives* : For $A = [a_{ij}]$,

define $$A' = [-a_{ij}] \in S^{m\times n}.$$

Then, $$A + A' = 0_{m\times n}.$$

Hence, $$A' = -A \in S^{m\times n}.$$

This proves the result.

Let us call a matrix with only one row (one column) a *row* (*column*) *matrix*. Thus, $[1 \quad 0 \quad 1]$ is a row-matrix and $\begin{bmatrix} -1 \\ 1 \end{bmatrix}$ is a column matrix.

6. *Matrix multiplication* : Let n be a positive integer. Consider a row matrix
$$A = [a_1, a_2, \ldots, a_n]$$
and a column matrix
$$B = \begin{bmatrix} b_1 \\ b_2 \\ \vdots \\ b_n \end{bmatrix},$$
both having n entries or components.

Then, the product AB is the 1×1-matrix $[a_1b_1 + a_2b_2 + \ldots + a_nb_n]$.

For example,
$$\begin{aligned} [1 \quad 2 \quad -3]\begin{bmatrix} 2 \\ 4 \\ 1 \end{bmatrix} &= [1\times 2 + 2\times 4 + (-3)\times 1] \\ &= [7] \end{aligned}$$

This type of multiplication has practical significance. For example, let a shopkeeper have 100 radios, 20 T.V sets and 10 VCR sets. Let him represent this stock by the row matrix $A = [100 \quad 20 \quad 10]$. Let the respective cost price of each of these three items be 15, 150 and 200 dollars, respectively. These cost prices can be represented by the column matrix
$$B = \begin{bmatrix} 15 \\ 150 \\ 200 \end{bmatrix}.$$

Then, $$AB = [100 \times 15 + 20 \times 150 + 10 \times 200]$$
$$= [6500]$$

is the total value of his stock.

We now define the product of two matrices. Let $A = [a_{ij}]$ be an $m \times p$-matrix and $B = [b_{ij}]$ be a $p \times n$-matrix over a ring S. Then, the *product* AB is the $m \times n$-matrix $[c_{ij}]$ such that

$$c_{ij} = a_{i1}b_{1j} + a_{i2}b_{2j} + \ldots + a_{ip}b_{pj}.$$

Observe that $$c_{ij} = A_i B'_j$$

where A_i is the ith row of A and B'_j is the jth column of B.

The ith row of AB is

$$[c_{i1}, c_{i2}, \ldots, c_{in}] = [A_i B'_1, A_i B'_2, \ldots, A_i B'_n]$$
$$A_i B.$$

Similarly, the jth column of AB is, AB'_j. If the number of columns of a matrix A is not equal to the number of rows of a matrix B, the product AB is not defined

Example 7.a. Let $$A = \begin{bmatrix} 1 & 0 & -1 \\ 2 & 3 & 1 \end{bmatrix},$$

and $$B = \begin{bmatrix} 0 & 1 & 0 \\ -1 & 0 & 0 \\ 1 & 2 & 5 \end{bmatrix}.$$

Here, the number of columns of A is equal to the number of rows of B. So, AB is defined. As A is a 2×3 and B is a 3×3-matrix, AB is a 2×3-matrix.

$$AB = \begin{bmatrix} c_{11} & c_{12} & c_{13} \\ c_{21} & c_{22} & c_{23} \end{bmatrix}$$

$$c_{11} = A_1 B'_1$$

$$= [1 \quad 0 \quad -1] \begin{bmatrix} 0 \\ -1 \\ 1 \end{bmatrix}$$

$$= 1 \times 0 + 0 \times (-1) + (-1) \times 1$$

$$= -1.$$

$$c_{12} = A_1 B'_2$$

$$= [1 \quad 0 \quad -1] \begin{bmatrix} 1 \\ 0 \\ 2 \end{bmatrix}$$

$$= 1\times 1+0\times 0+(-1)\times 2$$

$$= -1.$$

$$c_{13} = A_1 B_3'$$

$$= \begin{bmatrix} 1 & 0 & -1 \end{bmatrix} \begin{bmatrix} 0 \\ 0 \\ 5 \end{bmatrix}$$

$$= 1\times 0+0\times 0+(-1)\times 5$$

$$= -5$$

$$c_{21} = A_2 B_1'$$

$$= \begin{bmatrix} 2 & 3 & 1 \end{bmatrix} \begin{bmatrix} 0 \\ -1 \\ 1 \end{bmatrix}$$

$$= 2\times 0+3\times(-1)+1\times 1$$

$$= -2$$

$$c_{22} = \begin{bmatrix} 2 & 3 & 1 \end{bmatrix} \begin{bmatrix} 1 \\ 0 \\ 2 \end{bmatrix}$$

$$= 4$$

$$c_{23} = \begin{bmatrix} 2 & 3 & 1 \end{bmatrix} \begin{bmatrix} 0 \\ 0 \\ 5 \end{bmatrix}$$

$$= 5.$$

So that $$AB = \begin{bmatrix} -1 & -1 & -5 \\ -2 & 4 & 5 \end{bmatrix}, -5$$

BA is not defined as the number of columns of B is not equal to the number of rows of A.

Unlike numbers, even if both AB and BA may be defined, they need not be equal.

Example 7.b. Let $$A = \begin{bmatrix} 1 & 1 \\ 0 & 1 \end{bmatrix},$$

and $$B = \begin{bmatrix} 1 & 2 \\ 0 & 3 \end{bmatrix}.$$

Then, $$AB = \begin{bmatrix} 1 & 5 \\ 0 & 3 \end{bmatrix},$$

and $$BA = \begin{bmatrix} 1 & 3 \\ 0 & 3 \end{bmatrix}.$$

So that $AB \neq BA$.

Unlike numbers, the product of two non-zero matrices may be zero.

Example 7.c. Let $$A = \begin{bmatrix} 1 & 1 \\ 0 & 0 \end{bmatrix},$$

and $$B = \begin{bmatrix} 0 & 1 \\ 0 & -1 \end{bmatrix}.$$

Then, $$A \neq 0 \neq B.$$

However, $$AB = \begin{bmatrix} 0 & 0 \\ 0 & 0 \end{bmatrix} = BA.$$

Before we discuss some basic properties of matrix multiplication, we briefly discuss the *summation notation*. Suppose there is an expression, involving sum of many terms, and each of whose term is indexed by some symbol. Collect these symbols into a set I, called the *index set*.

For example, for $$s = a_1 + a_3 + a_7 + a_8 + a_{10} + a_{12} + a_{13}$$
$$I = \{1, 3, 7, 8, 10, 12, 13\}.$$

For this expression, we simply write
$$s = \sum_{i \in I} a_i .$$

Instead of i any other symbol can be used.

In fact, $$\sum_{i \in I} a_i = \sum_{j \in I} a_j = \sum_{\alpha \in I} a_\alpha .$$

Consider an expression $s' = b_5 + b_6 + b_7 + b_8 + b_9 + b_{10}$.

Here, the index set is $I' = \{5, 6, 7, 8, 9, 10\}$ which consists of all integers from 5 to 10. In such a situation, we write

$$s' = \sum_{i \in I'} b_i = \sum_{i=5}^{10} b_i = \sum_{j=5}^{10} b_j, \text{ etc.}$$

Thus, $$\sum_{i=1}^{5} c_i = c_1 + c_2 + c_3 + c_4 + c_5 .$$

In general, $$\sum_{i=1}^{n} d_i = \sum_{j=1}^{n} d_j = d_1 + d_2 + \ldots + d_n .$$

Keeping this in view, in the definition of the product $AB = [c_{ij}]$, we have

$$c_{ij} = a_{i1}\,b_{1j} + a_{i2}\,b_{2j} + \ldots + a_{ip}\,b_{pj}$$
$$= \sum_{k=1}^{p} a_{ik} b_{kj}.$$

Sometimes, we need more than one indexing set. For example, in

$$\sum_{i=1}^{3}\sum_{j=1}^{2} d_{ij},$$

i takes values 1, 2, 3. For each value of i, j takes values 1, 2.

So that
$$\sum_{i=1}^{3}\sum_{j=1}^{2} d_{ij} = \sum_{i=1}^{3}\left[\sum_{j=1}^{2} d_{ij}\right]$$
$$= d_{11} + d_{12} + d_{21} + d_{22} + d_{31} + d_{32}$$
$$= \sum_{j=1}^{2}\left[\sum_{i=1}^{3} d_{ij}\right].$$

In general,
$$\sum_{i=1}^{n}\left[\sum_{j=1}^{m} d_{ij}\right] = \sum_{j=1}^{m}\left[\sum_{i=1}^{n} d_{ij}\right].$$

Similar to the summation notation, we have the *product notation*.

Thus,
$$\prod_{i=2}^{5} a_i = a_2 a_3 a_4 a_5$$
$$\prod_{i\in I} b_i,$$

where $I = \{1, 5, 7, 9\}$ means $b_1 b_5 b_7 b_9$.

In the same way,
$$\prod_{j=1}^{3}\prod_{i=1}^{2} c_{ij} = c_{11}\,c_{21}\,c_{12}\,c_{22}\,c_{13}\,c_{23}, \text{etc.}$$

We now discuss same basic properties of matrix multiplication.

Theorem 8. Let A be an $m \times p$-matrix, B a $p \times q$-matrix and C a $q \times n$-matrix, over a ring S. Then, $(AB)C$ and $A(BC)$ both are defined and $(AB)C = A(BC)$.

Proof. Let $A = [a_{ij}]$, $B = [b_{ij}]$, and $C = [c_{ij}]$. As A is $m \times p$ and B is $p \times q$, by definition, AB is defined and it is an $m \times q$-matrix. Consequently, as C is $q \times n$, $(AB)C$ is an $m \times n$-matrix. Similarly, $A(BC)$ is defined and it is also an $m \times n$-matrix. To prove $(AB)C = A(BC)$, we need to prove that both the matrices have the same corresponding entries.

Now
$$AB = [u_{it}]_{m\times q}$$

with $$u_{it} = \sum_{s=1}^{p} a_{is}\, b_{st}.$$

Then, $$(AB)C = \left[v_{ij}\right]_{m\times n}$$

with $$v_{ij} = \sum_{t=1}^{q} u_{it}\, c_{tj} = \sum_{t=1}^{q}\left[\sum_{s=1}^{p} a_{is}\, b_{st}\right] c_{tj}. \qquad \text{... (1)}$$

Again, $$BC = \left[\omega_{sj}\right]_{p\times n}$$

with $$\omega_{sj} = \sum_{t=1}^{q} b_{st}\, c_{tj}.$$

So that $$A(BC) = \left[x_{ij}\right]_{m\times n}$$

with $$x_{ij} = \sum_{s=1}^{p} a_{is}\, \omega_{sj}$$

$$= \sum_{s=1}^{p} a_{is}\left[\sum_{t=1}^{q} b_{st}\, c_{tj}\right]$$

$$= \sum_{t=1}^{q}\left[\sum_{s=1}^{p} a_{is}\, b_{st}\right] c_{tj} \qquad \text{... (2)}$$

(1) and (2) give $$v_{ij} = x_{ij}.$$

Hence, $$(AB)C = A(BC).$$

For any positive integer *n*, let *I* denote the $n\times n$-identity matrix. $I_n = \left[\delta_{ij}\right]_{n\times n}$ such that $\delta_{ij} = 1$ if $i = j$ and $\delta_{ij} = 0$ for $i \neq j$; the function δ_{ij} is known as *Kronecker delta function*.

Theorem 9. For any $m\times n$-matrix A over a ring S, $I_m A = A = AI_n$.

Proof. By definition $$I_m A = \left[b_{ij}\right]_{m\times n}$$

such that $$b_{ij} = \sum_{s=1}^{m} \delta_{is}\, a_{sj}.$$

For $s \neq i$, $$\delta_{is} = 0.$$

So $$b_{ij} = \delta_{ii}\, a_{ij} = a_{ij}, \qquad \text{as } \delta_{ii} = 1.$$

Hence, $$I_m A = A.$$

Similarly, $$AI_n = A.$$

Theorem 10. (Distributive laws)

(*i*) Let A be an $m \times p$-matrix, and B and C be two $p \times n$-matrices, over a ring S. Then,

$$A(B+C) = AB + AC$$

(*ii*) Let B and C be two $m \times p$-matrices and A be a $p \times n$-matrix over a ring S.

Then, $$(B+C)A = BA + CA.$$

Proof. (*i*). Let $$A = \left[a_{ij}\right], B = \left[b_{ij}\right],$$

and $$C = \left[c_{ij}\right].$$

Then, $$B + C = \left[b_{ij} + c_{ij}\right]_{p \times m}.$$

Consequently, $$A(B+C) = \left[u_{ij}\right]$$

with $$u_{ij} = \sum_{s=1}^{p} a_{is}\left[b_{sj} + c_{sj}\right]$$

$$= \sum_{s=1}^{p} a_{is}\, b_{sj} + \sum_{s=1}^{p} a_{is}\, c_{sj}$$

which is the sum of the $(i, j)^{\text{th}}$ entries of AB and AC.

Consequently, $$A(B+C) = AB + AC.$$

Similarly, (*ii*) follows.

Theorem 11. For any ring S, and any positive integer n, $S^{n\times n}$ is a ring under matrix addition and multiplication operation. I is the identity in $S^{n\times n}$.

Proof. By Theorem 6, $S^{n\times n}$ is an abelian group under matrix addition. For any $A, B \in S^{n\times n}$ AB, being an $n \times n$-matrix, belongs to $S^{n\times n}$. So, matrix multiplication is a binary operation in $S^{n\times n}$. The multiplication is associative and it distributes over matrix addition (Theorem 8 and 10). Hence, $S^{n\times n}$ is a ring. By Theorem 9, I_n is the identity in $S^{n\times n}$.

Definition 12. For any $A = \left[a_{ij}\right] \in S^{m\times n}$ and $c \in S$, cA is the $m \times n$-matrix, whose $(i, j)^{\text{th}}$ entry is ca_{ij}, i.e., $cA = \left[ca_{ij}\right]_{m\times n}$; cA is called a *scalar multiple* of A.

This operation is analogous to that of multiplication of n-tuple, over S, by a member of S. If S is commutative, then for any $A, B \in S^{n\times n}$, $c \in S$, it can be easily seen that $c(AB) = (cA)B = A(cB)$. On similar lines as in Theorem 1, we have :

Theorem 13. Let S be any ring, and n and m be any two positive integers. Then,

(*I*). $S^{n\times m}$ is an abelian group under addition.

(*II*). For any $c \in S,\ A \in S^{n\times m},\ cA \in S^{n\times m}$

(*III*). For any $a,\ b \in S,\ A,\ B \in S^{n\times m}$,

(*i*) $a(A+B) = aA + aB$

(*ii*) $(a+b)A = aA + bA$

(*iii*) $(ab)A = a(bA)$

(*iv*) $1A = A$

(*IV*) If S is commutative, then, $c(AB) = (cA)B = A(cB)$, for any $A,\ B \in S^{n\times n}$ and $c \in S$.

Definition 14. For any $m \times n$ matrix, $A = \left[a_{ij}\right]$, its transpose A^T is the $n \times m$ matrix $\left[b_{ij}\right]$ such that $b_{ij} = a_{ji}$.

In order words, A^T is an $n \times m$-matrix, whose $(i, j)^{\text{th}}$ entry is the $(j, i)^{\text{th}}$ entry of A.

For $A = \left[a_{ij}\right]$, we usually write

$$A^T = \left[a_{ij}^T\right],$$

where

$$a_{ij}^T = a_{ji}$$

The ith row of A^T is $\left[a_{i1}^T\ a_{i2}^T \ldots a_{im}^T\right] = \left[a_{1i}\ a_{2i} \ldots a_{mi}\right]$

the transpose of the ith column of A.

Further, $a_{ii}^T = a_{ii}$.

Example 14.a. Let $A = \begin{bmatrix} 1 & 1 & 3 \\ 0 & -1 & 2 \end{bmatrix}$

As A is a 2×3-matrix, A^T is the 3×2-matrix $\begin{bmatrix} 1 & 0 \\ 1 & -1 \\ 3 & 2 \end{bmatrix}$.

Observe that

$$3A = \begin{bmatrix} 3 & 3 & 3 \\ 0 & -3 & 6 \end{bmatrix},$$

$$(3A)^T = \begin{bmatrix} 3 & 0 \\ 0 & -3 \\ 9 & 6 \end{bmatrix} = 3A^T$$

For

$$B = \begin{bmatrix} 3 & 0 & 0 \\ 0 & 1 & 0 \\ 0 & 0 & -1 \end{bmatrix},$$

$$B^T = B.$$

For the upper triangular matrix, $C = \begin{bmatrix} 1 & 2 \\ 0 & 1 \end{bmatrix}$.

$$C^T = \begin{bmatrix} 1 & 0 \\ 2 & 1 \end{bmatrix}$$

is a lower triangular matrix.

The following is immediately clear from the definition of transpose :

Lemma 15.

1. The transpose of a zero matrix is a zero matrix.
2. The transpose of a diagonal matrix A equals A.
3. The transpose of an upper triangular matrix is a lower triangular matrix.
4. The transpose of a column (row) matrix is a row (column) matrix.

Theorem 16. Let A and B be two $m \times n$-matrices over a ring S. Then,

1. $(A+B)^T = A^T + B^T$.
2. $(A^T)^T = A$.
3. For any $c \in S$, $(cA)^T = cA^T$.

Proof. Let $A = [a_{ij}]$, $B = [b_{ij}]$. Then, as $A + B$ is an $m \times n$-matrix, $(A+B)^T$ is an $n \times m$ matrix, whose (i, j)th entry is the (j, i)th entry $a_{ji} + b_{ji}$ of $A + B$; this is the same as the sum of the (i, j)th entries of A^T and B^T.

Consequently, $$(A+B)^T = A^T + B^T.$$

Again, $$A^T = \left[a_{ij}^T\right]_{n\times m},$$

gives $$(A^T)^T = \left[a_{ij}^{TT}\right]_{m\times n}$$

with $$a_{ij}^{TT} = a_{ji}^T = a_{ij},$$

so that $$(A^T)^T = A.$$

Finally, $$cA = \left[ca_{ij}\right]_{m\times n},$$

gives $$(cA)^T = \left[b_{ij}\right]_{n\times m}$$

with $$b_{ij} = ca_{ji} = ca_{ij}^T.$$

Hence, $$(cA)^T = cA^T.$$

Theorem 17. For any $m \times p$ matrix A and any $p \times n$-matrix B, over a commutative ring S, $(AB)^T = B^T A^T$.

Proof. Let $$A = [a_{ij}],\ B = [b_{ij}].$$

Then, $$AB = \left[c_{ij}\right]_{m\times n}$$

with $$c_{ij} = \sum_{k=1}^{p} a_{ik}\, b_{kj}.$$

Now, $$B^T A^T = \left[d_{ij}\right]_{n\times m}$$

with $$d_{ij} = \sum_{k=1}^{p} b_{ik}^T\, a_{kj}^T = \sum_{k=1}^{p} b_{ki}\, a_{jk}.$$

As S is commutative, $$b_{ki}\, a_{jk} = a_{jk}\, b_{ki}$$

so that $$d_{ij} = \sum_{k=1}^{p} a_{jk}\, b_{ki} = c_{ji} = c_{ij}^T.$$

Hence, $$(AB)^T = B^T A^T.$$

For any matrices $A_1, A_2, ..., A_n$, over a commutative ring S, If $A_1 A_2 ... A_n$ is defined, then by introduction on n, and by using the theorem, given above, it follows that

$$(A_1 A_2 ... A_n)^T = A_n^T A_{n-1}^T ... A_1^T.$$

It is obvious from the definition of a transpose that, if a matrix A is not a square matrix, then $A^T \neq A$. Even if A is a square matrix, A^T need not equal A.

For example, for $$A = \begin{bmatrix} 1 & 1 \\ 0 & 1 \end{bmatrix}.$$

$$A^T = \begin{bmatrix} 1 & 0 \\ 1 & 1 \end{bmatrix} \neq A.$$

However, for $$B = \begin{bmatrix} 1 & 2 \\ 2 & 0 \end{bmatrix},$$

$$B^T = B.$$

Another situation is illustrated by $D = \begin{bmatrix} 0 & 2 \\ -2 & 0 \end{bmatrix}$.

Here, $$D^T = \begin{bmatrix} 0 & -2 \\ 2 & 0 \end{bmatrix} = -D.$$

These examples lead to the following definition :

Definition 18. A matrix A over a ring S is called a *symmetric matrix* if $A^T = A$. A matrix A, over a ring S, is called a *skew-symmetric matrix* if $A^T = -A$.

The matrices B and D, given just above this definition, are symmetric and skew symmetric, respectively. Any diagonal matrix is symmetric. For any ring S with identity I, if the order of I under addition is infinite, S is said to be of *characteristic zero*; if the order of I is a positive integer n, then S is said to be of *characteristic n*,. Thus, $\mathbf{Z}$ is of characteristic zero. However, $\mathbf{Z}_5 = [0, 1, 2, 3, 4]$ is of characteristic 5. Any field F is either of characteristic zero or of characteristic p, a prime number. In a field F of characteristic zero, given $a \in F$, $n \in \mathbf{Z}, na = 0, n \neq 0$, gives $a = 0$. If a field F is of characteristic $p > 0$, then for any $a \neq 0$ in F and any $n \in \mathbf{Z}, na = 0$ iff p divides n (symbolically $p|n$). Consider any skew-symmetric matrix $A = [a_{ij}]$, over a field F of characteristic different from 2 (viz, $F = \mathbf{Q}$), $A^T = -A$ gives $a_{ii}^T = -a_{ii}$. However, $a_{ii}^T = a_{ii}$, so that $2a_{ii} = 0$. This gives $a_{ii} = 0$. Thus, in particular the main diagonal of any skew-symmetric matrix over $\mathbf{C}$ is zero. In $\mathbf{Z}_2, 1+1=0$ gives $1=-1$. Using this, it follows that $A = \begin{bmatrix} 1 & 1 \\ 1 & 1 \end{bmatrix}$ over $\mathbf{Z}_2$, is skew-symmetric as well as symmetric.

Theorem 19. Let A and B be two $n{\times}n$-symmetric matrices over a commutative ring S. Then,

1. $A + B$ is symmetric.
2. For any $a \in S$, aA is symmetric.
3. AB is symmetric iff $AB = BA$.

Proof. Now,
$$A^T = A,\ B^T = B,$$
so that
$$(A+B)^T = A^T + B^T = A + B.$$
This shows that $A + B$ is symmetric.

Again,
$$(aA)^T = aA^T = aA$$
shows that aA is symmetric.

Now,
$$(AB)^T = B^T A^T = BA.$$
Hence, AB is symmetric iff $(AB)^T = AB$ i.e., iff $BA = AB$.

Similarly, we have the following :

Theorem 20. Let A and B be two $n{\times}n$-skew symmetric matrices over a commutative ring S. Then,

1. $A + B$ is skew-symmetric.
2. For any $a \in S$, aA is skew-symmetric.
3. AB is skew symmetric iff $AB = -BA$.

Example 20.a. $A = \begin{bmatrix} 1 & 2 \\ 2 & 0 \end{bmatrix}$, and $B = \begin{bmatrix} 2 & 3 \\ 3 & 1 \end{bmatrix}$ are symmetric but $AB = \begin{bmatrix} 8 & 5 \\ 4 & 6 \end{bmatrix}$ is not symmetric.

Example 20.b. $A = \begin{bmatrix} 0 & -1 \\ 1 & 0 \end{bmatrix}$ is skew symmetric, but $AA = \begin{bmatrix} -1 & 0 \\ 0 & -1 \end{bmatrix}$ is not skew symmetric.

Theorem 21. Polynomial Expression. Consider $A \in S^{n\times n}$, S any ring. For any non-negative integer k, A^k is defined as follows :

$$A^0 = I_n$$

$$A^{m+1} = A^m . A, \qquad m \geq 0.$$

Thus, $$A^1 = A^0 A = I_n A = A,$$

$$A^2 = A^1 A = AA,$$

and $$A^3 = A^2 . A = (A.A).A = A.(A.A) = A.A^2.$$

As in rings, $$A^k . A^l = A^{k+l}$$

for any $k \geq 0$, $l \geq 0$. Consider $S = \mathbf{R}$, $A \in \mathbf{R}^{2\times 2}$,

say $$A = \begin{bmatrix} 1 & -2 \\ 3 & 4 \end{bmatrix}, \; f(X) = 2X^2 - X + 1.$$

We put
$$\begin{aligned} f(A) &= 2A^2 - A + I_2 \\ &= 2\begin{bmatrix} -5 & -10 \\ 15 & 10 \end{bmatrix} - \begin{bmatrix} 1 & -2 \\ 3 & 4 \end{bmatrix} + \begin{bmatrix} 1 & 0 \\ 0 & 1 \end{bmatrix} \\ &= \begin{bmatrix} -10 & -20 \\ 30 & 20 \end{bmatrix} - \begin{bmatrix} 1 & -2 \\ 3 & 4 \end{bmatrix} + \begin{bmatrix} 1 & 0 \\ 0 & 1 \end{bmatrix} \\ &= \begin{bmatrix} -10 & -18 \\ 27 & 17 \end{bmatrix}. \end{aligned}$$

We end this section with a brief discussion of *block matrices* and *submatrices*. Given an $m \times n$ matrix A over a ring S, any matrix obtained by deleting some rows or columns of A, and by keeping the other in respective orders is called a *submatrix* of A.

For example, consider $$A = \begin{bmatrix} 1 & 4 & 2 & 3 \\ 0 & 1 & 1 & 2 \\ 1 & 0 & 0 & 2 \end{bmatrix}$$

$$A_1 = \begin{bmatrix} 1 & 4 & 2 & 3 \\ 0 & 1 & 1 & 2 \end{bmatrix}$$

is obtained from A by deleting its third row.

So, A_1 is a submatrix of A.

But $$A_2 = \begin{bmatrix} 0 & 1 & 1 & 2 \\ 1 & 4 & 2 & 3 \end{bmatrix}$$

is not a submatrix of A, because the arrangement of rows in A_2 is not in the order in which they were in A.

Again,
$$B=\begin{bmatrix}1 & 4 & 3\\ 1 & 0 & 2\end{bmatrix}$$
is a submatrix of A obtained by deleting the second row and the third column of A.

Consider
$$B_{11}=\begin{bmatrix}1 & 4\\ 0 & 1\end{bmatrix},\ B_{12}=\begin{bmatrix}2 & 3\\ 1 & 2\end{bmatrix}$$
$$B_{21}=\begin{bmatrix}1 & 0\end{bmatrix},\ B_{22}=\begin{bmatrix}0 & 2\end{bmatrix}.$$

Now,
$$A=\begin{bmatrix}1 & 4 & : & 2 & 3\\ 0 & 1 & : & 1 & 2\\ .. & .. & .. & .. & ..\\ 1 & 0 & : & 0 & 2\end{bmatrix}.$$

One can recognize that A is partitioned into four blocks, the upper two blocks represent B_{11} and B_{12} and the lower two blocks represent B_{21} and B_{22}. For that matrix, we write
$$A=\begin{bmatrix}B_{11} & B_{12}\\ B_{21} & B_{22}\end{bmatrix}.$$

In this representation, we say A is a *block matrix*. A can be so represented in many ways.

For instance
$$A=\begin{bmatrix}C_{11} & C_{12} & C_{13}\\ C_{21} & C_{22} & C_{23}\end{bmatrix}$$
where
$$C_{11}=\begin{bmatrix}1\\ 0\end{bmatrix},\ C_{12}=\begin{bmatrix}4 & 2\\ 1 & 1\end{bmatrix},\ C_{13}=\begin{bmatrix}3\\ 2\end{bmatrix}$$
$$C_{21}=[1],\ C_{22}=\begin{bmatrix}0 & 0\end{bmatrix},\ C_{23}=[2].$$

The block matrix
$$\begin{bmatrix}I_2 & 0_{2\times 2}\\ 0_{2\times 2} & I_2\end{bmatrix}=\begin{bmatrix}1 & 0 & 0 & 0\\ 0 & 1 & 0 & 0\\ 0 & 0 & 1 & 0\\ 0 & 0 & 0 & 1\end{bmatrix}=I_4.$$

In any block representation of a matrix, the submatrices occurring in a row have to have the same number of rows and the submatrices occurring in any column have to have the same number of columns.

For example,
$$A=\begin{bmatrix}D_{11} & D_{12}\\ D_{21} & D_{22}\end{bmatrix}$$
with
$$D_{11}=\begin{bmatrix}1 & 4 & 2\\ 0 & 1 & 1\end{bmatrix},\ D_{12}=\begin{bmatrix}3\\ 2\end{bmatrix}$$
$$D_{21}=\begin{bmatrix}1 & 0\end{bmatrix},\ D_{22}=\begin{bmatrix}0 & 2\end{bmatrix}$$
is not a block representation of A.

Exercise 1.2

1. Let

$$A = \begin{bmatrix} 1 & -3 & 2 \\ 0 & 5 & 4 \end{bmatrix}, B = \begin{bmatrix} 2 \\ 5 \\ -1 \end{bmatrix}, C = \begin{bmatrix} 7 & 2 & 1 \\ 4 & 5 & 0 \\ 6 & -1 & 2 \end{bmatrix}, D = \begin{bmatrix} 1 & 0 & 1 \\ 0 & 1 & 1 \\ 1 & 0 & 1 \end{bmatrix}, E = \begin{bmatrix} 1 & 0 & 2 \end{bmatrix}.$$

If possible, compute

(i) $C+D$

(ii) $B+E$

(iii) $B^T C+E$

(iv) $A+D$

(v) $A^T A+C$

(vi) $5A$

(vii) $-3D+5C$

(viii) $(ED)B$

(ix) $(AC)D$

2. In the exercise, given above, verify that $(AC)^T = C^T A^T$

3. If $\begin{bmatrix} a+b & c+d \\ c-d & a-b \end{bmatrix} = \begin{bmatrix} 8 & 6 \\ 4 & 2 \end{bmatrix}$, find a, b, c, d

4. Verify that $A = \begin{bmatrix} 1 & 0 \\ 0 & -1 \end{bmatrix}$, and $B = \begin{bmatrix} 0 & 1 \\ 1 & 0 \end{bmatrix}$ are two symmetric matrices such that AB is skew symmetric.

5. If a real matrix A is symmetric as well as skew-symmetric, prove that $A = 0$.

6. Give an example of two 2×2 skew symmetric non-zero matrices A, B such that AB is symmetric

7. If an $m \times p$-matrix A has kth row a zero row, prove that for any $p \times n$-matrix B, the kth row of AB is a zero row.

8. For any $n \times n$-matrix A over a commutative ring S, prove the following :

(i) $A+A^T$ is symmetric.

(ii) $A-A^T$ is skew-symmetric.

(iii) AA^T and $A^T A$ are symmetric.

9. If an upper triangular matrix is symmetric, prove that it is a diagonal matrix.

10. Give an example of a non-zero real matrix A such that $A^2 \neq 0$, but $A^3 = 0$.

11. Let $A = \begin{bmatrix} 1 & 2 \\ 0 & 3 \end{bmatrix}$, find B such that $AB = I_2$.

12. Prove that the sum and the product of two $n \times n$-upper triangular matrices are upper triangular matrices.

1.3 Inverse of a Matrix

Definition 1. Let A be any $n \times n$-matrix over a ring S (with identity). Then, A is said to be *invertible* over S if there exists $a, B \in S^{n\times n}$ such that $AB = I_n = BA$; in this case, B is called an *inverse* of A.

Example 1.a. $A = \begin{bmatrix} 2 & 0 \\ 0 & 1 \end{bmatrix}$ is a matrix over **R**. It is invertible with inverse $\begin{bmatrix} \frac{1}{2} & 0 \\ 0 & 1 \end{bmatrix}$, but it is not invertible over Z.

An invertible matrix, over a field F, is called a *non-singular matrix*. An $n \times n$ non-invertible matrix over F is called a *singular matrix*. Observe that I_n has no row or column, consisting entirely of zeroes. Let $A \in S^{n\times n}$ have a row consisting of zeroes. Then, for any $B \in S^{n\times n}$, AB has the same numbered row consisting of zeroes, so that $AB \neq I_n$. This shows that A is not invertible. For similar reasons, if A has a column of zeroes, then A is not invertible. Finally, let $A \in S^{n\times n}$ be such that $AC = 0$ for some $C \neq 0$ in $S^{n\times n}$. If A has inverse B, we get $BAC = 0$, i.e., $I_n C = 0$, i.e., $C = 0$. Hence, A is not invertible.

Theorem 2. For any innvertible matrix $A \in S^{n\times n}$, its inverse is unique.

Proof. Let A have two inverses A', A''.

Then,
$$AA' = I_n = A'A,$$
$$AA'' = I_n = A''A.$$

Consequently, $A' = A'I_n = A'(AA'') = (A'A)A'' = I_nA'' = A''$.

Hence, A has unique inverse.

Because of the uniqueness of the inverse of an invertible matrix A, its inverse is denoted by A^{-1}

Theorem 3. Let A, B be two invertible matrices in $S^{n\times n}$. Then,

1. A^{-1} is invertible in $S^{n\times n}$, and $(A^{-1})^{-1} = A$.
2. AB is invertible in $S^{n\times n}$, and $(AB)^{-1} = B^{-1}A^{-1}$.

Proof.

(*i*) Now, $$AA^{-1} = I_n = A^{-1}A.$$

By definition, any $A' \in S^{n\times n}$ is inverse of A^{-1} if $A^{-1}A' = I_n = A'A^{-1}$.

Consequently, $A = (A^{-1})^{-1}$.

(*ii*) As
$$\begin{aligned}(AB)(B^{-1}A^{-1}) &= A\left[B(B^{-1}A^{-1})\right] \\ &= A\left[(BB^{-1})A^{-1}\right] \\ &= A\left[I_n A^{-1}\right] \\ &= AA^{-1} \\ &= I_n \\ &= (B^{-1}A^{-1})(AB),\end{aligned}$$

we get $(AB)^{-1} = B^{-1}A^{-1}$.

Corollary 4. For any positive integer n, the set $GL_n(S)$ of all invertible matrices in $S^{n\times n}$ is a group under matrix multiplication in $S^{n\times n}$.

Proof. The theorem given above, shows that for any $A, B \in GL_n(S)$, $AB \in GL_n(S)$ and $A^{-1} \in GL_n(S)$. Further, I_n is identity in $GL_n(S)$. As matrix multiplication in $S^{n\times n}$ is also associative, the result follows :

Observe that if $A \in S^{n\times n}$ is invertible and $B \in S^{n\times n}$ is not invertible, then AB and BA are not invertible. For, if AB is invertible, then by Theorem 3, $B = A^{-1}(AB)$ is also invertible.

5. Inverse of a 2×2 matrix over a field. Let F be any field.

Let $A = \begin{bmatrix} a & b \\ c & d \end{bmatrix}$ in $F^{2\times 2}$

have $A^{-1} = \begin{bmatrix} x & y \\ z & w \end{bmatrix}$.

Then $AA^{-1} = I_2$

gives
$$ax + bz = 1 \quad \text{... (1)}$$
$$cx + dz = 0 \quad \text{... (2)}$$
$$ay + bw = 0 \quad \text{... (3)}$$
$$cy + dw = 1. \quad \text{... (4)}$$

Elimination of x from (1) and (2) gives
$$(cb - ad)z = c. \quad \text{... (5)}$$

Elimination of z from (1) and (2) gives
$$(da - bc)x = d. \quad \text{... (6)}$$

If $\Delta = ad - bc = 0$, then, by (5) and (6), $c = 0 = d$, so that $A = \begin{bmatrix} a & b \\ 0 & 0 \end{bmatrix}$ has second row, a row of zeroes. Consequently, if $\Delta = ad - cb = 0$, A is singular. Let $\Delta \neq 0$. Then,

(5) and (6) give $z = -\frac{c}{\Delta}$, $x = \frac{d}{\Delta}$. In this case, (3) and (4) give $y = -\frac{b}{\Delta}$ and $w = \frac{a}{\Delta}$, so that A is invertible with

$$A^{-1} = \frac{1}{\Delta}\begin{bmatrix} d & -b \\ -c & a \end{bmatrix}.$$

$\Delta = ad - bc$ is called the *determinant* of A.

Example 5.a. For $A = \begin{bmatrix} 3 & 4 \\ 6 & 8 \end{bmatrix}$, $\Delta = 3\times 8 - 6\times 4 = 0$. Consequently, A is singular (Over **R**). For $B = \begin{bmatrix} 2 & 1 \\ -1 & 3 \end{bmatrix}$, determinant of B, $\Delta = 2\times 3 + 1\times 1 = 7 \neq 0$, so that B is invertible and

$$B^{-1} = \frac{1}{7}\begin{bmatrix} 3 & -1 \\ 1 & 2 \end{bmatrix}$$

$$= \begin{bmatrix} \frac{3}{7} & \frac{-1}{7} \\ \frac{1}{7} & \frac{2}{7} \end{bmatrix}.$$

Theorem 6. Let A be any $n\times n$-invertible matrix over a commutative ring S. Then, A^T is invertible and $(A^T)^{-1} = (A^{-1})^T$.

Proof. Now
$$AA^{-1} = I_n = A^{-1}A,$$
gives
$$(AA^{-1})^T = I_n^T = (A^{-1}A)^T,$$
i.e.,
$$(A^{-1})^T A^T = I_n = A^T (A^{-1})^T.$$

Hence, A^T is invertible and $(A^T)^{-1} = (A^{-1})^T$.

In later sections, we discuss general methods of finding inverse of non-singular matrices over a field F. Here, for illustration, we discuss a method of finding the inverse of a 3×3-upper triangular matrix.

Let
$$A = \begin{bmatrix} a_{11} & a_{12} & a_{13} \\ 0 & a_{22} & a_{23} \\ 0 & 0 & a_{33} \end{bmatrix}$$

be a non-singular 3×3-upper triangular matrix over a field F.

Let $B = \left[b_{ij}\right]$ be its inverse.

Then,
$$b_{23} = -a_{22}^{-1}a_{33}^{-1}a_{23}$$

$$b_{13} = -a_{11}^{-1}\left[-a_{22}^{-1}a_{33}^{-1}a_{12}a_{23} + a_{33}^{-1}a_{13}\right]$$

and $$b_{33} = \frac{1}{a_{33}}.$$

Now, $$u_{31} = 0,$$

gives $$a_{33}\, b_{31} = 0;\ u_{32} = 0,$$

gives $$a_{33}\, b_{32} = 0.$$

Consequently, $$b_{31} = 0 = b_{32}.$$

Again, $$u_{22} = 1,$$

gives $$a_{22}\, b_{22} = 1,$$

so that $$a_{22} \neq 0$$

and $$b_{22} = \frac{1}{a_{22}}.$$

As $$u_{21} = 0,$$

we get $$a_{22}\, b_{21} = 0,$$

so that $$b_{21} = 0.$$

Hence, $$B = \begin{bmatrix} b_{11} & b_{12} & b_{13} \\ 0 & b_{22} & b_{23} \\ 0 & 0 & b_{33} \end{bmatrix}$$

is an upper triangular matrix with $b_{33} \neq 0$, $b_{22} \neq 0$.

Again, $$u_{11} = 1$$

yields $$a_{11}\, b_{11} = 1.$$

Hence, $$b_{11} = \frac{1}{a_{11}}.$$

Thus, we find that if A is a non-singular 3×3-upper triangular matrix, then, each of its entries on the main diagonal is non-zero.

Further, $$b_{ii} = a_{ii}^{-1}, \qquad 1 \le i \le 3.$$

Now, $$u_{12} = 0$$

yields $$a_{11}\, b_{12} + a_{12}\, b_{22} = 0,$$

so that
$$\begin{aligned} b_{12} &= -a_{11}^{-1}\, a_{12}\, b_{22} \\ &= -a_{11}^{-1}\, a_{22}^{-1}\, a_{12}. \end{aligned}$$

Similarly,
$$\begin{aligned} b_{23} &= -a_{22}^{-1}\, a_{33}^{-1}\, a_{23}\, b_{13} \\ &= -a_{11}^{-1}\left[-a_{22}^{-1}\, a_{33}^{-1}\, a_{12}\, a_{23} - a_{33}^{-1}\, a_{13}\right] \end{aligned}$$

In particular, it gives that for the 3×3-diagonal matrix

$$A = \begin{bmatrix} a & 0 & 0 \\ 0 & b & 0 \\ 0 & 0 & c \end{bmatrix}$$

to be non-singular, we need $a \neq 0,\ b \neq 0,\ c \neq 0$.

In that case, $$A^{-1} = \begin{bmatrix} a^{-1} & 0 & 0 \\ 0 & b^{-1} & 0 \\ 0 & 0 & c^{-1} \end{bmatrix}.$$

Exercise 1.3

1. Show that $A = \begin{bmatrix} 2 & 4 \\ 5 & 7 \end{bmatrix}$ is non-singular over **R**. Show that this A is not invertible over Z.

2. Let A, B and C be three $n \times n$-matrices over a commutative ring S, such that A is invertible and $AB = AC$. Prove that $B = C$.

3. Find the inverse of each of the following matrices over **R**

$$\begin{bmatrix} 2 & 0 \\ 0 & 3 \end{bmatrix}, \begin{bmatrix} 1 & -1 \\ 0 & 1 \end{bmatrix}, \begin{bmatrix} 2 & 0 & 1 \\ 0 & 3 & 2 \\ 0 & 0 & 1 \end{bmatrix}$$

4. Prove that the product of two non-singular upper triangular matrices, over a field F, is a non-singular upper triangular matrix over F.

5. Write down all non-singular 2×2-matrices over the field Z_2.

6. If A and B are $n \times n$-non-singular matrices over a field F, prove the following :
   (*i*) $A + B$ need not be non-singular.
   (*ii*) Even if $A + B$ is non-singular, $(A + B)^{-1}$ need not equal $A^{-1} + B^{-1}$.
   (*iii*) For any $c \neq 0$ in F, $(cA)^{-1} = c^{-1}A^{-1}$.

7. A matrix $A \in S^{n \times n}$, S any ring, is said to be ***nilpotent***, if $A^k = 0$ for some $k \geq 1$. Prove the following :
   (*i*) Any 3×3-upper triangular matrix over a field F, with its main diagonal zero, is nilpotent.
   (*ii*) If $A \in F^{n \times n}$ is nilpotent, then, $I_n + A$ is non-singular.
   (*iii*) If $A,\ B \in F^{2 \times 2}$ are nilpotent, $A + B$ need not be nilpotent.
   (*iv*) If $A,\ B \in F^{n \times n}$ are nilpotent, and $AB = BA$, then, $A + B$ is nilpotent.

8. Give an example of a 2×2-matrix A such that $A \neq \pm I_2$, but $A^2 = I_2$.

9. Give an example of a 3×3-matrix A such that $A^3 = I_3$, but $A^2 \neq I_3$.

10. A subgroup H of a group G is called a *normal subgroup* if $x^{-1}ax \in H$ for every $a \in H$ and $x \in G$. Prove the following :

    The set H of all the 2×2-matrices over **R** with determinant $(A) = 1$, is a normal subgroup of $GL_2(\mathbf{R})$.

11. Let S be any ring with identity 1. An $a \in S$ is called a *unit* (or an invertible) element in case there exists an $a' \in S$ such that $aa' = 1 = a'a$; in such a case, a' is called an inverse of a, and is uniquely determined by a; a' is denoted by a^{-1}. Prove the following :

    (*i*) The set U of all invertible elements in S is a group under the multiplication operation in S.

    (*ii*) Find all units in Z.

    (*iii*) Find all units in the ring $Z[i] = \{a + bi : a,\ b \in Z\}$.

12. Let S be any ring. A subset J of S is called a *left ideal* of S, if J is a subgroup of S, under addition, and $ra \in J$ for every $r \in S$ and $a \in J$. Prove the following :

    (*i*) Any ideal of S is a left ideal.

    (*ii*) $A = \left\{\begin{bmatrix} a & 0 \\ b & 0 \end{bmatrix} : a,\ b \in \mathbf{R}\right\}$ is a left ideal of $\mathbf{R}^{2\times 2}$ but A is not an ideal of $\mathbf{R}^{2\times 2}$

1.4 Row Equivalent Matrices

Throughout, F is a field. A subset W of F^n is called a *subspace* of F^n, if W is non-empty and for any $\alpha,\ \beta \in W$, $c \in F$, we have $\alpha + \beta \in W$ and $c\alpha \in W$. For any subspace W of F^n, as some $\alpha \in W$, we get $0 = 0\ \alpha \in W$. For any $\beta \in W$. $-\beta = (-1)\beta \in W$. Observe that $W = \{(a, b, 0) : a, b \in \mathbf{R}\}$ is a subspace of $\mathbf{R}^3$.

Definition 1. Let $\alpha_1, \alpha_2, \ldots, \alpha_k$ be any finitely many members of F^n. Then, any $\alpha \in F^n$ is called a *linear combination* of $\alpha_1, \alpha_2, \ldots, \alpha_k$ if $\alpha = c_1\alpha_1 + c_2\alpha_2 + \ldots + c_k\alpha_k$ for some $c_1, c_2, \ldots, c_k \in F$.

Example 1.a. Consider
$$\alpha_1 = (1,\ 0),$$
$$\alpha_2 = (0,\ 2) \in \mathbf{R}^2$$
and
$$\alpha = (1,\ 1).$$
As
$$\alpha = (1,\ 1) = 1\alpha_1 + \frac{1}{2}\alpha_2,$$
α is a linear combination of α_1, α_2.

Example 1.b. Consider $\alpha_1 = (1, -1, 0)$

$$\alpha_2 = (0,\ 1,\ 0) \in \mathbf{R}^2.$$

Any linear combination $a_1\alpha_1 + a_2\alpha_2 = (a_1,\ a_2 - a_1, 0)$

has third component zero.

Consequently, $\alpha = (1,\ 0,\ 1)$

is not a linear combination of α_1 and α_2.

Theorem 2. Let $\alpha_1, \alpha_2, ..., \alpha_k \in F^n$, $k \geq 1$. Then, the set W of all linear combinations of these α_i's is a subspace of F^n with $\alpha_i \in W$.

Proof. Let $\delta_{ij} = 0$ for $i \neq j$ and $\delta_{ii} = 1$.

Then,
$$\alpha_1 = 1\alpha_1 + 0\alpha_2 + ... + 0\alpha_k$$
$$= \sum_{j=1}^{k} \delta_{1j}\ \alpha_j.$$

In general,
$$\alpha_i = \sum_{j=1}^{k} \delta_{ij}\ \alpha_j,$$

being a linear combination of $\alpha_1, \alpha_2, ..., \alpha_k$, belongs to W.

Consider any $\alpha, \beta \in W$ and $c \in F$.

By definition
$$\alpha = \sum_{i=1}^{k} c_i \alpha_i$$
$$\beta = \sum_{i=1}^{k} c_i' \alpha_i \qquad \text{for some } c_i,\ c_i' \in F.$$

Then,
$$\alpha + \beta = \sum_{i=1}^{k} (c_i + c_i')\ \alpha_i \in W$$
$$c\alpha = \sum_{i=1}^{k} (cc_i)\ \alpha_i \in W.$$

This proves the result.

Definition 3. Let F be any field. Any equation of the type $a_1x_1 + a_2x_2 + ... + a_kx_k = b$ where $a_1, a_2, ..., a_k, b$ are fixed elements of F, and $x_1, x_2, ..., x_k$ are unknowns, is called a *linear equation* over F; $a_1, a_2, ..., a_k$ are called its coefficients, $x_1, x_2, ..., x_k$ are called its variables and b is called its constant term. In case, $b = 0$, the equation is called a *homogeneous linear equation*.

Linear equations arise in several situations. Thus, $2x - 3y = 5$ is a linear equation in two variables x, y. It represents a straight line in the Euclidean plane. The equation $2x + y + z = 0$ is a homogeneous linear equation in three variables. It represents a plane in three dimensional Euclidean space, passing through the origin. One comes across a system consisting of more than one equation. By using matrix multiplication, we can write

$$a_1x_1 + a_2x_2 + \ldots + a_kx_k = b$$

as

$$[a_1a_2 \ldots a_k]\begin{bmatrix} x_1 \\ x_2 \\ \vdots \\ x_k \end{bmatrix} = [b]. \qquad \ldots (1)$$

Consider the following system of *m* linear equations in *n* variables

$$\left.\begin{array}{l} a_{11}x_1 + a_{12}x_2 + \ldots + a_{1n}x_n = b_1 \\ a_{21}x_1 + a_{22}x_2 + \ldots + a_{2n}x_n = b_2 \\ \ldots\ldots\ldots\ldots\ldots\ldots\ldots\ldots\ldots\ldots \\ \ldots\ldots\ldots\ldots\ldots\ldots\ldots\ldots\ldots\ldots \\ a_{m1}x_1 + a_{m2}x_2 + \ldots + a_{mn}x_n = b_m \end{array}\right] \qquad \ldots (2)$$

Keeping in view (1), we can write it as

$$AX = B$$

where $A = \left[a_{ij}\right]_{m \times n}$ is called the *co-efficient* matrix,

$$X = \begin{bmatrix} x_1 \\ x_2 \\ \vdots \\ x_n \end{bmatrix}$$

is called the *column of variables* and $B = \begin{bmatrix} b_1 \\ b_2 \\ \vdots \\ b_m \end{bmatrix}$

is called the *column of constant terms*. In case $B = 0$, the system (2) is called a system of homogeneous linear equations. Any $n \times 1$ vector

$$U = \begin{bmatrix} u_1 \\ u_2 \\ \vdots \\ u_n \end{bmatrix} \in F^{n \times 1}$$

is called a *solution of the system* (2) if

$$a_{i1}u_1 + a_{i2}u_2 + \ldots + a_{in}u_n = b_i$$

for $1 \le i \le m$; i.e., if $AU = B$. For example, $\begin{bmatrix} 1 \\ 1 \end{bmatrix}$ is a solution of the system of two equations

$$x + y = 2$$

$$3x - 2y = 1.$$

We aim to give a general method of solving a system like (2). For this, we discuss a simple example in detail.

Example 3.a. Consider a system of two linear equations

$$x - y = 3 \quad \text{... (1)}$$

$$2x - 3y = 5. \quad \text{... (2)}$$

Geometrically, this is a system of two straight lines in the Euclidean plane. The common solution is the point of intersection of these lines. The normal procedure of solving such a system consists in eliminating one of the two variables and, then, solve for the other. To do this, the following operations are used :

(*i*) Any of these equations could be multiplied by a non-zero constant.

(*ii*) A multiple of one equation could be added to the other.

(*iii*) One could write the equations in any other order, i.e., the second could be written as first and the first as the second.

The matrix form of the equations is

$$\begin{bmatrix} 1 & -1 \\ 2 & -3 \end{bmatrix} \begin{bmatrix} x \\ y \end{bmatrix} = \begin{bmatrix} 3 \\ 5 \end{bmatrix} \quad \text{... (3)}$$

Here, the co-efficient matrix $A = \begin{bmatrix} 1 & -1 \\ 2 & -3 \end{bmatrix}$, $X = \begin{bmatrix} x \\ y \end{bmatrix} B = \begin{bmatrix} 3 \\ 5 \end{bmatrix}$

We form another matrix $[A \; : \; B] = \begin{bmatrix} 1 & -1 & : & 3 \\ 2 & -3 & : & 5 \end{bmatrix}$ ... (4)

known as *augmented matrix*.

To eliminate *x*, we subtract 2 times (1) from (2). We get two equations

$$x - y = 3 \quad \text{... (5)}$$

$$0x - y = -1. \quad \text{... (6)}$$

The augmented matrix for (5) and (6) is

$$\begin{bmatrix} 1 & -1 & : & 3 \\ 0 & -1 & : & -1 \end{bmatrix}. \quad \text{... (7)}$$

It is immediately clear that this matrix is obtained from (4) by replacing the second row, by the row obtained by subtracting 2 times the first row from the second row in (4). To get *y* from (6), we need to multiply (6) by (–1). This gives

$$x - y = 3 \quad \text{... (8)}$$

$$0x + y = 1. \quad \text{... (9)}$$

Therefore, $y = 1.$

The corresponding augmented matrix

$$\begin{bmatrix} 1 & -1 & : & 3 \\ 0 & 1 & : & 1 \end{bmatrix} \qquad \dots (10)$$

is obtained from (7) by multiplying its second row by −1.

To eliminate y from (8) and (9), we add (9) to (8). This gives

$$x + 0.y = 4 \qquad \dots (11)$$

$$0x + y = 1. \qquad \dots (12)$$

that is, $x = 3,\ y = 1.$

Here, the augmented matrix $\begin{bmatrix} 1 & 0 & : & 4 \\ 0 & 1 & : & 1 \end{bmatrix}$

is obtained from (10) by adding the second row to the first row.

We now return to the system of m linear equations in n variables $AX = B$. The $m \times (n+1)$-matrix $[A \ : \ B]$, whose first n columns are the same as those of A and the last equals B, is called the *augmented matrix* of the system. The example, given above, shows that if we want to multiply a particular equation in the system by a non-zero constant (scalar), we multiply the corresponding row of the augmented matrix by the given constant. The interchange of the two equations amounts to the interchange of the corresponding rows of the augmented matrix. The addition of a certain multiple of one equation to the other amounts to a corresponding operation between the corresponding rows of the augmented matrix. This discussion leads us to the following operations in a matrix.

Definition 4. Any of the following three operations on an $m \times n$-matrix $A = [a_{ij}]$, over a field F, is called an *elementary row operation*.

I. Interchange k^{th} and l^{th} rows of A to obtain the $m \times n$-matrix $A' = [a'_{ij}]$ whose k^{th} row $[a'_{k1}, a'_{k2}, \dots, a'_{kn}]$ equals the lth row $[a_{l1}, a_{l2}, \dots, a_{ln}]$ of A, the lth row of A', $[a'_{l1}, a'_{l2}, \dots, a'_{ln}] = [a_{k1}, a_{k2}, \dots, a_{kn}]$, and any other row A' is the same as the corresponding row of A.

II. Multiplication of the kth row of A by an element $a \neq 0$ in F, to obtain an $m \times n$-matrix $A' = [a'_{ij}]$ whose k^{th} row $[a'_{k1}, a'_{k2}, \dots, a'_{kn}] = [aa_{k1}, aa_{k2}, \dots, aa_{kn}]$ and any other row of A' is the same as the corresponding row of A.

III. For any $t \neq s$, and $a \in F$ a times the tth row of A is added to the sth row of A, to obtain an $m \times n$-matrix $A' = [a'_{ij}]$ whose s^{th} row equal

$$a[a_{t1}, a_{t2}, \dots, a_{tn}] + [a_{s1}, a_{s2}, \dots, a_{sn}] = [aa_{t1} + a_{s1}, aa_{t2} + a_{s2}, \dots, aa_{tn} + a_{sn}],$$

and any other row of A' is the same as the corresponding row of A.

The elementary row operations given in (*I*), (*II*) and (*III*) are known to be of types *I*, *II* and *III*, respectively.

The row operations described in *I*, *II* and *III* are denoted by symbols $r_k \leftrightarrow r_l$, $r_k \to ar_k$ and $r_s \to r_s + ar_t$, respectively.

Example 4.a. Let
$$A = \begin{bmatrix} 2 & 3 & 1 \\ 0 & 2 & 1 \\ 1 & 0 & -1 \end{bmatrix}.$$

Apply $r_1 \leftrightarrow r_3$ to A, i.e., interchange first and third row of A. This gives
$$A_1 = \begin{bmatrix} 1 & 0 & -1 \\ 0 & 2 & 1 \\ 2 & 3 & 1 \end{bmatrix}.$$

Notice that the second row of A_1 is the same as the second row of A. Apply $r_2 \to \frac{1}{2} r_1$ to A_1, i.e., multiply the second row of A_1 by $\frac{1}{2}$. This gives
$$A_2 = \begin{bmatrix} 1 & 0 & -1 \\ 0 & 1 & \frac{1}{2} \\ 2 & 3 & 1 \end{bmatrix}.$$

Observe that only second row of A_1 is affected, the rows 1 and 3 of A_2 are the same as the rows 1 and 3, respectively of A_1. Apply $r_3 \to r_3 - 2r_1$ to A_2. This means the row 3 of A_2 is replaced by
$$\begin{aligned}(\text{row } 3) - 2(\text{row } 1) &= [2,\ 3,\ 1] - 2[1,\ 0,\ -1] \\ &= [0,\ 3,\ 3].\end{aligned}$$

This gives
$$A_3 = \begin{bmatrix} 1 & 0 & -1 \\ 0 & 1 & \frac{1}{2} \\ 0 & 3 & 3 \end{bmatrix}$$

Apply $r_3 \to r_3 - 3r_2$ to A_3.

This gives
$$A_4 = \begin{bmatrix} 1 & 0 & -1 \\ 0 & 1 & \frac{1}{2} \\ 0 & 0 & \frac{3}{2} \end{bmatrix}.$$

Apply $r_3 \to \frac{2}{3} r_3$ to A_4.

This gives
$$A_5 = \begin{bmatrix} 1 & 0 & -1 \\ 0 & 1 & \frac{1}{2} \\ 0 & 0 & 1 \end{bmatrix}$$

Apply $r_1 \to r_1 + r_3$ to A_5.

This gives $$A_6 = \begin{bmatrix} 1 & 0 & 0 \\ 0 & 1 & \frac{1}{2} \\ 0 & 0 & 1 \end{bmatrix}.$$

Apply $r_2 \to r_2 - \frac{1}{2} r_3$ to A_6.

This gives $$A_7 = \begin{bmatrix} 1 & 0 & 0 \\ 0 & 1 & 0 \\ 0 & 0 & 1 \end{bmatrix} = I_3.$$

Observe that A_1 is obtained from A by applying one elementary row operation. A_3 is obtained from A by applying three elementary row operations in succession. A_7 is obtained by applying a succession (or sequence) of seven elementary row operations to A.

Definition 5. Let A and B be two $m \times n$-matrices over a field F. Then, A is said to be *row equivalent* to B if B is obtained from A by applying a finite sequence of elementary row operations on A, in the sense that there is a finite sequence of $m \times n$-matrices

$$A = A_1, A_2, \ldots, A_s = B$$

such that either $A = B$ or each A_{i+1} for $1 \le i \le s-1$ is obtained from A_i by applying an elementary row operation.

In Example 4.a, A is row equivalent to $A_1, A_2, \ldots, A_7$. A_3 is row equivalent to A_6.

Remark 6. Consider an $m \times n$-matrix A.

(*i*) If A' is obtained by interchanging rows k and l of A, then, A can be obtained form A' by interchanging rows k and l.

(*ii*) If A' is obtained from A by multiplying k^{th} row of A by $a \neq 0$ in F, then, A can be obtained from A' by multiplying k^{th} row of A' by a^{-1}.

(*iii*) If A' is obtained from A by addiing a times l^{th} row to s^{th} row, then, A can be obtained from A' by adding $(-a)$ times l^{th} row to s^{th} row in A'.

(*iv*) Let A, B, C be three $m \times n$-matrices over a field F. We claim that the row-equivalence is an equivalence relation, i.e., it is reflexive, symmetric and transitive. By definition, A is row equivalent to A. This proves reflexivity. Let A be row equivalent to B and $A \neq B$. We get a finite sequence

$$A = A_1, A_2, \ldots, A_s = B$$

with $s \ge 2$, such that each A_{i+1} is obtained from A_i by applying an elementary row operation. (*i*), (*ii*) and (*iii*) show that A_i is obtained from A_{i+1} by an elementary row operation.

This, in $$B = A_s, A_{s-1}, \ldots, A_2, A_1 = A.$$

Each successive matrix is obtainable from its predecessor by an elementary row operation. Consequently, B is row equivalent to A. This proves symmetry. Finally, let A be row equivalent to B and B be row equivalent to C. clearly, by starting from A, we obtain C by a finite sequence of elementary row operations that first gives B from A, and then, C from B. Hence, A is row equivalent to C. This proves transitivity.

(*v*) The row equivalence divides $F^{m\times n}$ into equivalence classes. Any two members of an equivalence class are said to be *row equivalent*. Thus, in Example 4.a, A and I_3 are row equivalent, A_2 and A_5 are row equivalent.

Definition 7. Let A be any $m\times n$-matrix over a field F. Then, the subspace W of F^n, consisting of all linear combinations of the rows of A, is called the *row space* of A.

By Theorem 2, any row of a matrix A is a member of the row space W of A. [0 0...0] in F^n is in W. For any matrix A, A_i will the ith row of A.

Lemma 8. Let A and B be two $m\times n$-matrices over a field F, such that B is obtained from A by an elementary row operation on A. Then, A and B have the same row spaces.

Proof. Let W and W' be the row spaces of A and B, respectively.

Case I. B is obtained from A by interchanging two rows. Obviously, A and B have the same set of rows and, hence, $W = W'$

Case II. B is obtained from A by multiplying the kth row of A by $a \neq 0$ in F.

Let $\alpha \in W'$. Then,
$$\alpha = \sum_{i=1}^{m} a_i B_i.$$

However,
$$B_i = A_i \qquad \text{for } i \neq k$$
and
$$B_k = aA_k.$$

Consequently,
$$\alpha = \sum_{i\neq k} a_i A_i + a_k\, aA_k,$$
a linear combination of rows of A.

Consequently,
$$W' \subseteq W.$$

As A can be obtained from B, by multiplying the kth row of B by $a^{-1} \neq 0$ in F, we get $W \subseteq W'$. Hence, $W = W'$

Case III. B is obtained from A by adding a times the lth row to the sth row, $l \neq s$.

Therefore,
$$B_s = A_s + aA_l,\ a \in F,$$
$$B_i = A_i \qquad \text{for } i \neq s.$$

Let $\alpha \in W'$, then, by Theorem 2,
$$\alpha = \sum_{i\neq s} a_i B_i + a_s B_s$$
$$= \sum_{i\neq s} a_i A_i + a_s(A_s + aA_l) \in W.$$

Hence,
$$W' \subseteq W.$$

By Remark 6, A is obtainable from B by the elementary row operation of adding $(-a)$ times the lth row to the sth row.

So also, $$W \subseteq W'.$$

Hence, $$W = W'.$$

Theorem 9. Any two row equivalent matrices A and B have the same row spaces.

Proof. By the hypothesis, we have a sequence $A = A_0, A_1, ..., A_t = B$ such that either $A = B$ or each A_{i+1} is obtainable from A_i by an elementary row operation.

By Lemma 8,

$$\text{row space } (A) = \text{row space } (A_1) = ... = \text{row space } (A_t) = \text{row space } (B).$$

Definition 10. Any matrix E obtained from an identity matrix I_n by an elementary row operation is called an *elementary matrix*. Then, E is said to be of type (*I*), (*II*) or (*III*), according as the elementary row operation used to obtain E is of type (*I*), (*II*) or (*III*) respectively.

Example 10.a. Consider $$I_3 = \begin{bmatrix} 1 & 0 & 0 \\ 0 & 1 & 0 \\ 0 & 0 & 1 \end{bmatrix}.$$

Apply $r_2 \leftrightarrow r_3$. This gives $$E_1 = \begin{bmatrix} 1 & 0 & 0 \\ 0 & 0 & 1 \\ 0 & 1 & 0 \end{bmatrix},$$

an elementary matrix of type (*I*).

Apply $r_1 \to 5r_1$ to I_3. This gives $$E_2 = \begin{bmatrix} 5 & 0 & 0 \\ 0 & 1 & 0 \\ 0 & 0 & 1 \end{bmatrix},$$

an elementary matrix of type (*II*).

Apply $r_2 \to r_2 - 2r_3$ to I_3. This gives

$$E_3 = \begin{bmatrix} 1 & 0 & 0 \\ 0 & 1 & -2 \\ 0 & 0 & 1 \end{bmatrix},$$

an elementary matrix of type (*III*).

Theorem 11. Let A be an $m \times n$-matrix over a field F and B be obtained from A by an elementary row operation θ. Let E be the elementary matrix obtained from I_m by applying θ. Then, $EA = B$.

Proof. Let $A = [a_{ij}]$, $B = [b_{ij}]$, $EA = D = [d_{ij}]$, $E = [e_{ij}]$, $I_n = [\delta_{ij}]$.

Observe that $$[\delta_{i1}, \delta_{i2}, \ldots, \delta_{in}]A = A_i,$$
the ith row of A.

Case I. B is obtained from A by interchanging r^{th} and s^{th} rows.

Then, $B_r = A_s$, $B_s = A_r$, $B_i = A_i$ for $i \neq r, s$. For $i \neq r, s$, E_i is the same as the ith row of I_n, so that E_iA, the ith row of EA is the same as the ith row of $I_nA = A$, i.e., for $i \neq r, s$, $D_i = A_i = B_i$. Now, E_r being the sth row of I_n, gives the rth row of D,

$$D_r = E_rA = [\delta_{s1}, \delta_{s2}, \ldots, \delta_{sn}]A = A_s = B_r.$$

Similarly, $$D_s = B_s.$$

Hence, $$D = B.$$

Case II. B is obtained from A by multiplying the rth row of A by $a \neq 0$ in F.

Then, $$B_r = aA_r$$

and $$B_i = A_i \qquad \text{for } i \neq r.$$

For $i \neq r$, $$E_i = [\delta_{i1}, \delta_{i2}, \ldots, \delta_{in}]$$

gives $$D_i = E_iA = A_i = B_i.$$

As $$E_r = a[\delta_{r1}, \delta_{r2}, \ldots, \delta_{rn}],$$

$$D_r = E_rA = a[\delta_{r1}, \delta_{r2}, \ldots, \delta_{rn}]A = aA_r = B_r.$$

Hence, $$D = B.$$

Case III. B is obtained from A by adding c-times s^{th} row of A to the rth row of A.

Then, for $i \neq r$, $$B_i = A_i,$$

and as in case II, $$D_i = B_i.$$

Now, $$B_r = A_r + cA_s.$$

$$E_r = (I_n)_r + c(I_n)_s,$$

yields $$D_r = E_rA = (I_n)_rA + c(I_n)_sA = A_r + cA_s = B_r.$$

Once again, $$D = B.$$

Hence, $$EA = B.$$

Proposition 12. Any elementary matrix E, over a field F, is non-singular. Further, E and E^{-1} are of the same type.

Proof. Case I. If E is obtained from I_n by interchanging r^{th} and s^{th} rows, then, I_n can be obtained from E by interchanging r^{th} and s^{th} rows. So, by Theorem 11, $I_n = EE$. Hence, $E^{-1} = E$.

Case II. E is obtained from I_n by multiplying the rth row of I_n by a scalar $c \neq 0$. Then, I_n is obtained from E by multiplying the rth row of E by c^{-1}. Therefore, if E' is the elementary matrix obtained from I_n by muliplying the rth row of I_n by c^{-1}, then, $I_n = EE' = E'E$. Once again, E and $E' = E^{-1}$ are of the same type.

Case III. E is obtained from I_n by adding to the rth row, c times to the sth row, $r \neq s$. Then, I_n can be obtained from E by adding $(-c)$-times s^{th} row of E to the rth row of E. Let E' be the elementary matrix corresponding to the later elementary row operation E and E' are both of the same type.

By Theorem 11, $$I_n = E'E = EE'.$$

Remark 13. In analogy with elementary row operations, we can define corresponding elementary column operations. This, in turn, gives rise to elementary (column) matrices. Observe the following :

(i) If E' is obtained by interchanging the rth and the sth columns of I_n, then E' is the elementary matrix E, obtained by interchanging the rth and the sth rows of I_n.

(ii) If E' is obtained by multiplying the rth column of I_n by a scalar $c \neq 0$, then, E' is the elementary matrix E, obtained by multiplying the rth row of I_n by c.

(iii) If E' is obtained from I_n by adding c times the sth column to the rth column, then, E' is the elementary matrix E, obtained from I_n, by adding c times the rth row to the sth row.

(iv) If an $n \times n$-matrix B is obtained from A by an elementary column operation θ, and E' is the corresponding elementary (column) matrix, then, as in Theorem 11, $B = AE'$.

Theorem 14. Let A and B be two row equivalent $m \times n$-matrices over a field F. Then, $B=PA$ for some non-singular matrix P. Furthermore, if $m = n$, then A is non-singular if B is non-singular.

Proof. If $B = A$, $B = I_n A$. Let $B \neq A$.

We have a finite sequence of matrices

$$A = C_0,\ C_1, \ldots,\ C_s = B$$

where each $m \times n$ matrix C_{i+1} is obtained from C_i by an elementary row operation.

So by Theorem 11, $C_{i+1} = E_i C_i$ for some elementary matrix E_i.

This gives $$B = E_{s-1}\, E_{s-2} \ldots,\ E_0 A = PA$$

where $$P = E_{s-1}\, E_{s-2} \ldots\, E_0.$$

By Proposition 12, each E_i is non-singular. So, P is non-singular. The last part is now obvious.

Consider a matrix $A = \left[a_{ij}\right]_{m \times n}$ with $a_{11} = 1$.

Then, $$\begin{aligned} A_i - a_{i1}A_1 &= \left[a_{i1},\ a_{i2}, \ldots,\ a_{in}\right] - \left[a_{i1},\ a_{i1}a_{12}, \ldots, a_{i1}a_{1n}\right] \\ &= \left[0,\ a_{i2} - a_{i1}a_{12}, \ldots, a_{in} - a_{i1}a_{1n}\right]. \end{aligned}$$

Thus, by subtracting a_{i1} times A_1 from A_i, we get an n-tuple with the first component zero. Such a technique will be quite often adopted.

Theorem 15. An $n \times n$-matrix A over a field F is non-singular iff A is row equivalent to I_n.

Proof. Let A be non-singular. We prove that A is row equivalent to I_n. Let $A = [a_{ij}]$. As A is non-singular, its first column is non-zero. So, for some i, $a_{i1} \neq 0$. By interchanging the first and the ith rows of A, we get a matrix A_1, row equivalent to A, with $(1, 1)^{th}$ entry $a_{i1} \neq 0$. By operating $r_1 \to a_{i1}^{-1} r_1$ on A_1, we get matrix A_2, row equivalent to A, with $(1, 1)^{th}$ entry 1. Let $A_2 = [a'_{ij}]$. Here, $a'_{11} = 1$. For $i \geq 2$, apply operations $r_i \to r_i - a_{i1} r_1$ to A_2. As explained, just before this theorem, we get $n \times n$-matrix $B_1 = [b_{ij}]$ with $b_{11} = 1$ and $b_{i1} = 0$ for $i \geq 2$. Then, B_1 is row equivalent to A, so that B_1 is also non-singular (Theorem 14). B_1 looks like

$$B_1 = \begin{bmatrix} I_1 & X_1 \\ 0 & B'_1 \end{bmatrix}$$

where $I_1 = [1]$, $X_1 = [b_{12}, b_{13}, \ldots, b_{1n}]$ and B'_1 is the $(n-1) \times (n-1)$ submatrix, obtained from B by deleting the first row and the first column of B_1. Suppose for some k, $1 \leq k < n$, we obtain a matrix B_k, row equivalent to A, such that

$$B_k = \begin{bmatrix} I_k & X_k \\ 0 & B'_k \end{bmatrix}$$

where X_k is $k \times (n-k)$-matrix and B'_k is $(n-k) \times (n-k)$-matrix.

We claim that the first column of B'_k has a non zero entry. Suppose the contrary holds. Then, the $(k+1)^{th}$ column C_{k+1} of B_k is

$$\begin{bmatrix} b_1 \\ b_2 \\ \vdots \\ b_k \\ 0 \\ \vdots \\ 0 \end{bmatrix}.$$

For $1 \leq i \leq k$, the ith column C_i of B_k is

$$C_i = \begin{bmatrix} 0 \\ 0 \\ \vdots \\ 1 \\ 0 \\ \vdots \\ 0 \end{bmatrix} \text{..}i^{th} \text{ place.}$$

Then, $$C_{k+1}-\sum_{i=1}^{k}b_iC_i=0.$$

Thus, by applying to B_k, the elementary column operations $c_{k+1}\to c_{k+1}-b_ic_i$, $1\le i\le k$, in succession, we get an $n\times n$-matrix D column equivalent B_k, with its $(k+1)^{\text{th}}$ column zero. Therefore, D is singular. However, as B_k is non-singular, D must be non-singular (as in Theorem 14). This is a contradiction. Hence, B'_k has its first column non-zero. That means in B_k, for some $i\ge k+1$, the $(i,\ k+1)^{\text{th}}$ entry is non-zero. Apply $r_{k+1}\leftrightarrow r_i$. As all entries in B_k to the left of B'_k are zeros, we get a matrix L_k such that

$$L_k=\begin{bmatrix} I_k & X_k \\ 0 & B''_k \end{bmatrix}$$

with its $(k+1,\ k+1)^{\text{th}}$ entry, say $c\ne 0$.

By applying elementary row operation $r_{k+1}\to c^{-1}r_{k+1}$, we get L'_k, row equivalent to L_k (hence, to A), with $(k+1,k+1)^{\text{th}}$ entry 1. By subtracting appropriate multiples of the $(k+1)^{\text{th}}$ row of L'_k from other rows, we get a matrix B_{k+1}, such that, in its $(k+1)^{\text{th}}$ column, all the entries, except the $(k+1,\ k+1)^{\text{th}}$ are zeroes and the $(k+1,\ k+1)^{\text{th}}$ entry is 1. These operations do not change the first k columns.

Consequently, $$B_{k+1}=\begin{bmatrix} I_{k+1} & X_{k+1} \\ 0 & B'_{k+1} \end{bmatrix}.$$

Hence, by induction on k, we get $B_n=I_n$.

Therefore, A is row-equivalent to I_n. Conversely, if A is row equivalent to I_n. by Theorem 14, A is non-singular.

Corollary 16. An $n\times n$-matrix A, over a field F, is non-singular iff A is a product of elementary matrices.

Proof. Let A be non-singular. Then, I_n is row equivalent to A. As in the proof of Theorem 14,

$$A=E_tE_{t-1}\dots E_1I_n=E_tE_{t-1}\dots E_1$$

for some elementary matrices E_i, $1\le i\le t$. Conversely, if A is a product of elementary matrices, obviously, A is non-singular, as every elementary matrix is non-singular.

We can now strengthen Theorem 14.

Theorem 17. Let A and B be two $m\times n$-matrices over a field F. Then, A and B are row equivalent iff $B=PA$ for some non-singular matrix P.

Proof. In view of Theorem 14, we need to prove that if $B=PA$, then, A and B are row equivalent. By Corollary 16, $P=E_uE_{u-1}\dots E_1$ for some elementary matrices E_i. Put $C_0=A$, $C_1=E_1A,\dots,C_{i+1}=E_{i+1}C_i$. Then, $B=C_u$. Clearly, by Theorem 11, C_{i+1}, is row equivalent to C_i. Consequently, by using transitivity of row equivalence, we get, that B is row equivalent to A.

Given a non-singular ,matrix A, as A is row equivalent to I_n, $I_n = E_t E_{t-1} \dots E_1 A$ for some elementary matrices E_i, $1 \le i \le t$. Then, $A^{-1} = E_t E_{t-1} \dots E_1 I_n$ shows that if θ_1, θ_2, ..., θ_t is a finite sequence of elementary row operations on A, that gives I_n, then, the applications of the same sequence to I_n gives A^{-1}. This observation gives a practical method of finding A^{-1}. Start with any $n \times n$-matrix A. Form the $n \times 2n$-matrix $B = [A : I_n]$. On B, apply a sequence of elementary row operations, with the objective of getting I_n from A. These operations get applied to A as well as to I_n. If, on the way, in the part corresponding to A, we get row or column, entirely of zeroes, then, A is singular. On the other hand, we reach $[I_n : P]$, with $P = A^{-1}$. We shall illustrate this and Theorem 15 by the following example :

Example 17.a. Let
$$A = \begin{bmatrix} 0 & 2 & 1 \\ 3 & 1 & 4 \\ 2 & 4 & 6 \end{bmatrix}.$$

We form
$$B = \begin{bmatrix} 0 & 2 & 1 & : & 1 & 0 & 0 \\ 3 & 1 & 4 & : & 0 & 1 & 0 \\ 2 & 4 & 6 & : & 0 & 0 & 1 \end{bmatrix}.$$

Observe that in the first column of A, the first non-zero entry is in the second row. To get it to (1,1)-place, apply $r_1 \leftrightarrow r_2$.

This gives
$$B = \begin{bmatrix} 1 & 0 & 0 & -\frac{41}{18} & -\frac{53}{18} \\ 0 & 1 & 0 & \frac{11}{9} & \frac{4}{9} \\ 0 & 0 & 1 & -\frac{25}{18} & \frac{13}{18} \end{bmatrix}$$

(we could have used $r_1 \leftrightarrow r_3$, instead).

We want (1, 1)-entry 1. This we achieve by $r_1 \to \frac{1}{3} r_1$. We get

$$\begin{bmatrix} 1 & \frac{1}{3} & \frac{4}{3} & : & 0 & \frac{1}{3} & 0 \\ 0 & 2 & 1 & : & 1 & 0 & 0 \\ 2 & 4 & 6 & : & 0 & 0 & 1 \end{bmatrix}.$$

Now, (3, 1) entry is non-zero; it is 2. To get 0 instead, apply $r_3 \to r_3 - 2r_1$.

This gives
$$\begin{bmatrix} 1 & \frac{1}{3} & \frac{4}{3} & : & 0 & \frac{1}{3} & 0 \\ \cdots & \cdots & \cdots & \cdots & \cdots & \cdots & \cdots \\ 0 & 2 & 1 & : & 1 & 0 & 0 \\ 0 & \frac{10}{3} & \frac{10}{3} & : & 0 & -\frac{2}{3} & 1 \end{bmatrix}.$$

Look at the submatrix below the dotted line. In its first column, the very first entry is $\neq 0$. So to get 1, at the (2, 2)-place, apply $r_2 \to \frac{1}{2} r_2$.

This gives $\left[\begin{array}{ccc:ccc} 1 & \frac{1}{3} & \frac{4}{3} & 0 & \frac{1}{3} & 0 \\ 0 & 1 & \frac{1}{2} & \frac{1}{2} & 0 & 0 \\ 0 & \frac{10}{3} & \frac{10}{3} & 0 & -\frac{2}{3} & 1 \end{array}\right]$

(instead, one could have applied $r_3 \to \frac{3}{10} r_3$, followed by $r_2 \leftrightarrow r_3$).

We want all the entries in the second column, except the (2, 2)-entry to be zeroes. The (1, 2)-entry is $\frac{1}{3}$. Apply $r_1 \to r_1 - \frac{1}{3} r_2$; the (3, 2) entry is $\frac{10}{3}$. Apply $r_3 \to r_3 - \frac{10}{3} r_2$.

We get $\left[\begin{array}{ccc:ccc} 1 & 0 & \frac{7}{6} & -\frac{1}{6} & \frac{1}{3} & 0 \\ 0 & 1 & \frac{1}{2} & \frac{1}{2} & 0 & 0 \\ 0 & 0 & \frac{10}{6} & -\frac{10}{6} & -\frac{2}{3} & 1 \end{array}\right]$.

In the third row, the first non-zero entry is $\frac{10}{6}$. Apply $r_3 \to \frac{6}{10} r_3$.

This gives $\left[\begin{array}{ccc:ccc} 1 & 0 & \frac{7}{6} & -\frac{1}{6} & \frac{1}{3} & 0 \\ 0 & 1 & \frac{1}{2} & \frac{1}{2} & 0 & 0 \\ 0 & 0 & 1 & -1 & -\frac{2}{5} & \frac{6}{10} \end{array}\right]$.

Apply $r_1 \to r_1 - \frac{7}{6} r_3$, $r_2 \to r_2 - \frac{1}{2} r_3$.

This gives $\left[\begin{array}{ccc:ccc} 1 & 0 & 0 & 1 & \frac{4}{5} & -\frac{7}{10} \\ 0 & 1 & 0 & 1 & \frac{1}{5} & -\frac{3}{10} \\ 0 & 0 & 1 & -1 & -\frac{2}{5} & \frac{6}{10} \end{array}\right] = [I_3 : P]$

where $P = \begin{bmatrix} 1 & \frac{4}{5} & -\frac{7}{10} \\ 1 & \frac{1}{5} & -\frac{3}{10} \\ -1 & -\frac{2}{5} & \frac{6}{10} \end{bmatrix} = A^{-1}$.

Remark 18. Analogous results for elementary column operations are the following :

(*i*) An $n \times n$-matrix A, over a field F, is non-singular iff A is column equivalent to I_n.

(*ii*) Two $m \times n$-matrices A and B, over a field F, are column equivalent iff $B = AP$ for some non-singular matrix P iff A^T and B^T are row equivalent.

Remark 19. Let A and B be two $m \times n$-matrices over F, such that A and B are row equivalent. Let $\theta_1, \theta_2, \ldots, \theta_t$ be a finite sequence of elementary row operations that give B from A. Form the $m \times (n+m)$-matrix $C = [A : I_m]$.

Apply the sequence $\theta_1, \theta_2, \ldots, \theta_t$ to C. This will give $[B : P]$ where P is non-singular and $B = PA$.

20. Solutions of a system of linear equations. Consider a system of m linear equations in n variables, over a field F

$$\left.\begin{array}{l} a_{11}x_1 + a_{12}x_2 + \ldots + a_{1n}x_n = b_1 \\ a_{21}x_1 + a_{22}x_2 + \ldots + a_{2n}x_n = b_2 \\ \ldots\ldots\ldots\ldots\ldots\ldots\ldots\ldots\ldots\ldots \\ \ldots\ldots\ldots\ldots\ldots\ldots\ldots\ldots\ldots\ldots \\ a_{m1}x_1 + a_{m2}x_2 + \ldots + a_{mn}x_n = b_m \end{array}\right] \quad \ldots (1)$$

Let us name them as $E_1, E_2, \ldots, E_m$, respectively. So, E_i is the ith equation. Consider any $c_1, c_2, \ldots, c_m \in F$. Then, the equation E :

$$c_1\left[\sum_{j=1}^{n} a_{1j}x_j\right] + c_2\left[\sum_{j=1}^{n} a_{2j}x_j\right] + \ldots + c_m\left[\sum_{j=1}^{n} a_{mj}x_j\right] = \sum_{i=1}^{m} c_i b_i$$

that is
$$\left[\sum_{i=1}^{m} c_i a_{i1}\right]x_1 + \left[\sum_{i=1}^{m} c_i a_{i2}\right]x_2 + \ldots + \left[\sum_{i=1}^{m} c_i a_{in}\right]x_n = \sum_{i=1}^{m} c_i b_i \quad \ldots (2)$$

is called a *linear combination* of the system of linear equations, given above.

We write
$$E = \sum_{i=1}^{m} c_i E_i.$$

For any linear equation, $\quad a_1x_1 + a_2x_2 + \ldots + a_nx_n = b$,

let us call the row matrices $[a_1, a_2, \ldots, a_n]$ and $[a_1, a_2, \ldots, a_n, b]$ its *coefficient row* and its *augmented row*. It is clear from (2) that the augmented row of E is

$$\left[\sum_{i=1}^{m} c_i a_{i1}, \sum_{i=1}^{m} c_i a_{i2}, \ldots, \sum_{i=1}^{m} c_i a_{in}, \sum_{i=1}^{m} c_i b_i\right] = [c_1, c_2, \ldots, c_m][A : B],$$

and that E is simply
$$[c_1\ c_2\ \ldots\ c_m]AX = [c_1\ c_2\ \ldots\ c_m]B. \quad \ldots (3)$$

Therefore, if
$$U = \begin{bmatrix} u_1 \\ u_2 \\ \vdots \\ u_n \end{bmatrix}$$

is a solution of (1), then, $AU = B$ gives

$$[c_1\ c_2\ \ldots\ c_m]AU = [c_1, c_2, \ldots, c_m]B.$$

Consequently, any solution of (1) is a solution of any linear combination of the equations in (1). This discussion leads to the following.

Lemma 21. Let $AX = B$ be any system of m linear equations in n variables. Any linear combination of equations of this system is of the form

$$[c_1\ c_2 \ldots c_m]AX = [c_1\ c_2\ \ldots\ c_m]B.$$

Further, any solution of the system of linear equations is also a solution of any of its linear combination.

The following theorem is a step toward, giving an algorithm for finding the general solution of a system of linear equations.

Theorem 22. Let

$$AX = B \qquad \text{... (1)}$$

$$A_1X = B_1 \qquad \text{... (2)}$$

be two systems of linear equations, over a field F, such that their augmented matrices $[A : B]$ and $[A_1 : B_1]$ are row equivalent. Then, the two systems have the same sets of solutions.

Proof. As $[A : B]$ and $[A_1 : B_1]$ are row equivalent, by Theorem 9, they have the same row spaces. Therefore, any row of $[A_1 : B_1]$ is a linear combination of the rows of $[A : B]$. Any row of $[A_1 : B_1]$ is an augmented row of an equation in (2). Therefore, any equation in system (2) is a linear combination of equations in (1), and any solution of (1), by Lemma 21, is a solution of every equations in (2), i.e., it is a solution of (2). Similarly, every solution of (2) is a solution of (1).

Example 22.a. Consider the following system of three linear equations

$$\begin{aligned} x+y+\ z&=4 \\ x-y+2z&=1 \\ y+\ z&=3 \end{aligned}$$

It has unique solution $x = 1$, $y = 1$, $z = 1$. Its augmented matrix is

$$[A : B] = \begin{bmatrix} 1 & 1 & 1 & : & 4 \\ 1 & -1 & 2 & : & 1 \\ 0 & 1 & 1 & : & 3 \end{bmatrix}$$

Consider any non-singular 3×3-matrix P, say

$$P = \begin{bmatrix} 1 & 2 & 1 \\ -1 & 0 & 3 \\ 0 & 0 & 5 \end{bmatrix}$$

Then,

$$P[A : B] = \begin{bmatrix} 3 & 0 & 6 & : & 9 \\ -1 & 2 & 2 & : & 5 \\ 0 & 5 & 5 & : & 15 \end{bmatrix}$$

is row equivalent to $[A : B]$.

This gives a system of linear equations

$$\begin{aligned} 3x \qquad +6z&=9 \\ -x+2y+2z&=5 \\ 5y+5z&=15 \end{aligned}$$

This has the same solution as the previous system. viz., $x = 1$, $y = 2$, $z = 1$.

Example 22.b. Consider a system of linear equations with augmented matrix

$$[A:B]=\begin{bmatrix}1 & -1 & 0 & 0 & 2 & 1 & : & 3\\ 0 & 0 & 1 & 0 & 0 & -1 & : & 4\\ 0 & 0 & 0 & 1 & 1 & 1 & : & 5\end{bmatrix}$$

This system is

$$\begin{aligned} x_1 - x_2 \quad + 2x_5 + x_6 &= 3\\ x_3 \qquad\qquad - x_6 &= 4\\ x_4 \quad + x_5 + x_6 &= 5 \end{aligned}$$

This gives

$$\begin{aligned} x_1 &= x_2 - 2x_5 - x_6 + 3\\ x_3 &= \qquad\qquad x_6 + 4\\ x_4 &= \qquad - x_5 - x_6 + 5 \end{aligned}$$

Clearly x_2, x_5 and x_6 can be given any arbitrary values, there-after the above equations can be solved for x_1, x_3 and x_4. Thus, the given system gets immediately solved. This is so because of the special nature of the matrix.

Let A be a non-zero matrix. Its first non-zero column (counting from the left) is called its *pivotal column*. The first non-zero element in the pivotal column is called its pivot. In any non-zero row, the first non-zero entry is called its *leading entry*.

For example, in $A=\begin{bmatrix}0 & 0 & 2 & 1 & 5\\ 0 & -1 & 1 & 0 & 0\\ 0 & 0 & 2 & 0 & 0\end{bmatrix}$,

the second column is the pivotal column.

In the second column, the first non-zero element is −1. So, −1 is the pivot. In the third row, the leading entry is 2.

Definition 23. An $m\times n$-matrix A, over a field F, is said to be in *reduced row echelon* form if it satisfies the following :

1. If any row of A is a zero row, it must be below any non-zero row.
2. Leading entry of any non-zero row is 1.
3. If for any $i<m$, both the ith and the $(i+1)^{\text{th}}$ rows are non-zero, then, the leading entry of the $(i+1)^{\text{th}}$ row is to the right of the leading entry of the ith row, i.e., if the leading entries of the ith and the $(i+1)^{\text{th}}$ rows occur in $k_i{}^{\text{th}}$ and $k_{i+1}{}^{\text{th}}$ columns, respectively, then $k_i<k_{i+1}$.
4. If a column of A contains the leading entry of some non-zero row, then, all the other entries in that column are zeroes.

In Example 22.b, the augmented matrix is in reduced row echelon form. We are going to prove that any matrix is row-equivalent to a matrix in reduced row echelon form. To appreciate better the arguments in the proof, we discuss a specific example.

Example 23.a. Let $A=\begin{bmatrix}0 & 1 & 2 & 4 & -1\\ 1 & 3 & -1 & 0 & 5\\ 2 & 0 & 4 & 1 & 3\end{bmatrix}$

Observe that the leading entry of the first row is to the right of the leading entry of the second row. So, it does not satisfy condition (3) of the definition, given. Consequently, A is not in reduced row echelon form. Its pivotal column is the first column and its pivot is 1, that occurs in the second row. it must be brought to the first row. For this, apply $r_1 \leftrightarrow r_2$.

This yields $$\begin{bmatrix} 1 & 3 & -1 & 0 & 5 \\ 0 & 1 & 2 & 4 & -1 \\ 2 & 0 & 4 & 1 & 3 \end{bmatrix}$$

In the first column, we need all entries except the (1, 1)th entry zeros. Add suitable multiples of the first row to the other rows (see Theorem 15). Here, apply $r_3 \to r_3 - 2r_1$.

This gives $$A_1 = \begin{bmatrix} 1 & 3 & -1 & 0 & 5 \\ 0 & 1 & 2 & 4 & -1 \\ 0 & -6 & 6 & 1 & -7 \end{bmatrix}$$

Examine the submatrix, obtained by deleting the first row and the first column of the matrix, given above. Locates its pivot. It occurs at the (2, 2)th place in A_1 and it is already 1 (if it were some $c \neq 0$, we would have applied the operation $r_2 \to c^{-1} r_2$ to A_1, to get 1 at the (2, 2)th place). Apply elementary row operations to get all entries in the second column, other than the (2, 2)th entry zero, in such a way that the first column remains unaffected. This is possible by adding appropriate multiples of the second row to the other rows. The operations applied are $r_1 \to r_1 - 3r_2$, $r_3 \to r_3 + 6r_2$.

This gives $$\begin{bmatrix} 1 & 0 & -7 & -12 & 8 \\ 0 & 1 & 2 & 4 & -1 \\ 0 & 0 & 18 & 25 & -13 \end{bmatrix}$$

Leading entry in the third row is 18. Apply $r_3 \to \frac{1}{18} r_3$.

This gives $$\begin{bmatrix} 1 & 0 & -7 & -12 & 8 \\ 0 & 1 & 2 & 4 & -1 \\ 0 & 0 & 1 & \frac{25}{18} & -\frac{13}{18} \end{bmatrix}$$

Leading entry of the third row is in the third column. So, all the entries in the third column, except at the (3, 3)th place, must be made zeroes. For this, apply $r_1 \to r_1 + 7r_3$, $r_2 \to r_2 - 2r_3$.

This gives $$B = \begin{bmatrix} 1 & 0 & 0 & -\frac{41}{18} & -\frac{53}{18} \\ 0 & 1 & 0 & \frac{11}{9} & -\frac{4}{9} \\ 0 & 0 & 1 & -\frac{25}{18} & \frac{13}{18} \end{bmatrix}$$

Clearly, B is in reduced row echelon form. B is row equivalent to A.

Observe that any zero matrix and any identity matrix are in reduces row echelon form. The proof of the following is rather technical, but can be understood by keeping in view Example 23.a.

Theorem 24. Any matrix, over a field F, is row equivalent to a matrix in reduced row echelon form.

Proof. Let $A = [a_{ij}]$ be an $m \times n$-matrix over F. If $A = 0$, obviously, it is in reduced row echelon form. Let $A \neq 0$. Let k be the smallest positive integer such that the kth column of A is non-zero. Let i be the smallest integer such that $(i, k)^{\text{th}}$ entry $a_{ik} \neq 0$. Then, a_{ik} is the pivot of A. Apply the elementary row operation $r_1 \leftrightarrow r_i$, followed by $r_1 \to a_{ik}^{-1} r_1$. This gives a matrix A_1 whose pivotal column is the kth column and whose $(1, k)^{\text{th}}$ entry is 1. A_1 looks like

$$A_1 = \begin{bmatrix} & & k^{\text{th}} & & \\ 0 & 0 & 1 & * & .* \\ 0 & 0 & * & * & .* \\ \vdots & \vdots & * & * & .* \\ 0 & 0 & * & * & .* \end{bmatrix}$$

where * indicates arbitrary entries from F (not necessarily the same at different places of occurrences of *). As its first $k-1$ columns are zero columns, no elementary row operation on A_1 will change them. As in the proof of Theorem 15, subtraction of suitable multiples of the first row of A, from others, leads to a matrix A_2, whose k^{th} column has $(1, k)^{\text{th}}$ entry 1, and all the other entries zero. A_2 is row equivalent to A, and it looks like

$$\begin{bmatrix} 0\ldots & 0 & 1 & S \\ 0\ldots & 0 & 0 & \\ \vdots & \vdots & \vdots & B \\ \vdots & \vdots & \vdots & \\ 0 & 0 & 0 & \end{bmatrix}$$

where S is $1 \times (n - k_1)$-matrix and B is $(m-1) \times (n - k_1)$-matrix, with $k_1 = k$.

If $B = 0$, or $m = 1$, A_2 is in reduced row echelon form and we finish. Let $m > 1$ and $B \neq 0$. Observe that in the submatrix of A_2, obtained by deleting the first row, at least, the first $k_1 + 1$ columns are zero columns. As $B \neq 0$, we get the smallest k_2 such that the $k_2{}^{\text{th}}$ column of A_2 contains the first non-zero column of this submatrix. Then, $k_1 < k_2$, and we get an $i \geq 2$ such that in the ith row of A_2, the leading entry is at the $(i, k_2)^{\text{th}}$ place. Let this entry be $c \neq 0$. Then, the operation $r_2 \leftrightarrow r_i$, followed by $r_2 \to c^{-1} r_2$, gives a matrix A_3 such that A_2 and A_3 have first rows the same, first k_1 columns the same, all entries in A_3 below the first row to the left of kth column zeroes and the $(2, k_2)^{\text{th}}$ entry 1. By subtracting suitable multiples of second row from the others rows, all the others entries in the $k_2{}^{\text{th}}$ column can be reduced to zero. This gives matrix A_4. If $m = 2$, A_4 is in the reduced row echelon form. If $m > 2$, A_4 looks like

$$\begin{bmatrix} & & k_1^{\text{th}} & & k_2^{\text{th}} & & \\ 0\ldots & 0 & 1 & * & 0 & *\ldots & * \\ 0 & 0 & 0 & & 1 & & \\ \vdots & \ldots & & & 0 & & C \\ \vdots & \ldots & & & \vdots & & \\ 0 & 0 & 0 & . & 0 & & \end{bmatrix}$$

where C is an $(m-2)\times(n-k_2)$-submatrix. If $C=0$, we finish. If $c\neq 0$, continue the process as with B, and so on. As m is finite, this process finishes after a finite number of steps. We arrive at a matrix A' which is in the reduced row echelon form and which is row equivalent to A.

25. Gauss-Jordan Method. Let $AX = B$ be a system of linear equations. Form the augmented matrix $\left[A : B\right]$. Apply elementary row operations to this matrix to obtain a matrix $\left[A_1 : B_1\right]$ in the reduced row echelon form. Solve the system of linear equations $A_1X = B_1$. As $\left[A_1 : B_1\right]$ is row equivalent to $\left[A : B\right]$, the solution, if any, of $A_1X = B_1$, is a solution of $AX = B$. This procedure is called *Gauss-Jordan method.*

Example 25.a. Consider the system of two homogeneous linear equations

$$\left.\begin{array}{l} x-y+2z \quad\quad =0 \\ x \quad\quad + z + u = 0 \end{array}\right] \quad \text{... (1)}$$

in four variables x, y, z and u.

Its augmented matrix is $[A : 0] = \begin{bmatrix} 1 & -1 & 2 & 0 & : & 0 \\ 1 & 0 & 1 & 1 & : & 0 \end{bmatrix}$

It is not in the reduced row echelon form. Apply $r_2 \to r_2 - r_1$

This gives $\begin{bmatrix} 1 & -1 & 2 & 0 & : & 0 \\ 0 & 1 & -1 & 1 & : & 0 \end{bmatrix}$

Apply $r_1 \to r_1 + r_2$.

We get $\begin{bmatrix} 1 & 0 & 1 & 1 & : & 0 \\ 0 & 1 & -1 & 1 & : & 0 \end{bmatrix} = [A_1 : 0]$

in the reduced row echelon form.

This gives system $\left.\begin{array}{l} x+z+u=0 \\ y-z+u=0 \end{array}\right].$ ... (2)

Systems (1) and (2) have the same solutions. Observe that in $\left[A_1 : 0\right]$, the leading entries of the first row and the second row are in the first and the second columns, respectively. These columns are given by the coefficients of x and y, respectively. In (2), we shift the other variables z and u to the other side.

So that
$$x = -z-u$$
$$y = z-u$$

Assign to z and u any values, say $z = r$, $u = s$.

Then, $x=-r-s,\ y=r-s$, so that $X=\begin{bmatrix}-r-s\\ r-s\\ r\\ s\end{bmatrix}$

is a solution of (2) and, hence, of (1). This is the general solution of (1). It involves two parameters, r and s. By giving to r and s particular values, we get, the so called particular solutions. Thus, $r=1,\ s=-1$, gives the solution

$$\begin{bmatrix}0\\ 2\\ 1\\ -1\end{bmatrix}$$

Observe that the number of parameters is equal to the number of variables minus the number of non-zero rows in the matrix in the reduced row echelon form.

We now lay down an algorithm to write general solution of $AX=0$, in case A is in the reduced row echelon form.

Theorem 26. Let $AX=0$ be a system of m homogeneous linear equations in n variables, over a field F, such that the coefficient matrix $A\neq 0$, and is in the reduced row echelon form. Let its first s rows be non-zero and the remaining m-s rows be zero rows. For $1\le i\le s$, let the leading entry of the ith row be in the $k_i{}^{\text{th}}$ column. Then, the following hold

1. If $n=s$, the only solution is $x_i=0$ for $1\le i\le n$.
2. Let $s<n$. Then, there exist solutions with any pre-assigned values of the n-s variables, other than $x_{k_1}, x_{k_2},\ldots, x_{k_s}$. Further, if these n-s variables are named as $u_1, u_2,\ldots, u_{n-s}$, then, there exist $c_{ij}\in F$ such that

$$x_{k_i}=\sum_{j=1}^{n-s}c_{ij}u_j,\ 1\le i\le s$$

Proof. Let $A=\left[a_{ij}\right]_{m\times n}$. As the last m-s rows of A are zero rows, the last m-s equations have zero coefficients and, hence, they are redundant.

For $1\le i\le s$, consider $\qquad \displaystyle\sum_{j=1}^{n}a_{ij}x_j=0$ ... (1)

By definition, $\qquad a_{ik_i}=1$

and $\qquad a_{i'k_i}=0$ for $i'\neq i$

As $\qquad 1\le k_1<k_2<\ldots<k_s\le n,$

we get $\qquad s\le n.$

Case I. $\qquad s=n$

Then, $$1 \le k_1 < k_2 < \ldots < k_n \le n$$

gives $$k_j = j \qquad \text{for } i \le j \le n,$$

so that $$a_{jj} = 1$$

$$a_{ij} = 0 \qquad \text{for } i \ne j.$$

So, equation (1) gives $$a_{ii} x_i = 0$$

i.e. $$x_i = 0$$

Hence, $x_i = 0$, $1 \le i \le n$, is the only solution.

Case II. $$s < n$$

As $$a_{ik_i} = 1,$$

(1) gives $$x_{k_i} = -\sum_{j \ne x_{k_i}} a_{ij} x_j \qquad \ldots (2)$$

For $j = k_l \ne k_i$, $a_{ik_l} = 0$, so that the term with variable x_{k_l}, $l \ne i$, does not appear on the right hand side of (2). Consequently, if we rename the remaining *n-s* variables as $u_1, u_2, \ldots, u_{n-s}$, the right hand side of (2) is an expression, in these variables, so that (2) can be written as

$$x_{k_i} = \sum_{j=1}^{n-s} c_{ij} u_j \qquad \ldots (3)$$

This equation shows that, given any arbitrary values of $u_1, u_2, \ldots, u_{n-s}$, by using (3), we get corresponding values of x_{k_i}, $1 \le i \le s$. This completes the proof.

Remark 27. **In the notation of the theorem, given above, the general solution of $AX = 0$ is given by**

$$X = [x_1, x_2, \ldots, x_n]^T$$

such that for $j \ne k_i$, $1 \le i \le s$, x_j can take any arbitrary value, and

$$x_{k_i} = -\sum_{j \ne k_i} a_{ij} x_j \qquad \ldots (1)$$

Let $$S = \{1, 2, \ldots, n\} - \{k_1, k_2, \ldots, k_s\}$$

Then, (1) can be written as $$x_{k_i} = -\sum_{j \in S} a_{ij} x_j \qquad \ldots (2)$$

For each $j_0 \in S$, by taking $x_{j_0} = -1$ and $x_j = 0$ for $j \ne j_0$ in S, (2) gives

$$x_{k_i} = a_{ij_0}$$

This gives a solution $$U_{j_0} = \left[u_{1j_0}, u_{2j_0}, \ldots, u_{nj_0}\right]^T \qquad \ldots (3)$$

with k_i th component $u_{k_i j_0} = a_{ij_0}$, the j_0 th component -1, and all the other components (total $n-s-1$ in number) zeroes.

The set W of all solutions of $AX = 0$ is non-empty, as $0 \in F^{n\times 1}$ is in W.

For any $X_1, X_2 \in W,\ c \in F,\ AX_1 = 0,\ AX_2 = 0,$

gives $$A(X_1 + X_2) = AX_1 + AX_2 = 0,$$

so that $$X_1 + X_2 \in W$$

Again, $$A(cX_1) = cAX_1 = c0 = 0,$$

gives $$cX_1 \in W$$

Hence, W is a subspace of $F^{n\times 1}$. This W is called the *solution space*. Each $U_{j_0} \in W$. By using (2) and (3), we get that any $X = [x_1, x_2, \ldots, x_t]^T \in W$ satisfies

$$X = \sum_{j\in S} (-x_j)\, U_j$$

a linear combination of U_j's.

Let $$X = \sum_{j\in S} b_j U_j = 0 \qquad \ldots (4)$$

for some $b_j \in F$.

By comparing, the j_0 th component on both sides of (4) with $j_0 \in S$, we get

$$\sum_{j\in S} b_j u_{j_0 j} = 0 \qquad \ldots (5)$$

But by (3), $u_{j_0 j} = 0$ for $j \neq j_0$ and $u_{j_0 j_0} = -1$. We get $b_{j_0} = 0$. Hence, $b_j = 0$ for every $j \in S$. Given $X_1, X_2, \ldots, X_k \in F^{n\times 1}$, they are said to be *linearly independent* in $F^{n\times 1}$, if for any $a_1, a_2, \ldots, a_k \in F$, $a_1X_1 + a_2X_2 + \ldots + a_kX_k = 0$ implies $a_1 = 0 = a_2 = \ldots = a_k$. Thus, U_j's for $j \in S$ are linearly independent.

Remark 28.

(*i*) Consider a system of homogeneous linear equations $AX = 0$. Let A_1 be a matrix in the reduced row echelon form, row equivalent to A. If the number s of non-zero rows of A_1 is less than n, then, by Theorem 26, $n-s$ variables can be given any arbitrary values. So, by giving these variable some non-zero values, we get solutions $X = [x_1, x_2, \ldots, x_n]^T$ with some $x_i \neq 0$. Such solutions are called *non-trivial solutions*. If $s = n$, then, the system has only trivial solution.

(*ii*) Let $AX = 0$ be a system of m homogeneous linear equations in n variables such $n > m$. Then, $A = [a_{ij}]$ is an $m \times n$-matrix. If A_1, in reduced row echelon form, is row equivalent to A, then, the number s of rows of A satisfies $s \le m < n$. So, by (*i*), $AX = 0$ has a non-trivial solution. Consequently, if the number of equations in a system of homogeneous linear equations is less than the number of variables, the system has a non-trivial solution.

Example 28.a. Consider, the system of homogeneous linear equations

$$\begin{aligned} x_1+2x_2+\ x_3+2x_4+\ x_5&=0\\ x_1+2x_2+2x_3+\ x_4+2x_5&=0\\ 2x_1+2x_2+3x_3+3x_4+3x_5&=0\\ x_3-\ x_4-\ x_5&=0 \end{aligned}$$

The augmented matrix is

$$[A:0]=\begin{bmatrix} 1 & 2 & 1 & 2 & 1 & : & 0\\ 1 & 2 & 2 & 1 & 2 & : & 0\\ 2 & 4 & 3 & 3 & 3 & : & 0\\ 0 & 0 & 1 & -1 & -1 & : & 0 \end{bmatrix} \qquad \text{... (1)}$$

This is row equivalent to

$$[A_1:0]=\begin{bmatrix} 1 & 2 & 0 & 3 & 0 & : & 0\\ 0 & 0 & 1 & -1 & 0 & : & 0\\ 0 & 0 & 0 & 0 & 1 & : & 0\\ 0 & 0 & 0 & 0 & 0 & : & 0 \end{bmatrix} \qquad \text{... (2)}$$

Here, $n=5$, $s=3$, $n-s=2$. Leading entries are in the first, third and fifth columns, so variables x_2 and x_4 can take arbitrary values.

Now, (2) gives

$$\begin{aligned} x_1+2x_2+3x_4&=0\\ x_3-\ x_4&=0\\ x_5&=0 \end{aligned}$$

Therefore,

$$\begin{aligned} x_1&=-2x_2-3x_4\\ x_3&=\ 0x_2+\ x_4\\ x_5&=\ 0x_2+0x_4 \end{aligned}$$

So, the general solution is

$$X=\begin{bmatrix} x_1\\ x_2\\ x_3\\ x_4\\ x_5 \end{bmatrix}$$

$$=\begin{bmatrix} -2x_2-3x_4\\ x_2\\ x_4\\ x_4\\ 0 \end{bmatrix}$$

$$= (-x_2)\begin{bmatrix} 2 \\ -1 \\ 0 \\ 0 \\ 0 \end{bmatrix} + (-x_4)\begin{bmatrix} 3 \\ 0 \\ -1 \\ -1 \\ 0 \end{bmatrix}$$

$$= (-x_2)U_2 + (-x_4)U_4$$

where $U_2 = \begin{bmatrix} 2 \\ -1 \\ 0 \\ 0 \\ 0 \end{bmatrix}$, and $U_4 = \begin{bmatrix} 3 \\ 0 \\ -1 \\ -1 \\ 0 \end{bmatrix}$

Observe that in U_2 and U_4, the second and the fourth components, respectively, equal −1.

For any $b_2, b_4 \in R$ $\quad b_2U_2 + b_4U_4 = 0$

gives $\quad b_2 = 0,$

by comparing the second components, $b_4 = 0$, by comparing the fourth components.

Observe that U_2 and U_4 are non-trivial solutions.

Example 28.b. Consider the system of three linear equations

$$x - y + z = 2$$
$$x + y - z = 0$$
$$6x - 4y + 4z = 11$$

Its augmented matrix is $\begin{bmatrix} 1 & -1 & 1 & : & 2 \\ 1 & 1 & -1 & : & 0 \\ 6 & -4 & 4 & : & 11 \end{bmatrix}$

It is row equivalent to $\begin{bmatrix} 1 & 0 & 0 & : & 0 \\ 0 & 1 & -1 & : & 0 \\ 0 & 0 & 0 & : & 1 \end{bmatrix}$

The third row of this matrix gives $0x + 0y + 0z = 1$

This is impossible. Hence, the given system of linear equations does not have a solution.

Definition 29. A system of linear equations $AX = B$, over a field F, is said to be a *consistant system*, if it has solution; otherwise, it is said to be *inconsistant*.

Theorem 30. Let $A = [a_{ij}]$ and $B = [b_{ij}]$ be two $m \times n$-matrices over F in reduced row echelon form. If A and B have the same row spaces, then, $A = B$.

Proof. We apply induction on m. Let $m = 1$.

Then, $$A = [a_{11}, a_{12}, \dots, a_{1n}]$$

If $A = 0$, then obviously, $B = 0$ and $A = B$. Let $A \neq 0$ and its leading entry be in the kth column. Then, $a_{1k} = 1$ and $a_{ij} = 0$ for $j < k$.

Now, $$[b_{11}, b_{12}, \dots, b_{1n}] = c[a_{11}, a_{12}, \dots, a_{1n}]$$

As $a_{ij} = 0$ for $j < k, b_{1j} = ca_{1j} = 0$,

so that if the leading entry of B is $b_{1k'}$, then, $k' \geq k$

Similarly, $A = c'B$ for some $c' \in F$,

gives $k \geq k'$

Hence, $k = k'$

Then, $1 = b_{1k'} = ca_{1k} = c$,

so that $A = B$.

Let $m > 1$ and let the result hold in case the number of rows is less than m. Let A and B have s and s' non-zero rows, respectively. For $1 \leq i \leq s$, let the leading entry of the ith row A_i of A be in the $k_i{}^{\text{th}}$ column, and for $1 \leq j \leq s'$, let the leading entry of the jth row B_j of B be in the $k_i'{}^{\text{th}}$ column.

Now, $$B_1 = \sum_{j=1}^{m} c_{1j} A_j = \sum_{j=1}^{s} c_{1j} A_j \qquad \dots (1)$$

for some $c_{1j} \in F$.

In A_j, the first $k_1 - 1$ components are zero. So, (1) shows that, in B_1, the first $k_1 - 1$ components are zeros.

Consequently, $k_1 \leq k_1'$.

Similarly, $k_1' \leq k_1$.

Hence, $k_1 = k_1'$.

For $i \geq 2$ $$B_i = c_{i1} A_1 + \sum_{j=2}^{m} c_{ij} A_j \qquad \dots (2)$$

As $k_1 < k_i'$, $b_{ik_1} = 0$.

Also, for $j \geq 2$, $a_{jk_1} = 0$. So, by comparing the $k_1{}^{\text{th}}$ components in (2), we get

$$0 = c_{i1} a_{1k_1}$$

This yields $c_{i1} = 0$, as $a_{1k_1} = 1$.

Consequently, (2) becomes $$B_i = \sum_{j=2}^{m} c_{1j} A_j \qquad \dots (3)$$

Thus, if A' and B' are the submatrices of A and B, obtained by dropping their first rows, we find that A' and B' have the same row-spaces. Now, A' and B' each, is a matrix in the reduced row echelon form, and has $m-1$ rows.

Hence, by the induction hypothesis, $A' = B'$.

So, $$s = s',\ k_i = k_i',\ 1 \le i \le s.$$

Consider a fixed $$j_0,\ 2 \le j_0 \le s.$$

Then, $$b_{1k_{j_0}} = 0 = a_{jk_{j_0}} \qquad \text{for } j \ne j_0.$$

By (1), $$0 = b_{1k_{j_0}} = \sum_j c_{1j} a_{jk_{j_0}} = c_{1j_0} a_{j_0 k_{j_0}} = c_{1j_0},$$

so that $$B_1 = c_{11} A_1.$$

Then, $$1 = b_{1k_1} = c_{11} a_{1k_1} = c_{11}.$$

This gives $$B_1 = A_1.$$

As $$A' = B',$$

we have $$B_i = A_i \qquad \text{for } 2 \le i \le m.$$

Hence, $$A = B$$

This proves the theorem.

The following theorem improves Theorem 24.

Theorem 31. Any $m \times n$-matrix A, over a field F, is row equivalent to a unique $m \times n$-matrix over F in the reduced row echelon form.

Proof. By Theorem 24, A is row equivalent to a matrix B, in the reduced row echelon form. Let A be also row equivalent to a matrix B' in the reduced row echelon form. By Theorem 9, B and B' have the same row space. Hence, by Theorem 30, $B = B'$

Theorem 32. Any two $m \times n$-matrices A and B, over a field F, are row equivalent iff they have the same row space.

Proof. Let A and B be row equivalent, then, by Theorem 9, they have the same row space. Conversely, let A and B have the same row space. Let A_1 and B_1 be matrices in the reduced row echelon form row equivalent to A and B, respectively. Then, A_1 and B_1 have the same row space. By Theorem 30, $A_1 = B_1$. Hence, A and B are row equivalent.

Exercise 1.4

1. List the elementary row operation, used to obtain each of the following matrices, from an identity matrix

$$\begin{bmatrix} 0 & 1 \\ 1 & 0 \end{bmatrix}, \begin{bmatrix} 1 & 0 \\ 0 & -3 \end{bmatrix}, \begin{bmatrix} 1 & -2 \\ 0 & 1 \end{bmatrix}$$

$$\begin{bmatrix} 0 & 0 & 1 \\ 0 & 1 & 0 \\ 1 & 0 & 0 \end{bmatrix}, \begin{bmatrix} 1 & 0 & 0 \\ 0 & 2 & 0 \\ 0 & 0 & 1 \end{bmatrix}, \begin{bmatrix} 1 & 0 & 0 \\ 0 & 1 & 0 \\ 0 & 3 & 1 \end{bmatrix}.$$

2. Which of the following are elementary matrices? If any one is an elementary matrix, find its type and its inverse

$$\begin{bmatrix} 0 & 0 & 1 \\ 1 & 0 & 0 \\ 0 & 1 & 0 \end{bmatrix}, \begin{bmatrix} 3 & 0 & 0 \\ 0 & 1 & 0 \\ 0 & 0 & 1 \end{bmatrix}, \begin{bmatrix} 1 & 0 & 0 \\ 0 & 1 & 0 \\ 0 & 1 & -1 \end{bmatrix}, \begin{bmatrix} 1 & 5 \\ 0 & 1 \end{bmatrix}$$

3. Let
$$A = \begin{bmatrix} 1 & 2 & 3 & 0 \\ 0 & 1 & 5 & 7 \\ 1 & 1 & 0 & 4 \end{bmatrix}.$$

(*i*) Apply elementary row operations of different types to obtain four matrices row equivalent to A

(*ii*) Prove that $B = \begin{bmatrix} 1 & 1 & 0 & 4 \\ -1 & 0 & 5 & 3 \\ 1 & 4 & 11 & 3 \end{bmatrix}$ is row equivalent to A. Find a non-singular 3×3-matrix P such that $B = PA$.

4. Find matrix B in the reduced row echelon form, row equivalent to $A = \begin{bmatrix} 1 & 2 & -1 \\ 3 & 2 & 3 \\ 2 & 2 & 1 \end{bmatrix}$. Conclude that A is singular.

5. Find, if possible, the inverse of each of the following matrices

(*i*) $\begin{bmatrix} 2 & 1 \\ 5 & 3 \end{bmatrix}$

(*ii*) $\begin{bmatrix} 1 & 2 & 3 \\ 0 & 1 & 1 \\ 0 & 0 & 1 \end{bmatrix}$

(*iii*) $\begin{bmatrix} 1 & 1 & 1 & 1 \\ 1 & 2 & -1 & 2 \\ 1 & -1 & 2 & 1 \\ 1 & 3 & 3 & 2 \end{bmatrix}$

(*iv*) $\begin{bmatrix} 0 & 1 & 1 \\ 1 & 0 & 1 \\ 0 & 0 & 2 \end{bmatrix}$

(v) $$\begin{bmatrix} 2 & 1 & -1 & 3 \\ 0 & 1 & 5 & 2 \\ 1 & 0 & 1 & 0 \\ 0 & 0 & 1 & 1 \end{bmatrix}$$

6. If $A^{-1} = \begin{bmatrix} 5 & -1 \\ 2 & 3 \end{bmatrix}$, find A.

7. Find all those real numbers a such that

$$\begin{bmatrix} 1 & 2 & 0 \\ -1 & a & 3 \\ 1 & 2 & 1 \end{bmatrix}$$

is non-singular.

8. Find the general solution of the system of homogeneous linear equations

$$\begin{aligned} x + y + 3z - 3u &= 0 \\ 2y + z - 3u &= 0 \\ x + 2z - u &= 0 \end{aligned}$$

Write down a non-trivial solution of this system.

9. Let A be any $n \times n$-matrix over a field F. Prove that A is singular iff the matrix B is in the reduced row echelon form, row equivalent to A, has less than n non-zero rows.

10. By using the result of Question 9, prove the following :
A system of homogeneous linear equations

$$AX = 0,\ A = \left[a_{ij}\right]_{n\times n},$$

over a field F, has a non-trivial solution iff A is singular.

11. Solve the system of linear equations

$$\begin{aligned} 2x + y + z &= 9 \\ 2x + 4y + 3z &= 14 \\ x + y + z &= 6 \end{aligned}$$

12. Solve, if possible, the system :

$$\begin{aligned} x + y + 2z + 3w &= 12 \\ x - 2y + z + w &= 18 \\ 3x + z - w &= 1 \end{aligned}$$

13. If A is a non-singular $n \times n$-matrix, over a field F, prove that the system of linear equations $AX = B$ has unique solution.

14. Find all values of a for which the resulting system, given below, has (*i*) no solution (*ii*) a unique solution (*iii*) infinitely many solutions

$$\begin{aligned} x+2y- \quad z&= \quad 4 \\ 2x-3y- \quad 5z&= \quad 4 \\ x+2y+(a^2-5)z&=a+2 \end{aligned}$$

15. Let $A=\begin{bmatrix}3 & 2 & 1\\ 0 & 4 & 2\\ 0 & 0 & 4\end{bmatrix}$. Find the values of λ for which $\lambda I_3 - A$ is singular.

16. Consider $X_1=[2,\ 2,\ -3]$, $X_2=[2,\ -1,\ 2]$, $X_3=[4,\ -1,\ 0]$ in $\mathbf{R}^{1\times 3}$ Which of the following vectors in $\mathbf{R}^{1\times 3}$ are linear combinations of X_1, X_2, X_3?

    (*a*) $[4,\ 2,\ -5]$ (*b*) $[9,\ -1,\ 1]$ (*c*) $[2,\ 0,\ -2]$

17. Consider any system of linear equations $AX=B$. Let $[A_1 : B_1]$ be in the reduced row echelon form, row equivalent to $[A : B]$. Prove that $AX=B$ is inconsistant iff $[A_1 : B_1]$ has a row of the form $[0,\ 0,\ldots 0,\ 1]$.

Some Books on abstract algebra

1. Frank Ayres, Jr., Theory and Problems of Modern Algebra, Schaum's Outline Series, McGraw Hill, 1965.
2. David M. Burton, Abstract Algebra, WM. C Brown Publishers, Dubuque, Iowa, 1988.
3. P. B. Bhattacharya, S. K. Jain and S. R. Nagpaul, Basic Abstract Algebra, Cambridge University Press, 1986.
4. N. Block, Abstract Algebra with Applications, Englewood Cliffs, N. J., Prentice Hall, 1987.
5. G. Birkhoff and S. MacLane, A Survey of Modern Algebra, Macmillan, 1977.
6. P. M. Cohn, Algebra, Vols 1 and 2, John Wiley & Sons, 1982.
7. John B. Fraleigh, A First Course in Abstract Algebra, Addison Wesley Publishing Company, 1988.
8. R. A. Deans, Elements of Abstract Algebra, John Wiley & Sons, 1986.
9. N. Jacobson, Basic Algebra I, W. H. Freeman and Company, 1985.
10. S. Lang, Algebra, Addison-Wesley Publishing Company Inc. 1984.
11. S. Singh and Q. Zameeruddin, Modern Algebra, Vikas Publishing House Pvt Ltd, 1984.
12. B. L. Van der Waerden, Algebra, Valume I, Frederick Ungar Publishing Co., 1970.

Chapter 2

Vector Spaces

2.1 Vector Space

In the last chapter, we have studied F^n, the set of all n-tuples, over a field F. This set is an abelian group under component-wise addition. Further, for each $\alpha \in F^n$ and $a \in F$, there is $a\alpha \in F^n$; this operation is known as *scalar multiplication*. The basic properties of the two operations, in F^n, are listed in Theorem (1.2.1) [This means Theorem 1 in section (1.2)]. There are several other situations where we have a set V, and a field F, such that, under some binary operation denoted usually by addition, V becomes an abelian group, and there is a scalar multiple $a\alpha \in V$ for $\alpha \in V$ and $a \in F$, such that results, similar to Theorem (1.2.1), hold for V. For example $V = F^{n\times m}$ (Theorem (1.2.13)). Another such situation is the following. Consider two real numbers $a < b$. Let $V = C[a, b]$ be the set of all continuous real valued functions on the closed interval $[a, b]$. For $f, g \in V$, $a \in R$, $x \in [a, b]$, $f+g \in V$ and $af \in V$ are such that $(f+g)(x) = f(x)+g(x)$, $(af)(x) = af(x)$. Many more examples of such situations will be given. In all such situations, we are dealing with a vector space in the following sense. Throughout, F will denote an arbitrary field.

Definition 1. Let F be a field. A *vector space over F,* is a set V, together with operation $\oplus$ in V, called addition, and an operation $\odot$, called scalar multiplication, which for each $(a,\alpha) \in F \times V$ determines $a \odot \alpha \in V$, and it satisfies the following axioms:

I. (*a*) The addition is commutative,

i.e., $\alpha \oplus \beta = \beta \oplus \alpha$ for $\alpha, \beta \in V$

(*b*) The addition is associative,

i.e., $\alpha \oplus (\beta \oplus \gamma) = (\alpha \oplus \beta) \oplus \gamma$ for $\alpha, \beta, \gamma \in V$

(*c*) There exists an element $\overline{0} \in V$ such that

$$\alpha \oplus \overline{0} = \alpha \qquad \text{for } \alpha \in V$$

$\overline{0}$ is the identity element for addition.

(*d*) For each $\alpha \in V$, there exists $\alpha' \in V$ such that

$$\alpha \oplus \alpha' = \overline{0}.$$

II. For any $\alpha, \beta \in V$, $a, b \in F$ and the identity $1 \in F$, the following hold :

(*e*) $$a \odot (\alpha \oplus \beta) = (a \odot \alpha) \oplus (a \odot \beta)$$

(*f*) $$(a + b) \odot \alpha = (a \odot \alpha) \oplus (b \odot \alpha)$$

(*g*) $$(ab) \odot \alpha = a \odot (b \odot \alpha)$$

(*h*) $$1 \odot \alpha = \alpha$$

Axioms (*a*) through (*d*) tell that $\langle V, \oplus \rangle$ is an abelian group. Axioms (*e*) through (*h*) tell that for each $a \in F$, we have a function $\sigma_a : V \to V$ such that $\sigma_a(\alpha) = a \odot \alpha$. This functions has the following properties :

(*i*) $$\sigma_a(\alpha \oplus \beta) = \sigma_a(\alpha) \oplus \sigma_a(\beta)$$

(*ii*) $$\sigma_{a+b}(\alpha) = \sigma_a(\alpha) \oplus \sigma_b(\alpha)$$

(*iii*) $$\sigma_{ab}(\alpha) = (\sigma_a . \sigma_b)(\alpha)$$

(*iv*) σ_1 is the identity function on V.

In a situation described in Definition 1, we say V is a vector space over F, or V is an F-vector space. Further, V_F denotes that V is a vector space over F. In V_F, members of V are called *vectors* and the members of F are called *scalars*, $\overline{0} \in V$ is called the *zero vector*. As in a group, the identity element is unique, V has unique zero vector. For each $\alpha \in V$, there exists unique α', called the negative of α, such that $\alpha \oplus \alpha' = \overline{0}$; this is so, as the inverse of an element in a group is unique. This α' is denoted by $-\alpha$. For the sake of convenience, we shall denote $\alpha \oplus \beta$ and $a \odot \alpha$, simply by $\alpha + \beta$ and $a\alpha$, respectively, Further, $\alpha - \beta$ will denote $\alpha + (-\beta)$. Before we give examples of vector space, we prove a basic result on a vector space.

Lemma 2. Let V be a vector space over a field F. Then :

1. For each $\alpha \in V$, $\quad -(-\alpha) = \alpha$
2. For any α, β, γ, $\quad \alpha + \beta = \alpha + \gamma$ implies $\beta = \gamma$.
3. For any $a \in F$, $\alpha \in V$ $\quad a\alpha = \overline{0}$ iff $a = 0$ or $\alpha = \overline{0}$
4. For any $a \in F$, $\alpha \in V$ $\quad -(a\alpha) = (-a)\alpha = a(-\alpha)$
5. For any $\alpha \in V$, $\quad (-1)\alpha = -\alpha$.

Proof.

(1) As $\alpha + (-\alpha) = (-\alpha) + \alpha = \overline{0}$, by the definition of $-\alpha$, $-(-\alpha) = \alpha$.

(2) Let $$\alpha + \beta = \alpha + \gamma.$$

Then, $$(-\alpha) + (\alpha + \beta) = (-\alpha) + (\alpha + \gamma)$$

By axiom (*b*), $$[(-\alpha) + \alpha] + \beta = [(-\alpha) + \alpha] + \gamma$$

However, $$(-\alpha) + \alpha = \overline{0},$$

so that $$\overline{0} + \beta = \overline{0} + \gamma,$$

i.e., $$\beta = \gamma$$

by axiom (*c*)

(3) In F, $0 = 0 + 0$.

So, by axiom (*f*), $0\alpha = 0\alpha + 0\alpha$,

i.e., $0\alpha + \overline{0} = 0\alpha + 0\alpha$

By (2), $\overline{0} = 0\alpha$.

Similarly, $a\overline{0} = \overline{0}$ for any $a \in F$

So, for $a = 0$

or $\alpha = \overline{0}$,

we have $a\alpha = \overline{0}$.

Conversely, let $a\alpha = \overline{0}$.

If $a = 0$, we finish. Let $a \neq 0$.

Then, $a^{-1}(a\alpha) = a^{-1}\overline{0} = \overline{0}$, i.e.,

$(a^{-1}a)\alpha = \overline{0}$, by axiom (*g*), i.e.,

$1\alpha = \overline{0}$.

But $1\alpha = \alpha$ by axiom (*h*)

Hence, $\alpha = \overline{0}$.

(4) and (5), $a + (-a) = 0$

It gives $\overline{0} = 0\alpha = [a + (-a)]\alpha = a\alpha + (-a)\alpha$,

so that $-(a\alpha) = (-a)\alpha$.

Similarly, $\alpha + (-\alpha) = \overline{0}$

yields $-(a\alpha) = a(-\alpha)$.

Hence, $-(a\alpha) = (-a)\alpha = a(-\alpha)$.

In particular, for $a = 1$, $(-1)\alpha = -(1\alpha) = -\alpha$

As remarked in the introduction, F^n, $F^{m\times n}$ are vector spaces over F, under usual operations of addition and scalar multiplication. Also, $C[a, b]$ is a vector space over R.

Example 2.a. Let S be any non-empty set, and V be the set of all functions from S into F. For any f, $g \in V$ and $a \in F$, we define

$$f + g : S \to F,$$
$$af : S \to F$$

such that for any $x \in S$, $(f+g)(x) = f(x) + g(x)$

$$(af)(x) = af(x)$$

As $f(x), g(x) \in F$, clearly, $f(x) + g(x) \in F$, $af(x) \in F$, so that $f + g \in V$, $af \in V$. These two operations turn V into a vector space over F. We shall verify a few axioms, defining a vector space, for V. Let $f, g, h \in V$, $a, b \in F$, $x \in S$.

Then, $(f+g)(x) = f(x) + g(x)$,

$$(g+f)(x) = g(x) + f(x)$$

But in F, $f(x)+g(x)=g(x)+f(x)$.

So, $(f+g)(x)=(g+f)(x)$.

Hence, $f+g=g+f$

This proves that addition is commutative. Define $\overline{0}: S \to F$ such that $\overline{0}(x)=0$ for every $x \in S$.

Then, $(f+\overline{0})(x)=f(x)+\overline{0}(x)=f(x)+0=f(x)$,

gives $f+\overline{0}=f$.

This mapping $\overline{0}$ is called zero mapping. For $f \in V$, define $-f: S \to F$ such that

$$(-f)(x)=-f(x).$$

Then, $f+(-f)=\overline{0}$.

Now,
$$\begin{aligned}[a(f+g)](x) &= a[(f+g)(x)] \\ &= a[f(x)+g(x)] \\ &= af(x)+ag(x) \\ &= (af+ag)(x).\end{aligned}$$

This gives $a(f+g)=af+ag$

Obviously, $(1f)(x)=1f(x)=f(x)$;

gives $1f=f$

Example 2.b. In the foregoing example, a special case of interest is, when $S=N$ the set of natural numbers. Any function $f: N \to F$ is called a *sequence* over F. For any $k \in N$, $f(k)$ is called the kth term of f.

We write $f=\{f(0), f(1), f(2), \ldots, f(k), \ldots\}$

or, simply. $f=(s_k)_{k\geq 0}$,

$$s_k=f(k).$$

Thus $g=(1,\ -1,\ 1,\ -1,\ 1,\ -1, \ldots)$

is a sequence such that $g(0)=1$, $g(1)=-1$.

In general, $g(k)=(-1)^{k+1}$

The vector space V of all sequences over F consists of all $(s_k)_{k\geq 0}$, $s_k \in F$ such that for any $f=(s_k)$, $g=(s'_k)$, $a \in F$

$$f+g=(s_k+s'_k)_{k\geq 0}$$
$$af=(as_k)_{k\geq 0}$$

Example 2.c. Any vector space V over R is called a *real vector space* and any vector space over $\mathbf{C}$ is called a *complex vector space*. The real vector space $\mathbf{R}^2$ has special geometric significance. It is called a Euclidean plane and its members are called points. The set of all points $(x, 0) \in \mathbf{R}^2$, is called the X-axis and the set of points $(0, y) \in \mathbf{R}^2$, $y \in \mathbf{R}$, is called the Y-axis. The intersection of X-axis and the Y-axis consists of only one point (0, 0); this point is called the *origin*.

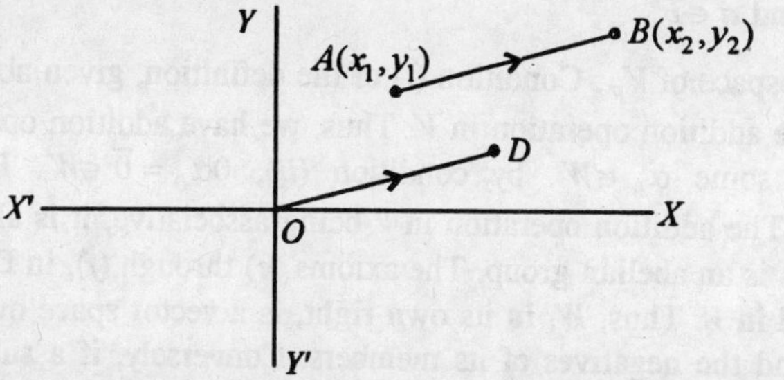

A geometric vector is a quantity that has magnitude and direction. It is represented by a directed line segment $\vec{AB}$, where $A(x_1, y_1)$, $B(x_2, y_2)$ are two points, such that the length $\sqrt{(x_1 - x_2)^2 + (y_1 - y_2)^2}$ of AB represents the magnitude of the vector, and θ the angle that $\vec{AB}$ makes with the X-axis, gives the direction of the vector. This vector can be represented by any other line segment $\vec{A'B'}$, provided $\vec{A'B'}$ has the same direction and length as $\vec{AB}$. In particular, draw $\vec{OD}$ such that $\vec{OD}$ has the same direction and magnitude as $\vec{AB}$. It can be easily seen that D is the point $(x_2 - x_1, y_2 - y_1)$. So, the vector $\vec{AB}$ is representable by

$$\begin{bmatrix} x_2 - x_1 \\ y_2 - y_1 \end{bmatrix} \in \mathbf{R}^{2\times 1}$$

The following diagram shows that the parallelogram law of addition of geometric vectors is the same as the addition operation in $\mathbf{R}^{2\times 1}$

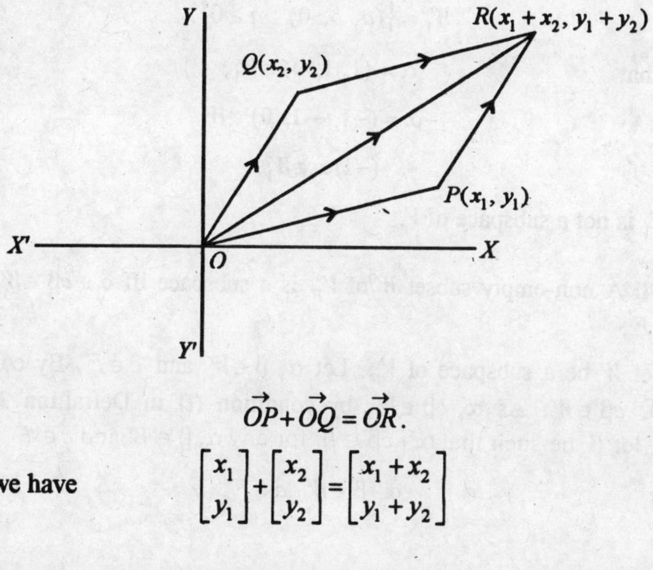

Here, $\vec{OP} + \vec{OQ} = \vec{OR}$.

In $\mathbf{R}^{2\times 1}$, we have $\begin{bmatrix} x_1 \\ y_1 \end{bmatrix} + \begin{bmatrix} x_2 \\ y_2 \end{bmatrix} = \begin{bmatrix} x_1 + x_2 \\ y_1 + y_2 \end{bmatrix}$

Definition 3. A non-empty subset W of a vector space V_F is called a *subspace* of V_F, if

(*i*) $\alpha + \beta \in W$

(*ii*) $a\alpha \in W$

for all $\alpha,\ \beta \in W$ and $a \in F$

Let W be a subspace of V_F. Condition (*i*) of the definition, given above, tells that W is closed under the addition operation in V. Thus, we have addition operation in W. As $W \neq \phi$, we have some $\alpha_0 \in W$. by condition (*ii*), $0\alpha_0 = \overline{0} \in W$. For any $\alpha \in W$, $-\alpha = (-1)\alpha \in W$. The addition operation in V being associative, it is also associative in W, so that $\langle W, +\rangle$ is an abelian group. The axioms (*e*) through (*f*), in Definition 1, hold in W, as they hold in V. Thus, W, in its own right, is a vector space over F. It contains the zero vector and the negatives of its members. Conversely, if a subset W of V is a vector space over F, under the operations in V, it has to be non-empty and $\alpha + \beta \in W$, $a\alpha \in W$ for all $\alpha,\ \beta \in W$ and $a \in F$.

$\{\overline{0}\}$ is a subspace of V, called the *zero subspace.* Further, V is its own subspace; these two are called *trivial subspace*. Any subspace W, other than $\{\overline{0}\}$ and V is called *proper subspaces* of V. Henceforth, we shall also write 0 for $\overline{0}$, whenever convenient.

Example 3.a. Let $V = \mathbf{R}^3$, $W = \{(a,b,0) : a,b \in \mathbf{R}\}$. $\overline{0} = (0,0,0) \in W$ so that $W \neq \phi$. Let $\alpha,\ \beta \in W,\ c \in \mathbf{R}$

Then,
$$\alpha = (a_1,\ b_1,\ 0),$$
$$\beta = (a_2,\ b_2,\ 0)$$

for some $a_i,\ b_i \in \mathbf{R}$.
$$\alpha + \beta = (a_1 + a_2,\ b_1 + b_2,\ 0) \in W,$$
$$c\alpha = (ca_1,\ cb_1,\ 0) \in W.$$

Hence, W is a subspace of V.

Consider
$$W_1 = \{(a,\ b,\ 0) : a \geq 0\}.$$

Observe that
$$\alpha = (1,\ 1,\ 0) \in W_1,$$

but
$$-\alpha = (-1,\ -1,\ 0) \notin W_1$$

i.e.
$$(-1)\alpha \notin W_1.$$

Hence, W_1 is not a subspace of V.

Lemma 4. A non-empty subset W of V_F is a subspace iff $\alpha + c\beta \in W$ for any $\alpha,\ \beta \in W$ and $c \in F$.

Proof. Let W be a subspace of V_F. Let $\alpha,\ \beta \in W$ and $c \in F$. By condition (*ii*) in Definition 3, $c\beta \in W$. As $\alpha,\ c\beta \in W$, by condition (*i*) in Definition 3, $\alpha + c\beta \in W$. Conversely, let W be such that $\alpha + c\beta \in W$ for any $\alpha,\ \beta \in W$ and $c \in F$.

Consider
$$\alpha,\ \beta \in W,\ c \in F.$$

As $-1 \in F,\ \alpha + (-1)\alpha \in W$

i.e., $0 \in W$.

Again, $\alpha + \beta = \alpha + 1\beta \in W$

As $0 \in W,\ \alpha \in W$,

we get $0 + c\alpha \in W$

i.e., $c\alpha \in W$.

Hence, W is a subspace of V.

Theorem 5. Let W_1 and W_2 be two subspaces of V_F. Then,

(*i*) $W_1 \cap W_2$ is a subspace of V

(*ii*) The sum $W_1 + W_2 = \{\alpha + \beta : \alpha \in W_1,\ \beta \in W_2\}$ is a subspace of V containing $W_1 \cup W_2$.

Proof.

(*i*) As $0 \in W_1 \cap W_2,\ W_1 \cap W_2 \neq \phi$.

Consider $\alpha,\ \beta \in W_1 \cap W_2,\ c \in F$.

Then, $\alpha,\ \beta \in W_1,\ c \in F$.

By Lemma 4, $\alpha + c\beta \in W_1$.

Similarly, $\alpha + c\beta \in W_2$.

So, $\alpha + c\beta \in W_1 \cap W_2$.

By Lemma 4, $W_1 \cap W_2$ is a subspace of V.

(*ii*) $0 = 0 + 0 \in W_1 + W_2$,

as $0 \in W_1$

and $0 \in W_2$,

so that $W_1 + W_2 \neq \phi$.

Let $\alpha, \beta \in W_1 + W_2$

and $c \in F$.

By definition $\alpha = \alpha_1 + \alpha_2,\ \beta = \beta_1 + \beta_2$

for some $\alpha_1,\ \beta_1 \in W_1$

and $\alpha_2,\ \beta_2 \in W_2$.

Consequently, $\alpha + c\beta = \alpha_1 + \alpha_2 + c(\beta_1 + \beta_2)$

$= (\alpha_1 + c\beta_1) + (\alpha_2 + c\beta_2) \in W_1 + W_2$, by Lemma 4.

Hence, $W_1 + W_2$ is a subspace of V.

Consider $\alpha_1 \in W_1$

and $\alpha_2 \in W_2$.

Then, $\alpha_1 = \alpha_1 + 0 \in W_1 + W_2,$ as $0 \in W_2$

$\alpha_2 = 0 + \alpha_2 \in W_1 + W_2,$ as $0 \in W_1$.

Hence, $W_1 \cup W_2 \subseteq W_1 + W_2$

Remark 6. By similar arguments as in the preceding theorem, we get that the intersection of any family of subspaces of a vector space V_F is a subspace of V_F.

Example 6.a. Let $V = \mathbf{R}^3$. The *XY*-plane is $W_1 = \{(x, y, 0) : x, y \in \mathbf{R}\}$ and the *XZ*-plane is $W_2 = \{(x, 0, z) : x, z \in \mathbf{R}\}$. These are subspaces of $\mathbf{R}^3$.

Then, $W_1 \cap W_2 = \{(x, 0, 0) : x \in \mathbf{R}\}$

is the *X*-axis.

Given any $\alpha = (x, y, z) \in V$,

$$\alpha = (x, y, 0) + (0, 0, z) \in W_1 + W_2,$$

as $(x, y, 0) \in W_1$

and $(0, 0, z) \in W_2$,

so that $V \subseteq W_1 + W_2 \subseteq V$

Hence, $V = W_1 + W_2$

i.e., $\mathbf{R}^3$ is a sum of the *XY*-plane and the *YZ*-plane.

Example 6.b. Let $V_{\mathbf{R}} = \mathbf{R}^{2\times 2}$. Let W_1 be the set of all symmetric matrices in V, and W_2 be the set of all skew-symmetric matrices in V.

As $0^T = 0 = -0$,

$0 \in W_1$

and $0 \in W_2$,

so that $W_1 \neq \phi$, $W_2 \neq \phi$.

Consider $A, B \in W_1$, $c \in \mathbf{R}$.

Then, $A^T = A$, $B^T = B$.

By Theorem (1.2.16), $(A + cB)^T = A^T + cB^T = A + cB$,

so that $A + cB \in W_1$.

By Lemma 4, W_1 is a subspace of V. Similarly, W_2 is a subspace of V.

Let $A \in W_1 \cap W_2$.

Then, $A \in W_1$,

gives $A^T = A$,

and $$A \in W_2$$

gives $$A^T = -A.$$

Thus $A = -A$ and, hence $A = 0$. This proves that $W_1 \cap W_2 = \{0\}$. Consider any $B \in V$.

We have $$B = \frac{1}{2}(B + B^T) + \frac{1}{2}(B - B^T)$$

Consider $$B_1 = \frac{1}{2}(B + B^T),\ B_2 = \frac{1}{2}(B - B^T).$$

By Theorem (1.2.16),

$$B_1^T = \frac{1}{2}\{B^T + (B^T)^T\} = \frac{1}{2}(B^T + B) = B_1$$

So that $$B_1 \in W_1.$$

Similarly, $$B_2 \in W_2.$$

Thus, $$B = B_1 + B_2$$

yields $$V \subseteq W_1 + W_2 \subseteq V.$$

Hence, $$V = W_1 + W_2,\ W_1 \cap W_2 = \{0\}.$$

Definition 7.

(*i*) Let $\alpha_1, \alpha_2, \ldots, \alpha_k$ be any $k \geq 1$ vectors in a vector space V_F. Then, any vector $\alpha \in V$ of the form

$$\alpha = a_1\alpha_1 + a_2\alpha_2 + \ldots a_k\alpha_k.$$

with $a_i \in F$, is called a *linear combination* of $\alpha_1, \alpha_2, \ldots, \alpha_k$.

(*ii*) Let S be any subset of V_F. Then, the *span* W of S is the set of those $\alpha \in V$ suc that either $\alpha = 0$ or α is a linear combination of some members $\alpha_1, \alpha_2, \ldots \alpha$ with $k \geq 1$, of S.

Remark 8.

(*i*) For $S = \{0\}$, as $a0 = 0$ for any $a \in F$, the span of $\{0\}$ is $\{0\}$

(*ii*) Let S be any subset of V. The span of S will be denoted by $\langle S\rangle$. For any $\alpha \in S$, $\alpha = 1\alpha_1$ with $\alpha_1 = \alpha$, shows that $\alpha \in \langle S\rangle$. Hence, the span $\langle S\rangle$ contains S.

(*iii*) Let W be a subspace of V. Let $\alpha \in \langle W\rangle$. For any $\alpha_1, \alpha_2, \ldots, \alpha_k \in W$, $\alpha_1 + \alpha_2 + \ldots + \alpha_k \in W$. If $\alpha = 0$, obviously $\alpha \in W$. Let $\alpha \neq 0$.

By definition, $$\alpha = c_1\beta_1 + c_2\beta_2 + \ldots + c_k\beta_k$$

for some $$\beta_i \in W,\ c_i \in F,$$

and $$k \geq 1.$$

Then, $$c_i\beta_i \in W_i.$$

Consequently, $$\alpha \in W.$$

Hence, $$W = \langle W \rangle,$$

i.e., the span of a subspace W of V is W itself.

(*iv*) Span of the empty subset ϕ is $\{0\}$.

Example 8.a. Let $$V = F^3$$

$$\in_1 = (1,\ 0,\ 0),\ \in_2 = (0,\ 1,\ 0)\ \in_2 = (0,\ 0,\ 1).$$

Let $$S = \{\in_1,\ \in_2,\ \in_3\}.$$

Given, $$\alpha = (a,\ b,\ c) \in V,$$

$$\alpha = (a,\ 0,\ 0) + (0,\ b,\ 0) + (0,\ 0,\ c)$$

$$= a \in_1 + b \in_2 + c \in_3$$

Thus, $$\alpha \in \langle S \rangle.$$

This gives $$V \subseteq \langle S \rangle \subseteq V,$$

i.e., $$V = \langle S \rangle.$$

Example 8.b. For any $m \times n$-matrix A in $F^{m \times n}$, the row space of A is the span of the set S of all the m rows of A. The *column space* W' of A is the span of the set S' of all the n columns of A.

Example 8.c. Let $$V = \mathbf{R}^3.$$

$$S = \{\alpha_1 = (1,\ 1,\ 0),\ \alpha_2 = (0,\ -1,\ 1),\ \alpha_3 = (1,\ 0,\ 1)\}.$$

Consider $\alpha = (a,\ b,\ c) \in V$. Let us determine the conditions on a, b, c so that $\alpha \in \langle S \rangle$. Let $\alpha \in \langle S \rangle$. Then, for some $a_1,\ a_2,\ a_3 \in \mathbf{R}$

$$(a,\ b,\ c) = a_1(1,\ 1,\ 0) + a_2(0,\ -1,\ 1) + a_3(1,\ 0,\ 1).$$

By comparing different components, we get

$$a_1 + \quad 0a_2 + 1a_3 = a$$

$$a_1 + (-1)a_2 + 0a_3 = b$$

$$0.a_1 + \quad 1.a_2 + 1.a_3 = c$$

These are three linear equations in three variables $a_1,\ a_2,\ a_3$.

The augmented matrix for this system is

$$[A : B] = \begin{bmatrix} 1 & 0 & 1 & : & a \\ 1 & -1 & 0 & : & b \\ 0 & 1 & 1 & : & c \end{bmatrix}$$

This is row equivalent to $\begin{bmatrix} 1 & 0 & 1 & : & a \\ 0 & 1 & 1 & : & a-b \\ 0 & 0 & 0 & : & c+b-a \end{bmatrix}$

This matrix gives $$0a_1 + 0a_2 + 0a_3 = c + b - a.$$

So, to have solution for a_1, a_2, a_3, we need $c+b-a=0$, i.e., $a=b+c$, so that $\alpha = (a, b, c) \in \langle S \rangle$ iff $a = b + c$. Consequently, $(3, 2, 1) \in \langle S \rangle$ but $(2, 2, 1) \notin \langle S \rangle$.

Given a finite non-empty subset $B = \{\alpha_1, \alpha_2, ..., \alpha_k\}$ of V_F, we say $\alpha \in V$ is a *linear combination* of B if α is a linear combination of its members $\alpha_1, \alpha_2, ..., \alpha_k$. Thus, for any subset S of V_F, any $\alpha \in \langle S \rangle$ iff $\alpha = 0$ or α is a linear combination of a finite non-empty subset of S.

Lemma 9. Let $B = \{\alpha_1, \alpha_2, ..., \alpha_k\}$ be a finite non-empty subset of V_F contained in a finite subset B' of V. Then, any α that is a linear combination of B is also a linear combination of B'.

Proof. If $B' = B$, there is nothing to prove.

Let $$B' \neq B.$$

Let $$B' - B = \{\beta_1, \beta_2, ..., \beta_s\}.$$

Consequently, $$B' = \{\alpha_1, \alpha_2, ..., \alpha_k, \beta_1, \beta_2, ..., \beta_s\}$$

By the hypothesis $$\alpha = a_1\alpha_1 + a_2\alpha_2 + ... + a_k\alpha_k,$$
$$a_i \in F.$$

Obviously, $$\alpha = a_1\alpha_1 + a_2\alpha_2 + ... + a_k\alpha_k + 0\beta_1 + 0\beta_2 + ... + 0\beta_s$$

a linear combination of B'.

Theorem 10. Let $B = \{\alpha_1, \alpha_2, ..., \alpha_k\}$ be a finite non-empty subset of V_F. Then, the span $\langle B \rangle$ is the set of all linear combinations of B. Further,

(*i*) $\langle B \rangle$ is a subspace of V

(*ii*) If a subspace W' contains B, then, $\langle B \rangle \subseteq W'$.

Proof. Clearly $0 = 0\alpha_1 + 0\alpha_2 + ... + 0\alpha_k$, is a linear combination of B. Let $\alpha \neq 0$ be in $\langle B \rangle$. By definition, α is a linear combination of a (finite) non-empty subset B' of B. By Lemma 9, as $B' \subseteq B$, α is a linear combination of B. Let $\alpha, \beta \in \langle B \rangle$ and $c \in F$.

Then $$\alpha = a_1\alpha_1 + a_2\alpha_2 + ... + a_k\alpha_k,$$
$$\beta = b_1\alpha_1 + b_2\alpha_2 + ... + b_k\alpha_k$$

for some $a_i, b_i \in F$.

Then, $$\alpha + c\beta = (a_1 + cb_1)\alpha_1 + (a_2 + cb_2)\alpha_2 + ... + (a_k + cb_k)\alpha_k$$

gives $$\alpha + c\beta \in \langle B \rangle.$$

By Lemma 4, $\langle B \rangle$ is a subspace of V. This proves (*i*). Let W be any subspace of V such that $B \subseteq W$. Then, for $1 \leq i \leq k$, $\alpha_i \in W$, gives $a_i\alpha_i \in W$, so that $\alpha = \sum_{i=1}^{k} a_i\alpha_i \in W$.

Hence $\langle B\rangle \subseteq W$.

The preceding theorem gives that $\langle B\rangle$ is the smallest subspace of V, that contains the finite subset B.

Theorem 11.

(*i*) For any subset S of V_F, the span $\langle S\rangle$ is the smallest subspace of V that contains S.

(*ii*) If $S_1 \subseteq S_2 \subseteq V$, then $\langle S_1\rangle \subseteq \langle S_2\rangle$.

(*iii*) For any two subsets S_1, S_2 of V_F,

$$\langle S_1 \cup S_2\rangle = \langle S_1\rangle + \langle S_2\rangle$$

Proof. By definition $0 \in \langle S\rangle$, so that $\langle S\rangle \neq \phi$. From Remark 8, $S \subseteq \langle S\rangle$. For $S = \phi, \langle S\rangle = \{0\}$ is clearly a subspace of V. Let $S \neq \phi$. Let $\alpha, \beta \in \langle S\rangle$ and $c \in F$. By the remark, given above, Lemma 9, there exist finite non-empty subsets B_1, B_2 of S such that $\alpha \in \langle B_1\rangle$ and $\beta \in \langle B_2\rangle$. Let $B = B_1 \cup B_2$. So, α in $\langle B_1\rangle$ is a linear combination of B by Lemma 9, i.e., $\langle B_1\rangle \subseteq \langle B\rangle$. The same way, $\langle B_2\rangle \subseteq \langle B\rangle$. Consequently, $\alpha, \beta \in \langle B\rangle$. As $\langle B\rangle$ is a subspace of V, by Theorem 10, we get $\alpha + c\beta \in \langle B\rangle$, and, hence, $\alpha + c\beta \in \langle S\rangle$. This proves that $\langle S\rangle$ is a subspace of V. Let W be any subspace of V such that $S \subseteq W$. Let $0 \neq \alpha \in \langle S\rangle$.

By definition, $$\alpha = \sum_{i=1}^{k} a_i \alpha_i$$

for some $\alpha_i \in S$, $a_i \in F$. As $\alpha_i \in W$, we get $\alpha \in W$.

Hence $\langle S\rangle \subseteq W$.

Thus $\langle S\rangle$ is the smallest subspace of V that contains S. This proves (*i*).

Let $S_1 \subseteq S_2 \subseteq V$.

Then, $S_1 \subseteq S_2 \subseteq \langle S_2\rangle$,

so that S_1 is a subset of the subspace $\langle S_2\rangle$.

Consequently, by (*i*) $\langle S_1\rangle \subseteq \langle S_2\rangle$. This proves (*ii*). Finally, consider any two subsets S_1, S_2 of V. Let $W_1 = \langle S_1\rangle$, $W_2 = \langle S_2\rangle$ and $W = W_1 + W_2$.

Now, $S_1 \subseteq W_1 \subseteq W$, $S_2 \subseteq W_2 \subseteq W$, by Theorem 5.

So, $S = S_1 \cup S_2 \subseteq W$.

Consequently, by (*i*) $\langle S\rangle \subseteq W$

Consider $\alpha \in W$.

Then, $\alpha = \alpha_1 + \alpha_2$

with $\alpha_1 \in \langle S_1\rangle, \alpha_2 \in \langle S_2\rangle$.

As $S_i \subseteq S$, $\langle S_i\rangle \subseteq \langle S\rangle$.

$$\alpha_1, \alpha_2 \in \langle S\rangle.$$

This in turn gives $\alpha = \alpha_1 + \alpha_2 \in \langle S \rangle$, $W \subseteq \langle S \rangle$.

Hence, $$\langle S \rangle = W = \langle S_1 \rangle + \langle S_2 \rangle.$$

Definition 12. A vector space V_F is said to be *spanned* by (or *generated* by) a subset S of V if the span $\langle S \rangle = V$. Further, V_F is said to be *finitely generated* if it is spanned by a finite subset.

Thus, if $V_F \neq \{0\}$ is finitely generated, we get finitely many elements $\alpha_1, \alpha_2, \ldots, \alpha_k$ in V such that any $\alpha \in V$ is a linear combination of these $\alpha_i's$.

Example 12.a. Let $V_F = F^2$, $S = \{(1, 0), (0, 1)\}$. Consider $\alpha \in V$

Then, $$\alpha = (a, b) = a(1, 0) + b(0, 1),$$

so that S spans V_F.

As S is finite, V is finitely generated. More generally, consider $V_F = F^n$, n any positive integer. Then, the n-tuples

$$\epsilon_1 = (1, 0, 0, \ldots, 0)$$
$$\epsilon_2 = (0, 1, 0, \ldots, 0)$$
$$\cdots\cdots\cdots\cdots\cdots\cdots$$
$$\epsilon_n = (0, 0, 0, \ldots, 1)$$

are in F^n.

Given, $$\alpha = (a_1, a_2, \ldots, a_n) \in F^n,$$

$$\alpha = a_1 \epsilon_1 + a_2 \epsilon_2 + \ldots + a_n \epsilon_n$$

gives that the finite subset $S = \{\epsilon_1, \epsilon_2, \ldots, \epsilon_n\}$ spans F^n, so that each F^n is a finitely generated vector space over F.

Example 12.b. Let V_F be the vector space of all sequences $(s_k)_{k \geq 0}$ over F, given in Example 2.b. Let W be the set of those sequences $f = (s_k)_{k \geq 0}$ in V such that for some integer $m \geq 0$, $s_i = 0$ for every $i \geq m$. For example, the sequence (1, 1, 1, ..., 1, ...) whose every term equals 1, is not in W. The zero sequence $\overline{0} = (0, 0, \ldots, 0, \ldots) \in W$. Thus, $W \neq \phi$. Consider any $f = (f(k))$, $g = (g(k))$ in W. By definition, there exist l, $m \in \mathbf{N}$, such that $f(k) = 0$ for $k \geq l$ and $g(k) = 0$ for $k \geq m$. Let $n = \max(k, l)$. Then, $f(k) = 0 = g(k)$ for $k \geq n$.

So, in $$f + cg = \{f(k) + cg(k)\},$$

$$f(k) + cg(k) = 0 + c.0 = 0, \quad \text{for } k \geq n$$

Hence $$f + cg \in W.$$

This proves that W is a subspace of V. We claim that W is not finitely generated. On the contrary, suppose that W is finitely generated. We get $S = \{f_1, f_2, \ldots, f_t\}$ a finite subset of W, that spans W. Now, $f_i = (f_i(k))$. There exists $m_i \in \mathbf{N}$ such that $f_i(k) = 0$ for $k \geq m_i$. Let m be the maximum among $m_1, m_2, \ldots, m_t$.

Then, for any $i,\ 1 \leq i \leq t,\ f_i(k) = 0$ for $k \geq m$

Consider $f \in W$.

Then, $f = a_1 f_1 + a_2 f_2 + \ldots + a_t f_t$,

gives $$f(k) = \sum_{i=1}^{t} a_i f_i(k) = 0 \qquad \text{for } k \geq m$$

Thus, for any $f = (f(k)) \in W,\ f(k) = 0$ for $k \geq m$

But we have sequence $g = (1, 1, \ldots, 1, 0, 0, \ldots)$

with $g(k) = 1$ for $1 \leq k \leq m+1$,

and $g(k) = 0$ for $k \geq m+2$, in W,

that does not satisfy $g(k) = 0$ for $k \geq m$.

This is a contradiction. Hence, W is not finitely generated.

13. Rings of formal power series and of polynomials. Closely related to the vector space considered in Example 12.b, is the vector space of *formal power series*. A *formal power series* $f(X)$, over a ring R, is an expression of the form

$$a_0 X^0 + a_1 X^1 + a_2 X^2 + \ldots + a_k X^k + \ldots..$$

with $a_i \in R$.

It is simply written as $\sum_{i=0}^{\infty} a_i X^i$.

Here, X is called an indeterminant, a_i is called the ith coefficient of $f(X)$.

Two power series $f(X) = \sum_{i=0}^{\infty} a_i X^i$ and $g(X) = \sum_{i=0}^{\infty} b_i X^i$ over R are equal iff $a_i = b_i$ for every i. For any $a \in R$, we define

$$f(X) + g(X) = \sum_{i=0}^{\infty} (a_i + b_i) X^i$$

$$af(X) = \sum_{i=0}^{\infty} (aa_i) X^i$$

$R[[X]]$ denotes the set of all formal power series, in the indeterminant X, over R. By taking $R = F$, a field, the two preceding operations, make $F[[X]]$ a vector space over F. This vector space is very much similar to the space V_F of Example 12.b (why?). $R[[X]]$ can be turned into a ring, by defining product

$$f(X)g(X) = \sum_{i=0}^{\infty} c_i X^i$$

where $$c_0 = a_0 b_0,$$

$$c_1 = a_0 b_1 + a_1 b_0;$$

in general, $$c_k = a_0 b_k + a_1 b_{k-1} + \ldots + a_k b_0 = \sum_{i+j=k} a_i b_j.$$

A *polynomial* over R is an expression of the form $f(X) = a_0 + a_1 X + \ldots + a_k X^k$ with $a_i \in R$, which, unlike a power series, involves only finitely many terms. However, we can regard any such polynomial as a power series

$$\sum_{i=0}^{\infty} a_i X^i$$

with $a_i = 0$ for $i \geq k+1$.

Consequently, the set $R[X]$ of all polynomials over R is a subset of $R[[X]]$. It is a subring of $R[[X]]$. In particular, $F[X]$ is a subspace of $F[[X]]$. This $F[X]$ is analogous to the subspace W of V_F in Example 12.b. Given a non-zero polynomial $f(X) = \sum_i a_i X^i$, we get largest integer n, such that $a_n \neq 0$. Then, a_n is called the *leading* coefficient of $f(X)$ and n is called the *degree* of $f(X)$; this is denoted by $\deg(f(X))$. The following result is going to be useful and can be easily proved.

Theorem 14. Let R be any ring, and $f(X)$, $g(X)$ be any two non-zero polynomials in $R[X]$. Then,

(*i*) $\deg(f(X)+g(X)) \leq \max(\deg f(X), \deg g(X))$ whenever $f(X)+g(X) \neq 0$

(*ii*) $\deg(f(X)g(X)) \leq \deg(f(X)) + \deg(g(X))$, whenever $f(X)g(X) \neq 0$

(*iii*) For any field F, given any non-zero $f(X), g(X) \in F[X]$

$$\deg(f(X)g(X)) = \deg(f(X)) + \deg(g(X)).$$

Example 14.a. Let n be any positive integer. For any field F, let

$$P_n(F) = \left\{ \sum_{i=0}^{n} a_i X^i \in F[X] : a_i \in F \right\}$$

$P_n(F)$ is the set of all polynomials in $F[X]$ of degree $\leq n$ and the zero polynomial. It is a subspace of $F[X]$. $S_n = \{1, X, X^2, \ldots, X^n\} \subseteq P_n(F)$ and $P_n(F)$ is spanned by S_n.

For $f(X) \in P_n(F)$,

$$f(X) = \sum_{i=0}^{n} a_i X^i = a_0 + a_1 X + \ldots + a_n X^n + 0 X^{n+1}$$

gives $$f(X) \in P_{n+1}(X),$$

so that $$P_n(X) \subseteq P_{n+1}(F).$$

However, $$X^{n+1} \in P_{n+1}(X) - P_n(X),$$

so $$P_n(X) \subset P_{n+1}(X).$$

Example 14. b. Let K be a field and F be a subfield of K. Now, $\langle K, +\rangle$ is an abelian group. For any $a \in F$, $\alpha \in K$, the product $a\alpha$ in K, defines a scalar multiplication. This makes K a vector space over F. Consider a specific example. Let $K = \mathbf{C}$, $F = \mathbf{R}$. Now, $1, i = \sqrt{-1} \in \mathbf{C}$. Any $z \in \mathbf{C}$ is of the form

$$z = a1 + bi,\ a,\ b \in \mathbf{R}$$

Thus, $\mathbf{C}$ as a vector space over $\mathbf{R}$, is spanned by $\{1, i\}$. Thus, $\mathbf{C}_{\mathbf{R}}$ is finitely generated.

Proposition 15. Let $\alpha_1, \alpha_2, \ldots, \alpha_k$ be some vectors that span a subspace W of V_F. If for some j, $1 \le j \le k$, α_j is a linear combination of the remaining $k-1$ vectors, then, $\alpha_1, \alpha_2, \ldots, \alpha_{j-1}, \alpha_{j+1}, \ldots, \alpha_k$ also span W.

Proof. By the hypothesis $\alpha_j = \sum_{i=1}^{j-1} a_i\alpha_i + \sum_{i=j+1}^{k} a_i\alpha_i;$

$$a_i \in F.$$

Consider only $\alpha \in W$. Then,

$$\alpha = \sum_{s=1}^{k} b_s\alpha_s = \sum_{s \ne j} b_s\alpha_s + b_j\left\{\sum_{i=1}^{j-1} a_i\alpha_i + \sum_{i=j+1}^{k} a_i\alpha_i\right\}$$

gives that α is a linear combination of the $k-1$ vectors $\alpha_1, \alpha_2, \ldots, \alpha_{j-1}, \alpha_{j+1}, \ldots, \alpha_k$.

Hence, these $k-1$ vectors span W.

Exercise 2.1

1. Let $V = \{(x, y) : x, y \in \mathbf{R}\}$. For any $\alpha = (x_1, y_1)$, $\beta = (x_2, y_2) \in V$, $c \in \mathbf{R}$, define

$$\alpha \oplus \beta = (x_1 + x_2 + 1,\ y_1 + y_2 + 1)$$
$$c \circledcirc \alpha = (cx_1,\ cy_1)$$

(*i*) Prove that $\langle V, \oplus\rangle$ is an abelian group

(*ii*) Verify that $c \circledcirc (\alpha \oplus \beta) \ne c \circledcirc \alpha \oplus (c \circledcirc \beta)$

(*iii*) Prove that V is not a vector space over $\mathbf{R}$ under the above two operations.

2. Let $V = \mathbf{R}^3$. For any $\alpha = (x_1, x_2, x_3)$, $\beta = (y_1, y_2, y_3) \in \mathbf{R}^3$, $c \in \mathbf{R}$, define

$$\alpha \oplus \beta = (x_1 + y_1 + 1,\ x_2 + y_2 + 1,\ x_3 + y_3 + 1)$$

$$c \circledcirc \alpha = (cx_1 + c - 1,\ cx_2 + c - 1,\ cx_3 + c - 1)$$

Prove that V is a vector space over R under these two operations.

3. Let $V = \{(x,1) : x \in \mathbf{R}\}$. For any $\alpha = (x, 1), \beta = (y, 1) \in V$ and $c \in \mathbf{R}$, define

$$\alpha \oplus \beta = (x + y, 1)$$

$$c \circledcirc \alpha = (cx, 1)$$

Verify that V is a vector space over R under these two operations.

4. Let $V = \mathbf{R}^3$. For any $\alpha = (x_1, x_2, x_3)$, $\beta = (y_1, y_2, y_3)$ and $c \in \mathbf{R}$, define

$$\alpha \oplus \beta = (x_1 + y_1, x_2 + y_2, x_3 + y_3)$$

$$c \circledcirc \alpha = (cx_1, cx_2, x_3)$$

Verify that V is not a vector space over R under these two operations. Which of the axioms of a vector space are not satisfied?

5. Let V = Set of all positive real numbers. For any $\alpha, \beta \in V$, $c \in \mathbf{R}$, define

$$\alpha \oplus \beta = \alpha\beta,$$

the product in R $\quad c \circledcirc \alpha = \alpha^c \in V$

Prove that V is a vector space over R under these two operations.

6. Let $V = C[0, 1]$, the set of all real valued continuous functions defined on the closed interval $[0, 1]$. Verify in details that V is a vector space over R under addition and scalar multiplication operations given as under.

Let $\quad f, g \in V,\ c \in \mathbf{R},\ x \in [0, 1],$

$$(f \oplus g)(x) = f(x) + g(x)$$

$$(c \circledcirc f)(x) = c.f(x)$$

(These two operations in V are called usual addition and scalar multiplication operations)

7. Let V be the vector space $\mathbf{R}^3$. Which of the following subsets of V are subspaces of V?

(*i*) $W_1 = \{(x, y, z) : x + y = 0\}$

(*ii*) $W_2 = \{(x, y, z) : x = 2y + 1\}$

(*iii*) $W_3 = \{(x, y, z) : x \geq 0\}$

(*iv*) $W_4 = \{(x, y, z) : x^2 = y^2\}$

8. Let $V = P_3(\mathbf{R}) = \{a_0 + a_1X + a_2X^2 + a_3X^3 : a_i \in \mathbf{R}\}$

(*i*) Verify that $W_1 = \{\alpha_0 + a_2X^2 : a_0, a_2 \in \mathbf{R}\}$ is a subspace of V.

(*ii*) Verify that $W_2 = \{a_0 + a_1X + a_2X^2 + a_3X^3 : a_0 = a_2 + a_3\}$ is a subspace of V.

(*iii*) Give an example of a subset W of V such that $0 \in W$, for any $p(X) \in W$, $-p(X) \in W$, but W is not a subspace of V.

9. Let $V = \mathbf{C}^{2\times 3}$, as a vector space over $\mathbf{C}$. Which of the following subsets of V are subspaces?

(*i*) $W_1 = \left\{\begin{bmatrix} a & b & c \\ d & 0 & 0 \end{bmatrix} : a, b, c, d \in \mathbf{C}\right\}$

(*ii*) $W_2 = \left\{\begin{bmatrix} a & b & c \\ d & e & f \end{bmatrix} : a + c = e + f\right\}$

(*iii*) $W_3 = \left\{\begin{bmatrix} a & b & c \\ d & e & f \end{bmatrix} : a \in \mathbf{R}, b, c, d, e, f \in \mathbf{C}\right\}$

(*iv*) $W_4 = \left\{\begin{bmatrix} a & b & c \\ d & e & f \end{bmatrix} : a > 0, b = c\right\}$

10. Let W be the set of all solutions in $\mathbf{R}^{3\times 1}$ of the system of linear equations

$$2x + y + z = 0$$
$$x - y - z = 0$$

Prove that W is a finitely generated subspace of $\mathbf{R}^{3\times 1}$.

11. Let $V_{\mathbf{R}} = C[0, 1]$.

(*i*) Prove that $W_1 = \{f \in V : f \text{ is differentiable over } [0, 1]\}$ is a subspace of V

(*ii*) Prove that $W_2 = \left\{f \in V : f\left(\frac{1}{2}\right) = 0\right\}$ is a subspace of V.

(*iii*) A function $f \in V$ is said to be an even *function* if $f(x) = f(-x)$. Prove that the set V_e of all even functions in V is a subspace of V

(*iv*) A function $f \in V$ is called on odd function if $f(-x) = -f(x)$. Prove that the set V_0 of all odd functions in V is a subspace of V

(*v*) Prove that $V_e \cap V_0 = \{0\}$

(*vi*) Is $V_{\mathbf{R}}$ finitely generated?

12. Which of the following vectors in $\mathbf{R}^3$ are linear combinations of the vectors

$$X_1 = (4, -2, 3), X_2 = (1, -1, 0), X_3 = (7, -3, 3)?$$

(*i*) $(5, -3, 3)$, (*ii*) $(8, -4, 4)$, (*iii*) $(15, -7, 6)$ (*iv*) $(1, 1, 1)$.

13. Prove that any vector $\alpha = (a, b) \in \mathbf{R}^2$ is a linear combination of $X_1 = (1, 1)$, $X_2 = (-1, 1)$.

14. Which of the following vectors in $\mathbf{R}^4$ are linear combinations of the four vectors

$$X_1 = (1, 2, 1, 0),\ X_2 = (1, 1, -1, 0),\ X_3 = (1, 1, 0, 0),\ X_4 = (2, 3, 0, 0)?$$

(*i*) (3, 4, 0, 0) (*ii*) (3, 4, 0, 1) (*iii*) (0, 0, 1, 0)

Further, prove that any $\alpha = (a, b, c, d) \in \mathbf{R}^4$ is a linear combinations of X_1, X_2, X_3, X_4 iff $d = 0$

15. Which of the following sets of vectors span $\mathbf{R}^2$?

(*i*) $\{(1, 1), (1, 2)\}$

(*ii*) $\{(1, 1), (-3, -3), (0, 0)\}$

(*iii*) $\{(1, 3), (3, 1), (2, 2)\}$

16. Which of the following polynomials in $P_2(\mathbf{R})$ are linear combinations of the polynomials

$$p_1(X) = X^2 + X + 1,\ p_2(X) = X^2 + 1,\ p_3(X) = X?$$

(*i*) $X^2 + 2X + 1$ (*ii*) $2X^2 - 3X + 3$ (*iii*) $X^2 + 2X + 2$

(*iv*) $3X^2 + 2X + 3$

17. Prove that the three polynomials $1, 1+X, (1+X)^2$ span $P_2(\mathbf{R})$.

18. Do the polynomials $X^3 + 2X + 1,\ X^2 - X + 2,\ X^3 + 2,\ -X^3 + X^2 - 5X + 2$ span $P_3(\mathbf{R})$?

19. For any field F, let W be the set of those polynomials in $F[X]$, in which the coefficients of odd powers of X are zeros. Prove that W is a subspace of $F[X]$ and that it is not finitely generated.

20. For any non-zero $f(X) = \sum_{i=0}^{\infty} a_i X^i \in R[[X]]$, R any ring, the order of $f(X)$, denoted by $0\{f(X)\}$ is the smallest suffix i such that $a_i \neq 0$

(*a*) For any two non-zero $f(X), g(X) \in R[[X]]$ prove that

(*i*) $0\{f(X) + g(X)\} \geq \min 0\{f(X)\}, 0\{g(X)\}$ if $f(X) + g(X) \neq 0$

(*ii*) $0\{f(X)g(X)\} \geq 0\{f(X)\} + 0\{g(X)\}$ if $f(X)g(X) \neq 0$

(*iii*) If $R = F$, a field, prove that $0\{f(X)g(X)\} = 0\{f(X)\} + 0\{g(X)\}$

(*b*) For any $n \geq 0$, let W_n be the set of those $f(X) \in F[[X]]$ such that either $f(X) = 0$ or $0\{f(X)\} \geq n$.

(*i*) Prove that W_n is a subspace of $F[[X]]$.

(*ii*) Prove that $W_{n+1} \subset W_n$.

2.2 Linear Independence and Basis

Definition 1. Let $\alpha_1, \alpha_2,..., \alpha_k$ be any $k \geq 1$ vectors in a vector space V_F. these k vectors are said to be *linearly independent* (over F) in V if for any $a_1, a_2,..., a_k \in F$ $a_1\alpha_1 + a_2\alpha_2 +...+ a_k\alpha_k = 0$ implies $a_1 = a_2 =...= a_k = 0$. If $\alpha_1, \alpha_2,..., \alpha_k$ are not linearly independent, they are said to be *linearly dependent*; that happens if for some a_1, $a_2,..., a_k \in F$ not all zeros. $a_1\alpha_1 + a_2\alpha_2 +...+ a_k\alpha_k = 0$

Example 1.a. Let $V = \mathbf{R}^2$. Let $\alpha_1 = (1, 2)$, $\alpha_2 = (2, 1)$ in V. Let for some $a_1, a_2 \in \mathbf{R}$, $a_1\alpha_1 + a_2\alpha_2 = 0$.

This gives
$$a_1(1, 2) + a_2(2, 1) = (0, 0)$$

so that
$$\begin{aligned} a_1 + 2a_2 &= 0 \\ 2a_1 + a_2 &= 0 \end{aligned}$$

The only solution for this system is $a_1 = 0 = a_2$. Hence, α_1, α_2 are linearly independent. Let $\alpha_3 = (1, 1)$. Notice that $1\alpha_1 + 1\alpha_2 + (-3)\alpha_3 = 0$. This shows that $\alpha_1, \alpha_2, \alpha_3$ are linearly dependent.

Consider
$$\beta_1 = (0, 0),$$
$$\beta_2 = (1, 3) \in V.$$

As
$$1.\beta_1 + 0\beta_2 = 0$$

we get β_1, β_2 are linearly dependent.

Definition 2. Let S be any subset of a vector space V_F. If any finitely many distinct vectors $\alpha_1, \alpha_2,..., \alpha_m \in S$ with $m \geq 1$, are linearly independent, then, S is said to be a *linearly independent subset.* If S is not linearly independent, it is said to be linearly dependent.

Remark 3. Let S be a subset of a vector space V_F. The following hold:

(*i*) If $\alpha_1, \alpha_2,..., \alpha_m \in V_F$ are linearly independent, then, for $1 \leq p \leq m$, $\alpha_1, \alpha_2,..., \alpha_p$ are linearly independent. For, if for some $a_i \in F$, $1 \leq i \leq p$, $a_1\alpha_1 + a_2\alpha_2 +...+ a_p\alpha_p = 0$, we get
$$a_1\alpha_1 + a_2\alpha_2 +...+ a_p\alpha_p + 0\alpha_{p+1} +...0\alpha_m = 0.$$
As $\alpha_1, \alpha_2,..., \alpha_m$ are linearly independent; in this equation, the coefficient of every α_i is zero, so that also $a_1 = a_2 =...= a_p = 0$

(*ii*) If S is linearly independent, then, any subset of S is linearly independent. To see this, let S' be a subset of S. Let $\alpha_1, \alpha_2,..., \alpha_m$ be any m distinct members of S'. Then, these $\alpha_i \in S$. By definition, $\alpha_1, \alpha_2,..., \alpha_m$ are linearly independent. Hence, S' is linearly independent.

(*iii*) The empty subset ϕ of V_F is linearly independent, as the condition defining linear independence holds vacuously for ϕ.

(*iv*) If $0 \in S$, S is linearly dependent.

Here, $\alpha_1 = 0 \in S$.

For $m = 1,\ 1.\alpha_1 = 0$,

so that $a_1\alpha_1 = 0$

with $a_1 = 1 \neq 0$.

Consequently, α_1 is linearly dependent. Hence, S is linearly dependent.

(*v*) If S is linearly dependent, then, any subset S' of V_F containing S is also linearly dependent. Here, by definition, there exist some $m \geq 1$ distinct members $\alpha_1, \alpha_2, \ldots, \alpha_m$ of S, which are linearly dependent. As these $\alpha_i \in S'$, we get that S' is linearly dependent.

(*vi*) For any $\alpha \neq 0$ in V_F, $\{\alpha\}$ is linearly independent.

Example 3.a. Let $S = \{\alpha_1, \alpha_2, \alpha_3\}$

with $\alpha_1 = (1,\ 2),\ \alpha_2 = (2,\ 1),\ \alpha_3 = (1,\ 1)$.

As seen in Example 1.a, $\alpha_1, \alpha_2, \alpha_3$ are linearly dependent, so S is a linearly dependent subset of $\mathbf{R}^2$. Now, $S' = \{\alpha_1, \alpha_2\}$ is linearly independent. So, a subset of linearly dependent set may not be linearly dependent. Observe that any subset S'' of S not equal to S is linearly independent.

Definition 4. A subset B of a vector space V_F is called a basis of V_F if

(*i*) B is linearly independent

(*ii*) B spans V_F

In Example 1.a, we saw that $S' = \{\alpha_1, \alpha_2\}$ with $\alpha_1 = (1,\ 2),\ \alpha_2 = (2,\ 1)$, is linearly independent in $\mathbf{R}^2$.

Given any $\alpha = (a,\ b) \in \mathbf{R}^2$, $\alpha = a_1\alpha_1 + a_2\alpha_2$,

$$a_1 = \frac{2b - a}{3},$$

$$a_2 = \frac{2a - b}{3}.$$

This gives that S' spans $\mathbf{R}^2$. Hence, S' is a basis of $\mathbf{R}^2$. In the same example, $S = \{\alpha_1, \alpha_2, \alpha_3\}$ is linearly dependent. However, as $S' \subseteq S$, S spans $\mathbf{R}^2$, but S is not a basis of $\mathbf{R}^2$.

Example 4.a. Let $V_F = F^n$. In Example (2.1.12.a), we had introduced n-tuples

$$\epsilon_1 = (1,\ 0, \ldots,\ 0,\ 0)$$

$$\epsilon_2 = (0,\ 1,\ 0, \ldots,\ 0,\ 0), \ldots \epsilon_n = (0,\ 0, \ldots,\ 0,\ 1) \quad \text{in } F^n$$

For any $a_1, a_2, \ldots, a_n \in F$, we saw that

$$a_1 \in_1 + a_2 \in_2 + \ldots + a_n \in_n = (a_1, a_2, \ldots, a_n) = \alpha$$

so that $\sum_{i=1}^{n} a_i \in_i = 0$ iff $a_i = 0$ for every i.

Hence, $B = \{\in_1, \in_2, \ldots, \in_n\}$

is a linearly independent subset of F^n.

Further, B also spans F^n. Hence, B is a basis of F^n. This basis is called the *standard basis* of F^n.

In particular in $\mathbf{R}^4$, the standard basis is

$$B = \{\in_1, \in_2, \in_3, \in_4\}$$

with
$$\in_1 = (1, 0, 0, 0),$$
$$\in_2 = (0, 1, 0, 0),$$
$$\in_3 = (0, 0, 1, 0),$$
$$\in_4 = (0, 0, 0, 1).$$

In $\mathbf{R}^2$, the standard basis is $\{(1, 0), (0, 1)\}$.

Example 4.b. Consider $P_2(\mathbf{R}) = \{a_0 + a_1 X + a_2 X^2 : a_i \in \mathbf{R}\}$

Now $B = \{1, X, X^2\} \subseteq P_2(\mathbf{R})$.

As any $f(X) \in P_2(\mathbf{R})$ is expressible as

$$f(X) = a_0 1 + a_1 X + a_2 X^2,$$

we get B spans this vector space.

Further, $a_0 1 + a_1 X + a_2 X^2 = 0$

gives $a_0 = a_1 = a_2 = 0$.

Thus, B is also linearly independent. This shows that B is a basis of $P_2(\mathbf{R})$. Similar arguments give that in $P_n(\mathbf{R})$ {indeed in $P_n(F)$}

$$B = \{1, X, X^2, \ldots, X^n\}$$

is a basis.

This is called the *standard basis* of $P_n(\mathbf{R})$ {or of $P_n(F)$}.

Given a set $B = \{\alpha_1, \alpha_2, \ldots, \alpha_k\}$ in V_F, the problem of determining whether B is linearly independent, gives rise to a system of homogeneous linear equations over F; if this system has non-trivial solution, then, B is linearly dependent. We illustrate this in the following:

Example 4.c. Let $V = \mathbf{R}^3$. Let $\alpha_1 = (4, 2, 1)$, $\alpha_2 = (2, 6, -5)$, $\alpha_3 = (1, -2, 3)$. Is $B = \{\alpha_1, \alpha_2, \alpha_3\}$ a basis of V? To answer this question, we need to check : (*i*) Is B linearly independent? (*ii*) Does B span V?

Let for some $a_1, a_2, a_3 \in \mathbf{R}$, $\quad a_1\alpha_1 + a_2\alpha_2 + a_3\alpha_3 = 0$

i.e., $\quad a_1(4, 2, 1) + a_2(2, 6, -5) + a_3(1, -2, 3) = (0, 0, 0)$

By comparing various components, we get the following system of homogeneous linear equations

$$\begin{aligned} 4a_1 + 2a_2 + a_3 &= 0 \\ 2a_1 + 6a_2 - 2a_3 &= 0 \\ a_1 - 5a_2 + 3a_3 &= 0 \end{aligned}$$

Its augmented matrix is $\begin{bmatrix} 4 & 2 & 1 & : & 0 \\ 2 & 6 & -2 & : & 0 \\ 1 & -5 & 3 & : & 0 \end{bmatrix}$

This is row equivalent to $\begin{bmatrix} 1 & 0 & \frac{1}{2} & : & 0 \\ 0 & 1 & -\frac{1}{2} & : & 0 \\ 0 & 0 & 0 & : & 0 \end{bmatrix}$

This shows that $a_3 = r$, arbitrary, $a_1 = -\frac{1}{2}r$, $a_2 = \frac{1}{2}r$.

For $r = -1$, we get $\quad a_1 = \frac{1}{2}, a_2 = \frac{-1}{2}, a_3 = -1,$

so that $\quad \frac{1}{2}\alpha_1 - \frac{1}{2}\alpha_2 - \alpha_3 = 0,$

i.e. $\quad \alpha_3 = \frac{1}{2}\alpha_1 - \frac{1}{2}\alpha_2 \quad$ ... (1)

Thus, $\alpha_1, \alpha_2, \alpha_3$ are linearly dependent. Consequently, B is not a basis of $\mathbf{R}^3$. Let W be the subspace of $\mathbf{R}^3$ spanned by B. Consider $\alpha \in W$.

Then, $\quad \alpha = b_1\alpha_1 + b_2\alpha_2 + b_3\alpha_3 \quad$ ... (2)

(1) and (2) give $\quad \alpha = b_1\alpha_1 + b_2\alpha_2 + b_3\left(\frac{1}{2}\alpha_1 - \frac{1}{2}\alpha_2\right)$

$$= \left(b_1 + \frac{1}{2}b_3\right)\alpha_1 + \left(b_2 - \frac{1}{2}b_3\right)\alpha_2$$

so that $B' = \{\alpha_1, \alpha_2\}$ spans W.

Now, α_1, α_2 are linearly independent. Hence, B' is a basis of W.

Remark 5. The only subsets of a zero space $\{0\}$ are ϕ and $\{0\}$. Now, $\{0\}$ is linearly dependent {see Remark 3(*iv*)} and ϕ is linearly independent. As ϕ spans $\{0\}$, ϕ is the only basis of a zero space.

Theorem 6. Let V_F be a vector space spanned by m vectors, m a positive integer. Then, for any $n > m$, any n vectors in V are linearly dependent.

Proof. Let $\alpha_1, \alpha_2, \ldots, \alpha_m$ span V. For an $n > m$, consider any n vectors $\beta_1, \beta_2, \ldots, \beta_n$ in V.

Now, $$\beta_j = \sum_{i=1}^{m} a_{ij}\alpha_i,\ a_{ij} \in F,\ 1 \le j \le n \qquad \ldots (1)$$

We want to find $c_1, c_2, \ldots, c_n \in F$, not all zeros (i.e., atleast, one of them not zero) such that

$$\sum_{j=1}^{n} c_j \beta_j = 0 \qquad \ldots (2)$$

i.e., $$\sum_{j=1}^{n} c_j \left[\sum_{i=1}^{m} a_{ij}\alpha_i\right] = 0$$

i.e., $$\sum_{i=1}^{m} \left[\sum_{j=1}^{n} a_{ij} c_j\right]\alpha_i = 0 \qquad \ldots (3)$$

Consider the $m \times n$-matrix $A = \left[a_{ij}\right]$. As $m < n$, the system of homogeneous linear equations

$$AX = 0$$

where $$X = \begin{bmatrix} x_1 \\ x_2 \\ \vdots \\ x_n \end{bmatrix} = \left[x_1, x_2, \ldots, x_n\right]^T$$

has a non-trivial solution, say $C = \left[c_1, c_2, \ldots, c_n\right]^T$ {Remark (1.4.28)}

Now, $$\sum_{j=1}^{n} a_{ij} c_j = 0,$$

and $$c_j \neq 0 \qquad \text{for some } j.$$

Then, (2) and (3) give $$\sum_{j=1}^{n} c_j \beta_j = 0.$$

Hence, $\beta_1, \beta_2, \ldots, \beta_n$ are linearly dependent.

Theorem 7. If a vector space V_F has a finite basis, then any two basis of V_F have the same number of vectors.

Proof. Let $B = \{\alpha_1, \alpha_2, ..., \alpha_n\}$ be a finite basis of V_F with n members. Let B' be another basis of V. By Theorem 6, more than n vectors in V are linearly dependent. So, if B' has m members, then, $m \le n$. Again as B' is a basis of V, with m vectors, more than m vectors in V must be linearly dependent. As $\alpha_1, \alpha_2, ..., \alpha_n$ are linearly independent, we get $n \le m$. Hence, $n = m$.

Theorem 8. Any finitely generated vector space V_F has a basis. Further, any two basis of V_F have the same number of vectors.

Proof. Let a finite subset S of V_F span V_F. It is possible that $0 \in S$. Consider $S' = S - \{0\}$. Then, S' is finite and it also spans V_F. If $S' = \phi$, then, $V = \{0\}$ and S' is the only basis of V_F. Let $S' = \{\alpha_1, \alpha_2, ..., \alpha_k\}$ with, $k \ge 1$ and every $\alpha_i \ne \{0\}$. Now, $\{\alpha_1\}$ is a linearly independent subset of S' {see Remark 3(*vi*)}. We get $B = \{\beta_1, \beta_2, ..., \beta_m\}$ a linearly independent subset of S', having largest number of members among the linearly independent subset of S'. Here, $m \ge 1$. Let W be the subspace of V spanned by B. Consider any $\alpha_i \in S$. If $\alpha_i \in B$, we get $\alpha_i \in W$. Let $\alpha_i \notin B$. Then, by the choice of B,

$$B' = \{\beta_1, \beta_2, ..., \beta_m, \alpha_i\}$$

is linearly dependent.

So, for some $a_1, a_2, ..., a_m, a$ in F, not all zeros

$$a_1\beta_1 + a_2\beta_2 + ... + a_m\beta_m + a\alpha_i = 0 \quad ... (1)$$

Suppose $a = 0$. We get $\quad a_1\beta_1 + a_2\beta_2 + ... + a_m\beta_m = 0 \quad ... (2)$

As $\quad \beta_1, \beta_2, ..., \beta_m$

are linearly independent, we get $a_1 = a_2 = ... = a_m = 0 = a$.

This is a contradiction. Hence, $a \ne 0$.

Then, (1) gives $\quad \alpha_i = -a^{-1}a_1\beta_1 - a^{-1}a_2\beta_2 - ... - a^{-1}a_m\beta_m$

i.e., α_i is linear combination of $\beta_1, \beta_2, ..., \beta_m$.

Consequently, $\quad \alpha_i \in W = \langle B \rangle$.

Hence, $\quad S' \subseteq W$.

Then $\quad V = \langle S' \rangle \subseteq W \subseteq V$

This gives $\quad V = \langle B \rangle$.

Hence, B is a basis of V. Thus, V has a finite basis. By Theorem 7, any two bases of V have the same number of vectors.

Given a finitely generated vector space V_F, the theorem, given above, associates a unique number to V_F, namely the number of members of a basis of V_F. This number is an invariant of V_F and a very important one at that.

Definition 9. Let V_F be any finitely generated vector space. The number of members of a basis of V_F is called the *dimension* of V_F; it is denoted by $\dim_F V$ {or simply by $\dim(V)$}.

We have seen in Example 4.a that F^n has a basis (standard basis) that has n members. Thus, $\dim_F F^n = n$. Example 4.b shows that $\dim_F P_n(F) = n+1$. The subspace W of $\mathbf{R}^3$ considered in Example 4.c has $\dim_{\mathbf{R}} W = 2$. A finitely generated vector space is also called a *finite dimensional* vector space. Observe that zero space is zero dimensional.

Corollary 10. If a finite set S spans a vector space V_F, then, $\dim V \le |S|$, the number of members of S. Further, S has a subset B that is a basis of V.

Proof. We saw in the proof of Theorem 8, that there exists a subset B of S that is a basis of V.

Consequently, $$\dim V = |B| \le |S|$$

Lemma 11. Let $\alpha_1, \alpha_2, \ldots, \alpha_m$ with $m \ge 1$, be non-zero and linearly dependent in V_F. Then,

(*i*) $m \ge 2$ and there exists some j such that $2 \le j \le m$, with α_j a linear combination of $\alpha_1, \alpha_2, \ldots, \alpha_{j-1}$.

(*ii*) If for some k, $1 \le k \le m$, $\alpha_1, \alpha_2, \ldots, \alpha_k$ are linearly independent, then, $k < j$.

Proof. As $\alpha_1 \ne 0$, $\{\alpha_1\}$ is linearly independent. So, $m \ge 2$. There exist $a_1, a_2, \ldots, a_m \in F$, not all zeros such that

$$a_1\alpha_1 + a_2\alpha_2 + \ldots + a_m\alpha_m = 0 \qquad \ldots (1)$$

As $a_i \ne 0$ for some $i \le m$, we can find largest j such that $a_j \ne 0$. Here, $j \le m$. Then, $a_i = 0$ for $i > j$. Then, (1) gives

$$a_1\alpha_1 + a_2\alpha_2 + \ldots + a_j\alpha_j = 0 \qquad \ldots (2)$$

If $j = 1$, this equation becomes $a_1\alpha_1 = 0$; this gives $a_1 = 0$, as $\alpha_1 \ne 0$. This is a contradiction. Hence, $j \ge 2$. Then, (2) gives

$$\alpha_j = \sum_{i=1}^{j-1} -a_j^{-1} a_i \alpha_i \qquad \ldots (3)$$

This proves that α_j is a linear combination of $\alpha_1, \alpha_2, \ldots, \alpha_{j-1}$. This proves (*i*). Let $\alpha_1, \alpha_2, \ldots, \alpha_k$ be linearly independent and $j \le k$. Then, $\alpha_1, \alpha_2, \ldots, \alpha_j$ are linearly independent by Remark 3(*i*). However, (2) gives $\alpha_1, \alpha_2, \ldots, \alpha_j$ are linearly dependent. This is a contradiction. Hence, $j > k$. This proves (*ii*).

Theorem 12. Let W be a subspace of a finitely generated vector space V_F.

Then, $\dim W \le \dim V$. Further, $W = V$ iff $\dim W = \dim V$.

Proof. Let $\dim V = n$. By Theorem 6, any more than n vectors in V are linearly dependent. So, as $W \subseteq V$, no linearly independent subset of W can have more than n members. W has, at least, one linearly independent subset, i.e., ϕ. So, we can find a linearly independent subset $B = \{\beta_1, \beta_2, ..., \beta_k\}$ of W such that B has largest number of members among all the linearly independent subsets of W. As seen above, $k \le n$. Consider any $\beta \ne 0$ in W.

Then, the $k+1$ vectors in W, $\quad \beta_1, \beta_2, ..., \beta_k, \beta$

are linearly dependent.

By Lemma 11(*i*), β is a linear combination of $\beta_1, \beta_2, ..., \beta_k$. Thus, B spans W. Hence, B is a basis of W. This proves that $\dim W = k \le n$. Finally, let $k = n$. Let $\alpha \ne 0$ in V.

Then, $n+1$ vectors $\quad \beta_1, \beta_2, ..., \beta_n, \alpha$

are linearly dependent.

So, by Lemma 11(*i*) α is a linear combination of

$$\beta_1, \beta_2, ..., \beta_n.$$

Hence, $\quad W = \langle B \rangle = V$

If $W = V$, obviously, $\quad \dim W = \dim V$.

Conversely, let $\quad \dim W = \dim V = n,$

and $\quad B = \{\alpha_1, \alpha_2, ..., \alpha_n\}$

be a basis of W.

As $\quad \dim V = n,$

given $0 \ne \alpha \in V$, the $n+1$ vectors $\alpha_1, \alpha_2, ..., \alpha_n, \alpha$ are linearly dependent. By Lemma 11, α is a linear combination of B. Hence, B is a basis of V and $V = \langle B \rangle = W$.

Theorem 13. Let V_F be an n dimensional vector space and S be any subset of V_F. Then, the following are equivalent :

(*i*) S is a basis

(*ii*) $|S| = n$ and S spans V

(*iii*) $|S| = n$ and S is linearly independent.

Proof. Let S be a basis.

Then, $\quad \dim V = |S| = n.$

By definition, S spans V and it is linearly independent. Hence, (*i*) implies (*ii*) and (*iii*). Let (*ii*) hold. By Corollary 10, S has a subset B that is a basis of V.

Then, $\quad \dim V = |B| = n = |S|.$

This gives $\quad B = S.$

Hence, (*ii*) implies (*i*). Finally, let (*iii*) hold. Let $W = \langle S \rangle$. Then, S being linearly independent, is a basis of W.

Hence, $$\dim W = |S| = n = \dim V.$$

By Theorem 12, $W = V$, so S is a basis of V. Hence, (*iii*) implies (*i*).

Lemma 14. **Let S be a linearly independent subset of a vector space V_F. Consider any $\alpha \neq 0$ in $V - S$. Then, $S \cup \{\alpha\}$ is linearly independent iff α does not belong to the span $\langle S \rangle$.**

Proof. **Let $S_1 = S \cup \{\alpha\}$ be linearly independent. Suppose that $\alpha \in W = \langle S \rangle$. Then, for some distinct members $\alpha_1, \alpha_2, \ldots, \alpha_k$ of S**

$$\alpha = a_1\alpha_1 + a_2\alpha_2 + \ldots + a_k\alpha_k,\ a_i \in F$$

This gives that the subset $\{\alpha_1, \alpha_2, \ldots, \alpha_k, \alpha\}$ of S_1 is linearly dependent. This contradicts the hypothesis that S_1 is linearly independent. Hence, $\alpha \notin \langle S \rangle$.

Conversely, let $\alpha \notin \langle S \rangle$, but S_1 be linearly dependent. So, S_1 has a finite non-empty subset B that is linearly dependent. As S is linearly independent, $B \not\subseteq S$.

However, $$S_1 = S \cup \{\alpha\},$$

we get $$\alpha \in B$$

$$B = \{\beta_1, \beta_2, \ldots, \beta_s, \alpha\},$$

For some distinct $$\beta_1, \beta_2, \ldots, \beta_s \in S.$$

As $$\beta_1, \beta_2, \ldots, \beta_s$$

are linearly independent, by Lemma 11.

$$\alpha = b_1\beta_1 + b_2\beta_2 + \ldots + b_s\beta_s,\ b_i \in F$$

So $$\alpha \in \langle S \rangle.$$

This is a contradiction. Hence, S_1 is linearly independent.

A vector space can have many bases. The following replacement theorem helps in constructing new bases from the others.

Theorem 15. **Let $\alpha_1, \alpha_2, \ldots, \alpha_k$, $k \geq 1$ be linearly independent members of a vector space V_F. Let a finite subset Y span V_F and that $0 \notin Y$. Then, there exists a subset A of Y containing k members such that**

$$(Y - A) \cup \{\alpha_1, \alpha_2, \ldots, \alpha_k\}$$

generates V_F.

Proof. **Let** $$Y = \{\beta_1, \beta_2, \ldots, \beta_m\}$$

We apply induction on k. Let $k = 1$.

Now, $$\alpha_1 = \sum_{i=1}^{m} a_i\beta_i,\ a_i \in F \qquad \text{... (1)}$$

As $\alpha_1 \neq 0$, some $a_i \neq 0$, say $a_1 \neq 0$.

Then, (1) gives $$\beta_1 = a_1^{-1}\alpha_1 - \sum_{i=2}^{m} a_1^{-1}\, a_i\beta_i$$

So, β_1 is in the subspace W spanned by

$$Y' = \{\alpha_1,\ \beta_2,\ \beta_3, \ldots,\ \beta_m\}$$

Clearly then, $$Y \subseteq W,$$

so $$V = \langle Y\rangle \subseteq W \subseteq V$$

Hence, $$V = \langle Y'\rangle,\ Y' = (Y - A)\cup\{\alpha_1\},$$

where $$A = \{\beta_1\}.$$

Thus, the result holds for $k = 1$. To apply induction, let $k > 1$ and the result hold for $k-1$ vectors. As $\alpha_1, \alpha_2, \ldots, \alpha_{k-1}$ are linearly independent, by the induction hypothesis, we get a subset A_1 of Y containing $k-1$ vectors such that

$$Y_1 = (Y - A_1)\cup\{\alpha_1,\ \alpha_2, \ldots,\ \alpha_{k-1}\}$$

spans V.

In particular, $$\alpha_k \in \langle Y_1\rangle.$$

By Lemma 14, $$\alpha_k \notin \langle \alpha_1,\ \alpha_2, \ldots,\ \alpha_{k-1}\rangle.$$

So, $$Y - A_1 \neq \phi.$$

Let $$Y - A_1 = \{\delta_1,\ \delta_2, \ldots,\ \delta_s\}.$$

Here, $$s \geq 1.$$

Now, $$\alpha_k = \sum_{i=1}^{s} d_i\delta_i + \sum_{j=1}^{k-1} b_j\alpha_j,$$

$$d_i,\ b_j \in F$$

If every $d_i = 0$, then, $\alpha_k \in$ span $\langle \alpha_1, \alpha_2, \ldots, \alpha_{k-1}\rangle$; which leads to a contradiction. Hence, some d_i, say $d_1 \neq 0$. As for the case for $k = 1$,

$$Y_2 = [Y_1 - \{\delta_1\}]\cup\{\alpha_k\} = (Y - A)\cup\{\alpha_1,\ \alpha_2, \ldots,\ \alpha_k\}$$

spans V, where $A = A_1 \cup \{\delta_1\}$ has k members.

Corollary 16. Let $\alpha_1, \alpha_2, \ldots, \alpha_k$ be linearly independent vectors in a finite dimensional vector space V_F. Given any basis B of V, there exists a subset A of B with k elements such that

$$B_1 = \{\alpha_1, \alpha_2, ..., \alpha_k\} \cup (B - A)$$

is a basis of V_F

Further, $$(B - A) \cap \{\alpha_1, \alpha_2, ..., \alpha_k\} = \phi$$

Proof. Let $\dim V = n$. In Theorem 15, take $Y = B$.

Then, $$B_1 = (B - A) \cup \{\alpha_1, \alpha_2, ..., \alpha_k\}$$

spans V.

As $$|B - A| = (n - k),$$

$$|B_1| \leq (n - k) + k = n.$$

Then, $$n = \dim V \leq |B_1| = n.$$

Hence, by Theorem 13, B_1 is a basis of V. The last part is obvious.

Example 16.a. Let $$V = \mathbf{R}^4$$

Now $$\alpha_1 = (1, 1, 0, 0),$$

$$\alpha_2 = (0, 1, 1, 0)$$

are linearly independent in V.

As $\dim V = 4$, by the above corollary, there exist $\beta_1, \beta_2 \in V$ such that

$$\{\alpha_1, \alpha_2, \beta_1, \beta_2\}$$

is a basis of V. We now determine such β_1, β_2.

Consider the standard basis $\epsilon_1, \epsilon_2, \epsilon_3, \epsilon_4$ of V.

Then, $$\alpha_1, \alpha_2, \epsilon_1, \epsilon_2, \epsilon_3, \epsilon_4 \quad ... (1)$$

are linearly dependent.

These vectors span V. For some $x_i \in \mathbf{R}$, $1 \leq i \leq 6$.

Let $$x_1\alpha_1 + x_2\alpha_2 + x_3 \epsilon_1 + x_4 \epsilon_2 + x_5 \epsilon_3 + x_6 \epsilon_4 = 0 = (0, 0, 0, 0) \quad ... (2)$$

By comparing different components, we get four homogeneous linear equations in the six variables x_i, $1 \leq i \leq 6$, with coefficient matrix

$$A = \begin{bmatrix} 1 & 0 & 1 & 0 & 0 & 0 \\ 1 & 1 & 0 & 1 & 0 & 0 \\ 0 & 1 & 0 & 0 & 1 & 0 \\ 0 & 0 & 0 & 0 & 0 & 1 \end{bmatrix} \text{ row equivalent to } B = \begin{bmatrix} 1 & 0 & 0 & 1 & -1 & 0 \\ 0 & 1 & 0 & 0 & 1 & 0 \\ 0 & 0 & 1 & -1 & 1 & 0 \\ 0 & 0 & 0 & 0 & 0 & 1 \end{bmatrix}$$

Leading entries of the rows of B are in column 1, 2, 3 and 6, so that x_4 and x_5 are arbitrary. Let $x_4 = r$, $x_5 = s$. Then, B gives

$$x_1 = -r + s$$

$$x_2 = -s$$
$$x_3 = r - s$$
$$x_6 = 0$$

For $r = -1$, $s = 0$, (2) yields $\quad \epsilon_2 = \alpha_1 - \epsilon_1$

For $s = -1$, $r = 0$, (2) yields $\quad \epsilon_3 = -\alpha_1 + \alpha_2 + \epsilon_1$

So, from (1), we get $\alpha_1, \alpha_2, \epsilon_1, \epsilon_4$ span V.

Hence,
$$B = \{\alpha_1, \alpha_2, \epsilon_1, \epsilon_4\}$$
is a basis of V.

Lemma. 17. Let S be any linearly independent subset of a vector space V_F and let $S = S_1 \cup S_2$ with $S_1 \cap S_2 = \phi$. If W, W_1 and W_2 are the subspaces of V spanned by S, S_1 and S_2 respectively, then,
$$W = W_1 + W_2$$
and
$$W_1 \cap W_2 = \{0\}$$

Proof. By Theorem (2.1.11) (*iii*),
$$\langle S_1 \cup S_2 \rangle = \langle S_1 \rangle + \langle S_2 \rangle$$
i.e.,
$$W = W_1 + W_2$$

Let $0 \neq \alpha \in W_1 \cap W_2$. Then, $\quad \alpha = \sum_{i=1}^{k} a_i \alpha_i = \sum_{j=1}^{s} b_j \alpha'_j$

for some $\quad \alpha_i \in S_1, \alpha'_j \in S_2$.

Here, $\quad k \geq 1, s \geq 1$.

This gives
$$\sum_{i=1}^{k} a_i \alpha_i - \sum_{j=1}^{s} b_j \alpha'_j = 0 \quad \text{... (1)}$$

As $\alpha \neq 0$, some $a_i \neq 0$ and some $b_j \neq 0$.

As
$$S_1 \cap S_2 = \phi,$$
$$B = \{\alpha_1, \alpha_2, ..., \alpha_k, \alpha'_1, \alpha'_2, ..., \alpha'_s\}$$
is a subset of S with $k + s$ elements.

By (1), B is linearly dependent. This contradicts the fact that S is linearly independent. Hence, $W_1 \cap W_2 = \{0\}$.

Lemma 18. Let W_1 and W_2 be two subspaces of a vector space V_F such that $W_1 \cap W_2 = \{0\}$. B_1 and B_2 be bases of W_1 and W_2, respectively. Then, $B = B_1 \cup B_2$ is a basis of $W_1 + W_2$ and $B_1 \cap B_2 = \phi$.

Proof. By Theorem (2.1.11), $W = W_1 + W_2$ is spanned by $B = B_1 \cup B_2$. To prove that B is a basis, we only need to prove that B is linearly independent. Consider any finite non-empty subset D of B. If $D \subseteq B_1$ or $D \subseteq B_2$, D is linearly independent, as each of B_1, B_2 is linearly independent. Let $D \not\subseteq B_1$ and $D \not\subseteq B_2$.

Then, $$D = D_1 \cup D_2,\ D_1 = B_1 \cap D,\ D_2 = B_2 \cap D$$

with $$D_1 \neq \phi,\ D_2 \neq \phi.$$

As $$W_1 \cap W_2 = \{0\}$$

and 0 cannot be in any linearly independent subset, $B_1 \cap B_2 = \phi$, so that $D_1 \cap D_2 = \phi$

Let $$D_1 = \{\alpha_1, \alpha_2, \ldots, \alpha_r\} \quad \ldots (1)$$

$$D_2 = \{\beta_1, \beta_2, \ldots, \beta_s\} \quad \ldots (2)$$

Observe that each D_i is linearly independent. Let D be linearly dependent. So, for some $a_i \in F$, $b_j \in F$, not all zeroes

$$\sum_{i=1}^{r} a_i\alpha_i + \sum_{j=1}^{s} b_j\beta_j = 0$$

So, $$\sum_{i=1}^{r} a_i\alpha_i = -\sum_{j=1}^{s} b_j\beta_j = \alpha \quad \text{(say)}$$

Then, $\alpha \neq 0$ and $\alpha \in W_1 \cap W_2$. This contradicts the hypothesis that $W_1 \cap W_2 = \{0\}$. Thus, B is linearly independent. Hence, B is a basis of W.

Theorem 19. Let W be any subspace of a finite dimensional vector space V_F. Then, there exists a subspace W' of V such that $V = W + W'$ and $W \cap W' = \{0\}$

Further, for any such subspace W',

$$\dim V = \dim W + \dim W'.$$

(Any such W' is called a *complement* of W in V).

Proof. Let $\dim V = n$ and $\dim W = m$. Then, $m \leq n$.

If $W = \{0\}$, then, $W' = V$.

If $W = V$, then, $W' = \{0\}$.

Let $W \neq \{0\}$ and $W \neq V$. Then, $1 \leq m < n$.

Let $$B_1 = \{\alpha_1, \alpha_2, \ldots, \alpha_m\}$$

be a basis of W

As $\alpha_1, \alpha_2, \ldots, \alpha_m$

are linearly independent,

Corollary 16, gives $n - m$ vectors $\beta_1, \beta_2, \ldots, \beta_{n-m}$

such that $$B = \{\alpha_1, \alpha_2, ..., \alpha_m, \beta_1, \beta_2, ..., \beta_{n-m}\}$$

is a basis of V.

Now, $$B = B_1 \cup B_2$$

with $$B_2 = \{\beta_1, \beta_2, ..., \beta_{n-m}\}.$$

Further, $$B_1 \cap B_2 = \phi.$$

Let $$W' = \langle B_2 \rangle.$$

As $$V = \langle B \rangle, \ W = \langle B_1 \rangle$$

by Lemma 17, $V = W + W'$ with $W \cap W' = \{0\}$.

Finally, let W'' be any subspace of V such that $V = W + W''$ and $W \cap W'' = \{0\}$. Let B_3 be any basis of W''. Then, $B_1 \cap B_3 = \phi$, by Lemma 18, $B' = B_1 \cup B_3$ is a basis of V, so that

$$\dim V = |B'| = |B_1| + |B_3| = \dim W + \dim W''$$

This proves the result.

Theorem 20. Let W_1 and W_2 be two finitely generated subspaces of a vector space V_F. Then, $W_1 + W_2$ is finitely generated and

$$\dim(W_1 + W_2) + \dim(W_1 \cap W_2) = \dim W_1 + \dim W_2$$

Proof. That $W_1 + W_2$ is finitely generated, is immediately clear from Theorem (2.1.11).

Let $$W = W_1 \cap W_2.$$

As $$W \subseteq W_2,$$

by Theorem 19, $$W_2 = W + W_3$$

with $$W \cap W_3 = \{0\} \qquad ... (1)$$

and $$\dim W_2 = \dim W + \dim W_3 \qquad ... (2)$$

Let $$\alpha \in W_1 \cap W_3.$$

As $$W_3 \subseteq W_2,$$

we get $$\alpha \in W_1 \cap W_2 = W,$$

so that $$\alpha \in W \cap W_3 = \{0\}.$$

Hence, $$W_1 \cap W_3 = \{0\}.$$

Thus, $$\dim(W_1 + W_3) = \dim W_1 + \dim W_3 \qquad ... (3)$$

But $$W_1 + W_2 = W_1 + W + W_3 = (W_1 + W) + W_3.$$

As $$W \subseteq W_1, \ W_1 + W = W_1.$$

Hence, $$W_1 + W_2 = W_1 + W_3 \qquad ... (4)$$

Consequently, by (3) and (4),

$$\dim(W_1 + W_2) = \dim W_1 + \dim W_3$$

By (2), $$\dim W_3 = \dim W_2 - \dim(W_1 \cap W_2).$$

Hence, $$\dim(W_1 + W_2) = \dim W_1 + \dim W_2 - \dim(W_1 \cap W_2),$$

i.e., $$\dim(W_1 + W_2) + \dim(W_1 \cap W_2) = \dim W_1 + \dim W_2$$

This situation in Theorem 19 leads to the following more general concept :

Definition 21. A vector space V_F is called an *internal direct sum* of its subspaces W_1, $W_2, \ldots, W_k$ if each α in V_F is uniquely expressible as $\alpha = \alpha_1 + \alpha_2 + \ldots + \alpha_k$, $\alpha_i \in W_i$, $1 \le i \le k$, in the sense that if also $\alpha = \beta_1 + \beta_2 + \ldots + \beta_k$, $\beta_i \in W_i$ then, $\alpha_i = \beta_i$ for $1 \le i \le k$; in this situation, we write

$$V = W_1 \oplus W_2 \oplus \ldots \oplus W_k$$

Example 21.a. Let V_F be any finitely generated vector space with $\dim V_F = k \ge 2$. Let $B = \{\alpha_1, \alpha_2, \ldots, \alpha_k\}$ be a basis of V_F. For each i, let $W_i = \langle \alpha_i \rangle = \{a\alpha_i : a \in F\}$. Consider $\alpha \in V_F$. As B is a basis of V_F

$$\alpha = a_1\alpha_1 + a_2\alpha_2 + \ldots + a_k\alpha_k,\ a_i \in F$$

Thus $$\alpha = \beta_1 + \beta_2 + \ldots + \beta_k,\ \beta_i = a_i\alpha_i \in W_i. \qquad \ldots (1)$$

Suppose also $$\alpha = \beta_1' + \beta_2' + \ldots + \beta_k',\ \beta_i' \in W_i.$$

Now, $$\beta_i' \in W_i$$

gives $$\beta_i' = a_i'\alpha_i$$

Hence, $$\alpha = a_1'\alpha_1 + a_2'\alpha_2 + \ldots + a_k'\alpha_k' \qquad \ldots (2)$$

(1) and (2) give $$\sum_{i=1}^{k} a_i\alpha_i = \sum_{i=1}^{k} a_i'\alpha_i$$

i.e., $$\sum_{i=1}^{k} (a_i - a_i')\alpha_i = 0$$

As $\alpha_1, \alpha_2, \ldots, \alpha_k$ are linearly independent, we get

$$a_i - a_i' = 0$$

i.e., $$a_i = a_i',$$

so that $$\beta_i = a_i\alpha_i = a_i'\alpha_i = \beta_i'.$$

Hence, $$V = W_1 \oplus W_2 \oplus \ldots \oplus W_k.$$

Thus, V is an internal direct sum of k subspaces, each one dimensional.

Example 21.b. Let $$V = \mathbf{R}^3.$$

Then, $$W_1 = \{(a, b, 0) : a, b \in \mathbf{R}\},$$
$$W_2 = \{(0, b, b) : b \in \mathbf{R}\},$$
$$W_3 = \{(0, b, c) : b, c \in \mathbf{R}\}$$
are subspaces of *V*.

Consider $$\alpha = (a, b, c) \in V.$$

Then, $$\alpha = (a, b-c, 0) + (0, c, c) = \alpha_1 + \alpha_2 \qquad \text{... (1)}$$

with $$\alpha_1 = (a, b-c, 0) \in W_1,$$
$$\alpha_2 = (0, c, c) \in W_2.$$

Let also $$\alpha = \beta_1 + \beta_2,\ \beta_1 \in W_1,\ \beta_2 \in W_2.$$

Then, $$\beta_1 = (a_1, b_1, 0),$$
$$\beta_2 = (0, b_2, b_2) \qquad \text{for some } a_1, b_1, b_2 \in \mathbf{R}.$$

So, $$\alpha = (a_1, b_1 + b_2, b_2)$$

This gives $$(a, b, c) = (a_1, b_1 + b_2, b_2), \text{ i.e.,}$$
$$a_1 = a,\ b_1 + b_2 = b,\ c = b_2.$$

Thus, $$b_1 = b - c.$$

This gives $$\alpha_1 = \beta_1,\ \alpha_2 = \beta_2.$$

Hence, $$V = W_1 \oplus W_2$$

Let us consider $W_1 + W_3$. Any $\alpha = (a, b, c)$ is expressible as
$$\alpha = (a, b, 0) + (0, 0, c) \qquad \text{... (3)}$$

Clearly, $(a, b, 0) \in W_1$ and $(0, 0, c) \in W_2$, so that $V = W_1 + W_2$

We can also write $$\alpha = (a, b+1, 0) + (0, -1, c) \qquad \text{... (4)}$$

Expressions (3) and (4) are different. Hence, *V* is not an internal direct sum of W_1 and W_3.

Theorem 22. Let $W_1, W_2, \ldots, W_k$ be any $k \geq 2$ subspaces of a vector space V_F. Then, the following are equivalent :

(*i*) $$V = W_1 \oplus W_2 \oplus \ldots \oplus W_k$$

(*ii*) $$V = W_1 + W_2 + \ldots + W_k$$

and for $$1 \leq i \leq k,\ W_i \cap V_i = \{0\}$$

where $$V_i = \sum_{j \neq i} W_j \qquad (\text{i.e., } V_1 = W_2 + W_3 + \ldots + W_k).$$

Proof.

Let $$V = W_1 \oplus W_2 \oplus \ldots \oplus W_k.$$

By definition, $V = W_1 + W_2 + ... + W_k$.

Let $\alpha \in W_i \cap V_i$

Then, $\alpha = \alpha_i \in W_i$ and $\alpha \in V_i$, gives

$$\alpha = \alpha_1 + ... + \alpha_{i-1} + \alpha_{i+1} + ... + \alpha_k,\ \alpha_j \in W_j$$

Thus, we can write α in two ways

$$\alpha = \beta_1 + \beta_2 + ... + \beta_k,\ \beta_j = 0 \qquad \text{for } j \neq i,\ \beta_i = \alpha_i$$

$$= \beta_1' + \beta_2' + ... + \beta_k',\ \beta_j' = \alpha_j \qquad \text{for } j \neq i,\ \beta_i' = 0$$

By comparing these two expressions, we get $\beta_j = \beta_j'$ for every j.

So, in particular, $\beta_i = \beta_i'$, i.e., $\alpha_i = 0$. Hence $\alpha = 0$. This proves that $W_i \cap V_i = \{0\}$. This proves that (*i*) implies (*ii*).

Let (*ii*) hold.

Let $\alpha \in V$.

As $V = W_1 + W_2 + ... + W_k$,

$$\alpha = \alpha_1 + \alpha_2 + ... + \alpha_k,\ \alpha_i \in W_i \qquad ... (1)$$

Let also $\alpha = \beta_1 + \beta_2 + ... + \beta_k,\ \beta_i \in W_i$ ... (2)

For any i, $1 \le i \le k$, (1) and (2) give

$$\alpha_i - \beta_i = \sum_{j \neq i} (\beta_j - \alpha_j) \in W_i \cap V_i = \{0\}.$$

Thus, $\alpha_i - \beta_i = 0$,

i.e., $\alpha_i = \beta_i$.

Hence, $V = W_1 \oplus W_2 \oplus ... \oplus W_k$

In the theorem, given above, for $k = 2$, $V_1 = W_2$, so that $W_1 \cap V_1 = \{0\}$ means $W_1 \cap W_2 = \{0\}$. Thus, the theorem gives the following :

Theorem 23. Let W_1 and W_2 be two subspaces of a vector space V_F. Then, the following are equivalent :

(*i*) $V = W_1 \oplus W_2$

(*ii*) $V = W_1 + W_2$

and $W_1 \cap W_2 = \{0\}$

Let W_1 and W_2 be two subspaces of V_F such that

$$W_1 \cap W_2 = \{0\}.$$

By Theorem 23, $W = W_1 + W_2 = W_1 \oplus W_2$.

By Theorem 19, $\dim W = \dim W_1 + \dim W_2$.

So, Theorem 19 can be restated to say that if a vector space V is an internal direct sum of two subspaces W and W', then,

$$\dim V = \dim W + \dim W'.$$

If $$V = W_1 \oplus W_2 \oplus \ldots \oplus W_k$$

for some $k \geq 2$, $$V_k = W_1 + W_2 + \ldots + W_{k-1}$$
$$= W_1 \oplus W_2 \oplus \quad \oplus W_{k-1}.$$

For any i, $1 \leq i \leq k-1$, V_i', the sum of W_j's for $j \neq i$, $1 \leq j \leq k-1$, $V_i' \subset V_i$, gives

$$W_j \cap V_i' = \{0\}.$$

The following theorem generalizes Theorem 19 :

Theorem 24. Let W_1, W_2,..., W_k be any $k \geq 1$, subspaces of a finitely generated vector space V_F such that

$$V_F = W_1 \oplus W_2 \oplus W_3 \oplus \ldots \oplus W_k$$

Further, for $1 \leq i \leq k$, let B_i be a basis of W_i. Then,

(*i*) B_1, B_2,..., B_k are pair-wise disjoint and $B = \bigcup_{i=1}^{k} B_i$ is a basis of V

(*ii*) $$\dim V = \sum_{i=1}^{k} \dim W_i$$

Proof. For $k = 1$, $V = W_1$, $B = B_1$. In this case, the result is obvious. Let $k > 1$ and that result holds for an internal direct sum of $k-1$ subspaces.

In this case, $$V = W_1 + \left[\sum_{i=2}^{k} W_1\right] = W_1 + V_1$$

where $$V_1 = \sum_{i=2}^{k} W_1.$$

By Theorem 22, $$W_1 \cap V_1 = \{0\}$$

so that $$V = W_1 \oplus V_1 \qquad \ldots (1)$$

By Theorem 19, $\dim V = \dim W_1 + \dim V_1$.

By Lemma 18, $B_1 \cup B'$ is a basis of V, for any basis B' of V_1.

However, $$V_1 = W_2 \oplus W_3 \oplus \ldots \oplus W_k \qquad \ldots (2)$$

is an internal direct sum of $k-1$ subspaces.

So, by the induction hypothesis, $$\dim V_1 = \sum_{i=2}^{k} \dim W_i \qquad \ldots (3)$$

Further, $B' = \bigcup_{i=1}^{k} B_i$ is a basis of V_1. Now $B_2, \ldots, B_k$ are pairwise disjoint. Also, $B_1 \cap B' = \phi$. Hence, $B_1, B_2, \ldots, B_k$ are pairwise disjoint.

Now, $$B = B_1 \cup B' = \overset{k}{\underset{i=1}{U}} B_i.$$

Finally, (1) and (3) give

$$\dim V = \dim W_1 + \sum_{i=2}^{k} \dim W_i = \sum_{i=1}^{k} \dim W_i$$

We now give another criterian for an internal direct sum.

Theorem 25. Let $W_1, W_2, \ldots, W_k$ be any $k \geq 1$, subspaces of a vector space V_F. Then, the following are equivalent :

(*i*) $$V = W_1 \oplus W_2 \oplus \ldots \oplus W_k$$

(*ii*) $$V = W_1 + W_2 + \ldots + W_k$$

and for any $\alpha_i \in W_i,\ 1 \leq i \leq k,\ \alpha_1 + \alpha_2 + \ldots + \alpha_k = 0$

gives $$\alpha_1 = \alpha_2 = \ldots = \alpha_k = 0$$

Proof. For $k = 1$, $V = W_1$ and the result is obvious. Let $k \geq 2$. Let (*i*) hold. Then, obviously, $V = W_1 + W_2 + \ldots + W_k$ and in the notation of Theorem 22, $W_i \cap V_i = \{0\}$.

Let for some $\alpha_j \in W_j$, $$\sum_{j=1}^{k} \alpha_j = 0,$$

Then, $$\alpha_i = -\sum_{j \neq i} \alpha_j \in W_i \cap V_i$$

yields, $\alpha_i = 0$ for every i.

Hence, (*i*) implies (*ii*). Let (*ii*) hold.

For some i, let $$\alpha \in W_i \cap V_i.$$

Then, $$\alpha = \beta_i \in W_i.$$

As $$\alpha \in V_i,$$

$$\alpha = \alpha_1 + \ldots + \alpha_{i-1} + \alpha_{i+1} + \alpha_{i+2} + \ldots + \alpha_k$$

for some $$\alpha_j \in W_j.$$

Consequently, $$\beta_i = \sum_{j \neq i} \alpha_j$$

i.e., $$\sum_{j=1}^{k} \alpha_j = 0,$$

where $$\alpha_i = -\beta_i./$$

By the hypothesis, $\alpha_j = 0$ for every j.

In particular, $\alpha_i = 0$

So $\alpha = -\alpha_i = 0.$

Hence, $W_i \cap V_i = \{0\}.$

By Theorem 22, $V = W_1 \oplus W_2 \oplus \ldots \oplus W_k$

Hence, (*ii*) implies (*i*).

Exercise 2.2

1. Let $V = R^3$. Which of the following sets of vectors in V are linearly independent?. If a set of vectors is linearly dependent, express one of the members of this set as a linear combination of the others.

   (*i*) $\{(1, 1, -2), (1, 0, 1)\}$

   (*ii*) $\{(2, 0, 1), (0, 1, 1), (3, 1, 0), (1, 1, 1)\}$

   (*iii*) $\{(2, 4, 8), (3, 9, 27), (1, 1, 1)\}$

   (*iv*) $\{(2, 3, -4), (-3, 2, 4), (1, -8, 4)\}$

2. Consider the set S of vectors in $\mathbf{R}^3$ consisting of

   $$\alpha_1 = (1, 1, 0), \alpha_2 = (0, 2, 3), \alpha_3 = (1, 2, 3), \alpha_4 = (1, -2, 3)$$

   (*i*) Prove that S is linearly dependent.

   (*ii*) Let $W = \langle S \rangle$, find a subset B of S such that B is a basis of W.

   (*iii*) Find dim W.

3. Which of the following subsets of $\mathbf{R}^4$ are linearly independent and which of them form a basis of $\mathbf{R}^4$?

   (*i*) $\alpha_1 = (1, 0, 0, 0), \alpha_2 = (1, 1, 0, 0), \alpha_3 = (1, 1, 1, 0), \alpha_4 = (1, 1, 1, 1)$

   (*ii*) $\alpha_1 = (1, 1, 1, 1), \alpha_2 = (1, 2, 2, 2), \alpha_3 = (1, 2, 3, 3), \alpha_4 = (1, 2, 3, 4)$

   (*iii*) $\alpha_1 = (1, 2, 1, 1), \alpha_2 = (1, 1, 2, 1), \alpha_3 = (1, 1, 1, 2), \alpha_4 = (1, 0, 1, 1)$,
   $\alpha_5 = (3, 0, 1, 1)$

   (*iv*) $\alpha_1 = (0, 0, 1, 1), \alpha_2 = (0, -1, 0, 0), \alpha_3 = (1, 2, 0, 0)$

4. Let $V = P_2(\mathbf{R})$. Which of the following sets of vectors form a basis of V? If a set does not form a basis, check if it spans V.

   (*i*) $\{-X^2 + X + 2, 2X^2 + 2X + 3, 4X^2 + 1\}$

   (*ii*) $\{1, 1 + X, 1 + X^2\}$

   (*iii*) $\{X^2 + 1, 3X^2 + X, 2X^2 - 2X + 1, X, X^2 - 1\}$

(*iv*) $\{3X^2+9X+1,\ -3X^2+8X+1\}$

Among these sets, which are linearly independent?

5. Let $V=\mathbf{R}^{2\times 2}$. Define

$$E_{11}=\begin{bmatrix}1&0\\0&0\end{bmatrix},\ E_{12}=\begin{bmatrix}0&1\\0&0\end{bmatrix},\ E_{21}=\begin{bmatrix}0&0\\1&0\end{bmatrix},\ E_{22}=\begin{bmatrix}0&0\\0&1\end{bmatrix}.$$

Prove that these four matrices form a basis of $\mathbf{R}^{2\times 2}$. What is $\dim V$?

6. Prove that the set W of all complex numbers of the form

$$a+2ai,\ a\in\mathbf{R}$$

is a subspace of $\mathbf{C}_{\mathbf{R}}$. Find a basis, and the dimension of W.

Let $W'=\{a-2ai : a\in\mathbf{R}\}$. Prove that $\mathbf{C}_{\mathbf{R}}=W\oplus W'$

7. Let $W=\{(a,\ b,\ c,\ d) : a=b+c,\ c=b+d\}$ in $\mathbf{R}^4$. Find a basis and the dimension of W.

8. Find a basis of $\mathbf{R}^3$ that includes one the subsets

(*i*) $\{(1,\ 2,\ 0)\}$

(*ii*) $\{(1,\ -2,\ 0),\ (0,\ 1,\ 1)\}$

9. Find a basis of $\mathbf{R}^4$ that contains the vectors $(1,\ 1,\ 0,\ 0)\ (0,\ 1,\ 0,\ -1)$

10. Let W be the set of those matrices A in $\mathbf{R}^{2\times 2}$ such that

$$A\begin{bmatrix}1&1\\2&2\end{bmatrix}=\begin{bmatrix}1&1\\2&2\end{bmatrix}A.$$

Prove that W is a subspace of $\mathbf{R}^{2\times 2}$. Find $\dim W$.

11. Let $V_{\mathbf{R}}=\mathbf{R}^{3\times 3}$, W be the set of symmetric matrices and W' be the set of skew symmetric matrices in $V_{\mathbf{R}}$. Prove that W and W' are subspaces. Further,

(*i*) Find $\dim W$ and $\dim W'$

(*ii*) Prove that $V=W\oplus W'$

12. Let $W=\{(a,\ b,\ 0) : a,b\in\mathbf{R}\}\subseteq\mathbf{R}^3$. Find its two different complements in $\mathbf{R}^3$.

13. Let $\alpha_1=(x_1,y_1),\ \alpha_2=(x_2,y_2)\in\mathbf{R}^2$ such that

$$x_1y_1+x_2y_2=0,$$

$$x_1^2+x_2^2=1=y_1^2+y_2^2$$

(*i*) Prove that $B=\{\alpha_1,\ \alpha_2\}$ is a a basis of $\mathbf{R}^2$

(*ii*) For any $\alpha=(a,\ b)\in\mathbf{R}^2$, find real numbers $a_1,\ a_2$ such that $\alpha=a_1\alpha_1+a_2\alpha_2$

14. Let $u \in \mathbf{R}$ be a transcendental number (e.g., $u = \pi$). Let W be the set of those real numbers in $\mathbf{R}$ which are of the type

$$a_0 + a_1 u + a_2 u^2 + ... + a_k u^k, \; a_i \in \mathbf{Q}$$

and $$k \geq 0.$$

Prove that W is an infinite dimensional subspace of $\mathbf{R}$ over $\mathbf{Q}$.

15. Let a vector space V_F be an internal direct sum of $W_1, W_2, ..., W_k$ with $k \geq 2$. Let A and B be two subsets of $\{1, 2, 3, ..., k\}$ such that $A \neq \phi$, $B \neq \phi$, $A \cap B = \phi$. Let W_A be the sum of those W_i such that $i \in A$, and W_B be the sum of those W_j such that $j \in B$. Prove the following :

 (*i*) $W_A + W_B = W_A \oplus W_B$

 (*ii*) If $A \cup B = \{1, 2, ..., k\}$, then $V = W_A \oplus W_B$

 (*iii*) $W_A = \sum_{i \in A} \oplus W_i$

16. Express $\mathbf{R}^5$ as a direct sum of two subspaces W_1, W_2 with $\dim W_1 = 2$, $\dim W_2 = 3$.

17. Let $V_F = W_1 + W_2$ for some finite dimensional subspaces W_1 and W_2 of V. If $\dim V = \dim W_1 + \dim W_2$, prove that $V = W_1 \oplus W_2$.

18. If $B = \{\alpha_1, \alpha_2, \alpha_3, \alpha_4\}$ is a basis of a vector space V_F, prove that each of the following is a bases of V_F.

 (*i*) $\{\alpha_1, \alpha_2 + \alpha_1, \alpha_3 + \alpha_1, \alpha_4\}$

 (*ii*) $\{\alpha_1, \alpha_1 + \alpha_2, \alpha_1 + \alpha_2 + \alpha_3, \alpha_1 + \alpha_2 + \alpha_3 + \alpha_4\}$

 (*iii*) $\{c_1\alpha_1, c_2\alpha_2, c_3\alpha_3, c_4\alpha_4\}$, where c_1, c_2, c_3, c_4 are any four non-zero members of F.

19. Find four matrices A_1, A_2, A_3, A_4 in $\mathbf{R}^{2\times2}$ such that $A_i^2 = A_i$ for $1 \leq i \leq 4$, and $\{A_1, A_2, A_3, A_4\}$ is a basis of $\mathbf{R}^{2\times2}$

20. Let V_F be a finite dimensional vector space and $V = W_1 \oplus W_2$. For any subspace L of V, prove that $2\dim L - \dim V \leq \dim(W_1 \cap L) + \dim(W_2 \cap L) \leq \dim L$.

21. Give an example of a vector space $V_{\mathbf{R}}$ having three different non-zero subspaces V_1, V_2 and V_3 such that $V = V_1 \oplus V_2 = V_2 \oplus V_3 = V_3 \oplus V_1$.

2.3 Rank of a Matrix

Let $A = [a_{ij}]$ be any $m \times n$-matrix over a field F. Recall that the row space of A is the subspace of $F^{1\times n}$ (or just F^n)spanned by the m rows of A. The dimension of the row space of A is called the *row-rank* of A. The column space of A is the subspace of $F^{m\times1}$, spanned by the n columns of A, and the dimension of this space is called the *column-rank* of A.

Proposition 1. Let A and B be two row equivalent matrices over F. Then,

(*i*) row-rank $(A) \le \min(m, n)$

(*ii*) row-rank $(A) =$ row-rank (B)

Proof. Let $W =$ row-space (A). As W is spanned by m rows of A, $\dim W \le m$. As $W \subseteq F^n$ and $\dim F^n = n$, $\dim W \le n$.

Hence, row-rank $(A) = \dim W \le \min(m, n)$

As A and B are row equivalent,

row-space $(A) =$ row-space (B) {Theorem (1.4.9)}.

Hence, row-rank $(A) =$ row-rank (B).

Proposition 2. Let $A = \left[a_{ij}\right]$ be a non-zero $m \times n$-matrix over F, in reduced row echelon form. Then, row-rank (A) equals the number s of non-zero rows of A. The set of non-zero rows of A is a basis of the row space of A.

Proof. By definition, each of the first s rows of A is non-zero and the other rows are zero rows. So, the row space W of A is spanned by the first s rows A_i, $1 \le i \le s$. For $1 \le i \le s$, let the leading entry of A_i be in k_ith column.

Then, $$k_1 < k_2 < \ldots < k_s,\ a_{ik_i} = 1,$$

but $$a_{jk_i} = 0 \qquad \text{for } j \ne k_i.$$

For some $c_j \in F$, let $$\sum_{i=1}^{s} c_j A_j = 0 \qquad \ldots (1)$$

So, for $1 \le l \le n$, the lth component in (1)

$$\sum_{i=1}^{s} c_j a_{jl} = 0 \qquad \ldots (2)$$

For $l = k_i$, $a_{jl} = 0$ for $j \ne k_i$. So (2) gives

$$c_i a_{ik_i} = 0,$$

i.e., $$c_i = 0,$$

as $$a_{ik_i} = 1.$$

Hence, in (1) every coefficient $c_j = 0$. This proves that $A_1, A_2, \ldots, A_s$ are linearly independent. Hence, W has a basis $\{A_1, A_2, \ldots, A_s\}$ with s members.

Consequently, row-rank $(A) = \dim W = s$.

Observe that given a matrix A, the rows of A^T are the columns of A, so that the row-rank $(A^T) =$ column-rank (A). However, we prove a more general result.

Theorem 3. For any $m \times n$-matrix A over a field F, row-rank (A) = column-rank (A).

Proof. Let row-rank $(A) = s$ and column-rank $(A) = t$. Let R_i and C_j denote the ith row and the jth column of $A = [a_{ij}]$. Let $\{\alpha_1, \alpha_2, ..., \alpha_s\}$ be a basis of the row space W of A.

As $$\alpha_j \in F^{1\times n}, \alpha_j = [b_{j1}, b_{j2}, ..., b_{jn}] \quad ... (1)$$

Now, $$R_i = \sum_{j=1}^{s} r_{ij}\alpha_j, \; r_{ij} \in F, \text{ i.e.,}$$

$$[a_{i1}, a_{i2}, ..., a_{in}] = \sum_{j=1}^{s} r_{ij}[b_{j1}, b_{j2}, ..., b_{jn}].$$

This gives $$a_{iu} = \sum_{j=1}^{s} r_{ij}b_{ju},$$

$$1 \le u \le n,$$

so that the uth column $$C_u = \begin{bmatrix} a_{1u} \\ a_{2u} \\ \vdots \\ a_{mu} \end{bmatrix} = \sum_{j=1}^{s} \begin{bmatrix} r_{1j} \\ r_{2j} \\ \vdots \\ r_{mj} \end{bmatrix} b_{ju}$$

This shows that the column space W' of A is a subspace of *a* subspace of $F^{m\times 1}$, spanned by s vectors, so that

$$\text{Column-rank } (A) \le s = \text{row-rank } (A)$$

Similarly, $$\text{row-rank } (A) \le \text{column-rank } (A)$$

Hence, $$\text{row-rank } (A) = \text{column-rank } (A)$$

Theorem 4. Let A and B be two non-zero row equivalent matrices over F, such that B is in reduced row echelon form. Let B have s non-zero rows and let the leading entry of the ith row, for $1 \le i \le s$, of B be in k_i^{th} column. Then, the set consisting of the $k_1^{th}, k_2^{th}, ..., k_s^{th}$ columns of A form a basis of the column space of A.

Proof. Let A_j and B_j denote the jth column of A and B, respectively.

Then, for any $$X = \begin{bmatrix} x_1 \\ x_2 \\ \vdots \\ x_n \end{bmatrix}$$

$$AX = \sum_{j=1}^{n} A_j x_j \quad ... (1)$$

$$BX = \sum_{j=1}^{n} B_j x_j \quad \text{... (2)}$$

Now, row-rank $(B) = s$, by Proposition 2.

By Theorem 3, column-rank $(B) = s$.

Now, the k_1^{th} column, k_2^{th} column,..., k_s^{th} column, of B are

$$\begin{bmatrix} 1 \\ 0 \\ 0 \\ \vdots \\ 0 \end{bmatrix}, \begin{bmatrix} 0 \\ 1 \\ 0 \\ \vdots \\ 0 \end{bmatrix}, \ldots, \begin{bmatrix} 0 \\ \vdots \\ 0 \\ 1 \\ 0 \\ \vdots \\ 0 \end{bmatrix} \text{-}s^{\text{th}}$$

These are linearly independent, i.e.,

$$B_{k_1}, B_{k_2}, \ldots, B_{k_s}$$

are linearly independent.

Let for some $c_{k_1}, c_{k_2}, \ldots, c_{k_s} \in F$

$$\sum_{i=1}^{s} A_{k_i} c_{k_i} = 0$$

This gives solution $X' = [c_1, c_2, \ldots, c_n]^T$

of $AX = 0$, with $c_j = 0$ for $j \neq k_i$, $1 \leq i \leq s$.

This is also a solution of $BX = 0$ {Theorem (1.4.22)}, so that

$$\sum_{j=1}^{n} B_j c_j = 0$$

i.e.,

$$\sum_{i=1}^{s} B_{k_i} c_{k_i} = 0,$$

as $c_j = 0$ for $j \neq k_i$.

The linear independence of B_{k_i}'s, $1 \leq i \leq s$ gives

$$c_{k_i} = 0.$$

Hence, the s columns $A_{k_1}, A_{k_2}, \ldots, A_{k_s}$ of A are linearly independent.

By Proposition 1 and Theorem 3, column-rank $(A) = s$. Hence, these s columns form a basis of the column space of A.

Remark 5.

(*i*) Let A and B be two row equivalent matrices over F. As $AX = 0$ and $BX = 0$ have, same sets of the solutions, in the notation of the above theorem, if some B_j is expressible as

$$B_j = \sum_{i=1}^{u} B_{t_i} r_i, \qquad r_i \in F$$

then,

$$A_j = \sum_{i=1}^{u} A_{t_i} r_i$$

(*ii*) To find a basis of the row-space of A and a basis of the column space of A, first of all find a matrix B in reduced row echelon form that is row equivalent to A. Now, row-space (A) = row-s[ace (B). By Proposition 2, the set of non-zero rows of B is a basis of the row-space (B) and, hence, of row-space (A). Use Theorem 4 to find a basis of the column-space (A).

(*iii*) Another method of finding a basis of the column space of a matrix A is the following. Form A^T. Find matrix B in reduced row echelon form row-equivalent to A^T. Form B^T. Then, the set of non-zero column of B^T is a basis of the column-space of A.

Example 5.a. Let
$$A = \begin{bmatrix} 0 & 2 & 5 & -1 & 5 \\ 0 & 0 & 2 & 3 & 4 \\ 4 & 2 & -11 & 11 & 11 \\ 2 & 0 & -6 & 9 & 7 \end{bmatrix}$$

By applying row operations, we get

$$B = \begin{bmatrix} 1 & 0 & 0 & 9 & \frac{19}{2} \\ 0 & 1 & 0 & -\frac{17}{4} & -\frac{5}{2} \\ 0 & 0 & 1 & \frac{3}{2} & 2 \\ 0 & 0 & 0 & 0 & 0 \end{bmatrix}$$

in reduced row echelon form row equivalent to A.

As B has three non-zero row,

$$\text{row-rank } (B) = 3 = \text{row-rank } (A) = \text{column-rank } (A).$$

Further, the set of non-zero rows of B is a basis of the row-space of A.

So
$$B = \left\{\left(1\ 0\ 0\ 9\ \frac{19}{2}\right), \left(0\ 1\ 0\ -\frac{17}{4}\ -\frac{5}{2}\right), \left(0\ 0\ 1\ \frac{3}{2}\ 2\right)\right\}$$

is a basis of the row-space (A).

The leading entries of rows of B are in columns 1, 2 and 3. So, columns numbered 1, 2, 3 of A form a basis of the column-space (A).

Thus, $$B' = \left\{ \begin{bmatrix} 0 \\ 0 \\ 4 \\ 2 \end{bmatrix}, \begin{bmatrix} 2 \\ 0 \\ 2 \\ 0 \end{bmatrix}, \begin{bmatrix} 5 \\ 2 \\ -11 \\ -6 \end{bmatrix} \right\}$$

is a basis of the column-space (A).

Observe that in B, column 4

$$B_4 = \begin{bmatrix} 9 \\ -\frac{17}{4} \\ \frac{3}{2} \\ 0 \end{bmatrix} = 9 \begin{bmatrix} 1 \\ 0 \\ 0 \\ 0 \end{bmatrix} - \frac{17}{4} \begin{bmatrix} 0 \\ 1 \\ 0 \\ 0 \end{bmatrix} + \frac{3}{2} \begin{bmatrix} 0 \\ 0 \\ 1 \\ 0 \end{bmatrix} = 9B_1 - \frac{17}{4} B_2 + \frac{3}{2} B_3$$

In A, $$A_4 = \begin{bmatrix} -1 \\ 3 \\ 11 \\ 9 \end{bmatrix} = 9 \begin{bmatrix} 0 \\ 0 \\ 4 \\ 2 \end{bmatrix} - \frac{17}{4} \begin{bmatrix} 2 \\ 0 \\ 2 \\ 0 \end{bmatrix} + \frac{3}{2} \begin{bmatrix} 5 \\ 2 \\ -11 \\ -6 \end{bmatrix}$$

Once again, $$A^T = \begin{bmatrix} 0 & 0 & 4 & 2 \\ 2 & 0 & 2 & 0 \\ 5 & 2 & -11 & -6 \\ -1 & 3 & 11 & 9 \\ 5 & 4 & 11 & 7 \end{bmatrix}$$

It is row-equivalent to the following matrix in reduced row echelon form

$$B' = \begin{bmatrix} 1 & 0 & 0 & -\frac{1}{2} \\ 0 & 1 & 0 & 1 \\ 0 & 0 & 1 & \frac{1}{2} \\ 0 & 0 & 0 & 0 \\ 0 & 0 & 0 & 0 \end{bmatrix}$$

Then, $$(B')^T = \begin{bmatrix} 1 & 0 & 0 & 0 & 0 \\ 0 & 1 & 0 & 0 & 0 \\ 0 & 0 & 1 & 0 & 0 \\ -\frac{1}{2} & 1 & \frac{1}{2} & 0 & 0 \end{bmatrix}$$

It has three non-zero columns; they form a basis of the column space (A).

Thus, $$B'' = \left\{ \begin{bmatrix} 1 \\ 0 \\ 0 \\ -\frac{1}{2} \end{bmatrix}, \begin{bmatrix} 0 \\ 1 \\ 0 \\ 1 \end{bmatrix}, \begin{bmatrix} 0 \\ 0 \\ 1 \\ \frac{1}{2} \end{bmatrix} \right\}$$

is a basis of the column-space (A).

Example 5.b. Let W be the subspace of $\mathbf{R}^3$ spanned by the vectors

$$(1, 2, 3), (2, 1, 4), (-1, -1, 2), (0, 1, 2)$$

We form matrix A with these vectors as its rows, i.e.,

$$A = \begin{bmatrix} 1 & 2 & 3 \\ 2 & 1 & 4 \\ -1 & -1 & 2 \\ 0 & 1 & 2 \end{bmatrix}$$

so that
$$W = \text{row-space } (A).$$

Now, A is row equivalent to the matrix

$$B = \begin{bmatrix} 1 & 0 & 0 \\ 0 & 1 & 0 \\ 0 & 0 & 1 \\ 0 & 0 & 0 \end{bmatrix}$$

Thus,
$$\{(1, 0, 0), (0, 1, 0), (0, 0, 1)\}$$

is a basis of W.

As this is also a basis of $\mathbf{R}^3$, we get $W = \mathbf{R}^3$.

Consider
$$A^T = \begin{bmatrix} 1 & 2 & -1 & 0 \\ 2 & 1 & -1 & 1 \\ 3 & 4 & 2 & 2 \end{bmatrix}$$

Observe that given vectors spanning W are occurring as columns of A^T, so that the column-space (A) is the set of transposes of members of W.

Now, A^T is row equivalent to

$$B' = \begin{bmatrix} 1 & 0 & 0 & \frac{10}{13} \\ 0 & 1 & 0 & -\frac{3}{13} \\ 0 & 0 & 1 & \frac{4}{13} \end{bmatrix}$$

in reduced row echelon form.

Leading entries of rows of B' are in columns 1, 2, 3. So, in A^T, its first three columns, form a basis of column-space (A^T). Their transposes give a basis of W. That means

$$B = \{(1, 2, 3), (2, 1, 4), (-1, -1, 2)\}$$

is a basis of W.

This B is a subset of the set of four vectors that are given in the beginning.

Definition 6. For any matrix A over a field F, as row-rank (A) = column-rank (A), this common number is called the *rank* of A.

Theorem 7. For any $m \times n$-matrix A over a field F, the dimension of the solution space W of the system of homogeneous linear equations $AX = 0$ is n-rank(A)

Proof. If $A = 0$, then, $W = F^n$ has dimension $n = n -$ rank (A), as rank (A) = 0. Let $A \neq 0$, and B in reduced row echelon form be row equivalent to A. Then, solution space of $BX = 0$ is W itself. If B has s non-zero rows, then, s = rank (A). By Remark (1.4.27), W is spanned by $n - s$ vectors, which are linearly independent.

Hence, $$\dim W = n - s = n - \text{rank } (A).$$

Theorem 8. For any $n \times n$-matrix A over F, the following are equivalent

(*i*) rank (A) = n

(*ii*) A is non-singular

(*iii*) The solution space of the system of homogeneous linear equations $AX = 0$ is the zero space.

Proof. Let W be the solution space of $AX = 0$.

Then, $$\dim \text{W} = n - \text{rank } (A).$$

So, $$W = \{0\}$$

iff $$n - \text{rank } (A) = 0$$

iff $$n = \text{rank } (A)$$

This proves the equivalence of (*i*) and (*iii*). Let B be the $n \times n$ matrix in reduced row echelon form that is row equivalent to A, so that rank (A) = n iff rank (B) = n iff B has n non-zero rows. The only $n \times n$-matrix in reduced row echelon form, having n non-zero rows, is I_n. Hence, rank (A) = n iff A is row equivalent to I_n, i.e., iff A is non-singular {Theorem (1.4.15)}.

Theorem 9. For a system of linear equations $AX = B$, the following are equivalent :

(*i*) $AX = B$ is consistent

(*ii*) B is a linear combination of columns of A

(*iii*) column-space (A) = column-space $[A : B]$

(*iv*) rank (A) = rank $[A : B]$

Proof. Let A be an $m \times n$-matrix. Let $X = \begin{bmatrix} x_1 \\ x_2 \\ \vdots \\ x_n \end{bmatrix}$. Then, $AX = \text{B}$ is equivalent to

$$A_1 x_1 + A_2 x_2 + \ldots + A_n x_n = B \quad \ldots(1)$$

where A_j is the jth column of A.

Let W and W' be the column-spaces of A and $[A : B]$, respectively. Obviously, $W \subseteq W'$. Let $AX = B$ be consistent.

So, there exists a solution $X' = [c_1, c_2, ..., c_n]^T$.

Then, (1) gives $\sum A_i c_i = B$

Thus, B is a linear combination of column of A.

Hence, (*i*) implies (*ii*).

Now, $W' = \langle A_1, A_2, ..., A_n, B \rangle$

If B is a linear combination of columns $A_1, A_2, ..., A_n$, then,

$$W' = \langle A_1, A_2, ..., A_n \rangle = W$$

This proves (*ii*) implies (*iii*). Now (*iii*) implies (*iv*) is obvious. Let (*iv*) hold.

Thus, $W' = W = \langle A_1, A_2, ..., A_n \rangle$

As $B \in W$, B is a linear combination of $A_1, A_2, ..., A_n$ i.e., $B = \sum_{i=1}^{n} A_i c_i'$, $c_i' \in F$

Consequently, $X'' = [c_1', c_2', ..., c_n']^T$
is a solution of $AX = B$.

So, the system is consistent. Hence, (*iv*) implies (*i*).

Remark 10. Any polynomial $p(X) = \sum_{i=0}^{n} a_i X^i \in P_n(F)$ determines the n + 1-tuple $(a_0, a_1, ..., a_n) \in F^{n+1}$

It can be easily seen that polynomials $p_1(X), p_2(X), ..., p_k(X)$ in $P_n(F)$ are linearly dependent iff their corresponding $(n + 1)$-tuples in F^{n+1} are linearly dependent. This information can be used to find bases and dimensions of subspaces of $P_n(F)$. Similar situation can also arise while dealing with subspaces of $F^{m \times n}$. We illustrate this by the following example :

Example 10.a. Let W be the subspace of $P_3(\mathbf{R})$ spanned by the following four polynomials

$$p_1(X) = X^3 + X^2 - 2X + 1$$

$$p_2(X) = X^2 + 1 = 0X^3 + X^2 + 0X + 1$$

$$p_3(X) = X^3 - 2X = X^3 + 0X^2 - 2X + 0$$

$$p_4(X) = 2X^3 + 3X^2 - 4X + 3$$

We find a basis of W among these four polynomials. These polynomials give the following four vectors in $\mathbf{R}^4$

$$\alpha_1 = (1, -2, 1, 1)$$

$$\alpha_2 = (1, 0, 1, 0)$$

$$\alpha_3 = (0, -2, 0, 1)$$

$$\alpha_4 = (3, -4, 3, 2)$$

Form the matrix $A = \begin{bmatrix} 1 & 1 & 0 & 3 \\ -2 & 0 & -2 & -4 \\ 1 & 1 & 0 & 3 \\ 1 & 0 & 1 & 2 \end{bmatrix} \begin{matrix} - \text{co eff}. X^0 \\ - \text{co eff}. X \\ - \text{co eff}. X^2 \\ - \text{co eff}. X^3 \end{matrix}$

Its column are linearly dependent iff the given polynomials are linearly dependent, so that column-rank $(A) = \dim W$.

Now, A is row equivalent to $B = \begin{bmatrix} 1 & 0 & 1 & 2 \\ 0 & 1 & -1 & 1 \\ 0 & 0 & 0 & 0 \\ 0 & 0 & 0 & 0 \end{bmatrix}$

In B, leading entries are in columns 1 and 2. So, that columns 1 and 2 of A form a basis of the column-space of A. This, in turn, gives that $\{p_1(X), p_2(X)\}$ is a basis of W. By looking at columns 3 and 4, we get

$$p_3(X) = p_1(X) - p_2(X)$$

$$p_4(X) = 2p_1(X) + p_2(X)$$

Let us find another basis of W.

Consider $A^T = \begin{bmatrix} 1 & -2 & 1 & 1 \\ 1 & 0 & 1 & 0 \\ 0 & -2 & 0 & 1 \\ 3 & -4 & 3 & 2 \end{bmatrix} \begin{matrix} - \text{co eff. of } p_1(X) \\ - \text{co eff. of } p_2(X) \\ - \text{co eff. of } p_3(X) \\ - \text{co eff. of } p_4(X) \end{matrix}$

This matrix is row equivalent to

$$B' = \begin{bmatrix} 1 & 0 & 1 & 0 \\ 0 & 1 & 0 & -\frac{1}{2} \\ 0 & 0 & 0 & 0 \\ 0 & 0 & 0 & 0 \end{bmatrix}$$

Thus, row-space (A^T) has a basis consisting of

$$\beta_1 = (1, 0, 1, 0)$$

$$\beta_2 = \left(0, 1, 0, -\frac{1}{2}\right)$$

These correspond to the polynomials

$$q_1(X) = 1 + X^2$$

$$q_2(X) = X - \frac{1}{2}X^3$$

Then, $$\left\{1+X^2, X - \frac{1}{2}X^3\right\}$$

is a basis of W.

Theorem 11. Let $A = [a_{ij}]$ and $B = [b_{ij}]$ be two $m \times n$-matrices over F. Then,

(*i*) row-space $(A + B) \subseteq$ row-space (A) + row-space (B)

(*ii*) column-space $(A + B) \subseteq$ column-space (A) + column-space (B)

(*iii*) rank $(A + B) \leq$ rank (A) + rank (B)

Proof. Let A_i and B_i stand for the ith rows of A and B, respectively.

Then, $$A_i = [a_{i1}\ a_{i2} \ldots\ a_{in}]$$

$$B_i = [b_{i1}\ b_{i2} \ldots\ b_{in}]$$

$$A_i + B_i = [a_{i1} + b_{i1}\ a_{i2} + b_{i2} \ldots\ a_{in} + b_{in}]$$

is the ith row of $A + B$, so that any row of $A + B$ is in row-space (A) + row-space (B)

Hence, row-space $(A + B) \subseteq$ row-space (A) + row-space (B).

This, in turn, gives $\quad \text{rank}\ (A + B) \leq \dim(W_1 + W_2)$,

where $\quad W_1 =$ row-space (A),

$W_2 =$ row-space (B)

But $$\dim(W_1 + W_2) \leq \dim W_1 + \dim W_2 = \text{rank}(A) + \text{rank}(B)$$

Hence, rank $(A + B) \leq$ rank (A) + rank (B).

This proves (*i*) and (*iii*). Similar to that of (*i*), is the proof of (*ii*).

Theorem 12. Let $A = [a_{ij}]$ be an $m \times p$-matrix and $B = [b_{ij}]$ be a $p \times n$-matrix over F. Then,

(*i*) row-space $(AB) \subseteq$ row -space (B)

(*ii*) column-space $(AB) \subseteq$ column-space (A)

(*iii*) rank $(AB) \leq$ max {rank (A), rank (B)}

Proof. For any $$[x_1\ x_2 \ldots x_p] \in F^{1 \times p}$$

$$[x_1\ x_2 \ldots x_m]B = x_1R_1 + x_2R_2 + \ldots + x_pR_p$$

where R_i is the ith row of B, so that in AB, the ith row

$$[a_{i1}\ a_{i2} \ldots a_{ip}]B = \sum_{j=1}^{p} a_{ij}R_j$$

is in row-space (B).

Hence, row-space $(AB) \subseteq$ row-space (B)

This proves (*i*) and gives $\text{rank}(AB) \leq \text{rank}(B)$ ... (1)

Again, $$A\begin{bmatrix} x_1 \\ x_2 \\ \vdots \\ x_n \end{bmatrix} = C_1x_1 + C_2x_2 + \ldots + C_px_p,$$

where C_j is the jth column of A, gives

$$\text{column-space}(AB) \subseteq \text{column-space}(A)$$

This proves (*ii*) and also gives $\text{rank}(AB) \leq \text{rank}(A)$ ... (2)

(1) and (2) give $\text{rank}(AB) \leq \min\{\text{rank}(A), \text{rank}(B)\}$.

Theorem 13. Let A be any $m \times n$-matrix over F. Let P and Q be non-singular $n \times n$ and $m \times m$-matrices, respectively, over F.

Then, $$\text{rank}(A) = \text{rank}(QA) = \text{rank}(AP)$$

Proof.

By the above given theorem, $\text{rank}(AP) \leq \text{rank}(A)$

again $$\text{rank}(A) = \text{rank}(\text{APP}^{-1}) \leq \text{rank}(AP),$$

so that $$\text{rank}(A) = \text{rank}(AP).$$

The same way $$\text{rank}(QA) = \text{rank}(A).$$

Exercise 2.3

1. Find a basis of the row-space and a basis of the column-space of each of the following matrices. Find their ranks

(*i*) $$\begin{bmatrix} 1 & 0 & 2 & 0 & 0 \\ 0 & 1 & -3 & 0 & 1 \\ 0 & 0 & 0 & 1 & 0 \\ 0 & 0 & 0 & 0 & 0 \end{bmatrix}$$

(*ii*) $$\begin{bmatrix} 1 & 2 & 3 & 1 \\ 3 & 1 & -5 & 1 \\ -1 & 1 & 0 & 2 \end{bmatrix}$$

(*iii*) $$\begin{bmatrix} 1 & -2 & -1 \\ 2 & -1 & 3 \\ 7 & -8 & 3 \\ 5 & -7 & 0 \end{bmatrix}$$

2. Use the method of finding a basis of a row-space, described in Remark 5, to find a basis of each of the subspaces spanned by the following sets of vectors

   (*i*) (2, 1, 4), (–1, –1, 2), (0, 1, 2), (1, 1, 1) in $\mathbf{R}^3$

   (*ii*) (2, 2, 3), (–1, –2, 1), (4, 6, 1) in $\mathbf{R}^3$

   (*iii*) (1, 2, 2, 1), (4, 6, 8, 6), (0, 3, 2, 1), (1, 0, 0, 2) in $\mathbf{R}^4$

   (*iv*) $X^2+1, X-2, X+3$ in $P_2(\mathbf{R})$.

   (*v*) $X^3+2X^2+X+3,\ 2X^3+X^2+3X+2,\ X^3+X^2+2X+2$ in $P_3(\mathbf{R})$

3. Use the method of finding a basis of row-space to find a basis of the column-space of

$$\begin{bmatrix} 1 & 0 & 1 & 2 & 0 \\ 1 & 1 & -2 & 3 & 3 \\ -2 & 0 & 0 & -4 & -2 \\ 1 & 1 & 0 & 3 & 1 \end{bmatrix}$$

4. Find dimension and a basis of the solution space W of

$$\begin{bmatrix} 1 & 3 & 1 & 3 & 2 \\ 1 & 3 & 2 & 4 & 3 \\ 1 & 4 & 3 & 5 & 4 \\ 0 & 0 & 1 & 1 & 1 \end{bmatrix}\begin{bmatrix} x_1 \\ x_2 \\ x_3 \\ x_4 \\ x_5 \end{bmatrix} = \begin{bmatrix} 0 \\ 0 \\ 0 \\ 0 \end{bmatrix}$$

5. Find the dimension of the solution space of the system $AX = 0$,

   where $$A = \begin{bmatrix} 1 & 3 & 0 & 5 & -1 & 0 \\ 0 & 0 & 1 & 6 & 1 & 0 \\ 0 & 0 & 0 & 0 & 0 & 0 \\ 0 & 0 & 0 & 0 & 0 & 0 \end{bmatrix}$$

6. Find the ranks of the coefficient matrix and the augmented matrix for

$$\begin{bmatrix} 1 & 2 & 5 & -2 \\ 2 & 3 & -2 & 4 \\ 5 & 1 & 0 & 2 \end{bmatrix}\begin{bmatrix} x_1 \\ x_2 \\ x_3 \\ x_4 \end{bmatrix} = \begin{bmatrix} 7 \\ 1 \\ 1 \end{bmatrix}$$

   Is this system consistent?

7. Answer the following :

   (*i*) If A is a 3×4 real matrix, what is the largest possible value of rank (A)

   (*ii*) If A is a 5×3-matrix, what is the largest possible value of rank (A)

   (*iii*) For a 5×5-matrix A over F, what must be rank (A), for A to be non-singular

8. Give examples of two non-zero 3×3-matrices A, B such that

$$\text{rank}(A+B) = \text{rank}(A) + \text{rank}(B)$$

9. Give examples of two non-zero 2×3-matrices A, B such that $A+B \neq 0$ and rank $(A+B) < \text{rank}(A) + \text{rank}(B)$

10. Give examples of two non-zero 3×3-matrices A, B such that $AB \neq 0$ and rank $(AB) < \min\{\text{rank}(A), \text{rank}(B)\}$

11. Let A be any non-singular $m \times m$-matrix over F and B be any $m \times n$-matrix over F, prove that

$$\text{rank}(AB) = \text{rank}(B)$$

[Hint : Use Theorem 12]

2.4 Change of Basis

Let V_F be a finitely generated vector space. We take $V_F \neq \{0\}$.

Let $B = \{\alpha_1, \alpha_2, \ldots, \alpha_k\}$ be a basis of V_F. It was seen in Example (2.2.21.a) that for each $\alpha \in V$, there exist unique $a_1, a_2, \ldots, a_k \in F$ such that $\alpha = a_1\alpha_1 + a_2\alpha_2 + \ldots + a_k\alpha_k$

If we fix an order of arrangement of members of B, $\alpha \in V$ determines a unique $k \times 1$-matrix $\begin{bmatrix} a_1 \\ a_2 \\ \vdots \\ a_k \end{bmatrix} \in F^{k \times 1}$. This $k \times 1$–matrix is going to play a significant role in the following chapters. To make the concept more precise, we introduce the concept of an ordered basis, and co-ordinate vectors.

Definition 1. Let V_F be a vector space of dimension $k \geq 1$. Then, any k-tuple $B = (\alpha_1, \alpha_2, \ldots, \alpha_k)$ of members of V is called an *ordered basis* of V if the set $\{\alpha_1, \alpha_2, \ldots, \alpha_k\}$ is a basis of V. For any $\alpha \in V$, if $\alpha = a_1\alpha_1 + a_2\alpha_2 + \ldots + a_k\alpha_k$, $a_i \in F$ then, the $k \times 1$-matrix $\begin{bmatrix} a_1 \\ a_2 \\ \vdots \\ a_k \end{bmatrix}$ is called the *co-ordinate vector* of α relative to the ordered basis B.

We write
$$[\alpha]_B = \begin{bmatrix} a_1 \\ a_2 \\ \vdots \\ a_k \end{bmatrix}$$

As an ordered basis $B = (\alpha_1, \alpha_2, \ldots, \alpha_k)$ is essentially a set of k distinct members of V_F, as an abuse of notation, we simply write B as a set $B = \{\alpha_1, \alpha_2, \ldots, \alpha_k\}$. In F^k, the *standard ordered basis* is $(\epsilon_1, \epsilon_2, \ldots, \epsilon_k)$ where ϵ_i is the k-tuple with i^{th} component 1, and all the other components zeros.

Thus, we write this ordered basis as $\{\epsilon_1, \epsilon_2, ..., \epsilon_k\}$

Thus, for $\mathbf{R}^3$, ordered basis

$$B_1 = \{\epsilon_1, \epsilon_2, \epsilon_3\} = \{(1, 0, 0), (0, 1, 0), (0, 0, 1)\}$$

is the standard ordered basis.

But the ordered basis $$B_2 = \{\epsilon_2, \epsilon_1, \epsilon_3\}$$

is not the standard ordered basis of $\mathbf{R}^3$.

For any $$\alpha = (a, b, c)$$

$$\alpha = a\epsilon_1 + b\epsilon_2 + c\epsilon_3$$

gives $$[\alpha]_{B_1} = \begin{bmatrix} a \\ b \\ c \end{bmatrix}.$$

As $$\alpha = b\epsilon_2 + a\epsilon_1 + c\epsilon_3,\ [\alpha]_{B_2} = \begin{bmatrix} b \\ a \\ c \end{bmatrix}.$$

Thus, for the same $\alpha \in V$, its co-ordinate vector changes with different ordered bases. To find relationship between different co-ordinate vectors, we, first of all, discuss relation between different ordered bases.

Definition 2. Let $B = \{\alpha_1, \alpha_2, ..., \alpha_k\}$ and $B' = \{\beta_1, \beta_2, ..., \beta_k\}$ be two ordered bases of a vector space V_F of dimension $k \geq 1$. The $k \times k$-matrix $P = [p_{ij}]$ over F such that

$$\beta_j = \sum_{i=1}^{k} p_{ij}\alpha_i, \qquad 1 \leq j \leq k$$

is called the *matrix of* B' relative to B.

Observe that in the above definition $[\beta_j]_B = \begin{bmatrix} p_{1j} \\ p_{2j} \\ \vdots \\ p_{kj} \end{bmatrix}$, so that in P, the jth column is the co-ordinate vector of β_j relative to B.

Example 2.a. Let $$V = \mathbf{R}^3.$$

Let $$B_1 = \{\epsilon_1, \epsilon_2, \epsilon_3\},\ B_2 = \{\alpha_1, \alpha_2, \alpha_3\}$$

with $$\alpha_1 = (1, 1, 0),\ \alpha_2 = (1, -1, 0),\ \alpha_3 = (0, 1, 1).$$

Consider $$\alpha = (1, 2, 3) \in \mathbf{R}^3.$$

As $\alpha = 1 \in_1 + 2 \in_2 + 3 \in_3$, $[\alpha]_{B_1} = \begin{bmatrix} 1 \\ 2 \\ 3 \end{bmatrix}$.

We, now, find $[\alpha]_{B_2} = \begin{bmatrix} a_1 \\ a_2 \\ a_3 \end{bmatrix}$.

By definition, $a_1\alpha_1 + a_2\alpha_2 + a_3\alpha_3 = \alpha$

i.e., $a_1(1, 1, 0) + a_2(1, -1, 0) + a_3(0, 1, 1) = (1, 2, 3)$.

By comparing the three components, we get linear equations

$$\begin{aligned} a_1 + a_2 + 0a_3 &= 1 \\ a_1 - a_2 + a_3 &= 2 \\ 0a_1 + 0a_2 + a_3 &= 3 \end{aligned}$$

For this system, the augmented matrix is

$$\begin{bmatrix} 1 & 1 & 0 & : & 1 \\ 1 & -1 & 1 & : & 2 \\ 0 & 0 & 1 & : & 3 \end{bmatrix}$$

Solving the system, we get $a_1 = 0, a_2 = 1, a_3 = 3$,

so that $[\alpha]_{B_2} = \begin{bmatrix} 0 \\ 1 \\ 3 \end{bmatrix}$.

Once again observe that $[\alpha_1]_{B_1} = \begin{bmatrix} 1 \\ 1 \\ 0 \end{bmatrix}$,

$$[\alpha_2]_{B_1} = \begin{bmatrix} 1 \\ -1 \\ 0 \end{bmatrix},$$

$$[\alpha_3]_{B_1} = \begin{bmatrix} 0 \\ 1 \\ 1 \end{bmatrix}$$

So, the matrix of B_2 relative to B_1 is

$$P = \begin{bmatrix} 1 & 1 & 0 \\ 1 & -1 & 1 \\ 0 & 0 & 1 \end{bmatrix}$$

P is non-singular with $$P^{-1} = \begin{bmatrix} \frac{1}{2} & \frac{1}{2} & -\frac{1}{2} \\ \frac{1}{2} & -\frac{1}{2} & \frac{1}{2} \\ 0 & 0 & 1 \end{bmatrix}.$$

One can easily verify that P^{-1} is the matrix of B_1 relative to B_2.

Further, $$[\alpha]_{B_2} = P^{-1}[\alpha]_{B_1}$$

Theorem 3. Let B and B' be two ordered bases of a vector space V_F with $\dim V_F = k \geq 1$. Let P be the matrix of B' relative to B. Then, P is non-singular and P^{-1} is the matrix of B relative to B'.

Proof. Let $P = [p_{ij}]$. Let $P' = [q_{ij}]$ be the matrix of B relative to B'.

If $$B = \{\alpha_1, \alpha_2, \ldots, \alpha_k\}$$
$$B' = \{\beta_1, \beta_2, \ldots, \beta_k\}$$

by Definition 2 $$\beta_j = \sum_{i=1}^{k} p_{ij}\alpha_i, \qquad 1 \leq j \leq k$$

$$\alpha_s = \sum_{i=1}^{k} q_{is}\beta_i, \qquad 1 \leq s \leq k$$

This gives $$\alpha_s = \sum_{j=1}^{k} q_{js}\beta_j = \sum_{j=1}^{k} q_{js}\left[\sum_{i=1}^{k} p_{ij}\alpha_i\right],$$

i.e., $$\alpha_s = \sum_{i=1}^{k}\left[\sum_{j=1}^{k} q_{js}p_{ij}\right]\alpha_i = \sum_{i=1}^{k}\left[\sum_{j=1}^{k} p_{ij}q_{js}\right]\alpha_i, \qquad \ldots (1)$$

as in F, $$q_{js}p_{ij} = p_{ij}q_{js}$$

Recall that Kronecker delta function is :

$$\delta_{ij} = 1 \qquad \text{for } i = j$$
$$= 0 \qquad \text{for } i \neq j$$

So, $$\alpha_s = \sum_{i=1}^{k} \delta_{is}\alpha_i \qquad \ldots (2)$$

(1) and (2) give $$\sum_{i=1}^{k} \delta_{is}\alpha_i = \sum_{i=1}^{k}\left[\sum_{j=1}^{k} p_{ij}q_{js}\right]\alpha_i$$

By comparing on both sides the coefficients of α_i, we get

$$\delta_{is} = \sum_{j=1}^{k} p_{ij}q_{js}$$

so that $$I_k = [\delta_{is}] = PP'$$

Similarly, $$P'P = I_k.$$

Hence, P is non-singular and $P' = P^{-1}$.

Theorem 4. Let B and B' be two ordered bases of a vector space V_F of dimension $k \geq 1$.

Then for any $\alpha \in V$ $$[\alpha]_B = P[\alpha]_{B'}$$

where P is the matrix of B' relative to B.

Proof. Let $$B = \{\alpha_1, \alpha_2, \ldots, \alpha_k\}$$

$$B' = \{\beta_1, \beta_2, \ldots, \beta_k\}$$

and $$P = [p_{ij}].$$

Then, $$\beta_j = \sum_{i=1}^{k} p_{ij}\alpha_i, \qquad 1 \leq j \leq k$$

Let $[\alpha]_B = \begin{bmatrix} a_1 \\ a_2 \\ \vdots \\ a_k \end{bmatrix}$, $[\alpha]_{B'} = \begin{bmatrix} b_1 \\ b_2 \\ \vdots \\ b_k \end{bmatrix}$

$$\alpha = \sum_{j=1}^{k} b_j\beta_j = \sum_{j=1}^{k} b_j \left[\sum_{i=1}^{k} p_{ij}\alpha_i\right],$$

i.e., $$\alpha = \sum_{i=1}^{k} \left[\sum_{j=1}^{k} p_{ij}b_j\right]\alpha_i \qquad \ldots (1)$$

Also, $$\alpha = \sum_{i=1}^{k} a_i\alpha_i \qquad \ldots (2)$$

By comparing the coefficient of α_i in (1) and (2) we get

$$a_i = \sum_{j=1}^{k} p_{ij}b_j, \qquad 1 \leq i \leq k$$

This gives $$[\alpha]_B = \begin{bmatrix} a_1 \\ a_2 \\ \vdots \\ a_k \end{bmatrix} = \begin{bmatrix} \sum_{j=1}^{k} p_{1j}b_j \\ \sum_{j=1}^{k} p_{2j}b_j \\ \vdots \\ \vdots \end{bmatrix} = P[\alpha]_{B'}$$

The following result will provide an important example of the concept of vector space isomorphism, that will be discussed in the next chapter.

Theorem 5. Let B be an ordered basis of a vector space V_F of dimension $k \geq 1$.

Then, the mapping $$\sigma : V \to F^{k\times 1}$$

given by $\sigma(\alpha) = [\alpha]_B$, $\alpha \in V$ satisfies the following conditions :

For any $a, \beta \in V, a \in F$

(*i*) $$\sigma(\alpha + \beta) = \sigma(\alpha) + \sigma(\beta)$$

(*ii*) $$\sigma(a\alpha) = a\sigma(\alpha)$$

(*iii*) σ is one-to-one and onto

Proof. Let $B = \{\alpha_1, \alpha_2, ..., \alpha_k\}$. Consider $\alpha, \beta \in V$ and $a \in F$.

Let $[\alpha]_B = \begin{bmatrix} a_1 \\ a_2 \\ \vdots \\ a_k \end{bmatrix}, [\beta]_B = \begin{bmatrix} b_1 \\ b_2 \\ \vdots \\ b_k \end{bmatrix}$.

Then, $$\alpha = \sum_{i=1}^{k} a_i \alpha_i \quad \text{... (1)}$$

$$\beta = \sum_{i=1}^{k} b_i \alpha_i \quad \text{... (2)}$$

Then, $$\alpha + \beta = \sum_{i=1}^{k} (a_i + b_i)\alpha_i$$

so that $$[\alpha + \beta]_B = \begin{bmatrix} a_1 + b_1 \\ a_2 + b_2 \\ \vdots \\ a_k + b_k \end{bmatrix} = \begin{bmatrix} a_1 \\ a_2 \\ \vdots \\ a_k \end{bmatrix} + \begin{bmatrix} b_1 \\ b_2 \\ \vdots \\ b_k \end{bmatrix} = [\alpha]_B + [\beta]_B$$

i.e., $$\sigma(\alpha + \beta) = \sigma(\alpha) + \sigma(\beta)$$

Again, (1) gives $$a\alpha = \sum_{i=1}^{k} a a_i \alpha_i$$

so that $$[a\alpha]_B = \begin{bmatrix} aa_1 \\ aa_2 \\ \vdots \\ aa_k \end{bmatrix} = a \begin{bmatrix} a_1 \\ a_2 \\ \vdots \\ a_k \end{bmatrix} = a[\alpha]_B$$

i.e., $$\sigma(a\alpha) = a\sigma(\alpha)$$

Now, $\sigma(\alpha) = \sigma(\beta)$

implies $[\alpha]_B = [\beta]_B$

i.e.,
$$\begin{bmatrix} a_1 \\ a_2 \\ \vdots \\ a_k \end{bmatrix} = \begin{bmatrix} b_1 \\ b_2 \\ \vdots \\ b_k \end{bmatrix}$$

This gives $a_i = b_i$ for $1 \le i \le k$.

Then (1) and (2) give $\alpha = \beta$

Hence, σ is one-to-one.

For any $\alpha' = \begin{bmatrix} c_1 \\ c_2 \\ \vdots \\ c_k \end{bmatrix} \in F^{k\times 1}$, $\lambda = \sum_{i=1}^{k} c_i\alpha_i \in V$

such that $\sigma(\lambda) = \alpha'$.

Hence, σ is also onto.

Remark 6. Consider two ordered bases $B = \{\alpha_1, \alpha_2, \ldots, \alpha_k\}$ and $B' = \{\beta_1, \beta_2, \ldots, \beta_k\}$ of F^k, and let $\alpha \in F^k$. A practical way for finding $[\alpha]_B$, and the matrix of B' relative to B is the following. Let A and B be two $k \times k$-matrices, such that the jth column of A and B are α_j and β_j, respectively, put in column forms.

Let $\alpha = (a_1, a_2, \ldots, a_k)$.

Take
$$C = \begin{bmatrix} a_1 \\ a_2 \\ \vdots \\ a_k \end{bmatrix}$$

(*i*) Form $[A : C]$. This is a $k \times (k+1)$-matrix. Apply elementary row operations to this matrix to get a matrix in reduced row echelon form. The result is $[I_k : C']$. Then $[\alpha]_B = C'$

(*ii*) Form $[A : B]$. It is row equivalent to some $[I_k : P]$, in reduced row echelon form. Then, P is the matrix of B' relative to B.

The method described in the above remark can be adapted to other situations, in particular, to $V = P_k(F)$.

Example 6.a. Let $V = \mathbf{R}^3$

Let
$$B = \{(1, 1, 1), (1, 2, 3), (0, 1, 0)\}$$
$$B' = \{(1, 2, 2), (2, 1, 3), (1, 1, 0)\}$$

We get
$$A=\begin{bmatrix}1&1&0\\1&2&1\\1&3&0\end{bmatrix},$$
$$C=\begin{bmatrix}1&2&1\\2&1&1\\2&3&0\end{bmatrix}$$

for B and B'.

Consider any $\alpha=(a,\ b,\ c)\in\mathbf{R}^3$. We find $[\alpha]_B$.

Consider the matrix
$$\begin{bmatrix}1&1&0&:&a\\1&2&1&:&b\\1&3&0&:&c\end{bmatrix}$$

This is row equivalent to
$$\begin{bmatrix}1&0&0&:&\frac{(3a-c)}{2}\\0&1&0&:&\frac{c-a}{2}\\0&0&1&:&\frac{2b-a-c}{2}\end{bmatrix}$$

so that
$$[\alpha]_B=\begin{bmatrix}\frac{(3a-c)}{2}\\\frac{c-a}{2}\\\frac{2b-a-c}{2}\end{bmatrix}$$

To find the matrix P of B' relative to B, we consider
$$[A:C]=\begin{bmatrix}1&1&0&:&1&2&1\\1&2&1&:&2&1&1\\1&3&0&:&2&3&0\end{bmatrix}$$

This is row equivalent to
$$\begin{bmatrix}1&0&0&:&\frac{1}{2}&\frac{3}{2}&\frac{3}{2}\\0&1&0&:&\frac{1}{2}&\frac{1}{2}&-\frac{1}{2}\\0&0&1&:&\frac{1}{2}&-\frac{3}{2}&\frac{1}{2}\end{bmatrix}$$

Thus,
$$P=\begin{bmatrix}\frac{1}{2}&\frac{3}{2}&\frac{3}{2}\\\frac{1}{2}&\frac{1}{2}&-\frac{1}{2}\\\frac{1}{2}&-\frac{3}{2}&\frac{1}{2}\end{bmatrix}$$

Theorem 7. Let $B=\{\alpha_1,\ \alpha_2,\ldots,\ \alpha_k\}$ be an ordered basis of a non-zero finite dimensional vector space V_F and $P=[p_{ij}]$ be any $k\times k$ non-singular matrix over F. For $1\le j\le k$, if $\beta_j=\sum_{i=1}^{k}p_{ij}\alpha_i$, then, $B'=\{\beta_1,\ \beta_2,\ldots,\ \beta_k\}$ is an ordered basis of V_F.

Proof. As $\dim V = k$ and B' has at most k members, it is enough to prove that B' spans V. Let $W = \langle B' \rangle$ and $P^{-1} = [q_{ij}]$.

Then, for $1 \le t \le k$,

$$\sum_{j=1}^{k} q_{jt}\beta_j = \sum_{j=1}^{k} q_{jt}\left[\sum_{i=1}^{k} p_{ij}\alpha_i\right]$$

$$= \sum_{i=1}^{k}\left[\sum_{j=1}^{k} p_{ij}q_{jt}\right]\alpha_i$$

$$= \sum_{i=1}^{k} \delta_{it}\alpha_i$$

$$= \alpha_t.$$

This gives $\alpha_t \in W$.

Hence, $B \subseteq W$.

Consequently, $V \subseteq W \subseteq V$,

gives, V is spanned by B'

Exercise 2.4

1. Show that $B' = \{(1, 1), (1, -2)\}$ is an ordered basis of $\mathbf{R}^2$.

   (*i*) For $\alpha = (-1, 1)$, find the co-ordinate vector $[\alpha]_{B'}$.

   (*ii*) Find the co-ordinate vector of any $\alpha = (a, b) \in \mathbf{R}^2$ relative to B'.

   (*iii*) Let B be the standard ordered basis of $\mathbf{R}^2$. Find the matrix P of B' relative to B, and the matrix P' of B relative to B'.

   (*iv*) For $\alpha = (1, 0)$, verify that $[\alpha]_B = P[\alpha]_{B'}$

2. Let $V = P_2(\mathbf{R})$. Consider the following three ordered bases of V

$$B_1 = \{1, 1+X, 1+X+X^2\}$$

$$B_2 = \{X, 1, 1+X^2\}$$

$$B_3 = \{X, 1+X^2, 1\}$$

   (*i*) For $p(X) = 1 - X + 2X^2$, find

$$[p(X)]_{B_1}, [p(X)]_{B_2}, [p(X)]_{B_3}$$

   (*ii*) Find matrix P_1 of B_3 relative to B_2

   (*iii*) Find matrix P_2 of B_2 relative to B_1

   (*iv*) Find matrix P of B_3 relative to B_1 and verify that $P = P_2P_1$

3. Find the co-ordinate vector of (a, b, c) relative to the ordered basis $\{(2, 4, 6), (2, 5, 1), (2, 4, 1)\}$ in $\mathbf{R}^3$.

4. Let B, B', B'' be any three ordered bases of a vector space V_F with dimension $k \geq 1$. If P_1 is the matrix of B' relative to B, and P_2 is the matrix of B'' relative to B', prove that P_2P_1 is the matrix of B'' relative to B.

5. Find the co-ordinate vector of $(1, 0, 0)$ in $\mathbf{C}^3$, as a vector space over $\mathbf{C}$, relative to the ordered basis $\{(2i, 1, 0), (2, -1, 1), (0, 1+i, 1-i)\}$.

6. Let W be the subspace of $\mathbf{C}^3$ over $\mathbf{C}$ spanned by

$$\alpha_1 = (1, 0, i), \alpha_2 = (i, 0, 1).$$

Prove the following :

(*i*) α_1, α_2 form a basis of W

(*ii*) $\beta_1 = (1+i, 0, 1+i), \beta_2 = (1-i, 0, i-1)$ form a basis of W

Find the matrix of the ordered basis $B' = \{\beta_1, \beta_2\}$ relative to the ordered basis $B = \{\alpha_2, \alpha_1\}$.

7. Let
$$A = \begin{bmatrix} 0 & 1 & 1 & -1 \\ 1 & 0 & 1 & -5 \\ 2 & 1 & -6 & 1 \\ 1 & -1 & -3 & 0 \end{bmatrix}.$$

Find an ordered basis B of the column-space (A). Further, find the co-ordinate vector of each column of A relative to B.

8. For any two vectors
$$\alpha = (a_1, a_2, \ldots, a_k),$$
$$\beta = (b_1, b_2, \ldots, b_k),$$
define
$$\alpha.\beta = a_1b_1 + a_2b_2 + \ldots + a_kb_k.$$

In $\mathbf{R}^3$, consider any three non-zero vectors $\alpha_1 = (a_1, a_2, a_3)$, $\alpha_2 = (b_1, b_2, b_3)$, $\alpha_3 = (c_1, c_2, c_3)$ such that $\alpha_i.\alpha_j = 0$ for $i \neq j$. Prove that $B = \{\alpha_1, \alpha_2, \alpha_3\}$ is an ordered basis of $\mathbf{R}^3$. Further, find co-ordinate vector of each of the vector in the standard basis, relative to B.

Chapter 3

Linear Transformations

3.1 Linear Transformations

To study relationships between different vector spaces over the same field, we introduce the concept of a linear transformation (or a vector space homomorphism, in analogy with the concept of a homomorphism of a group into another group).

Definition 1. Let V and W be any two vector spaces over the same field F. Then, a mapping $T : V \to W$ is called a *linear transformation* (or a vector space homomorphism) if it satisfies the following two conditions :

(*i*) $$T(\alpha+\beta) = T(\alpha)+T(\beta)$$

(*ii*) $$T(a\alpha) = aT(\alpha)$$

for any $\alpha,\ \beta \in V$ and $a \in F$

Given a linear transformation $T : V_F \to W_F$, in order to emphasise the field F of scalars, T is also called an *F-linear transformation*. If T is one-to-one, it is called a *monomorphism*. If T is onto, it is called an *epimorphism*. A linear transformation $T : V_F \to W_F$, which is one-to-one and onto, is called an *F-vector space isomorphism* (or simply an *isomorphism*). One can observe that a linear transformation $T : V_F \to W_F$ is a mapping that commutes with addition and scalar multiplication operations. Observe that $T(\alpha+c\beta) = T(\alpha)+T(c\beta) = T(\alpha)+cT(\beta)$

Example 1.a. Let $V = F^{m\times 1}$, $W = F^{n\times 1}$. Let $A = \left[a_{ij}\right]$ be any fixed $m\times n$-matrix over F. Define $T : F^{n\times 1} \to F^{m\times 1}$ such that $T(X) = AX$ for $X \in F^{n\times 1}$.

Observe that as A is $m\times n$ and X is an $n\times 1$-matrix, the product AX is an $m\times 1$-matrix. So, T is a well-defined mapping. Consider any $X_1, X_2 \in F^{n\times 1}$. By the property of distributivity of matrix multiplication over matrix addition

$$A(X_1+X_2) = AX_1+AX_2 \qquad \text{... (1)}$$

Also, by the property of multiplication of a matrix by a scalar

$$A(aX_1) = a(AX_1) \qquad \text{... (2)}$$

(1) and (2) give
$$T(X_1+X_2)=T(X_1)+T(X_2)$$
$$T(aX_1)=aT(X_1)$$

Hence, T is a linear transformation. To be specific, let $m=2$, $n=3$. Choose any 2×3 -matrix, say $A=\begin{bmatrix}1 & 2 & 0\\ 0 & 3 & 4\end{bmatrix}$ over **R**. This gives the linear transformation

$$T:\mathbf{R}^{3\times 1}\to\mathbf{R}^{2\times 1}$$

such that
$$T\left[\begin{bmatrix}a\\ b\\ c\end{bmatrix}\right]=\begin{bmatrix}1 & 2 & 0\\ 0 & 3 & 4\end{bmatrix}\begin{bmatrix}a\\ b\\ c\end{bmatrix}=\begin{bmatrix}a+2b\\ 3b+4c\end{bmatrix}$$

Example 1.b. Consider $T:\mathbf{R}^2\to\mathbf{R}^2$ such that

$$T(a,\,b)=(a^2,\,b)\qquad\text{for any } (a,\,b)\in\mathbf{R}^2$$

Let
$$\alpha=(a_1,\,b_1),$$
$$\beta=(a_2,\,b_2)\in\mathbf{R}^2.$$

Then,
$$T(\alpha+\beta)=T(a_1+a_2,\,b_1+b_2)=\left\{(a_1+a_2)^2,\,b_1+b_2\right\}$$
$$T(\alpha)+T(\beta)=\left(a_1^2,\,b_1\right)+\left(a_2^2,\,b_2\right)=\left(a_1^2+a_2^2,\,b_1+b_2\right).$$

It is clear that $T(\alpha+\beta)\neq T(\alpha)+T(\beta)$ unless $a_1a_2=0$. Hence, T is not a linear transformation.

We start with some basic properties of linear transformations.

Proposition 2. Let $T:V_F\to W_F$ be a linear transformation. Then,

(*i*) $T(0_v)=0_w$, where 0_v and 0_w are the zero vectors in V and W, respectively.

(*ii*) For any $\alpha\in V$ $\quad T(-\alpha)=-T(\alpha)$.

(*iii*) $\left\{\alpha\in V:\,T(\alpha)=0_w\right\}$ is a subspace of V.

(*iv*) For any finitely many $\alpha_1,\,\alpha_2,\ldots,\,\alpha_k$, $k\geq 1$ vectors in V,

$$T\left[\sum_{i=1}^{k}\alpha_i\right]=\sum_{i=1}^{k}T(\alpha_i)$$

Proof.

(*i*) As
$$0_v=0_v+0_v$$
$$T(0_v)+0_w=T(0_v)=T(0_v+0_v)$$

i.e.,
$$T(0_v)+0_w=T(0_v)+T(0_v)$$

By the cancellation property for addition {Lemma (2.1.2)}

$$0_w=T(0_v)$$

(*ii*) Now $\alpha + (-\alpha) = 0_v$

so that $T[\alpha + (-\alpha)] = T(0_v)$

i.e., $T(\alpha) + T(-\alpha) = 0_w$, by using (*i*).

Consequently, $T(-\alpha) = -T(\alpha)$

(*iii*) Let $W' = \{\alpha \in V : T(\alpha) = 0_w\}$

By (*i*) $0_v \in W'$

So, $W' \neq \phi$.

Let $\alpha, \beta \in W'$

and $c \in F$.

Then, $T(\alpha) = 0_w, T(\beta) = 0_w$

and, hence, $T(\alpha + c\beta) = T(\alpha) + cT(\beta) = 0_w + c0_w = 0_w$.

This yields $\alpha + c\beta \in W'$

Hence, by Lemma (2.1.4), W' is a subspace of V. Finally, we prove (*iv*) by induction on k. For $k = 1$, the result obviously holds. Let $k > 1$ and let the result hold for any $k - 1$ vectors.

Now, $$\sum_{i=1}^{k} \alpha_i = \sum_{i=1}^{k-1} \alpha_i + \alpha_k = \beta_1 + \beta_2$$

where $$\beta_1 = \sum_{i=1}^{k-1} \alpha_i,$$

$$\beta_2 = \alpha_k.$$

By the induction hypothesis,

$$T(\beta_1) = \sum_{i=1}^{k-1} T(\alpha_i).$$

By definition, $$T\left[\sum_{i=1}^{k} \alpha_i\right] = T(\beta_1) + T(\beta_2)$$

$$= \sum_{i=1}^{k-1} T(\alpha_i) + T(\alpha_k)$$

$$= \sum_{i=1}^{k} T(\alpha_i)$$

Definition 3. Let $T : V_F \to W_F$ be a linear transformation. Then, the subspace $\{\alpha \in V : T(\alpha) = 0_w\}$ is called the *kernel of* T, and is denoted by ker(T). (The kernel of T is also called the *null space* of T). The dimension of ker(T) is called the *nullity* of T, which is denoted by $\nu(T)$.

Before we give some examples of kernels, we prove the following :

Proposition 4. Let $T : V_F \to W_F$ be a linear transformation. Then,

(*i*) $$T(\alpha - \beta) = T(\alpha) - T(\beta) \qquad \text{for all } \alpha, \beta \in V.$$

(*ii*) T is one-to-one iff $\ker(T) = \{0_v\}$.

Proof.

(*i*) $$\alpha - \beta = \alpha + (-\beta),$$

gives $$T(\alpha - \beta) = T[\alpha + (-\beta)] = T(\alpha) + T(-\beta).$$

However, $$T(-\beta) = -T(\beta), \qquad \text{by Proposition}$$

Consequently, $T(\alpha - \beta) = T(\alpha) + [-T(\beta)] = T(\alpha) - T(\beta)$

(*ii*) Let T be one-to-one. Let $\alpha \in \ker(T)$.

Then, $$T(\alpha) = 0_w.$$

Also, $$T(0_v) = 0_w,$$

so that $$T(\alpha) = T(0_v).$$

This gives $$\alpha = 0_v,$$

as T is one-to-one.

Hence, $$\ker T = \{0_v\}.$$

Conversely, let $$\ker(T) = \{0_v\}.$$

For some $\alpha, \beta \in V$, let $$T(\alpha) = T(\beta).$$

Then, $$T(\alpha - \beta) = T(\alpha) - T(\beta) = 0_w,$$

$$\alpha - \beta \in \ker T = \{0_v\},$$

gives $$\alpha - \beta = 0_v,$$

i.e., $$\alpha = \beta.$$

Hence, T is one-to-one.

Proposition 5. Let $T : V_F \to W_F$ be a linear transformation. Then, the following hold :

(*i*) For any subspace V_1 of V,

$$T(V_1) = \{T(\alpha) : \alpha \in V_1\}$$

is a subspace of W.

(*ii*) For any subspace W_1 of W,

$$T^{-1}(W_1) = \{\alpha \in V : T(\alpha) \in W_1\}$$

is a subspace of V containing $\ker(T)$

Proof.

(*i*) As $$0_v \in V_1,\ 0_w = T(0_v) \in T(V_1),$$

so that $T(V_1) \neq \phi$.

Let $\alpha', \beta' \in T(V_1)$ and $c \in F$.

By definition, $\alpha' = T(\alpha), \beta' = T(\beta)$ for some $\alpha, \beta \in V_1$.

As V_1 is a subspace of V, we get $\alpha + c\beta \in V_1$.

So that, $T(\alpha + c\beta) = T(\alpha) + cT(\beta) = \alpha' + c\beta' \in T(V_1)$.

Hence, $T(V_1)$ is a subspace of W. (lemma 2.1.4).

(*ii*) Let W_1 be a subspace of W.

As $T(0_v) = 0_w \in W_1$, $0_v \in T^{-1}(W_1)$.

Hence, $T^{-1}(W_1) \neq \phi$.

Let $\alpha, \beta \in T^{-1}(W_1)$ and $c \in F$.

By definition $T(\alpha) \in W_1$, $T(\beta) \in W_1$.

So that $T(\alpha) + cT(\beta) \in W_1$,

i.e., $T(\alpha + c\beta) \in W_1$.

Consequently, $\alpha + c\beta \in T^{-1}(W_1)$.

This proves that $T^{-1}(W_1)$ is a subspace of V. Consider any $\alpha \in \ker(T)$.

Then $T(\alpha) = 0_w \in W_1$,

yields $\alpha \in T^{-1}(W_1)$.

This proves that $\ker(T) \subseteq T^{-1}(W_1)$.

In particular the above result gives that range(T) = $T(V)$ is a subspace of W. Hence forth we shall not be using distinguishing symbols for the zero vectors in different vector spaces. A linear transformation $T : V_F \to W_F$ for which range(T) $\neq \{0\}$ is called a *non-zero linear transformation*.

Definition 6. Let $T : V_F \to W_F$ be a linear transformation. Then the dimension of range(T) is called the *rank* of T and is denoted by $\rho(T)$.

Example 6.a. Let $T : \mathbf{R}^3 \to \mathbf{R}^2$ be given by $T(a, b, c) = (a+b, b+c)$. It can be easily checked that T is a linear transformation.

Any $\alpha = (a, b, c) \in \ker(T)$

iff $T(\alpha) = 0$

iff $(a+b, b+c) = 0 = (0, 0)$,

i.e., iff
$$a + b = 0$$
$$b + c = 0$$

This is a system of two homogeneous linear equations. The general solution for this system is $a = -r$, $b = r$, $c = -r$, r arbitrary.

Hence, $\ker(T) = \{(-r, r, -r) : r \in \mathbf{R}\} = \{r(-1, 1, -1) : r \in \mathbf{R}\}$.

So that $\ker(T)$ is a subspace of $\mathbf{R}^3$ spanned by the vector $(-1, 1, -1)$.

Hence, $\dim\{\ker(T)\} = 1$, i.e., nullity $(T) = 1$.

Now, $$\text{Rang}(T) = \{(a+b, b+c) : a, b, c \in \mathbf{R}\}$$

$$(a+b, b+c) = a(1, 0) + b(1, 1) + c(0, 1) = a\beta_1 + b\beta_2 + c\beta_3 \quad \text{... (1)}$$

where $$\beta_1 = (1, 0), \beta_2 = (1, 1), \beta_3 = (0, 1)$$

For $a = 1, b = 0, c = 0$, the right hand side in (1) becomes β_1.

So, $$\beta_1 \in \text{range}(T).$$

Similarly, $$\beta_2, \beta_2 \in \text{range}(T).$$

Further, (1) shows that $\beta_1, \beta_2, \beta_3$ span range(T).

However, $$\beta_2 = \beta_1 + \beta_3$$

so that $$\text{range}(T) = \langle \beta_1, \beta_3 \rangle.$$

However, β_1, β_2 are linearly independent. Hence, range(T) has $\{\beta_1, \beta_2\}$ as its basis.

So that $$\text{rank}(T) = 2.$$

As $$\text{range}(T) \subseteq \mathbf{R}^2,$$

and $$\dim\{\text{range}(T)\} = \dim(\mathbf{R}^2),$$

we get $$\text{range}(T) = \mathbf{R}^2$$

This proves that T is into, i.e., T is an epimorphism. As $\ker T \neq \{0\}$, T is not one-to-one.

Example 6.b. Let A be any $m \times n$-matrix over a field F. It gives a linear transformation

$$T : F^{n\times 1} \to F^{m\times 1}$$

such that $$T(X) = AX,$$

for every $$X = \begin{bmatrix} x_1 \\ x_2 \\ \vdots \\ x_n \end{bmatrix} \in F^{n\times 1}$$

Now, $$X \in \ker(T) \quad \text{iff } AX = 0.$$

Hence, $\ker(T)$ is precisely the solution space of the system of homogeneous linear equations $AX = 0$

So, by Theorem (2.3.7) $$\dim\{\ker(T)\} = n - \text{rank}(A)$$

i.e., $$\text{nullity}(T) = n - \text{rank}(A) \quad \text{... (1)}$$

If $A_1, A_2, \ldots, A_n$ are the n columns of A,

$$T(X) = A_1x_1 + A_2x_2 + \ldots + A_nx_n.$$

This shows that the range(T), consists of all linear combinations of the column of A.

That gives $\qquad$ range(T) = column-space (A) $\qquad$ ... (2)

so that $\qquad$ rank(T) = rank(A) $\qquad$ ... (3)

(1) and (3) give $\qquad$ nullity$(T) = n -$ rank(T)

i.e., $\qquad$ rank(T) + nullity$(T) = n$

Example 6.c. Let
$$A = \begin{bmatrix} 1 & 0 & 2 \\ 0 & 1 & 3 \\ 1 & 1 & 0 \\ 0 & 0 & 1 \end{bmatrix}$$

This gives a linear transformation $T : \mathbf{R}^{3\times 1} \to \mathbf{R}^{4\times 1}$

such that
$$T\begin{bmatrix} x \\ y \\ z \end{bmatrix} = \begin{bmatrix} 1 & 0 & 2 \\ 0 & 1 & 3 \\ 1 & 1 & 0 \\ 0 & 0 & 1 \end{bmatrix}\begin{bmatrix} x \\ y \\ z \end{bmatrix} = \begin{bmatrix} x+2z \\ y+3z \\ x+y \\ z \end{bmatrix}$$

If
$$X = \begin{bmatrix} x \\ y \\ z \end{bmatrix} \in \ker(T),$$

we get
$$\begin{bmatrix} x+2z \\ y+3z \\ x+y \\ z \end{bmatrix} = \begin{bmatrix} 0 \\ 0 \\ 0 \\ 0 \end{bmatrix}.$$

By solving for x, y, z, we get $x = 0, y = 0, z = 0$. Hence, $\ker(T) = \{0\}$, so that T is one-to-one and $\nu(T)$ = nullity$(T) = 0$

Now, $\qquad$ range(T) = column-space(A),

so that $\qquad$ rank(T) = rank$(A) = 3 < \dim(\mathbf{R}^{4\times 1})$.

This gives $\qquad \dim\{\text{range}(T)\} < \dim(\mathbf{R}^{4\times 1})$

i.e., $\qquad$ range$(T) \neq \mathbf{R}^{4\times 1}$

so that T is one-to-one, but T is not onto.

Lemma 7. Let $T : V_F \to W_F$ be a linear transformation, where V_F is spanned by a finite subset $B = \{\alpha_1, \alpha_2, \ldots, \alpha_m\}$. Then, (*i*) Range($T$) is spanned by

$$T(B)=\{T(\alpha_1),\, T(\alpha_2),\ldots,\, T(\alpha_m)\}$$

(*ii*) Rank(T) $\le \dim V$

Proof. If $B=\phi$, $V=\{0\}$. Then, $T(V)=\{T(0)\}=\{0\}$ is spanned by $\phi=T(\phi)$: in this case, rank(T) $=0=\dim V$. So, the result holds in case $B=\phi$. Let $B\ne\phi$. Consider any $\alpha\in V$

Then, $$\alpha=\sum_{i=1}^{m} a_i\alpha_i, \qquad a_i\in F$$

By Proposition 2, $$T(\alpha)=\sum_{i=1}^{m} T(a_i\alpha_i)=\sum_{i=1}^{m} a_i T(\alpha_i).$$

This shows that range(T) is spanned by

$$T(B)=\{T(\alpha_1),\, T(\alpha_2),\ldots,\, T(\alpha_m)\}.$$

This proves (*i*). To prove (*ii*), let $\dim V=k$.

Then, V has a basis $$B'=\{\alpha_1',\, \alpha_2',\ldots,\, \alpha_k'\}.$$

By part(*i*), range(T) is spanned by $T(B')$, so that

$$\text{rank}(T)=\dim\{\text{range}(T)\}\le |T(B')|\le k$$

Hence, $$\text{rank}(T)\le \dim V.$$

Proposition 8. Let $T: V_F\to W_F$ be a linear transformation. Let V' be a finite dimensional subspace of V such that $V'\cap\ker(T)=\{0\}$. Then, the following hold :

(*i*) If $B=\{\alpha_1,\, \alpha_2,\ldots,\, \alpha_k\}$ is a basis of V', then, $\{T(\alpha_1),\, T(\alpha_2),\ldots,\, T(\alpha_k)\}$ is a basis of $T(V')$

(*ii*) $\dim V'=\dim T(V')$

(*iii*) If T is one-to-one and V is finite dimensional, then, rank(T) $=\dim V$.

Proof. By Lemma 7, $T(B)$ spans $T(V')$. Let for some $a_i\in F$, $1\le i\le k$

$$\sum_{i=1}^{k} a_i T(\alpha_i)=0$$

i.e., $$T\left[\sum_{i=1}^{k} a_i\alpha_i\right]=0$$

This gives $$\sum_{i=1}^{k} a_i\alpha_i\in\ker(T)\cap V'=\{0\}.$$

So that $$\sum_{i=1}^{k} a_i\alpha_i=0$$

However, $\alpha_1, \alpha_2, \ldots, \alpha_k$ are linearly independent. Hence, $a_i = 0$ for every i. This gives that $T(B)$ has k members, and they are linearly independent. Hence, $T(B')$ form a basis of $T(V')$, so that $\dim T(V') = k = \dim V'$. Finally, let $\dim V = k$, and T be one-to-one.

Then, $$\ker(T) = \{0\}.$$

Let $$V = V'.$$

Obviously, $$V' \cap \ker(T) = \{0\}.$$

So, by (*ii*) $$\dim V = \dim T(V) = \text{rank}(T)$$

Theorem 9 (Sylvester Law). Let T be a linear transformation of a finite dimensional vector space V_F into a vector space W_F.

Then, $$\dim V = \text{rank}(T) + \text{nullity}(T)$$

Proof. Let $L = \ker(T)$, $\dim L = \text{nullity}\ (T)$. Then, there exists a subspace W' of V such that

$$V = L + W' \text{ and } L \cap W' = \{0\} \quad \ldots (1)$$

Consider any $\alpha \in V$.

Then, $$\alpha = \beta + \beta'$$

with $$\beta \in L,\ \beta' \in W'$$

Now, $$T(\beta) = 0,$$

as $$\beta \in \ker(T),$$

so that $$T(\alpha) = T(\beta + \beta') = T(\beta) + T(\beta') = T(\beta')$$

is in $T(W')$.

Hence, $$\text{range}(T) \subseteq T(W') \subseteq \text{range}(T).$$

This gives $$\text{range}(T) = T(W'),$$

so that $$\text{rank}(T) = \dim T(W') = \dim W', \quad \text{by Proposition 8.}$$

By Theorem (2.2.19), $$\dim V = \dim L + \dim W'$$

i.e., $$\dim V = \text{nullity}(T) + \text{rank}(T)$$

By definition, isomorphism $T : V_F \to W_F$ has to be one-to-one. In certain situations, it is enough for a linear transformation to be isomorphism, that it is one-to-one or onto. This is shown in the following theorem.

Theorem 10. Let V_F and W_F be finite dimensional with $\dim V = \dim W$. Then, for a linear transformation $T : V_F \to W_F$, the following are equivalent :

(*i*) T is an isomorphism

(*ii*) T is onto

(*iii*) T is one-to-one

Proof. Obviously, (*i*) implies (*ii*) and (*iii*). Let (*ii*) hold.

Then, $$\text{range}(T) = W$$

gives $$\text{rank}(T) = \dim W = \dim V.$$

However, by Theorem 9 $$\dim V = \text{nullity}(T) + \text{rank}(T) \quad \text{... (1)}$$

This gives $\text{nullity}(T) = 0$, so that T is one-to-one. Hence, T is an isomorphism. Consequently, (*ii*) implies (*i*). Let (*iii*) hold. Then, $\text{nullity}(T) = 0$.

So, (1) gives $$\text{rank}(T) = \dim V = \dim W.$$

As $$\text{rank}(T) = \dim\{\text{range}(T)\} = \dim W,$$

we get $$W = \text{range}(T).$$

Consequently, T is onto. Hence, T is an isomorphism. This proves that (*iii*) implies (*i*).

Example 10.a. Let $V_F = F[X]$. This vector space is not finite dimensional

Define $$T : F[X] \to F[X]$$

such that $$T(a_0 + a_1X + a_2X^2 + ... + a_kX^k) = a_0 + a_1X^2 + a_2X^4 + ... + a_kX^{2k}$$

for any $$f(X) = a_0 + a_1X + ...\ a_kX^k \in F[X]$$

T is a linear transformation. It is one-to-one. The range of T is $F[X^2] \neq F[X]$, so T is not onto. Thus, T is not an isomorphism.

The following result is extremely useful in the construction of linear transformations.

Theorem 11. Let V_F be a finite dimensional vector space with $\dim V = k \geq 1$. Let $B = \{\alpha_1, \alpha_2, ..., \alpha_k\}$ be a basis of V. Let W_F be any vector space. Then, for any sequence of k vectors $\beta_1, \beta_2, ..., \beta_k$ in W, there exists a unique linear transformation $T : V_F \to W_F$ such that $T(\alpha_i) = \beta_i$, $1 \leq i \leq k$.

Proof. Let $\alpha, \beta \in V_F$ and $c \in F$.

Then, $$\alpha = \sum_{i=1}^{k} a_i\alpha_i$$

$$\beta = \sum_{i=1}^{k} b_i\alpha_i$$

for some $a_i, b_i \in F$.

As $\alpha_1, \alpha_2, ..., \alpha_k$ are linearly independent, $\alpha = \beta$ iff $a_i = b_i$, for $1 \leq i \leq k$.

Define $$T : V_F \to W_F$$

such that $$T\left[\sum_{i=1}^{k} a_i\alpha_i\right] = \sum_{i=1}^{k} a_i\beta_i$$

for any $$\alpha = \sum_{i=1}^{k} a_i \alpha_i \in V.$$

For $\alpha = \beta$, we get $$a_i = b_i,$$

so that $$T(\alpha) = \sum_{i=1}^{k} a_i \beta_i = \sum_{i=1}^{k} b_i \beta_i = T(\beta).$$

Consequently, T is well-defined.

Now, $$\alpha + \beta = \sum_{i=1}^{k} (a_i + b_i)\alpha_i$$

Gives $$T(\alpha+\beta) = \sum_{i=1}^{k} (a_i + b_i)\alpha_i = \sum_{i=1}^{k} a_i \alpha_i + \sum_{i=1}^{k} b_i \alpha_i = T(\alpha) + T(\beta)$$

Again, $$c\alpha = \sum_{i=1}^{k} (ca_i)\alpha_i,$$

gives $$T(c\alpha) = \sum_{i=1}^{k} (ca_i)\beta_i = c\sum_{i=1}^{k} a_i \beta_i = cT(\alpha)$$

Hence, T is a linear transformation.

Now $$\alpha_i = \sum_{j=1}^{k} \delta_{ij} \alpha_j, \qquad 1 \le i \le k$$

where δ_{ij} is the Kronecker delta {defined after Theorem (1.2.8)}, so that

$$T(\alpha_i) = \sum_{j=1}^{k} \delta_{ij} \beta_j = \beta_i, \qquad 1 \le i \le k$$

Let T' be another linear transformation from V_F into W_F such that $T'(\alpha_i) = \beta_i$, $1 \le i \le k$.

Then, $$T'(\alpha) = \sum_{i=1}^{k} a_i T'(\alpha_i) = \sum_{i=1}^{k} a_i \beta_i = T(\alpha)$$

so that $T' = T$.

Hence T is unique.

Example 11.a. Consider $V = \mathbf{R}^3$, $W = \mathbf{R}^4$. Now, $B = \{\epsilon_1, \epsilon_2, \epsilon_3\}$ is the standard basis of $\mathbf{R}^3$.

Choose any three vectors in $\mathbf{R}^4$, not necessarily distinct, say $\beta_1 = (0,1,-1,0)$, $\beta_2 = (2,1,1,0)$, $\beta_3 = (1,1,0,1)$. The above theorem gives a linear transformation

$$T : V \to W$$

such that for any $$\alpha = a\,\epsilon_1 + b\,\epsilon_2 + c\,\epsilon_3$$

$$T(\alpha) = a\beta_1 + b\beta_2 + c\beta_3$$

i.e., $$T(a, b, c) = a(0, 1, -1, 0) + b(2, 1, 1, 0) + c(1, 1, 0, 1)$$
$$= (2b + c, a + b + c, -a + b, c)$$

Example 11.b. Let $T : \mathbf{R}^2 \to \mathbf{R}^2$ be a linear transformation such that

$$T(1, 1) = (1, 3), T(-1, 1) = (3, 1).$$

Let us find $T(a, b)$ for any $\alpha = (a, b) \in \mathbf{R}^2$. Let $\alpha_1 = (1, 1), \alpha_2 = (-1, 1)$. The first step is to express α as a linear combination of α_1 and α_2.

Clearly $$\alpha = \left[\frac{a+b}{2}\right]\alpha_1 + \left[\frac{b-a}{2}\right]\alpha_2,$$

so that $$T(\alpha) = \left[\frac{a+b}{2}\right]T(\alpha_1) + \left[\frac{b-a}{2}\right]T(\alpha_2)$$

i.e., $$T(a, b) = \left[\frac{a+b}{2}\right](1, 3) + \left[\frac{b-a}{2}\right](3, 1) = (2b - a, a + 2b)$$

Theorem 12. V_F be a non-zero finite dimensional vector space, and W_F be any vector space. Let $\alpha_1, \alpha_2, \ldots, \alpha_k$ be any k linearly independent vectors in V, and $\beta_1, \beta_2, \ldots, \beta_k$ be any k vectors in W. Then, there exists a linear transformation $T : V \to W$ such that $T(\alpha_i) = \beta_i, 1 \le i \le k$.

Proof. Let $\dim V = m$. Then $m \ge k$. By Corollary (2.2.16), we can find $m - k$ vectors $\alpha_{k+1}, \alpha_{k+2}, \ldots, \alpha_m$ such that $\alpha_1, \alpha_2, \ldots, \alpha_m$ is a basis of V. Put $\beta_i = 0$ for $k + 1 \le i \le m$. So, by Theorem 11, there exists a linear transformation $T : V_F \to W_F$ such that $T(\alpha_i) = \beta_i, 1 \le i \le m$. In particular $T(\alpha_i) = \beta_i, 1 \le i \le k$.

Theorem 13. A mapping $T : V_F \to W_F$ is a linear transformation

iff $$T(\alpha + c\beta) = T(\alpha) + cT(\beta)$$

for all $\alpha, \beta \in V$, and $c \in F$.

Proof. Consider any $\alpha, \beta \in V$ and $c \in F$. Let T be a linear transformation.

So, $$T(c\beta) = cT(\beta).$$

Consequently, $$T(\alpha + c\beta) = T(\alpha) + T(c\beta) = T(\alpha) + cT(\beta).$$

Conversely, let T satisfy the given condition.

Then, $$T(\alpha + \beta) = T(\alpha + 1.\beta) = T(\alpha) + 1.T(\beta) = T(\alpha) + T(\beta).$$

By using this and, then, following the arguments in Proposition 2(i), we get

$$T(0) = 0,$$

so that $$T(c\beta) = T(0 + c\beta) = T(0) + cT(\beta) = 0 + cT(\beta)$$

i.e., $$T(c\beta) = cT(\beta)$$

Hence, T is a linear transformation.

The theorem, given above, provides a useful criterian for a linear transformation.

Exercise 3.1

1. Which of the following functions from $\mathbf{R}^2$ to $\mathbf{R}^2$ are linear transformations?

   (i) $T(x_1, x_2) = (x_1 + 1, y_1)$

   (ii) $T(x_1, x_2) = (x_1 + x_2, x_1 - x_2)$

   (iii) $T(x_1, x_2) = \left(x_1^2, x_2\right)$

   (iv) $T(x_1, x_2) = (ax_1 + bx_2, cx_1 + dx_2)$ for some fixed $a, b, c, d \in \mathbf{R}$

2. Which of the following are linear transformations?

   (i) $T : \mathbf{R}^3 \to \mathbf{R}^3$, $T(x_1, x_2, x_3) = (x_1, x_2, 0)$

   (ii) $T : \mathbf{R}^3 \to P_2(\mathbf{R})$, $T(x_1, x_2, x_3) = (x_1 - x_2)X + x_3X^2$

   (iii) $T : \mathbf{R}^{2\times 2} \to \mathbf{R}$, $T\left[\begin{bmatrix} a & b \\ c & d \end{bmatrix}\right] = ad - bc$

   (iv) $T : \mathbf{R}^3 \to \mathbf{R}$, $T(a_1, a_2, a_3) = \int_0^1 (a_1 + a_2X + a_3X^2)\, dx$

3. Let $T : \mathbf{R}^3 \to \mathbf{R}^2$ be the linear transformation such that

   $$T(1, 2, 3) = (1, 0, 0),\ T(1, 2, 0) = (0, 1, 0),\ T(1, -1, 0) = (0, 1, 0).$$

   Find $T(a, b, c)$ for any $\alpha = (a, b, c) \in \mathbf{R}^3$

4. Does there exist a linear transformation $T : \mathbf{R}^4 \to \mathbf{R}^3$ such that

   $$T(1, 0, 0, 0) = (1, 0, 0),\ T(0, 1, 0, 0) = (1, 1, 0),$$
   $$T(0, 0, 1, 0) = (1, 0, -1),\ T(1, 1, 1, 0) = (2, 1, 3)$$

   Explain your answer.

5. Let $T : \mathbf{R}^{4\times 1} \to \mathbf{R}^{3\times 1}$ be defined by $T\begin{bmatrix} x_1 \\ x_2 \\ x_3 \\ x_4 \end{bmatrix} = \begin{bmatrix} 1 & 0 & -1 & 3 \\ 2 & 0 & -1 & 5 \\ 3 & 0 & -2 & 8 \end{bmatrix}\begin{bmatrix} x_1 \\ x_2 \\ x_3 \\ x_4 \end{bmatrix}$.

   (i) Does $\begin{bmatrix} 1 \\ 0 \\ 1 \\ 0 \end{bmatrix} \in \ker(T)$?

   (ii) Does $\begin{bmatrix} 3 \\ 5 \\ 8 \end{bmatrix} \in \text{range}(T)$?

   (iii) What are nullity(T) and rank(T)?

6. Let T be a non-zero linear transformation from $\mathbf{R}^5$ into $\mathbf{R}^2$, such that T is not onto. Find rank(T) and nullity(T).

7. Let $D : P_3(\mathbf{R}) \to P_3(\mathbf{R})$ be given by

$$D(a_0 + a_1X + a_2X^2 + a_3X^3) = a_1 + 2a_2X + 3a_3X^2$$

This D is called a *differentiation operator*.

(*i*) Prove that D is a linear transformation

(*ii*) Find ker(D) and range(D)

(*iii*) Verify that nullity(D) + rank(D) = 4

8. Let $T : \mathbf{R}^{3\times 1} \to \mathbf{R}^{3\times 1}$ be given by

$$T\begin{bmatrix} x \\ y \\ z \end{bmatrix} = \begin{bmatrix} 0 & 0 & 1 \\ 1 & 0 & -1 \\ 0 & 1 & 2 \end{bmatrix}\begin{bmatrix} x \\ y \\ z \end{bmatrix}$$

(*i*) Prove that T is a an isomorphism

(*ii*) Find the coordinate vector $[\alpha]_\beta$ of $\alpha = \begin{bmatrix} 1 \\ 2 \\ -1 \end{bmatrix}$ relative to the ordered basis

$$B = \left\{ \begin{bmatrix} 1 \\ 1 \\ 0 \end{bmatrix}, \begin{bmatrix} 0 \\ 1 \\ 1 \end{bmatrix}, \begin{bmatrix} -1 \\ 0 \\ 1 \end{bmatrix} \right\}$$

(*iii*) Find the co-ordinate vector $[T(\alpha)]_B$

(*iv*) Verify that $[T(\alpha)]_B = P^{-1}A\alpha$,

where $P = \begin{bmatrix} 1 & 0 & -1 \\ 1 & 1 & 0 \\ 0 & 1 & 1 \end{bmatrix}$, $A = \begin{bmatrix} 0 & 0 & 1 \\ 1 & 0 & -1 \\ 0 & 1 & 2 \end{bmatrix}$.

9. Let $V_1, V_2, \ldots, V_n$ be any $n \geq 2$, finite dimensional vector spaces over a field F

Let $T_i : V_i \to V_{i+1}$ $(1 \leq i \leq n-1)$ be any linear transformations, such that

(*i*) $\ker(T_1) = \{0\}$

(*ii*) $\ker(T_{i+1}) = \text{range}(T_i)$ for $1 \leq i \leq n-2$

(*iii*) $\text{range}(T_{n-1}) = V_n$

Prove that $\sum_{i=1}^{n} (-1)^i \dim V_i = 0$

10. Prove that $T : P_3(\mathbf{R}) \to P_3(\mathbf{R})$ such that $T\{f(X)\} = f(X+1)$, $f(X) \in P_3(\mathbf{R})$ is an isomorphism.

3.2 Algebra of Linear Transformations

In this section, we discuss certain algebraic operations among linear transformations. These operations help us in constructing examples of various kinds of groups, rings, fields and several other algebraic systems. They also find applications in other areas of pure and applied mathematics.

For any two vector spaces V_F and W_F, $\text{Hom}_F(V, W)$ (or simply Hom(V, W), in case there is no ambiguity concerning the underlying field of scalars) will denote the set of all linear transformations from V_F to W_F. We start with the following :

Lemma 1. Let T and T' be two linear transformations from V_F to W_F. Then,

(*i*) $$T+T' : V_F \to W_F$$

such that $(T+T')(\alpha) = T(\alpha)+T'(\alpha),\ \alpha \in V$

is a linear transformation

(*ii*) For any $a \in F$, $aT : V_F \to W_F$ such that $(aT)(\alpha) = aT(\alpha)$, $\alpha \in V$ is a linear transformation.

Proof. Let $\alpha,\ \beta \in V$ and $c \in F$

Then
$$\begin{aligned}(T+T')(\alpha+c\beta) &= T(\alpha+c\beta)+T'(\alpha+c\beta)\\ &= [T(\alpha)+cT(\beta)]+[T'(\alpha)+cT'(\beta)]\\ &= [T(\alpha)+T'(\alpha)]+c[T(\beta)+T'(\beta)]\\ &= (T+T')(\alpha)+c(T+T')(\beta)\end{aligned}$$

Hence, by Theorem (3.1.13), $T+T'$ is a linear transformation. This proves (*i*)

Again,
$$\begin{aligned}(aT)(\alpha+c\beta) &= a[T(\alpha+c\beta)]\\ &= a[T(\alpha)+cT(\beta)]\\ &= aT(\alpha)+acT(\beta)\\ &= aT(\alpha)+caT(\beta), \qquad \text{as } ac = ca\\ &= (aT)(\alpha)+c[(aT)(\beta)]\end{aligned}$$

Hence, aT is a linear transformation.

Theorem 2. For any two vector spaces V_F and W_F, $\text{Hom}_F(V, W)$ is a vector space over F under the addition and scalar multiplication operations given below.

For any $T,\ T' \in \text{Hom}_F(V, W)$, and $a \in F$, $T+T'$ and aT are given by the formula :

$$(T+T')(\alpha) = T(\alpha)+T'(\alpha)$$
$$(aT)(\alpha) = a[T(\alpha)]$$

for any $\alpha \in V$

Proof. Let $T_1,\ T_2,\ T_3 \in \text{Hom}_F(V, W)$, $a,\ b \in F$ and $\alpha \in V$. By Lemma 1, T_1+T_2 and aT are in $\text{Hom}_F(V, W)$.

I. **$Hom_F(V, W)$ is an abelian group under addition :**

(*i*) $$(T_1+T_2)(\alpha) = T_1(\alpha)+T_2(\alpha),$$
$$(T_2+T_1)(\alpha) = T_2(\alpha)+T_1(\alpha)$$

But in W, $$T_1(\alpha)+T_2(a) = T_2(\alpha)+T_1(\alpha)$$

This gives $$(T_1+T_2)(\alpha) = (T_2+T_1)(\alpha)$$

Hence, $$T_1+T_2 = T_2+T_1$$

This proves that addition is commutative.

(*ii*) $$[T_1+(T_2+T_3)](\alpha) = T_1(\alpha)+(T_2+T_3)(\alpha)$$
$$= T_1(\alpha)+[T_2(\alpha)+T_3(\alpha)]$$
$$= [T_1(\alpha)+T_2(\alpha)]+T_3(\alpha),$$

as addition is associative in W.

So that $$[T_1+(T_2+T_3)](\alpha) = [(T_1+T_2)(\alpha)]+T_3(\alpha)$$
$$= [(T_1+T_2)+T_3](\alpha)$$

Hence, $$T_1+(T_2+T_3) = (T_1+T_2)+T_3,$$

thus addition is associative.

(*iii*) Define $\overline{0} : V \to W$ such that $\overline{0}(\alpha) = 0$, zero vector in W, for every $\alpha \in V$. Then, for any $\alpha, \beta \in V$

$$\overline{0}(\alpha+c\beta) = 0 = 0+c0 = \overline{0}(\alpha)+c\overline{0}(\beta).$$

This proves that $\overline{0}$ is a linear transformation, called the *zero linear transformation*.

As $$(T_1+\overline{0})(\alpha) = T_1(\alpha)+\overline{0}(\alpha) = T_1(\alpha)+0 = T_1(\alpha),$$
$$T_1+\overline{0} = T_1.$$

So, $\overline{0}$ is the identity under addition.

(*iv*) For any $$T \in Hom_F(V, W),$$

as $$[T+(-1)T](\alpha) = T(\alpha)+[(-1)T(\alpha)]$$
$$= T(\alpha)+(-1)T(\alpha)$$
$$= T(\alpha)-T(\alpha)$$
$$= 0$$
$$= \overline{0}(\alpha),$$

we get $$T+(-1)T = \overline{0}.$$

Hence, $$(-1)T = -T$$

This proves that $Hom_F(V, W)$ is an abelian group under addition.

II. **Properties of scalar multiplication :**

(*v*) $$[a(T_1+T_2)](\alpha) = a[(T_1+T_2)(\alpha)]$$

$$= a[T_1(\alpha) + T_2(\alpha)]$$
$$= aT_1(\alpha) + aT_2(\alpha)$$
$$= (aT_1 + aT_2)(\alpha).$$

This gives $\quad a(T_1 + T_2) = aT_1 + aT_2.$

(*vi*) $$[(a+b)T_1](\alpha) = (a+b)[T_1(\alpha)].$$

In W, $(a+b)[T_1(\alpha)] = aT_1(\alpha) + bT_1(\alpha) = (aT_1 + bT_1)(\alpha)$

Hence, $\quad (a+b)T_1 = aT_1 + bT_1.$

(*vii*) $$[(ab)T_1](\alpha) = (ab)T_1(\alpha).$$

In W, $\quad (ab)T_1(\alpha) = a[bT_1(\alpha)] = [a(bT_1)](\alpha),$

so that $\quad (ab)T_1 = a(bT_1).$

(*viii*) $$(1T_1)(\alpha) = 1T_1(\alpha) = T_1(\alpha).$$

So that $\quad 1T_1 = T_1.$

This proves that $\text{Hom}_F(V, W)$ is a vector space over F.

Theorem 3. Let U, V and W be any three vector spaces over the same field F. Then, the following hold :

(*i*) For any linear transformations $T : U_F \to V_F, T' : V_F \to W_F$, the composite $T'T : U_F \to W_F$ is a linear transformation

(*ii*) For any three linear transformations

$$T_1 : U_F \to V_F,\ T_2 : U_F \to V_F,\ T : V_F \to W_F$$
$$T(T_1 + T_2) = TT_1 + TT_2$$

(*iii*) For any three linear transformations

$$S_1 : V_F \to W_F,\ S_2 : V_F \to W_F,\ S : U_F \to V_F,$$
$$(S_1 + S_2)S = S_1S + S_2S$$

Proof. Let $\alpha, \beta \in U$, $c \in F$. As T is a linear transformation,

$$T(\alpha + c\beta) = T(\alpha) + cT(\beta)$$

This gives $\quad (T'T)(\alpha + c\beta) = T'[T(\alpha + c\beta)] = T'[T(\alpha) + cT(\beta)]$

As $\quad T(\alpha), T(\beta) \in V$

and $\quad T' : V \to W$

is a linear transformation

$$T'[T(\alpha) + cT(\beta)] = T'[T(\alpha)] + cT'[T(\beta)] = (T'T)(\alpha) + c[(T'T)(\beta)]$$

Hence, $\quad (T'T)(\alpha + c\beta) = (T'T)(\alpha) + c(T'T)(\beta)$

This proves that $T'T$ is a linear transformation. This proves (*i*).

For (*ii*)
$$\begin{aligned}[T(T_1+T_2)](\alpha) &= T[(T_1+T_2)(\alpha)]\\ &= T[T_1(\alpha)+T_2(\alpha)]\\ &= T[T_1(\alpha)]+T[T_2(\alpha)]\\ &= (TT_1)(\alpha)+(TT_2)(\alpha)\\ &= (TT_1+TT_2)(\alpha)\end{aligned}$$

i.e.,
$$[T(T_1+T_2)](\alpha) = (TT_1+TT_2)(\alpha).$$

Hence,
$$T(T_1+T_2) = TT_1+TT_2$$

For (*iii*)
$$\begin{aligned}[(S_1+S_2)S](\alpha) &= (S_1+S_2)[S(\alpha)]\\ &= S_1[S(\alpha)]+S_2[S(\alpha)]\\ &= (S_1S)(\alpha)+(S_2S)(\alpha)\\ &= [S_1S+S_2S](\alpha).\end{aligned}$$

So that
$$[(S_1+S_2)S](\alpha) = (S_1S+S_2S)(\alpha)$$

Hence,
$$(S_1+S_2)S = S_1S+S_2S$$

Theorem 4. Let $T : U_F \to V_F$ and $T' : V_F \to W_F$ be any two linear transformations.

Then, for any $a \in F$ $\quad a(T'T) = (aT')T = T'(aT)$

Proof. Let
$$\alpha \in U$$

Then
$$[a(T'T)](\alpha) = a[(T'T)(\alpha)] \quad \text{... (1)}$$
$$[(aT')T](\alpha) = (aT')[T(\alpha)] = a[T'\{T(\alpha)\}] = a[(T'T)(\alpha)] \quad \text{... (2)}$$
$$[T'(aT)](\alpha) = T'[(aT)(\alpha)] = T'[a(T\alpha)]$$

As $T' : V_F \to W_F$ is a linear transformation and $T(\alpha) \in V$, we get
$$T'[a(T\alpha)] = a[T'\{T(\alpha)\}] = a[(T'T)(\alpha)]$$

So that
$$[T'(aT)](\alpha) = a[(T'T)(\alpha)] \quad \text{... (3)}$$

(1), (2) and (3) show that
$$[a(T'T)](\alpha) = [(aT')T](\alpha) = [T'(aT)](\alpha)$$

Hence,
$$a(T'T) = (aT')T = T'(aT)$$

Theorem 5. For any isomorphism $T : V_F \to W_F$, $T^{-1} : W_F \to V_F$ is an isomorphism

Proof. Let $\alpha, \beta \in W$ and $c \in F$. As T is one-to-one and onto, there exist unique α', β' $\in V$ such that $T(\alpha') = \alpha$, $T(\beta') = \beta$.

Then,
$$T^{-1}(\alpha) = \alpha', \; T^{-1}(\beta) = \beta'.$$

As T is a linear transformation

$$T(\alpha' + c\beta') = T(\alpha') + cT(\beta') = \alpha + c\beta$$

This gives $$T^{-1}(\alpha + c\beta) = \alpha' + c\beta' = T^{-1}(\alpha) + cT^{-1}(\beta)$$

Hence, T^{-1} is a linear transformation. As T is one-to-one and onto, T^{-1} is also one-to-one and onto. Hence, T^{-1} is an isomorphism.

Definition 6. A vector space V_F is said to be *isomorphic* to a vector space W_F, in symbol $V_F \cong W_F$, if there exists an isomorphism $T : V_F \to W_F$

Let $$\dim V_F = k \geq 1$$

and $$B = \{\alpha_1, \alpha_2, \ldots, \alpha_k\}$$

be any ordered basis of V_F.

Then, the mapping $$\sigma : V \to F^{k\times 1}$$

such that $$\sigma(\alpha) = [\alpha]_B, \ \alpha \in V$$

is an isomorphism {see Theorem (2.4.5)}.

Hence, $$V \cong F^{k\times 1}$$

We now prove that the relation of isomorphism is an equivalence relation.

Theorem 7. The relation of isomorphism in any set of vector spaces over a field F is an equivalence relation.

Proof. Consider any three vector spaces U_F, V_F and W_F

(*i*) *Reflexivity* : The identity function

$$I_V : V \to V$$

such that $$I_V(\alpha) = \alpha \qquad \text{for all } \alpha \in V,$$

is an isomorphism of V onto V.

Hence, $$V_F \cong V_F.$$

(*ii*) *Symmetry* : Let $$U_F \cong V_F.$$

So there exists an isomorphism

$$T : U_F \to V_F.$$

By Theorem 5, $$T^{-1} : V_F \to U_F$$

is an isomorphism.

Hence, $$V_F \cong U_F$$

(*iii*) *Transitivity* : Let $$U_F \cong V_F, V_F \cong W_F.$$

So, we have isomorphisms $$T : U_F \to V_F$$

and $$T' : V_F \to W_F.$$

By Theorem 3 $\qquad T'T : U_F \to W_F$

is a linear transformation.

As T' and T both are one-to-one and onto, the composite $T'T$ is also one-to-one and onto.

Consequently, $\qquad T'T : U_F \to W_F$

is an isomorphism.

This proves that $\qquad U_F \cong W_F$.

Two vector spaces V_F and W_F are said to be isomorphic if $V_F \cong W_F$ (or, equivalently, $W_F \cong V_F$). We now strengthen Theorem (2.4.5)

Theorem 8. Two finite dimensional vector spaces V_F and W_F are isomorphic iff $\dim V = \dim W$.

Proof. Let $V_F \cong W_F$. So, there exists an isomorphism $T : V_F \to W_F$. Let $B = \{\alpha_1, \alpha_2, ..., \alpha_k\}$ be a basis of V, $k = \dim V$; if $V = \{0\}$, $B = \phi$.

By Proposition (3.1.8), $T(B)$ is a basis of $T(V) = W$. As $T(B)$ also has k members, obviously, $\dim W = k = \dim V$.

Conversely, let $\qquad \dim V = \dim W = k$.

If $\qquad k = 0, V = W = \{0\}$

then, obviously, $\qquad V_F \cong W_F$.

Let $k \geq 1$. Then, $\qquad V_F \cong F^{k\times 1}$

and $\qquad F^{k\times 1} \cong W_F$

By the property of transitivity of isomorphism,

$$V_F \cong W_F.$$

Example 8.a. Let $T : \mathbf{R}^{3\times 1} \to \mathbf{R}^{2\times 1}$ and $T' : \mathbf{R}^{3\times 1} \to \mathbf{R}^{2\times 1}$ be given by

$$T\begin{bmatrix} x \\ y \\ z \end{bmatrix} = \begin{bmatrix} x+y \\ y-z \end{bmatrix}$$

$$T'\begin{bmatrix} x \\ y \\ z \end{bmatrix} = \begin{bmatrix} x-y \\ z-y \end{bmatrix}$$

for any $\qquad X = \begin{bmatrix} x \\ y \\ z \end{bmatrix} \in \mathbf{R}^{3\times 1}$

Here $$\ker(T) = \left\{ \begin{bmatrix} -r \\ r \\ r \end{bmatrix} : r \in \mathbf{R} \right\}$$

gives $$\text{nullity}(T) = 1$$

So, $$\text{rank}(T) = \dim(\mathbf{R}^{3\times1}) - 1 = 2.$$

Similarly, $$\text{rank}(T') = 2$$

$$(T+T')\begin{bmatrix} x \\ y \\ z \end{bmatrix} = T\begin{bmatrix} x \\ y \\ z \end{bmatrix} + T'\begin{bmatrix} x \\ y \\ z \end{bmatrix} = \begin{bmatrix} 2x \\ 0 \end{bmatrix}$$

One can see that $$\text{rank}(T+T') = 1.$$

$$3T : \mathbf{R}^{3\times1} \to \mathbf{R}^{2\times1}$$

is given by $$(3T)\begin{bmatrix} x \\ y \\ z \end{bmatrix} = 3T\begin{bmatrix} x \\ y \\ z \end{bmatrix} = \begin{bmatrix} 3x+3y \\ 3y-3z \end{bmatrix}$$

Observe that $$\text{rank}(3T) = 2.$$

Example 8.b. Let A be any non-singular 3×3-matrix, say $A = \begin{bmatrix} 1 & 5 & 7 \\ 0 & 1 & 9 \\ 0 & 0 & 1 \end{bmatrix}$. We get two linear transformations $T : \mathbf{R}^{3\times1} \to \mathbf{R}^{3\times1}$, $T' : \mathbf{R}^{3\times1} \to \mathbf{R}^{3\times1}$ such that for any $X \in \mathbf{R}^{3\times1}$.

$$T(X) = AX, \; T'(X) = A^{-1}X$$

Then, $$(T'T)(X) = T'[T(X)] = T'(AX) = A^{-1}(AX) = X = I(X),$$

where I is the identity transformation on $\mathbf{R}^{3\times1}$, so that $T'T = I$.

Similarly, $TT' = I$. Hence, $T' = T^{-1}$, so that T is an isomorphism of $\mathbf{R}^{3\times1}$ onto $\mathbf{R}^{3\times1}$.

Definition 9. Any linear transformation T from a vector space V_F into V_F itself is called a *linear operator* on V_F. A linear operator on V_F that is an isomorphism is called an automorphism of V_F

In Example 8.b., T and T' are two automorphisms of $\mathbf{R}^{3\times1}$. The symbol $L_F(V)$ {or simply $L(V)$} will denote the set of all linear operators on V.

Clearly, $$L_F(V) = \text{Hom}_F(V, V)$$

The symbol $GL_F(V)$ {or simply $GL(V)$} will denote the set of all automorphisms of V_F. Let $T, T' \in L(V)$. Then, by Theorem 3, $T'T : V \to V$ is a linear transformation, so that $T'T \in L(V)$. Hence, $L(V)$ admits two binary operations viz. addition and multiplication. For any $a \in F$, $aT \in L(V)$. By Theorem 2, $L(V)$ is an abelian group. The composition of functions, is associative in general, so that the product operation in $L(V)$ is associative. Further, Theorem 3 shows that multiplication distributes over addition. Hence, $L(V)$ is a ring. This along with the result in Theorem 4, gives the following :

Theorem 10. For any vector space V_F, the set $L_F(V)$ of all linear operators on V is a ring under the addition and multiplication operations on linear operators. Further,

(*i*) $L_F(V)$ is a vector space over F

(*ii*) For any $T,\ T' \in L_F(V)$ and $a \in F$

$$a(TT') = (aT)T' = T(aT')$$

An *algebra* over a field F is a ring A along with a scalar multiplication giving $a\alpha \in A$ for any $a \in F$ and $\alpha \in A$, which makes $\langle A, +\rangle$ a vector space over F, such that for any $\alpha,\ \beta \in A$ and $a \in F$

$$a(\alpha\beta) = (a\alpha)\beta = \alpha(a\beta)$$

Theorem 10 gives that $L(V)$ is an algebra over F. Any subset S of an algebra R over a field F is called a *subalgebra* of R, if S itself is an algebra under the operations in R.

Example 10.a. Let F be any field and $R = F^{n\times n}$, the set of all $n\times n$-matrices over F. By Theorem (1.2.11), $F^{n\times n}$ is a ring under matrix addition and multiplication operations.

By Theorem (1.2.13),
$$a(AB) = (aA)B = A(aB) \qquad \text{... (1)}$$
for any $a \in F$, $A,\ B \in F^{n\times n}$

Hence, $F^{n\times n}$ is an algebra over F.

Let S be the set of all upper triangular matrices in $F^{n\times n}$. As sums and products of upper triangular matrices are upper triangular, {see Exercise (1.2.11)}, it is immediate that S is a ring under matrix addition and multiplication. If A is upper triangular and $a \in F$, then, aA is upper triangular. Further, (1) holds in S, as $S \subseteq F^{n\times n}$, so that S, in its own right, is an algebra over F. We see that S is a subalgebra of $F^{n\times n}$. Similarly, the set S' of all diagonal matrices in $F^{n\times n}$ is a subalgebra of $F^{n\times n}$.

Theorem 11. Let V_F and W_F be any two finite dimensional vector spaces with $\dim V = n \geq 1$, $\dim W = m \geq 1$. Then, $\mathrm{Hom}_F(V,\ W)$ is of dimension the product mn. Indeed given any basis $B = \{\alpha_1,\ \alpha_2,\ldots,\ \alpha_n\}$ of V and any basis $B' = \{\beta_1,\ \beta_2,\ldots,\ \beta_m\}$ of W, there exists a basis

$$B'' = \{E_{ij} : 1 \leq i \leq m,\ 1 \leq j \leq n\}$$

depending on B and B', of $\mathrm{Hom}_F(V,\ W)$, such that

$$\begin{aligned} E_{ij}(\alpha_k) &= \beta_i && \text{if } j = k \\ &= 0 && \text{if } j \neq k \end{aligned}$$

Proof. Given any n vectors $\gamma_1,\ \gamma_2,\ldots,\ \gamma_n$ (not necessarily distinct) in W, by Theorem (3.1.11), there exists a unique linear transformation $T : V_F \to W_F$ such that $T(\alpha_i) = \gamma_i$,

$1 \le i \le m$. Consider any i, j such that $1 \le i \le m$, $1 \le j \le n$. Depending upon i, j, we get $\gamma_1, \gamma_2, \ldots, \gamma_n$ in W such that $\gamma_j = \beta_i$ and $\gamma_k = 0$ for $k \ne j$. This choice gives a linear transformation T that we denote by E_{ij}.

Clearly, $E_{ij}(\alpha_j) = \gamma_j = \beta_i$

and $E_{ij}(\alpha_k) = \gamma_k = 0 \qquad k \ne j.$

As $i = 1, 2, \ldots, m$ $j = 1, 2, \ldots, n$, we get mn E_{ij}'s.

We claim that $B'' = \{E_{ij} : 1 \le i \le m, 1 \le j \le n\}$

is a basis of $\mathrm{Hom}_F(V, W)$.

This will prove that $\dim\{\mathrm{Hom}_F(V, W)\} = mn$.

Consider any $T \in \mathrm{Hom}_F(V, W)$.

For $1 \le j \le n$, as $T(\alpha_j) \in W$ and $\beta_1, \beta_2, \ldots, \beta_m$ constitute a basis of W, we get

$$T(\alpha_j) = \sum_{i=1}^{m} a_{ij}\beta_i, \; a_{ij} \in F \qquad \ldots (1)$$

Consider
$$T' = \sum_{i=1}^{m}\sum_{j=1}^{n} a_{ij}E_{ij} \qquad \ldots (2)$$

in $\mathrm{Hom}_F(V, W)$.

For $1 \le k \le n$
$$T'(\alpha_k) = \sum_{i=1}^{m}\sum_{j=1}^{n} a_{ij}E_{ij}(\alpha_k) = \sum_{i=1}^{m} a_{ik}\beta_i \qquad \ldots (3)$$

by using the fact that $E_{ij}(\alpha_k) = 0$ for $j \ne k$ and that $E_{ik}(\alpha_k) = \beta_i$.

Then, (1) and (3) give $T(\alpha_k) = T'(\alpha_k)$ for $1 \le i \le k$.

By Theorem (3.1.11), $T = T'$,

a linear combination of members of B'', so that B'' spans $\mathrm{Hom}_F(V, W)$.

It only remains to be shown that B'' is linearly independent.

Let for some $b_{ij} \in F$
$$\sum_{i=1}^{m}\sum_{j=1}^{n} b_{ij}E_{ij} = \overline{0}$$

Then, for $1 \le k \le n$
$$\sum_{i=1}^{m}\sum_{j=1}^{n} b_{ij}E_{ij}(\alpha_k) = \overline{0}(\alpha_k) = 0$$

i.e.,
$$\sum_{i=1}^{m} b_{ik}\beta_i = 0$$

The linear independence of $\beta_1, \beta_2, \ldots, \beta_m$ yield $b_{ik} = 0$ for $1 \le i \le m$, $1 \le k \le n$. Hence, B'' is linearly independent.

Example 11.a. Let $V = \mathbf{R}^3$, $W = \mathbf{R}^2$. Consider the standard basis $B = \{\epsilon_1, \epsilon_2, \epsilon_3\}$ and $B' = \{\eta_1, \eta_2\}$ of V and W, respectively.

Here, $\dim V = 3$, $\dim W = 2$,

gives dimension of Hom(V, W) is 6

We get the basis $B'' = \{E_{ij} : 1 \le i \le 2, 1 \le j \le 3\}$

of Hom(V, W).

Now, $E_{11}(\epsilon_1) = \eta_1$, $E_{11}(\epsilon_2) = 0$, $E_{11}(\epsilon_3) = 0$.

So, for any $\alpha = (a, b, c) \in V$, $\alpha = a\,\epsilon_1 + b\,\epsilon_2 + c\,\epsilon_3$,

gives $E_{11}(\alpha) = aE_{11}(\epsilon_1) + bE_{11}(\epsilon_2) + cE_{11}(\epsilon_3) = a\eta_1 = a(1, 0) = (a, 0)$

i.e., $E_{11}(a, b, c) = (a, 0)$

Again, $E_{12}(\alpha) = aE_{12}(\epsilon_1) + bE_{12}(\epsilon_2) + cE_{12}(\epsilon_3) = a0 + b\eta_1 + c.0 = (b, 0)$

Similarly, we get

$$E_{13}(\alpha) = (c, 0)$$
$$E_{21}(\alpha) = (0, a)$$
$$E_{22}(\alpha) = (0, b)$$
$$E_{23}(\alpha) = (0, c)$$

Theorem 12. Let U, V and W be any three finite dimensional vector spaces over a field F. Let $T_1, T_2 : U_F \to V_F$ and $T_3 : V_F \to W_F$ be three linear transformations. Then,

(*i*) $\rho(T_1 + T_2) \le \rho(T_1) + \rho(T_2)$

(*ii*) $\rho(T_3 T_1) \le \min\{\rho(T_1), \rho(T_3)\}$

Proof.

(*i*) For any $\alpha \in U$, $(T_1 + T_2)(\alpha) = T_1(\alpha) + T_2(\alpha) \in \text{range}(T_1) + \text{range}(T_2)$,

so that $\text{range}(T_1 + T_2) \subseteq \text{range}(T_1) + \text{range}(T_2)$.

This gives $\rho(T_1 + T_2) \le \dim\{\text{range}(T_1) + \text{range}(T_2)\}$

However, for any two subspaces V_1, V_2 of V,

$$\dim(V_1 + V_2) \le \dim V_1 + \dim V_2$$

so that $\rho(T_1 + T_2) \le \rho(T_1) + \rho(T_2)$.

Let $\rho(T_1) = k$. Consequently, range(T_1) is spanned by a subset B with k vectors, so that $T_3\{\text{range}(T_1)\}$ is spanned by $T(B)$, that has $\le k$ vectors.

Hence, $\dim[T_3\{\text{range}(T_1)\}] \le k = \rho(T_1)$

i.e., $\dim\{\text{range}(T_3 T_1)\} \le \rho(T_1)$

i.e., $\rho(T_3 T_1) \le \rho(T_1)$ ... (1)

Again, as $\quad T_3\{\text{range}(T_1)\} \subseteq \text{range}(T_3)$

we get $\quad \rho(T_3T_1) \le \rho(T_3)$ ... (2)

Then, (1) and (2) give $\quad \rho(T_3T_1) \le \min\{\rho(T_1),\ \rho(T_3')\}$

Remark 13.

(*i*) For any linear operator T on V_F, T^k, for any $k \ge 0$, is defined as follows :

$$T^0 = I, \qquad \text{the identity operator on } V$$

$$T^{k+1} = T^k T$$

It can be easily proved that $T^m T^n = T^{n+m}$ for any $n \ge 0$, $m \ge 0$

(*ii*) For any finite dimensional vector space V_F,

$$(\dim V_F)^2 = \dim_F L(V),$$

as $\quad L(V) = \text{Hom}_F(V, V),$

the result follows from Theorem 11.

(*iii*) For any vector space V_F, the set $GL(V)$ of all automorphism of V_F is a group, under the composition of mappings. To see this, notice that, for any $T, T' \in GL(V)$, $T'T \in GL(V)$, as seen in the proof of Theorem 7. The composition is associative, $I \in GL(V)$ and $T^{-1} \in GL(V)$. This group is called a *general linear group.*

(*iv*) Let T be a linear operator on a finite dimensional vector space V_F.

If $\quad \text{rank}(T) = \dim V,$

then, $\quad \text{nullity}(T) = 0,$

gives T is one-to-one and, hence, it is an automorphism of V_F {Theorem (3.1.10)}.

(*v*) Let $T \in L(V_F)$. For any polynomial $f(X) = a_0 + a_1X + \ldots + a_nX^n$ we define

$$f(T) = a_0I + a_1T + a_2T^2 + \ldots + a_nT^n \in L(V)$$

It can be seen that given any two polynomials

$$f(X),\ g(X) \in F(X)$$

if $\quad h(X) = f(X) + g(X)$

and $\quad p(X) = f(X)g(X),$

then $\quad h(T) = f(T) + g(T)$

and $\quad p(T) = f(T)g(T).$

Exercise 3.2

1. Let $T : \mathbf{R}^3 \to \mathbf{R}^3$ be given by

$$T(x, y, z) = (y, -x, 2z)$$

(*a*) Find the formulae for each of the following linear operators

$$T^3,\ T^2 + 2I,\ T^3 - 2T^2 + T - 2I$$

(*b*) Is any of the operators in (*a*) equal to zero operator?

(*c*) Verify that T is invertible. Find the formula for T^{-1}.

2. Let $T_1 : \mathbf{R}^4 \to P_2(\mathbf{R})$, $T_2 : P_2(\mathbf{R}) \to \mathbf{R}^4$ be such that

$$T_1(a, b, c, d) = a + (b+c)X + dX^2$$

$$T_2(a + bX + cX^2) = (a, 2b + a, -b - a, c)$$

for any $\alpha = (a, b, c, d) \in \mathbf{R}^4$, and $p(X) = a + bX + cX^2 \in P_2(\mathbf{R})$.

(*i*) Find $T_2T_1 : \mathbf{R}^4 \to \mathbf{R}^4$ and $T_1T_2 : P_2(\mathbf{R}) \to P_2(\mathbf{R})$

(*ii*) Is T_1T_2 the identity operator on $P_2(\mathbf{R})$?

(*iii*) Is $\text{range}(T_2) \cap \ker(T_1) = \{0\}$?

3. Let T_1 and T_2 be linear operators on $\mathbf{R}^2$ such that

$$T_1(x, y) = (0, x), \; T_2(x, y) = (y, 0)$$

Prove the following :

(*i*) $T_2^2 = T_1^2 = 0$

(*ii*) $T_1T_2 \neq 0$

(*iii*) $(T_1T_2)^2 = T_1T_2$

(*iv*) $T_1 + T_2$ is invertible

4. Let $D : P_3(\mathbf{R}) \to P_3(\mathbf{R})$ be the differentiation operator.

(*i*) Find a positive integer k such that $D^k = 0$

(*ii*) Prove that $I + D$ is invertible. Find the formula for $(I + D)^{-1}$.

5. Let T be the integration operator : $P_3(\mathbf{R}) \to P_4(\mathbf{R})$ given by

$$T(a_0 + a_1X + a_2X^2 + a_3X^3) = a_0X + \frac{a_1}{2}X^2 + \frac{a_2}{3}X^3 + \frac{a_3}{4}X^4$$

(*i*) Prove that T is a monomorphism, but T is not an isomorphism

(*ii*) If $D : P_4(\mathbf{R}) \to P_3(\mathbf{R})$ is the differentiation operator, prove that DT is the identity operator on $P_3(\mathbf{R})$.

6. Prove that $\mathbf{R}^{2\times 2} \cong \mathbf{R}^4$. Give two different isomorphisms of $\mathbf{R}^{2\times 2}$ onto $\mathbf{R}^4$.

7. Let T be a linear operator on a finite dimensional vector space V_F. Prove the following :

(*i*) $\ker(T^s) \subseteq \ker(T^{s+1})$ for any $s \geq 0$

(*ii*) $\text{range}(T^{s+1}) \subseteq \text{range}(T^s)$ for any $s \geq 0$

(*iii*) There exists a positive integer s such that $\ker(T^s) = \ker(T^{s+1})$.

(*iv*) If for a positive integer s,

$$\ker(T^s) = \ker(T^{s+1}),$$

then, $$\text{range}(T^s) \cap \ker(T^s) = \{0\}$$

8. Let $\dim V_F = n$ and $\dim W_F = m$. Let $T_1 : V_F \to W_F$ and $T_2 : W_F \to V_F$ be any two linear transformations.

   (*i*) If $n > m$, prove that T_2T_1 can never be an automorphism of V_F

   (*ii*) Prove that $$\rho(T_1) + \rho(T_2) - n \le \rho(T_1T_2)$$

9. Let T_1 and T_2 be two linear operators on a finite dimensional vector space V_F, such that one of them is an automorphism of V_F.

   Prove that $$\rho(T_1T_2) = \min\{\rho(T_1), \rho(T_2)\}$$

10. Give an example of two non-zero linear operators T_1, T_2 on a finite dimensional vector space V_F such that

$$\rho(T_1 + T_2) = \rho(T_1) + \rho(T_2)$$

11. A linear operator T on a vector space V_F is called an *idempotent operator* if $T^2 = T$

   (*i*) Prove that $T : \mathbf{R}^3 \to \mathbf{R}^3$ such that $T(x, y, z) = (x, y, 0)$ is an idempotent operator

   (*ii*) Let T and T' be two idempotent linear operators on a vector space V_F. Prove that if $T + T'$ is idempotent, then, $TT' = -T'T$

   (*iii*) If T is an idempotent linear operator on a vector space V_F, prove that

$$\ker(T) \cap \text{range}(T) = \{0\}$$

12. Let E_1 and E_2 be two idempotent linear operators on a vector space V_F such that $E_1E_2 = E_2E_1$. Prove that $E = E_1 + E_2 - E_1E_2$ is an idempotent linear operator. Further, prove that range(E) = range(E_1) + range(E_2).

13. Let T be a non-zero linear operator on $\mathbf{R}^2$ such that $T^2 = 0$. Prove that there exists an $\alpha \in \mathbf{R}^2$ such that $\{\alpha, T(\alpha)\}$ is a basis of $\mathbf{R}^2$.

14. Let $\dim V_F = n \ge 1$ and T be a linear operator on V_F such that $T^n = 0$, but for some $\alpha \in V$, $T^{n-1}\alpha \ne 0$. Prove that $\alpha, T(\alpha), \ldots, T^{n-1}(\alpha)$ form a basis of V. Any linear operator T' on a vector space V_F is called a *nilpotent operator* if $(T')^k = 0$ for some $k \ge 1$.

15. Let T be a nilpotent linear operator on a vector space V_F. Prove that $I + T$ is invertible.

   [**Hint :** If $$T^k = 0,$$
   then, $$I - T + T^2 - \ldots + (-1)^{k-1} T^{k-1}$$ is the inverse of $I + T$]

3.3 Matrix of a Linear Transformation

All the general vector spaces considered in this section are finite dimensional, and F is a field.

Definition 1. Let V_F and W_F be any two vector spaces, with $\dim V = n$ and $\dim W = m$, $B = \{\alpha_1, \alpha_2, ..., \alpha_n\}$ and $B' = \{\beta_1, \beta_2, ..., \beta_m\}$ be any two ordered bases of V and W, respectively. Let $T : V_F \to W_F$ be any linear transformation.

For $1 \le j \le n$, if $$T(\alpha_j) = \sum_{i=1}^{m} a_{ij}\beta_i, \qquad a_{ij} \in F$$

then, the $m \times n$-matrix $A = [a_{ij}]$ is called the *matrix of T relative to B and B'*.

We write $$[T]_{(B, B')} = [a_{ij}]$$

Observe that in the above definition, the jth column of A is the co-ordinate vector $[T(\alpha_j)]_{B'}$. If T is a linear operator on a vector space V_F, then, for any ordered basis B of V_F, the matrix $[T]_{(B,B)}$ is called the *matrix of T relative to B*, and is simply denoted by $[T]_B$.

Example 1.a. Let $T : \mathbf{R}^3 \to \mathbf{R}^2$ be the linear transformation given by

$$T(x, y, z) = (x - y, z)$$

$$B = \{(1, 1, 0), (0, 1, 1), (1, 0, 1)\}$$

and $$B' = \{(1, 2), (2, 1)\}$$

are ordered bases of $\mathbf{R}^3$ and $\mathbf{R}^2$, respectively.

Write $$\alpha_1 = (1, 1, 0), \alpha_2 = (0, 1, 1), \alpha_3 = (1, 0, 1),$$

$$\beta_1 = (1, 2), \beta_2 = (2, 1).$$

The jth column of $[T]_{(B, B')}$ is the co-ordinate vector $[T(\alpha_j)]_{B'}$.

Now, $$T(\alpha_1) = T(1, 1, 0) = (0, 0) = 0\beta_1 + 0\beta_2$$

Consequently, $$[T(\alpha_1)]_{B'} = \begin{bmatrix} 0 \\ 0 \end{bmatrix}$$

$$T(\alpha_2) = T(0, 1, 1) = (-1, 1) = \beta_1 - \beta_2$$

Thus, $$[T(\alpha_2)]_{B'} = \begin{bmatrix} 1 \\ -1 \end{bmatrix}.$$

Similarly $$[T(\alpha_3)]_{B'} = \begin{bmatrix} \frac{1}{3} \\ \frac{1}{3} \end{bmatrix}.$$

Hence, $$[T]_{(B, B')} = \begin{bmatrix} 0 & 1 & \frac{1}{3} \\ 0 & -1 & \frac{1}{3} \end{bmatrix}.$$

Remark 2. Given any linear transformation $T : F^n \to F^m$ and any ordered bases B, B' of F^n and F^m, respectively, we describe a practical way of finding $[T]_{(B,B')}$. This method can be easily justified by the results on solutions of systems of linear equations.

Let
$$B = \{\alpha_1, \alpha_2, ..., \alpha_n\},$$
$$B' = \{\beta_1, \beta_2, ..., \beta_m\}.$$

Form the $m \times m$-matrix C, whose j^{th} column is the vector β_j, written in column form, i.e., the co-ordinate vector of β_j relative to the standard ordered basis of F^m. Let D be the $m \times n$-matrix whose j^{th} column is the vector $T(\alpha_j)$ in F^m written in column form. Form the $m \times (m+n)$-matrix $[C : D]$. As C is non-singular, we get a sequence of elementary row operations, $\theta_1, \theta_2, ..., \theta_k$ which, when applied to C, gives I_m. Apply this sequence of elementary row operations to $[C : D]$.

We get a matrix
$$[I_m : A].$$

Then,
$$A = [T]_{(B,B')}.$$

Example 2.a. Let $T : \mathbf{R}^{3\times 1} \to \mathbf{R}^{3\times 1}$ be the linear operator given by

$$T\begin{bmatrix} a \\ b \\ c \end{bmatrix} = \begin{bmatrix} 1 & 2 & 1 \\ 2 & -1 & 0 \\ 0 & 2 & 1 \end{bmatrix}\begin{bmatrix} a \\ b \\ c \end{bmatrix}$$

Consider the ordered basis
$$B = \left\{\begin{bmatrix} 1 \\ 2 \\ 0 \end{bmatrix}, \begin{bmatrix} 0 \\ 1 \\ -1 \end{bmatrix}, \begin{bmatrix} 0 \\ 0 \\ 1 \end{bmatrix}\right\}$$

Here, matrix
$$C = \begin{bmatrix} 1 & 0 & 0 \\ 2 & 1 & 0 \\ 0 & -1 & 1 \end{bmatrix}$$

$$T\begin{bmatrix} 1 \\ 2 \\ 0 \end{bmatrix} = \begin{bmatrix} 1 & 2 & 1 \\ 2 & -1 & 0 \\ 0 & 2 & 1 \end{bmatrix}\begin{bmatrix} 1 \\ 2 \\ 0 \end{bmatrix} = \begin{bmatrix} 5 \\ 0 \\ 4 \end{bmatrix}$$

$$T\begin{bmatrix} 0 \\ 1 \\ -1 \end{bmatrix} = \begin{bmatrix} 1 \\ -1 \\ 1 \end{bmatrix},$$

$$T\begin{bmatrix} 0 \\ 0 \\ 1 \end{bmatrix} = \begin{bmatrix} 1 \\ 0 \\ 1 \end{bmatrix}$$

so that
$$D=\begin{bmatrix}5 & 1 & 1\\ 0 & -1 & 0\\ 4 & 1 & 1\end{bmatrix}.$$

Then,
$$[C:D]=\begin{bmatrix}1 & 0 & 0 & : & 5 & 1 & 1\\ 2 & 1 & 0 & : & 0 & -1 & 0\\ 0 & -1 & 1 & : & 4 & 1 & 1\end{bmatrix}$$

Apply to this 3×6-matrix elementary row operations, to bring it in reduced row echelon form.

This gives
$$\begin{bmatrix}1 & 0 & 0 & : & 5 & 1 & 1\\ 0 & 1 & 0 & : & -10 & -3 & -2\\ 0 & 0 & 1 & : & -6 & -2 & -1\end{bmatrix}$$

so that
$$[T]_B=\begin{bmatrix}5 & 1 & 1\\ -10 & -3 & -2\\ -6 & -2 & -1\end{bmatrix}$$

Theorem 3. Let $\dim V_F = n$, $\dim W_F = m$. Let B and B' any ordered basis of V and W, respectively

Then, the mapping
$$\lambda : \text{Hom}_F(V, W) \to F^{m\times n}$$
such that
$$\lambda(T)=[T]_{(B,\,B')},\ T\in \text{Hom}_F(V, W),$$
is a vector space isomorphism.

Proof. Let
$$B=\{\alpha_1, \alpha_2,\ldots, \alpha_n\}$$
and
$$B'=\{\beta_1, \beta_2,\ldots, \beta_m\},$$
T_1 and T_2 be in $\text{Hom}_F(V, W)$ and $a\in F$.

Further let
$$[T_1]_{(B,\,B')}=A=[a_{ij}]$$
$$[T_2]_{(B,\,B')}=B=[b_{ij}]$$

By definition,
$$T_1(\alpha_j)=\sum_{i=1}^{m} a_{ij}\beta_i$$
$$T_2(\alpha_j)=\sum_{i=1}^{m} b_{ij}\beta_i$$

Consequently, $(T_1+T_2)(\alpha_j)=T_1(\alpha_j)+T_2(\alpha_j)=\sum_{i=1}^{m}(a_{ij}+b_{ij})\beta_i$

$$(aT_1)(\alpha_j)=a\{T_1(\alpha_j)\}=\sum_{i=1}^{m} aa_{ij}\beta_i$$

so that
$$[T_1+T_2]_{(B,\,B')}=[a_{ij}+b_{ij}]=A+B$$
$$[aT_1]_{(B,\,B')}=[aa_{ij}]=aA$$

Hence,
$$\lambda(T_1+T_2)=\lambda(T_1)+\lambda(T_2)$$
$$\lambda(aT_1)=a\lambda(T_1)$$

Thus, λ is a linear transformation.

If
$$\lambda(T_1)=0,$$
then
$$T_1(\alpha_j)=0 \qquad 1\le j\le n.$$

Consequently, T_1 being zero on the basis B, gives $T_1=0$. This proves that $\ker(\lambda)=\{0\}$. Hence, λ is one-to-one. Consider any $A'=[a'_{ij}]\in F^{m\times n}$

We get
$$\alpha'_j=\sum_{i=1}^{m}a'_{ij}\beta_i\in W, \qquad 1\le j\le n$$

By Theorem (3.1.11), there exists a unique linear transformation $T':V\to W$ such that
$$T'(\alpha_j)=\alpha'_j=\sum_{i=1}^{m}a'_{ij}\beta_i$$
so that
$$\lambda(T')=[T']_{(B,\,B')}=[a'_{ij}]=A'.$$

Hence, λ is onto. This all proves that λ is a vector space isomorphism.

As dimension of $\text{Hom}_F(V,W)=mn$, the theorem, given above once again shows that the dimension of $F^{m\times n}$ is mn.

Theorem 4. Let B_1, B_2, B_3 be any ordered bases of U_F, V_F and W_F respectively. Then, for any linear transformations
$$T:U_F\to V_F,$$
$$T':V_F\to W_F$$
$$[T'T]_{(B_1,\,B_3)}=[T']_{(B_2,\,B_3)}.[T]_{(B_1,\,B_2)}$$

Proof. Let
$$B_1=\{\alpha_1,\alpha_2,\ldots,\alpha_n\},$$
$$B_2=\{\beta_1,\beta_2,\ldots,\beta_p\},$$
$$B_3=\{\gamma_1,\gamma_2,\ldots,\gamma_m\}.$$

Further, let
$$A=[a_{ij}]_{p\times n}=[T]_{(B_1,\,B_2)}$$
and
$$B=[b_{ij}]_{m\times p}=[T']_{(B_2,\,B_3)}.$$

Then,
$$T(\alpha_j) = \sum_{k=1}^{p} a_{kj}\beta_k$$

$$T'(\beta_k) = \sum_{i=1}^{m} b_{ik}\gamma_i$$

Consequently,
$$(T'T)(\alpha_j) = T'\{T(\alpha_j)\}$$

$$= T'\left(\sum_{k=1}^{p} a_{kj}\beta_k\right)$$

$$= \sum_{k=1}^{p} a_{kj}T'(\beta_k)$$

$$= \sum_{k=1}^{p} a_{kj}\left[\sum_{i=1}^{m} b_{ik}\gamma_i\right]$$

$$= \sum_{i=1}^{m} c_{ij}\gamma_i,$$

$$c_{ij} = \sum_{k=1}^{p} b_{ik}a_{kj}$$

so that
$$[T'T]_{(B_1, B_3)} = [c_{ij}]_{m\times n} = BA$$

i.e.,
$$[T'T]_{(B_1, B_3)} = [T']_{(B_2, B_3)}[T]_{(B_1, B_2)}.$$

Consider an ordered basis B of V_F. By taking $V_F = W_F$, $B = B'$ in Theorem 3 and by taking $U_F = V_F$, $B_1 = B_2 = B$ in Theorem 4, we get the following result :

Theorem 5. Let $\dim V_F = n$ and B be an ordered basis of V.

Then, the mapping $\lambda : \text{Hom}_F(V; V) \to F^{n\times n}$

such that $\lambda(T) = [T]_B,\ T \in \text{Hom}_F(V, V)$

is one-to-one and onto, and is such that for any $T_1, T_2 \in \text{Hom}_F(V, V)$, $a \in F$,

(i) $\lambda(T_1 + T_2) = \lambda(T_1) + \lambda(T_2)$

(ii) $\lambda(T_1T_2) = \lambda(T_1).\lambda(T_2)$

(iii) $\lambda(aT_1) = a\lambda(T_1)$

Given two algebras R and R' over a field F, a one-to-one mapping σ of R onto R' is called an *algebra isomorphism*, if for any $x, y \in R$, $a \in F$,

$$\sigma(x + y) = \sigma(x) + \sigma(y),$$

$$\sigma(xy) = \sigma(x)\sigma(y),$$

$$\sigma(ax) = a\sigma(x),$$

i.e., if σ is a ring isomorphism as well as a vector space isomorphism.

Further R and R' are said to be isomorphic if there exists an algebra isomorphism $\sigma : R \to R'$. Thus, Theorem 5, shows that algebra $\text{Hom}_F(V, V)$ is isomorphic to $F^{n\times n}$ and that λ is an algebra isomorphism. This gives $[I]_B$ and $[0]_B$ are I_n and 0 respectively.

We now study the effect of change of bases on a matrix of a linear transformation. For any function $T : V_F \to W_F$, we shall write $T\alpha$ for $T(\alpha)$.

Theorem 6. Let $\dim V_F = n$, $\dim W_F = m$. Let B and B' be two ordered bases of V_F, and B_1, B_1' be two ordered bases of W_F.

Then, for any linear transformation $T : V_F \to W_F$,

$$[T]_{(B',\,B_1')} = S^{-1}[T]_{(B,\,B_1)}P$$

where P is the matrix of B' relative to B, and S is the matrix of B_1' relative to B_1.

Proof. Let $B = \{\alpha_1, \alpha_2, ..., \alpha_n\}$

$$B' = \{\alpha_1', \alpha_2', ..., \alpha_n'\}$$

$$B_1 = \{\beta_1, \beta_2, ..., \beta_m\}$$

and $B_1' = \{\beta_1', \beta_2', ..., \beta_m'\}$.

Let $P = [p_{ij}],\ S = [s_{ij}]$.

By definition, $$\alpha_j' = \sum_{k=1}^{n} p_{kj}\alpha_k,\ 1 \le j \le n \qquad \text{... (1)}$$

$$\beta_u' = \sum_{k=1}^{m} s_{ku}\beta_k,\ 1 \le u \le m \qquad \text{... (2)}$$

Further, let $$[T]_{(B,\,B_1)} = A = [a_{ij}] \qquad \text{... (3)}$$

$$[T]_{(B',\,B_1')} = D = [b_{ij}] \qquad \text{... (4)}$$

(3) gives $$T\alpha_j = \sum_{i=1}^{m} a_{ij}\beta_i$$

Consequently,

$$T\alpha_j' = \sum_{k=1}^{n} p_{kj}T\alpha_k = \sum_{k=1}^{n} p_{kj}\left[\sum_{i=1}^{m} a_{ik}\beta_i\right] = \sum_{i=1}^{m}\left[\sum_{k=1}^{n} a_{ik}p_{kj}\right]\beta_i \qquad \text{... (5)}$$

Let $$S^{-1} = \left[s'_{ij}\right].$$

Then S^{-1} is the matrix of B_1 relative to B'_1.

so that $$\beta_i = \left[\sum_{l=1}^{m} s'_{li}\beta'_l\right] \qquad \text{... (6)}$$

(5) and (6) yield

$$T\alpha'_j = \sum_{i=1}^{m}\left[\sum_{k=1}^{n} a_{ik}p_{kj}\right]\left[\sum_{l=1}^{m} s'_{li}\beta'_l\right] = \sum_{l=1}^{m}\left[\sum_{k=1}^{n}\sum_{i=1}^{m} s'_{li}a_{ik}p_{kj}\right]\beta'_l \qquad \text{... (7)}$$

By (4), $$T\alpha'_j = \sum_{l=1}^{m} b_{lj}\beta'_l \qquad \text{... (8)}$$

By comparing (7) and (8), we get $b_{lj} = \sum_{k=1}^{n}\sum_{i=1}^{m} s'_{li}a_{ik}p_{kj}$.

so that $$D = S^{-1}AP$$

i.e., $$[T]_{(B',\, B'_1)} = S^{-1}[T]_{(B,\, B_1)}P$$

In the above theorem, take $V_F = W_F$, $B = B_1$, $B' = B'_1$.

Then, $$[T]_{(B,\, B_1)} = [T]_B,\ [T]_{(B',\, B'_1)} = [T]_{B'},\ S = P.$$

So, we get the following :

Corollary 7. For any finite dimensional non-zero vector space V_F, given any linear operator T on V_F, and any two ordered bases B and B' of V_F, $[T]_{B'} = P^{-1}[T]_B P$ where P is the matrix of B' relative to B.

The two results, given above, lead to the following concepts for matrices :

Definition 8.

(*i*) Let A and B be any two $m \times n$-matrices over a field F. Then, A is said to be *equivalent* to B over F if $B = QAP$ for some non-singular $m \times m$-matrix Q over F and some non-singular $n \times n$-matrix P over F.

(*ii*) Let A and B be two $n \times n$-matrices over a field F. Then, A is said to be *similar to* B over F if there exists a non-singular matrix P over F such that $B = P^{-1}AP$

The relation of equivalence, among matrices in an $F^{m\times n}$, is an equivalence relation, and so is the relation of similarity among matrices of an $F^{n\times n}$. It is clear from the definition, given above, that two similar matrices are equivalent.

Example 8.a. Let $$A = \begin{bmatrix} 0 & 1 \\ 1 & 0 \end{bmatrix}$$

and $$P=\begin{bmatrix}1 & -1\\ 1 & 1\end{bmatrix}$$

over R.

Then, P is non-singular and $P^{-1}AP=\begin{bmatrix}1 & 0\\ 0 & -1\end{bmatrix}$.

This shows that A is similar to the diagonal matrix

$$\begin{bmatrix}1 & 0\\ 0 & -1\end{bmatrix}.$$

Suppose $\begin{bmatrix}0 & 1\\ 1 & 0\end{bmatrix}$ is similar to $I_2=\begin{bmatrix}1 & 0\\ 0 & 1\end{bmatrix}$

Then, for some non-singular 2×2-matrix S over R.

$$\begin{bmatrix}0 & 1\\ 1 & 0\end{bmatrix}=S^{-1}I_2S=I_2.$$

This is a contradiction.

Hence, $\begin{bmatrix}0 & 1\\ 1 & 0\end{bmatrix}$ is not similar to $\begin{bmatrix}1 & 0\\ 0 & 1\end{bmatrix}$.

However, $$I_2=Q\begin{bmatrix}0 & 1\\ 1 & 0\end{bmatrix}P,$$

where $$Q=\begin{bmatrix}0 & 1\\ 1 & 0\end{bmatrix},\ P=I_2,$$

so that I_2 and $\begin{bmatrix}0 & 1\\ 1 & 0\end{bmatrix}$ are equivalent.

This shows that two equivalent matrices need not be similar.

Let $T: V_F \to W_F$ be a linear transformation. Theorem 6 shows that matrices of T relative to two different sets of bases are equivalent. The following result proves the converse :

Theorem 9.

(i) Let $T: V_F \to W_F$ be a linear transformation. Let B and B_1 be ordered bases of V and W, respectively, and let $A=[T]_{(B,\,B_1)}$. Let A be equivalent to D. Then, there exist ordered bases B' and B_1' of V and W, respectively, such that $D=[T]_{(B',\,B_1')}$.

(ii) Let T be a linear operator on V_F. For some ordered bases B of V_F, let $A=[T]_B$. Let A be similar to D. Then, there exists an ordered basis B' of V such that $[T]_{B'}=D$.

Proof.

(*i*) Let $\dim V_F = n$, $\dim W_F = m$. Then, A is an $m \times n$-matrix and $D = QAP$ for some non-singular $n \times n$-matrix *P*, and a non-singular $m \times m$-matrix *Q*. Let $S = Q^{-1}$. Then, $D = S^{-1}AP$. By Theorem (2.4.7), there exists an ordered bases B' of *V* such that the matrix of B' relative to *B* is *P*, and there exists an ordered basis B_1' of *W* such that *S* is the matrix of B_1' relative to B_1.

Then, by Theorem 6 $\quad [T]_{(B', B_1')} = S^{-1}AP = D.$

This proves (*i*). Finally, (*ii*) is a special case of (*i*).

Theorem 10. Let *A* be any $m \times n$-matrix over *F*, *B* and B' be the standard bases of $F^{n\times 1}$ and $F^{m\times 1}$, respectively.

Then, for
$$T : F^{n\times 1} \to F^{m\times 1}$$
given by
$$T(X) = AX,$$
one has
$$[T]_{(B, B')} = A.$$
For $m = n$,
$$A = [T]_B.$$

Proof. Let $B = \{\epsilon_1, \epsilon_2, \ldots, \epsilon_n\}, B' = \{\epsilon_1', \epsilon_2', \ldots, \epsilon_m'\}$. Then, for any $m \times n$-matrix *A* over *F*, $A \epsilon_j$ is the *j*th column of *A*.

So, if $A = [a_{ij}]$,
$$T \epsilon_j = A \epsilon_j = \begin{bmatrix} a_{1j} \\ a_{2j} \\ \vdots \\ \vdots \\ a_{mj} \end{bmatrix} = \sum_{i=1}^{m} a_{ij} \epsilon_i'.$$

This gives
$$[T \epsilon_j]_{B'}$$
is precisely the *j*th column of *A*.

Hence,
$$A = [T]_{(B, B')}.$$
For $m = n$,
$$B' = B.$$
So
$$A = [T]_B.$$

The next theorem gives an extremely useful relationship between a linear transformation and its matrix.

Theorem 11. Let $T : V_F \to W_F$ be a linear transformation and *B*, B' be ordered bases of *V* and *W* respectively, and $A = [T]_{(B, B')}$. Then,

(*i*) For any $\alpha \in V$, $\quad [T\alpha]_{B'} = A[\alpha]_B$

(*ii*) $\quad \operatorname{rank}(T) = \operatorname{rank}(A)$

Proof. Let $B=\{\alpha_1, \alpha_2,\ldots, \alpha_n\}$, $B'=\{\beta_1, \beta_2,\ldots, \beta_m\}$

and $A=[T]_{(B,\,B')}=[a_{ij}]$.

By definition, $T\alpha_j=\sum_{i=1}^{m} a_{ij}\beta_i$.

Consider $\alpha\in V$. Let $[\alpha]_B=\begin{bmatrix} u_1\\ u_2\\ \vdots\\ \vdots\\ u_n\end{bmatrix}$.

Then, $\alpha=\sum_{j=1}^{\eta} u_j\alpha_j$

gives $T\alpha=\sum_{j=1}^{n} u_jT\alpha_j=\sum_{j=1}^{n} u_j\left[\sum_{i=1}^{m} a_{ij}\beta_i\right]$

i.e., $T\alpha=\sum_{i=1}^{m}\left[\sum_{j=1}^{n} a_{ij}u_j\right]\beta_i$

Thus, in $[T\alpha]_{B'}$, the ith component $\sum_{j=1}^{m} a_{ij}u_j$ is the ith component in the product $A[\alpha]_B$.

Consequently $[T\alpha]_{B'}=A[\alpha]_B$

This proves (*i*).

Now, $\sigma: V\to F^{n\times 1}$

such that $\sigma(\alpha)=[\alpha]_B,\ \alpha\in V$,

is an isomorphism {Theorem(2.4.5)}.

Let W' be the solution space of the system of linear equation $AX=0$.

Then, $\dim W'=n-\text{rank}(A)$ ... (1)

Now, $\alpha\in\ker(T)$

iff $T\alpha=0$

iff $[T\alpha]_{B'}=0$

iff $A[\alpha]_B=0$

iff $[\alpha]_B\in W'$.

Consequently, as $\sigma(\alpha)=[\alpha]_B$,

we get $\alpha \in \ker(T)$

iff $\sigma(\alpha) \in W'$,

so that $\ker(T) = \sigma^{-1}(W')$;

$$\text{nullity}(T) = \dim\{\sigma^{-1}(W')\}$$

As σ^{-1} is an isomorphism, $\dim(W') = \dim\{\sigma^{-1}(W')\}$.

Hence, $\text{nullity}(T) = \dim(W') = n - \text{rank}(A)$.

By the Sylvester law, {Theorem (3.1.9)},

$$\text{nullity}(T) = \dim V - \text{rank}(T) = n - \text{rank}(T).$$

Hence, $n - \text{rank}(T) = n - \text{rank}(A)$,

$$\text{rank}(T) = \text{rank}(A).$$

Corollary 12. Let T be a linear operator on V_F and let B be an ordered basis of V_F. Then, T is an automorphism of V_F iff $A = [T]_B$ is non-singular.

Proof. Now, T is an automorphism iff $\rho(T) = \dim V = n$ iff $\text{rank}(A) = n$ iff A is non-singular {Theorem (2.3.8)}. Hence, T is an automorphism iff A is non-singular.

Theorem 13. Let A be any $m \times n$-matrix over a field F. Then, $\text{rank}(A) = k$ iff A is equivalent to the matrix $\begin{bmatrix} I_k & 0 \\ 0 & 0 \end{bmatrix}_{m \times n}$

Proof. Let $T : F^{n\times 1} \to F^{m\times 1}$ be given by $T(X) = AX$, $X \in F^{n\times 1}$. Let B and B' be the standard bases of $F^{n\times 1}$ and $F^{m\times 1}$, respectively.

Then, $A = [T]_{(B,\, B')}$ by Theorem 10.

By Theorem 11, $\text{rank}(T) = \text{rank}(A)$.

Let $\text{rank}(A) = k$,

so that $\text{rank}(T) = k$

and $F^{n\times 1} = W \oplus W'$

where $W' = \ker(T)$,

so that $\dim W' = n - \text{rank}(T) = n - k$,

and $\dim W = k$.

We get an ordered basis $B_1 = \{\alpha_1, \alpha_2, \ldots, \alpha_n\}$

of $F^{n\times 1}$ such that $B_2 = \{\alpha_1, \alpha_2, \ldots, \alpha_k\}$

is a basis of W and $B_3 = \{\alpha_{k+1}, \alpha_{k+2}, \ldots, \alpha_n\}$

is a basis of W'.

Then, range(T) = $T(W)$ has basis

$$T(B_2) = \{T(\alpha_1), T(\alpha_2), \ldots, T(\alpha_k)\}$$

{see the proof of Theorem (3.1.9)}:

As $\dim F^{m\times 1} = m$,

we get an ordered basis $B_1' = \{\beta_1, \beta_2, \ldots, \beta_m\}$

of $F^{m\times 1}$ such that $\beta_j = T\alpha_j$ for $1 \le j \le k$

Thus, for $1 \le j \le k$ $\quad T\alpha_j = \beta_j = \sum_{i=1}^{m} \delta_{ij}\beta_i,$

gives

$$[T\alpha_j]_{B_1'} = \begin{bmatrix} 0 \\ 0 \\ \vdots \\ 0 \\ 1 \\ 0 \\ \vdots \\ 0 \end{bmatrix} \; j^{\text{th}}$$

For $k+1 \le j \le n$, $\quad T\alpha_j = 0,$

gives $\quad [T\alpha_j]_{B'} = 0$

Hence,

$$B = [T]_{(B_1, B_1')} = \begin{bmatrix} 1 & 0\ldots0 & 0\ldots0 \\ 0 & 1\ldots0 & 0\ldots0 \\ 0 & 0\ldots0 & 0\ldots0 \\ \vdots & \vdots & 0\ldots0 \\ \vdots & \vdots & \vdots \\ 0 & 0\ldots1 & 0\ldots0 \\ 0 & 0\ldots0 & 0\ldots0 \\ \vdots & \vdots & \vdots \\ 0 & 0\ldots0 & 0\ldots0 \end{bmatrix} = \begin{bmatrix} I_k & 0 \\ 0 & 0 \end{bmatrix}_{m\times n}$$

By Theorem 6, A is equivalent to B. Conversely, let A be equivalent to $B = \begin{bmatrix} I_k & 0 \\ 0 & 0 \end{bmatrix}$. This B is in reduced row-echelon form and it has k non-zero rows. So rank(B) = k. Now, $A = QBP$ for some non-singular matrices Q and P.

By Theorem (2.3.13), rank(B) = rank(QB) = rank(QBP).

So that $\quad$ rank(A) = rank(B) = k.

Theorem 14. Two $m\times n$-matrices A and B over F are equivalent iff rank(A) = rank(B).

Proof. Let A and B be equivalent, then, $A = QBP$ for some non-singular matrices P and Q

So by Theorem (2.3.13), $\operatorname{rank}(A) = \operatorname{rank}(B)$.

Conversely, let $\operatorname{rank}(A) = \operatorname{rank}(B) = k$.

Then, by Theorem 13, A and B both are equivalent to the same matrix $\begin{bmatrix} I_k & 0 \\ 0 & 0 \end{bmatrix}_{m\times n}$

Hence, A and B are equivalent.

Lemma 15. Let T be an idempotent linear operator on any vector space V_F.

Then, $V = \operatorname{range}(T) \oplus \ker(T)$.

For any $\alpha \in \operatorname{range}(T)$, $T\alpha = \alpha$.

Proof. By definition $T^2 = T$.

So, for any $\alpha \in V$, $T^2\alpha = T\alpha$,

gives $T(\alpha - T\alpha) = 0$.

Consequently, $\alpha - T\alpha \in \ker(T)$.

Also, $T\alpha \in \operatorname{range}(T)$.

Hence, $\alpha = T\alpha + (\alpha - T\alpha) \in \operatorname{range}(T) + \ker(T)$.

Hence, $V = \operatorname{range}(T) + \ker(T)$.

It only remains to prove that $\operatorname{range}(T) \cap \ker(T) = \{0\}$.

Let $\alpha \in \operatorname{range}(T) \cap \ker(T)$.

Then, $T\alpha = 0$

and $\alpha = T\beta$,

for some $\beta \in V$.

This yields $0 = T\alpha = TT\beta = T\beta = \alpha$

i.e., $\alpha = 0$.

Hence, $\operatorname{range}(T) \cap \ker(T) = \{0\}$

and $V = \operatorname{range}(T) \oplus \ker(T)$

The last part is obvious.

Theorem 16. Let T be an idempotent linear operator on a finite dimensional vector space V_F such that $\operatorname{rank}(T) = k \geq 1$. Then, there exists an ordered basis B of V such that

$$[T]_B = \begin{bmatrix} I_k & 0 \\ 0 & 0 \end{bmatrix}_{n\times n}$$

where $n = \dim V$.

Proof. Let $W = \text{range}(T),\ W' = \ker(T)$.

By the lemma, given above, $V = W \oplus W'$

Further, $\dim W = k,$

and $\dim W' = n - k.$

So, we get an ordered basis

$$B = \{\alpha_1, \alpha_2, \ldots, \alpha_k, \alpha_{k+1}, \ldots, \alpha_n\}$$

such that $B_1 = \{\alpha_1, \alpha_2, \ldots, \alpha_k\}$

and $B_2 = \{\alpha_{k+1}, \alpha_{k+2}, \ldots, \alpha_n\}$

are ordered bases of W and W', respectively.

For $1 \le j \le k$ $$T\alpha_j = \alpha_j = \sum_{i=1}^{n} \delta_{ij}\alpha_i$$

so that the jth column of $[T]_B$ is

$$\begin{bmatrix} \delta_{1j} \\ \delta_{2j} \\ \vdots \\ \vdots \\ \delta_{nj} \end{bmatrix} = \begin{bmatrix} 0 \\ 0 \\ \vdots \\ 1 \\ 0 \\ \vdots \\ 0 \\ \vdots \\ 0 \end{bmatrix} \begin{matrix} \\ \\ \\ \leftarrow j^{\text{th}} \\ \\ \\ \\ \\ \\ \end{matrix}$$

For $k + 1 \le j \le n,$ $T\alpha_j = 0,$

gives that the jth column of $[T]_B$ is zero.

Hence, $$[T]_B = \begin{bmatrix} I_k & 0 \\ 0 & 0 \end{bmatrix}_{n\times n}$$

A square matrix is called an *idempotent matrix,* if $A^2 = A$.

Theorem 17. Let A be any non-zero $n\times n$-matrix over a field F. Then, A is idempotent iff A is similar to $\begin{bmatrix} I_k & 0 \\ 0 & 0 \end{bmatrix}_{n\times n}$ where $k = \text{rank}(A)$.

Proof. Let A be idempotent, i.e., $A^2 = A$.

Then, the linear operator $T : F^{n\times 1} \to F^{n\times 1}$

such that $T(X) = AX,\ X \in F^{n\times 1}$

is idempotent.

Now, $\text{rank}(T) = \text{rank}(A) = k.$

By Theorem 16, there exists an ordered basis B' of $F^{n\times 1}$ such that

$$[T]_{B'} = \begin{bmatrix} I_k & 0 \\ 0 & 0 \end{bmatrix}_{n\times n}$$

By Theorem 10, $A = [T]_B$

where B is the standard ordered basis of $F^{n\times 1}$.

Hence, by Corollary 7, A and $\begin{bmatrix} I_k & 0 \\ 0 & 0 \end{bmatrix}$ are similar. Conversely, let A be similar to $B = \begin{bmatrix} I_k & 0 \\ 0 & 0 \end{bmatrix}$.

So $\text{rank}(A) = \text{rank}(B) = k$.

Observe that $B^2 = B$.

Now, $A = P^{-1}BP$

for some non-singular matrix P.

Then, $A^2 = P^{-1}BP\,P^{-1}BP = P^{-1}B^2P = P^{-1}BP = A$

Hence, A is idempotent.

Theorem 18. Let $\dim V_F = k \geq 1$ and T be a linear operator on V such that $T^k = 0$ but $T^{k-1} \neq 0$. Then, there exists an ordered basis B of V such that

$$[T]_B = \begin{bmatrix} 0 & 0 & 0 & \ldots & 0 & 0 \\ 1 & 0 & 0 & \ldots & 0 & 0 \\ 0 & 1 & 0 & \ldots & 0 & 0 \\ \vdots & \vdots & \vdots & \ldots & \vdots & \vdots \\ \vdots & \vdots & \vdots & \ldots & \vdots & \vdots \\ 0 & 0 & 0 & \ldots & 1 & 0 \end{bmatrix}$$

Proof. As $T^{k-1} \neq 0$, we can find an $\alpha \in V$ such that $T^{k-1}\alpha \neq 0$. Consider the k vectors $\alpha, T\alpha, T^2\alpha, \ldots, T^{k-1}\alpha$. Let for some $a_0, a_1, a_2, \ldots, a_{k-1} \in F$.

$$a_0\alpha + a_1T\alpha + a_2T^2\alpha + \ldots + a_{k-1}T^{k-1}\alpha = 0 \qquad \ldots (1)$$

We prove that every $a_i = 0$.

As $T^k = 0,\ T^j = 0$ for every $j \geq k$.

Apply T^{k-1} on (1), we get $a_0T^{k-1}\alpha + a_1T^k\alpha + \ldots + a_{k-1}T^{2k-1}\alpha = 0$

i.e., $a_0T^{k-1}\alpha = 0$

as all the other terms are zeros.

However, $T^{k-1}\alpha \neq 0$,

so $a_0 = 0$.

To apply induction, let for some i, $0 \le i \le k-1$, we have proved $a_0 = a_1 = \ldots = a_i = 0$. Then, (1) becomes

$$a_{i+1}T^{i+1}\alpha + a_{i+2}T\alpha^{i+2} + \ldots + a_{k-1}T^{k-1}\alpha = 0 \qquad \ldots (2)$$

Apply T^{k-i-2} on this equation, we get

$$a_{i+1}T^{k-1}\alpha + a_{i+2}T^{k} + \ldots + a_{k-1}T^{2k-i-3} = 0$$

i.e., $$a_{i+1}T^{k-1}\alpha = 0$$

This gives $$a_{i+1} = 0.$$

So that $$\alpha,\ T\alpha, \ldots,\ T^{k-1}\alpha$$

are linearly independent.

As $\dim V = k$, we get $$B = \{\alpha,\ T\alpha,\ T^2\alpha, \ldots,\ T^{k-1}\alpha\}$$

is an ordered basis of V.

Write $$\alpha_i = T^{i-1}\alpha \qquad \text{for } 1 \le i \le k.$$

Then, $$B = \{\alpha_1,\ \alpha_2, \ldots, \alpha_k\}.$$

For $1 \le i \le k-1$, $$T\alpha_i = \alpha_{i+1} = 0\alpha_1 + 0\alpha_2 + \ldots + 0\alpha_i + 1\alpha_{i+1} + 0\alpha_{i+1} + \ldots$$

gives that in $[T]_B$, ith column has $(i+1)$th entry 1, and all the other entries zeros.

Further, $$T\alpha_k = TT^{k-1}\alpha = T^k\alpha = 0$$

gives that the kth column of $[T]_B$ is zero.

Hence, $$[T]_B = \begin{bmatrix} 0 & 0 & 0 & \ldots & 0 & 0 \\ 1 & 0 & 0 & \ldots & 0 & 0 \\ 0 & 1 & 0 & \ldots & 0 & 0 \\ 0 & 0 & 1 & \ldots & 0 & 0 \\ \vdots & \vdots & \vdots & \ldots & \vdots & \vdots \\ 0 & 0 & 0 & \ldots & 1 & 0 \end{bmatrix}_{k \times k}$$

By using the theorem, given above, and arguments similar to those in Theorem 17, we get the following :

Theorem 19. Let A be any $k \times k$-matrix over a field F such that $A^k = 0$, but $A^{k-1} \ne 0$. Then, A is similar to the matrix

$$\begin{bmatrix} 0 & 0 & 0 & \ldots & 0 & 0 \\ 1 & 0 & 0 & \ldots & 0 & 0 \\ 0 & 1 & 0 & \ldots & 0 & 0 \\ 0 & 0 & 1 & \ldots & 0 & 0 \\ \vdots & \vdots & \vdots & \ldots & \vdots & \vdots \\ 0 & 0 & 0 & \ldots & 1 & 0 \end{bmatrix}$$

We now give examples illustrating Theorem 17 and Theorem 19

Example 19.a $$A = \begin{bmatrix} 1 & 1 \\ 0 & 0 \end{bmatrix}$$

satisfies $A^2 = A$.

It defines $T : \mathbf{R}^{2\times 1} \to \mathbf{R}^{2\times 1}$ such that

$$T\begin{bmatrix} a \\ b \end{bmatrix} = \begin{bmatrix} 1 & 1 \\ 0 & 0 \end{bmatrix}\begin{bmatrix} a \\ b \end{bmatrix} = \begin{bmatrix} a+b \\ 0 \end{bmatrix}.$$

$$W' = \ker(T) = \left\{\begin{bmatrix} a \\ b \end{bmatrix} : b = -a\right\}$$

It is one-dimensional, and has a basis $B_2 = \left\{\begin{bmatrix} 1 \\ -1 \end{bmatrix}\right\}$.

$$\text{range}(T) = \left\{\begin{bmatrix} x \\ 0 \end{bmatrix} : x \in R\right\}$$

has basis $$B_1 = \left\{\begin{bmatrix} 1 \\ 0 \end{bmatrix}\right\}$$

So, we get ordered basis $B' = \left\{\begin{bmatrix} 1 \\ 0 \end{bmatrix}, \begin{bmatrix} 1 \\ -1 \end{bmatrix}\right\} = \{X_1, X_2\}$

of $\mathbf{R}^{2\times 1}$, where $$X_1 = \begin{bmatrix} 1 \\ 0 \end{bmatrix}, X_2 = \begin{bmatrix} 1 \\ -1 \end{bmatrix}.$$

Then $$TX_1 = X_1 = 1X_1 + 0X_2$$
$$TX_2 = 0 = 0X_1 + 0X_2$$

so that $[T]_{B'} = \begin{bmatrix} 1 & 0 \\ 0 & 0 \end{bmatrix}$ which is similar to A.

The matrix of B' relative to the standard ordered basis is

$$P = [X_1 X_2] = \begin{bmatrix} 1 & 1 \\ 0 & -1 \end{bmatrix}.$$

One can easily see that $$P^{-1}AP = \begin{bmatrix} 1 & 0 \\ 0 & 0 \end{bmatrix}.$$

Example 19.b. Let $$A = \begin{bmatrix} 0 & 1 & 2 \\ 0 & 0 & -3 \\ 0 & 0 & 0 \end{bmatrix}.$$

Then, $$A^2 = \begin{bmatrix} 0 & 0 & -3 \\ 0 & 0 & 0 \\ 0 & 0 & 0 \end{bmatrix} \neq 0$$

but $$A^3 = 0.$$

We need an $X_1 \in \mathbf{R}^{3\times1}$ such that $A^2X_1 \neq 0$. This could be any vector, not in the solution space of $A^2X = 0$. One such is $X_1 = \begin{bmatrix} 0 \\ 0 \\ 1 \end{bmatrix}$; this corresponds to $T^2X_1 \neq 0$, where T is the linear operator on $\mathbf{R}^{3\times1}$ given by A.

Now, $$AX_1 = \begin{bmatrix} 2 \\ -3 \\ 0 \end{bmatrix},$$

$$A^2X_1 = \begin{bmatrix} -3 \\ 0 \\ 0 \end{bmatrix}.$$

Then, $$B' = \left\{ \begin{bmatrix} 0 \\ 0 \\ 1 \end{bmatrix}, \begin{bmatrix} 2 \\ -3 \\ 0 \end{bmatrix}, \begin{bmatrix} -3 \\ 0 \\ 0 \end{bmatrix} \right\}$$

is an ordered basis of $\mathbf{R}^{3\times1}$ whose matrix relative to the standard ordered basis is

$$p = \begin{bmatrix} 0 & 2 & -3 \\ 0 & -3 & 0 \\ 1 & 0 & 0 \end{bmatrix}$$

Then, $$P^{-1}AP = \begin{bmatrix} 0 & 0 & 0 \\ 1 & 0 & 0 \\ 0 & 1 & 0 \end{bmatrix}$$ (check ?)

Remark 20. (*i*)Let A be any $m \times n$-matrix over a field F. If an $m \times n$-matrix B is equivalent to A, then, $B = QAP$ for some non-singular matrices P and Q This means B is obtained from A by applying a finite sequence of elementary row and column operations. Thus, Theorem 16 simply says that by applying an appropriate finite sequence of elementary row and column operations to A, we get

$$\begin{bmatrix} I_k & 0 \\ 0 & 0 \end{bmatrix}_{m \times n}$$

Example 20.a. Let $$A = \begin{bmatrix} 1 & 2 & -2 & 3 \\ -1 & -4 & 4 & -3 \\ 3 & -1 & 0 & 2 \\ 5 & 5 & -6 & 8 \end{bmatrix}$$

Apply $r_2 \to r_2 + r_1$, $r_3 \to r_3 - 3r_1$, $r_4 \to r_4 - 5r_1$.

We get $$A_1 = \begin{bmatrix} 1 & 2 & -2 & 3 \\ 0 & -2 & 2 & 0 \\ 0 & -7 & 6 & -7 \\ 0 & -5 & 4 & -7 \end{bmatrix}$$

Apply $c_2 \to c_2 - 2c_1$, $c_3 \to c_3 + 2c_1$, $c_4 \to c_4 - 3c_1$.

We get $$A_2 = \begin{bmatrix} 1 & 0 & 0 & 0 \\ 0 & -2 & 2 & 0 \\ 0 & -7 & 6 & -7 \\ 0 & -5 & 4 & -7 \end{bmatrix}$$

Apply $r_2 \to -\frac{1}{2}r_2$.

We get $$A_3 = \begin{bmatrix} 1 & 0 & 0 & 0 \\ 0 & 1 & -1 & 0 \\ 0 & -7 & 6 & -7 \\ 0 & -5 & 4 & -7 \end{bmatrix}$$

Apply $r_3 \to r_3 + 7r_2$, $r_4 \to r_4 + 5r_2$.

We get $$A_4 = \begin{bmatrix} 1 & 0 & 0 & 0 \\ 0 & 1 & -1 & 0 \\ 0 & 0 & -1 & -7 \\ 0 & 0 & -1 & -7 \end{bmatrix}$$

Apply $c_3 \to c_3 + c_2$.

We get $$A_5 = \begin{bmatrix} 1 & 0 & 0 & 0 \\ 0 & 1 & 0 & 0 \\ 0 & 0 & -1 & -7 \\ 0 & 0 & -1 & -7 \end{bmatrix}$$

Apply $r_3 \to (-1)r_3$ and follow it by $r_4 \to r_4 - r_3$.

We get $$A_6 = \begin{bmatrix} 1 & 0 & 0 & 0 \\ 0 & 1 & 0 & 0 \\ 0 & 0 & 1 & 7 \\ 0 & 0 & 0 & 0 \end{bmatrix}$$

Apply $c_4 \to c_4 - 7c_3$.

We get $$B = \begin{bmatrix} 1 & 0 & 0 & 0 \\ 0 & 1 & 0 & 0 \\ 0 & 0 & 1 & 0 \\ 0 & 0 & 0 & 0 \end{bmatrix} = \begin{bmatrix} I_3 & 0 \\ 0 & 0 \end{bmatrix}_{4\times 4}$$

Thus, $$\text{rank}(A) = 3.$$

Exercises 3.3

1. Let $T : \mathbf{R}^2 \to \mathbf{R}^2$ be the linear operator such that
$$T(x_1, x_2) = (x_1, x_1 + x_2).$$
Let $$B = \{\epsilon_1, \epsilon_2\}, B' = \{(1, 2), (1, 1)\}.$$
(*i*) Find $[T]_B, [T]_{B'}, [T]_{(B, B')}, [T]_{(B', B)}$
(*ii*) Find the matrix P of B' relative to B and verify that $[T]_{B'} = P^{-1}[T]_B P$
(*iii*) Find the formula for T^{-1}, and find $[T^{-1}]_{B'}$. Verify that $[T^{-1}]_{B'} = [T]_{B'}^{-1}$.

2. Let $T : \mathbf{R}^2 \to P_2(\mathbf{R})$ be given by $T(a, b) = bX + aX^2$. Consider the following bases $B = \{(1, -1), (2, 0)\}$ of $\mathbf{R}^2$ and $B' = \{1, X, X^2\}$ of $P_2(\mathbf{R})$
(*i*) Find $A = [T]_{(B, B')}$
(*ii*) Verify that $[T(1, 0)]_{B'} = [T]_{(B, B')}[(1, 0)]_B$
(*iii*) Find a basis of the column space of A. Use this basis to find a basis of the range(T).

3. Let $T : \mathbf{R}^3 \to \mathbf{R}^3$ be the linear operator whose matrix relative to the standard ordered basis B is $A = \begin{bmatrix} 1 & 2 & 1 \\ 0 & 1 & 1 \\ -1 & 3 & 4 \end{bmatrix}$
(*i*) Find $T\epsilon_1, T\epsilon_2, T\epsilon_3$
(*ii*) Find the matrix of each of the following relative to the standard ordered basis
$$T^2, T^2 + I, (2T)^3 + 5T - I$$
(*iii*) Verify that $T^3 - 6T^2 + 7T = 0$ and that $E = \left[\frac{6}{7}I - \frac{T}{7}\right]$ is idempotent
(*iv*) Find a basis of $\ker(E)$ and a basis of range(T).

4. Let $D : P_3(\mathbf{C}) \to P_3(\mathbf{C})$ be the differentiation operator. Let B be the standard ordered basis $\{1, X, X^2, X^3\}$ and $B' = \{1, 1+X, (1+X)^2, (1+X)^3\}$
(*i*) Find , $[D]_B, [D]_{B'}$ and $[D]_{(B', B)}$
(*ii*) For $A = [D]_{B'}$ verify that $A^4 = 0$, but $A^3 \neq 0$
(*iii*) For any $c \neq 0$ in $\mathbf{C}$, prove that $cI + D$ is invertible.

5. Let $A = \begin{bmatrix} 1 & 6 & 4 & 2 \\ 2 & 3 & -1 & -5 \\ -1 & 0 & 2 & 4 \\ 3 & 15 & 7 & -1 \end{bmatrix}$ and let $T : \mathbf{R}^{4\times 1} \to \mathbf{R}^{4\times 1}$ be given by $T(X) = AX$.

(*i*) Find $[T]_{B'}$, where $B' = \{\in_2, \in_4, \in_1, \in_3\}$

(*ii*) Find a basis B_1 of $\ker(T)$

(*iii*) Find an ordered basis B' of $\mathbf{R}^{4\times1}$ that contains B_1 as its final segment.

(*iv*) Find an ordered basis B'' of $\mathbf{R}^{4\times1}$ such that

$$[T]_{(B', B'')} = \begin{bmatrix} 1 & 0 & 0 & 0 \\ 0 & 1 & 0 & 0 \\ 0 & 0 & 0 & 0 \\ 0 & 0 & 0 & 0 \end{bmatrix}$$

6. Let $T: R^{2\times1} \to R^{2\times1}$ be the linear operator, whose matrix relative to the ordered basis $B' = \left\{\begin{bmatrix} 1 \\ -1 \end{bmatrix}, \begin{bmatrix} 0 \\ 1 \end{bmatrix}\right\}$ is $A = \begin{bmatrix} 1 & 4 \\ 1 & -2 \end{bmatrix}$.

(*i*) Find $T\begin{bmatrix} x \\ y \end{bmatrix}$ and the matrix of T relative to the standard ordered basis of $\mathbf{R}^{2\times1}$

(*ii*) Find an $X \neq 0$ in $\mathbf{R}^{2\times1}$ such that $(T+3I)X \neq 0$. For such an X, prove that $T'X = 2X$, where $T' = T - 3I$

(*iii*) Find, if possible, an ordered basis B'' of $\mathbf{R}^{2\times1}$ such that $[T]_{B''}$ is a diagonal matrix

7. Let $T : \mathbf{R}^{2\times1} \to \mathbf{R}^{2\times1}$ be given by $T\begin{bmatrix} x_1 \\ x_2 \end{bmatrix} = \begin{bmatrix} -x_2 \\ x_1 \end{bmatrix}$

(*i*) Find the matrix of T relative to the ordered basis $B = \left\{\begin{bmatrix} 1 \\ 2 \end{bmatrix}\begin{bmatrix} 1 \\ -1 \end{bmatrix}\right\}$

(*ii*) For any ordered basis B' of $\mathbf{R}^{2\times1}$ if $[T]_{B'} = \begin{bmatrix} a_{11} & a_{12} \\ a_{21} & a_{22} \end{bmatrix}$, prove that $a_{12}a_{21} \neq 0$.

8. For any real number θ, prove that $\begin{bmatrix} \cos\theta & -\sin\theta \\ \sin\theta & \cos\theta \end{bmatrix}$ and $\begin{bmatrix} e^{i\theta} & 0 \\ 0 & e^{-i\theta} \end{bmatrix}$ are similar over the field of complex numbers.

9. Let
$$A = \begin{bmatrix} 1 & 1 & -2 & 0 & 0 \\ 1 & 2 & 3 & 6 & 7 \\ 2 & 1 & 3 & 6 & 5 \end{bmatrix}.$$

(*i*) Find $\text{rank}(A)$.

(*ii*) Find matrix B in the form $\begin{bmatrix} I_k & 0 \\ 0 & 0 \end{bmatrix}_{3\times5}$ equivalent to A.

10. Prove that $\begin{bmatrix} 0 & 1 \\ 0 & 0 \end{bmatrix}$ and $\begin{bmatrix} 0 & 0 \\ 1 & 0 \end{bmatrix}$ are similar over **R**.

11. Prove that $A = \begin{bmatrix} 0 & 0 & 1 \\ 0 & 0 & 0 \\ 0 & 0 & 0 \end{bmatrix}$ and $B = \begin{bmatrix} 0 & 1 & 0 \\ 0 & 0 & 1 \\ 0 & 0 & 0 \end{bmatrix}$ are not similar over **R**, but B is similar to $\begin{bmatrix} 0 & 0 & 0 \\ 1 & 0 & 0 \\ 0 & 1 & 0 \end{bmatrix}$

12. Let E_1 and E_2 be any two idempotent linear operators on a finite dimensional vector space V_F. Prove that range(E_1) = range(E_2) iff $E_1E_2 = E_2$ and $E_2E_1 = E_1$.

13. Let A be any upper triangular $k \times k$-matrix over a field F such that the main diagonal of A consists of zeros. Prove that $A^k = 0$.

14. Give an example of a 3×3-matrix A that is equivalent to $B = \begin{bmatrix} 1 & 0 & 0 \\ 0 & 1 & 0 \\ 0 & 0 & 0 \end{bmatrix}$, but is not similar to B.

3.4 Dual Spaces

For any two vector spaces V_F and W_F, we had discussed the vector space $\text{Hom}_F(V, W)$, in the previous two sections. Now, F is a one-dimensional vector space over F. In the present section, we shall be discussing $\text{Hom}_F(V, F)$ in further details, as this vector space is of great significance in many areas of mathematics. Throughout, V_F is a finite dimensional vector space of dimension $n \geq 1$, unless otherwise stated.

Definition 1. Any linear transformation f of a vector space V_F into F is called a *linear functional* on V_F. The F-vector space $\text{Hom}_F(V, F)$ is called the *dual space* of V and is denoted by V^*.

As $$\dim_F V = n$$

and $$\dim_F F = 1,$$ by Theorem (3.2.11),

$$\dim V^* = \dim \text{Hom}_F(V, F) = n.1 = n.$$

Further, let $$B = \{\alpha_1, \alpha_2, ..., \alpha_n\}$$

be any basis of V.

Now $$B' = \{\beta_1\},$$

where $\beta_1 = 1$, is a basis of V_F.

In the notation of the same theorem, we get a basis

$$B'' = \{E_{11}, E_{12}, ..., E_{1n}\}$$

of V^* such that $\qquad E_{1j}(\alpha_j) = \beta_1 = 1$

and $\qquad E_{1j}(\alpha_k) = 0 \qquad$ for $j \neq k$

Let us write $\qquad \alpha_j^* = E_{1j}$

so that $\qquad B'' = \{\alpha_1^*, \alpha_2^*, ..., \alpha_n^*\}$

is such that $\qquad \alpha_j^*(\alpha_i) = \delta_{ij}$

where δ_{ij} is the Kronecker delta; to remind, $\delta_{ij} = 1$ if $i = j$ and $\delta_{ij} = 0$ for $i \neq j$. Thus, we have the following :

Theorem 2. Let $\dim V_F = n \geq 1$ and $B = \{\alpha_1, \alpha_2, ..., \alpha_n\}$ be a basis of V_F.

Then, there exists a basis $\qquad B^* = \{\alpha_1^*, \alpha_2^*, ..., \alpha_n^*\}$

of V^* such that $\qquad \alpha_j^*(\alpha_i) = \delta_{ij}$

(This basis B^* is called *the dual basis* of B).

Example 2.a. Let $V = \mathbf{R}^3$. Consider any three numbers a_1, a_2, a_3 in **R**; to be specific, let $a_1 = 1, a_2 = 2, a_3 = -2$.

Then, we get function $\qquad f: \mathbf{R}^3 \to \mathbf{R}$

such that $\qquad f(x_1, x_2, x_3) = a_1x_1 + a_2x_2 + a_3x_3$

This f is a linear functional on $\mathbf{R}^3$.

Consider any $\qquad \alpha = (x_1, x_2, x_3),$

$$\beta = (y_1, y_2, y_3) \in \mathbf{R}^3$$

and $\qquad c \in \mathbf{R}.$

Then, $\qquad \alpha + c\beta = (x_1 + cy_1, x_2 + cy_2, x_3 + cy_3)$

gives
$$\begin{aligned} f(\alpha + c\beta) &= a_1(x_1 + cy_1) + a_2(x_2 + cy_2) + a_3(x_3 + cy_3) \\ &= (a_1x_1 + a_2x_2 + a_3x_3) + c(a_1y_1 + a_2y_2 + a_3y_3) \\ &= f(\alpha) + cf(\beta) \end{aligned}$$

Consider $\qquad B = \{\alpha_1, \alpha_2, \alpha_3\},$

where $\qquad \alpha_1 = (1, 1, 2), \alpha_2 = (0, 2, 1), \alpha_3 = (0, 0, 5)$

We determine the dual basis $\quad B^* = \{\alpha_1^*, \alpha_2^*, \alpha_3^*\}.$

To understand the method, consider any F^n. Let f be any linear functional of F^n.

Let $$f(\in_i) = a_i,\ 1 \le i \le n,$$

where $\{\in_1, \in_2, \ldots, \in_n\}$ is the standard basis of F^n.

Any $\alpha \in F^n$ is given by $\alpha = (x_1, x_2, \ldots, x_n) = \sum_{i=1}^{n} x_i \in_i$

so that $$f(\alpha) = \sum_{i=1}^{n} x_i f(\in_i) = a_1x_1 + a_2x_2 + \ldots + a_nx_n$$

We come back to finding $\alpha_1^*, \alpha_2^*, \alpha_3^*$

Now, $$\alpha_1^*(x_1, x_2, x_3) = a_1x_1 + a_2x_2 + a_3x_3$$

for some a_1, a_2, a_3.

By definition $\alpha_1^*(\alpha_1) = 1, \alpha_1^*(\alpha_2) = 0, \alpha_1^*(\alpha_3) = 0$.

This gives a system of three linear equations

$$a_1 + a_2 + 2a_3 = 1$$
$$0a_1 + 2a_2 + 1a_3 = 0$$
$$0a_1 + 0a_2 + 5a_3 = 0$$

i.e., $$A\begin{bmatrix} a_1 \\ a_2 \\ a_3 \end{bmatrix} = \begin{bmatrix} 1 \\ 0 \\ 0 \end{bmatrix},$$

where $$A = \begin{bmatrix} 1 & 1 & 2 \\ 0 & 2 & 1 \\ 0 & 0 & 5 \end{bmatrix}.$$

By solving this system of linear equations, we get

$$a_1 = 1,\ a_2 = 0,\ a_3 = 0,$$

so that $$\alpha_1^*(x_1, x_2, x_3) = x_1$$

Similarly, α_2^* is given by solving $A\begin{bmatrix} a_1' \\ a_2' \\ a_3' \end{bmatrix} = \begin{bmatrix} 0 \\ 1 \\ 0 \end{bmatrix}$

which gives $$a_1' = -\frac{1}{2},\ a_2' = \frac{1}{2},\ a_3' = 0,$$

so that $$\alpha_2^*(x_1, x_2, x_3) = -\frac{1}{2}x_1 + \frac{1}{2}x_2$$

α_3^* is obtained by solving $A\begin{bmatrix} b_1 \\ b_2 \\ b_3 \end{bmatrix} = \begin{bmatrix} 0 \\ 0 \\ 1 \end{bmatrix}$

This gives $$b_1 = -\frac{3}{10},\ b_2 = -\frac{1}{10},\ b_3 = \frac{1}{5},$$

so that $$\alpha_3^*(x_1, x_2, x_3) = -\frac{3}{10}x_1 - \frac{1}{10}x_2 + \frac{1}{5}x_3$$

Example 2.b. Let $W_F = F^{n\times n}$. For any $A = [a_{ij}] \in W_F$, define the trace of A to be $a_{11} + a_{22} + \ldots + a_{nn}$. Let us denote this trace by $\text{tr}(A)$. Clearly, $\text{tr}(A) \in F$.

Let $$A = [a_{ij}],\ B = [b_{ij}] \in F^{n\times n}.$$

Then, $$A + cB = [a_{ij} + cb_{ij}],$$

gives
$$\begin{aligned}\text{tr}(A+cB) &= \sum_{i=1}^{n}(a_{ii} + cb_{ii}) \\ &= \sum_{i=1}^{n} a_{ii} + c\sum_{i=1}^{n} b_{ii} \\ &= \text{tr}(A) + c\,\text{tr}(B)\end{aligned}$$

Hence, $$\text{tr} : F^{n\times n} \to F$$
is a linear functional.

Example 2.c. Let $$A = [a_{ij}] \in F^{n\times n}.$$

$$U = [a_1, a_2, \ldots, a_n] \in F^{1\times n}$$

For any $$X = [x_1, x_2, \ldots, x_n] \in F^{1\times n},$$

define $$f_u(X) = UAX^T = \sum_{j=1}^{n}\sum_{i=1}^{n} a_{ij} a_i x_j .$$

This f_u is a linear functional on $F^{1\times n}$.

Example 2.d. Let $C[a, b]$ be the real vector space of all continuous real valued functions defined on the closed interval $[a, b]$.

For any $f \in C[a, b]$, define $\sigma(f) = \int_a^b f(t)\, dt \in \mathbf{R}$.

For any $f, g \in C[a, b]$ and $c \in \mathbf{R}$

$$\begin{aligned}\sigma(f + cg) &= \int_a^b \{f(t) + cg(t)\}\, dt \\ &= \int_a^b f(t)\, dt + c\int_a^b g(t)\, dt\end{aligned}$$

$$=\sigma(f)+c\sigma(g).$$

This shows that σ is a linear functional on $C[a, b]$.

Example 2.e. Let $V = P_2(\mathbf{R})$. Let a_1, a_2, a_3 be any three distinct real numbers.

For $i = 1, 2, 3$, define $\quad f_i : V \to \mathbf{R}$

such that for any $p(X) \in V$, $\quad f_i\{p(X)\} = p(a_i)$;

f_i is a linear functional on V, i.e., $\quad f_i \in V^*$.

Suppose for some real numbers b_1, b_2, b_3,

$$b_1 f_1 + b_2 f_2 + b_3 f_3 = 0$$

Then, as $\quad 1 \in V, X \in V, X^2 \in V$

$$b_1 f_1(1) + b_2 f_2(1) + b_3 f_3(1) = 0$$

$$b_1 f_1(X) + b_2 f_2(X) + b_3 f_3(X) = 0$$

$$b_1 f_1(X^2) + b_2 f_2(X^2) + b_3 f_3(X^2) = 0$$

i.e.,

$$b_1 + b_2 + b_3 = 0$$

$$b_1 a_1 + b_2 a_2 + b_3 a_3 = 0$$

$$b_1 a_1^2 + b_2 a_2^2 + b_3 a_3^2 = 0$$

This gives a system of three homogeneous linear equations in b_1, b_2, b_3 with coefficient matrix

$$A = \begin{bmatrix} 1 & 1 & 1 \\ a_1 & a_2 & a_3 \\ a_1^2 & a_2^2 & a_3^2 \end{bmatrix}$$

$$A^T = \begin{bmatrix} 1 & a_1 & a_1^2 \\ 1 & a_2 & a_2^2 \\ 1 & a_3 & a_3^2 \end{bmatrix}$$

Apply $r_2 \to r_2 - r_1,\ r_3 \to r_3 - r_1$.

This gives $\quad \begin{bmatrix} 1 & a_1 & a_1^2 \\ 0 & a_2 - a_1 & a_2^2 - a_1^2 \\ 0 & a_3 - a_1 & a_3^2 - a_1^2 \end{bmatrix}$

Apply $r_2 \to \dfrac{1}{(a_2 - a_1)} r_2,\ r_3 \to \dfrac{1}{(a_3 - a_1)} r_3$.

This gives
$$\begin{bmatrix} 1 & a_1 & a_1^2 \\ 0 & 1 & a_2 + a_1 \\ 0 & 1 & a_3 + a_1 \end{bmatrix}.$$

Continue this process, and this finally gives I_3, so that A^T is non-singular. Hence, A is non-singular.

Consequently,
$$b_1 = b_2 = b_3 = 0,$$
so that f_1, f_2, f_3 are linearly independent.

As
$$\dim V^* = \dim V = 3,$$
we get
$$B = \{f_1, f_2, f_3\}$$
is a basis of V^*.

Theorem 3. Let $B = \{\alpha_1, \alpha_2, \ldots, \alpha_n\}$ be an ordered basis of V_F and $B^* = \{\alpha_1^*, \alpha_2^*, \ldots, \alpha_n^*\}$ be the ordered basis of V^*, which is the dual of B. Then, the following hold :

(*i*) For any $f \in V^*$,
$$f = \sum f(\alpha_i)\,\alpha_i^*$$

(*ii*) The mapping
$$\eta : V^* \to F^n$$
such that
$$\eta(f) = \big(f(\alpha_1), f(\alpha_2), \ldots, f(\alpha_n)\big)$$
is an isomorphism

(*iii*) For any $\alpha \neq 0$ in V, there exists an $f \in V^*$ such that $f(\alpha) \neq 0$.

Proof.

(*i*) As B^* is a basis of V^*, given
$$f \in V^*,\ f = \sum_{i=1}^{n} a_i \alpha_i^*,\ a_i \in F.$$

For $1 \le j \le n$
$$f(\alpha_j) = \sum a_i \alpha_i^*(\alpha_j) = \sum_{i=1}^{n} a_i \delta_{ij} = a_j.$$

Hence,
$$f = \sum_{i=1}^{n} f(\alpha_i)\,\alpha_i^*$$

(*ii*) For any $f, g \in V^*$, $c \in F$ (*i*) gives
$$f + cg = \sum_{i=1}^{n} (f + cg)(\alpha_i)\,\alpha_i^*$$

But
$$(f + cg)(\alpha_i) = f(\alpha_i) + cg(\alpha_i),$$
so that
$$f + cg = \sum_{i=1}^{n} \big[f(\alpha_i) + cg(\alpha_i)\big]\,\alpha_i.$$

Thus, $\eta(f+cg) = (f(\alpha_1)+cg(\alpha_1),\dots,f(\alpha_n)+cg(\alpha_n))$

$$= \eta(f)+c\eta(g).$$

Hence, η is a linear transformation. Now, as B is a basis of V, any $f \in V^*$ is zero iff $f(\alpha_j)=0$ for every $j = 1, 2,\dots, n$, so

$$\eta(f)=0$$

iff
$$(f(\alpha_1), f(\alpha_2),\dots, f(\alpha_n))=0$$

iff
$$f(\alpha_j)=0$$

for $1 \le j \le n$ iff $f=0$.

Hence, $\ker(\eta)=\{0\}$.

This proves that η is one-to-one.

As $\dim V^* = n = \dim F^n$, we get η is also onto. This proves (*ii*).

(*iii*) Consider $\alpha \neq 0$ in V. We can find a basis $B=\{\alpha_1, \alpha_2,\dots, \alpha_n\}$ of V such that $\alpha_1 = \alpha$.

Then, $f = \alpha_1^*$

satisfies $f(\alpha)=\alpha_1^*(\alpha_1)=1 \neq 0.$

Corollary 4.

(*i*) Let $V = F^n$. Then, any $f \in V^*$ is such that for any $\alpha = (x_1, x_2,\dots, x_n)$

$$f(x_1, x_2,\dots, x_n) = a_1x_1 + a_2x_2 + \dots + a_nx_n$$

for a uniquely determined $(a_1, a_2,\dots, a_n) \in F^n$

(*ii*) The mapping $\eta : V^* \to F^n$ such that $\eta(f)=(a_1, a_2,\dots, a_n)$, whenever f is given by the formula (*i*), is an isomorphism

Proof. Let $B = \{\in_1, \in_2,\dots, \in_n\}$ be the standard ordered basis of F^n. Let $f \in V^*$ and $f(\in_i)=a_i$.

Then, for any $\alpha = (x_1, x_2,\dots, x_n) \in F^n$, $\alpha = \sum_{i=1}^{n} x_i \in_i$, gives

$$f(\alpha) = \sum_{i=1}^{n} x_i f(\in_i) = \sum_{i=1}^{n} a_i x_i.$$

That η is an isomorphism follows from part (*ii*) of the theorem, given above.

Theorem 5. Let $\dim V_F = n$, $\alpha_1, \alpha_2,\dots, \alpha_n$ be any n members of V and $f_1, f_2, \dots, f_n$ be any n members of V^*, such that

$$f_i(\alpha_j) = \delta_{ij} \qquad \text{for } 1 \le i \le n, 1 \le j \le n$$

Then, $B = \{\alpha_1, \alpha_2, ..., \alpha_n\}$ is a basis of V and $B' = \{f_1, f_2, ..., f_n\}$ is a basis of V^* such that B' is the dual of B. Further, for any $\alpha \in V$, $\alpha = \sum_{j=1}^{n} f_j(\alpha)\alpha_j$.

Proof. Let for some $\quad a_1, a_2, ..., a_n \in F, \sum_{i=1}^{n} a_i f_i = 0$

Then, $$\left[\sum_{i=1}^{n} a_i f\right](\alpha_j) = 0,$$

gives $$0 = \sum_{i=1}^{n} a_i f_i(\alpha_j) = \sum a_i \delta_{ij} = a_j$$

Hence, $f_1, f_2, ..., f_n$ are linearly independent. As $\dim V^* = n$, we get

$$B' = \{f_1, f_2, ..., f_n\}$$

is a basis of V^*

Let for some $\quad b_1, b_2, ..., b_n \in F, \sum_{j=1}^{n} b_j \alpha_j = 0.$

Then, $$f_i\left[\sum_{j=1}^{n} b_j \alpha_j\right] = 0,$$

gives $$0 = \sum_{j=1}^{n} b_j f_i(\alpha_j) = \sum_{j=1}^{n} b_j \delta_{ij} = b_i.$$

Thus, $\alpha_1, \alpha_2, ... \alpha_n$ are linearly independent, so that $B = \{\alpha_1, \alpha_2, ..., \alpha_n\}$ is a basis of V. That B' is the dual basis of B is obvious. Consider $\alpha \in V$.

Then, $$\alpha = \sum_{j=1}^{n} c_j \alpha_j, \; c_j \in F.$$

Then for $1 \le i \le n$, $$f_i(\alpha) = \sum_{j=1}^{n} c_j f_i(\alpha_j) = \sum_{j=1}^{n} c_j \delta_{ij} = c_i.$$

Hence, $$\alpha = \sum_{j=1}^{n} f_j(\alpha)\alpha_j$$

Theorem 6. Let $\quad V = P_{n-1}(F), \; n \ge 2.$

(*i*) For any $a \in F$, the function $L_a : V \to F$

given by $$L_a\{p(X)\} = p(a)$$

is a linear functional on V.

(*ii*) Let $a_1, a_2, ..., a_n$ be any n distinct members of F.

For $1 \le i \le n$, define $$p_i(X) = \frac{\left[\prod_{j \ne i}(X - a_j)\right]}{\prod_{j \ne i}(a_i - a_j)}.$$

Then, $p_1(X), p_2(X), ..., p_n(X)$ form a basis of V and for any $p(X) \in V$,

$$p(X) = \sum_{i=1}^{n} p(a_i) p_i(X)$$

Proof. Before we give the formal proof, let us illustrate the concept. Consider $P_3(\mathbf{R})$. Here, $n = 4$. Let $a = 5$.

Then, $$L_5(a + bX + cX^2 + dX^3) = a + 5b + 5^2 c + 5^3 d = a + 5b + 25c + 125d.$$

Let $$a_1 = 1, a_2 = 3, a_3 = 7, a_4 = 10.$$

Then, $$p_1(X) = \frac{(X - a_2)(X - a_3)(X - a_4)}{(a_1 - a_2)(a_1 - a_3)(a_1 - a_4)} = \frac{(X-3)(X-7)(X-10)}{(-2)(-6)(-9)} = -\frac{1}{108}(X-3)(X-7)(X-10),$$

a polynomial of degree three.

Again, $$p_2(X) = \frac{(X - a_1)(X - a_3)(X - a_4)}{(a_2 - a_1)(a_2 - a_3)(a_2 - a_4)}.$$

and so on.

We return to the proof.

For any $p(X), q(X) \in F[X]$ and $a, c \in F$, if $h(X) = p(X) + cq(X)$, it is immediately clear that

$$h(a) = p(a) + cq(a)$$

so that $$L_a\{p(X) + cq(X)\} = L_a\{p(X)\} + cL_a\{q(X)\}.$$

Hence, L_a is a linear functional on V.

Now, $$\dim V = n, \; L_{a_i}(X) = a_i.$$

As $a_i \ne a_j$ for $i \ne j$, we get that $L_{a_1}, L_{a_2}, ..., L_{a_n}$ are n distinct members of V^*

For $i \ne j$, $$P_j(X) = \frac{\left[\prod_{i \ne j}(X - a_i)\right]}{\prod_{i \ne j}(a_j - a_i)}$$

is a polynomial of degree $(n-1)$, and $p_j(a_i) = 0$.

Further, $$p_j(a_j) = 1.$$

So that $$L_{a_i}\{p_j(X)\} = p_j(a_i) = \delta_{ij}.$$

Hence, by Theorem 5, $$B' = \{L_{a_1}, L_{a_2}, \ldots, L_{a_n}\}$$

and $$B = \{p_1(X), p_2(X), \ldots, p_n(X)\}$$

are bases of V^* and V, respectively.

Further, given any $p(X) \in V$,

$$p(X) = \sum_{i=1}^{n} L_{a_i}\{p(X)\} p_i(X) = \sum_{i=1}^{n} p(a_i) p_i(X).$$

(This formula for $p(X)$ is called *Lagrange's interpolation formula*).

Example 6.a. Let $$V = P_2(\mathbf{R}).$$

Now $$\dim V = 3.$$

Consider $$a_1 = 2,\ a_2 = -1,\ a_3 = 5.$$

So that $$p_1(X) = \frac{(X-a_2)(X-a_3)}{(a_1-a_2)(a_1-a_3)} = \frac{-1}{9}(X+1)(X-5)$$

$$p_2(X) = \frac{(X-a_1)(X-a_3)}{(a_2-a_1)(a_2-a_3)} = \frac{1}{18}(X-2)(X-5)$$

$$p_3(X) = \frac{(X-a_1)(X-a_2)}{(a_3-a_1)(a_3-a_2)} = \frac{1}{18}(X-2)(X+1)$$

Consider $$p(X) = X.$$

Then, $$p(a_1) = 2,\ p(a_2) = -1,\ p(a_3) = 5.$$

We have $$p(X) = p(a_1)p_1(X) + p(a_2)p_2(X) + p(a_3)p_3(X)$$

i.e., $$X = \frac{-2}{9}(X+1)(X-5) - \frac{1}{18}(X-2)(x-5) + \frac{5}{18}(x-2)(X+1).$$

One can easily check that this identity is true.

Theorem 7. For any finite dimensional vector space V_F, the mapping $\sigma : V \to (V^*)^*$ such that $[\sigma(\alpha)](f) = f(\alpha)$, $\alpha \in V$, $f \in V^*$, is an isomorphism.

Proof. For any $\alpha \in V$, define $$L_\alpha : V^* \to F$$

such that $$L_\alpha(f) = f(\alpha),\ f \in V^*$$

For any $f, g \in V^*$, $c \in F$,

$$L_\alpha(f+cg)=(f+cg)(\alpha)=f(\alpha)+cg(\alpha)=L_\alpha(f)+cL_\alpha(g).$$

This shows that L_α is a linear functional on V^*, i.e., $L_\alpha \in (V^*)^*$. We define $\sigma(\alpha)=L_\alpha$ that way $[\sigma(\alpha)](f)=f(\alpha)$.

For any $\alpha, \beta \in V$, $c \in F$

$$L_{(\alpha+c\beta)}(f)=f(\alpha+c\beta)=f(\alpha)+cf(\beta)=(L_\alpha+cL_\beta)(f)$$

This gives $\sigma(\alpha+c\beta)=\sigma(\alpha)+c\sigma(\beta)$.

Hence, σ is a linear transformation.

For $\alpha \neq 0$, by Theorem 3, there exists $f \in V^*$

such that $f(\alpha)\neq 0$

This gives $[\sigma(\alpha)](f)\neq 0$,

so that $\sigma(\alpha)\neq 0$ and, hence, $\alpha \notin \ker\sigma$.

Consequently, $\ker\sigma=\{0\}$.

Hence, σ is one-to-one.

Now $\dim V=\dim V^*=\dim(V^*)^*$

Hence, σ being one-to-one, must also be onto. This proves that σ is an isomorphism.

We shall write V^{**} for $(V^*)^*$ and call it the *double dual* of V.

Remark 8. Observe that the isomorphism $\sigma: V \to V^{**}$ given by the theorem, given above, is not defined by specifying bases of V and V^{**}. This σ is known as a *canonical isomorphism*. We shall see that this isomorphism helps in giving a natural relation between V and V^*.

Theorem 9. Let $V_F = n \geq 1$ and let $B'=\{f_1, f_2,..., f_n\}$ be any basis of V^*. Then there exists a basis $B=\{\alpha_1, \alpha_2,..., \alpha_n\}$ of V such that B' is a dual basis of B.

Proof. As V^{**} is the dual of V^*, we get a basis

$$B''=\{g_1, g_2,..., g_n\}$$

of V^{**} that is dual of B' (Theorem 2).

So, by definition $g_i(f_j)=\delta_{ij}$.

Now, consider the canonical isomorphism $\sigma: V \to V^{**}$

For $1 \leq j \leq n$, we get $\alpha_j \in V$

such that $\sigma(\alpha_j)=g_j$.

Then, $B=\{\alpha_1, \alpha_2,..., \alpha_n\}$

is a basis of V_F.

Further, $$f_i(\alpha_j) = [\sigma(\alpha_j)](f_i) = g_j(f_i) = \delta_{ji}.$$

This gives that B' is the dual basis of B.

Definition 10. For any subset S of V_F, its annihilator $S^\perp$ in V^* is the set of all those $f \in V^*$ such that $f(a) = 0$ for any $\alpha \in S$. For any subset A of V^*, its annihilator $A^\perp$ in V is the set of all those $\alpha \in V$ such that $f(\alpha) = 0$ for every $f \in A$.

In other words, given $S \subseteq V_F$, $S^\perp$ is the set of those $f \in V^*$ such that $S \subseteq \ker(f)$. For any $A \subseteq V^*$, $A^\perp$ is the intersection of all the $\ker(f)$, for $f \in A$. As $0 \in \ker(f)$ for every $f \in V^*$. Similarly, for $0 \in V^*$, $\{0\}^\perp = V$. For V_F finite dimensional, given $\alpha \neq 0$ in V, there exists $f \in V^*$ such that $f(\alpha) \neq 0$. Consequently, $V^\perp$ contains no non-zero functional i.e., $V^\perp = \{0\}$. We get the following :

Lemma 11. Let V_F be finite dimensional. Then

(i) $$\{0\}^\perp = V^*$$

(ii) For $0 \in V^*$, $$\{0\}^\perp = V$$

(iii) $$V^\perp = \{0\}$$

(iv) $$(V^*)^\perp = \{0\}.$$

Lemma 12. Let S_1, S_2 be two subsets of V_F. Then,

(i) $S_1^\perp$ is a subspace of V^* and $S_1 \subseteq (S_1^\perp)^\perp$

(ii) If $S_1 \subseteq S_2$, then $S_2^\perp \subseteq S_1^\perp$

(iii) If W is the subspace of V spanned by S_1, then $W^\perp = S_1^\perp$.

Proof.

(i) For any $\alpha \in V$, $0(\alpha) = 0$, so that $0 \in S_1^\perp$. Let $f, g \in S_1^\perp$ and $c \in F$.

Then, for any $\alpha \in S_1$, $\quad f(\alpha) = 0 = g(\alpha)$,

gives $$(f + cg)(\alpha) = f(\alpha) + cg(\alpha) = 0$$

so that $$f + cg \in S_1^\perp.$$

Hence, $S_1^\perp$ is a subspace of V^*. As $f(\alpha) = 0$ for every $f \in S_1^\perp$, we get
$$S_1 \subseteq (S_1^\perp)^\perp.$$

(ii) Let $S_1 \subseteq S_2$. Let $f \in S_2^\perp$. Then, for any $\alpha \in S_1$, as $\alpha \in S_2$, we get $f(\alpha) = 0$.

Consequently, $$f \in S_1^\perp.$$

Hence, $$S_2^\perp \subseteq S_1^\perp.$$

(iii) As $S_1 \subseteq W$, by *(ii)* $$W^\perp \subseteq S_1^\perp.$$

Consider an $f \in S_1^{\perp}$ and $\alpha \in W$.

If $\alpha = 0$, obviously, $\quad f(\alpha) = 0.$

Let $\alpha \neq 0$. Then, $\quad \alpha = \sum_{i=1}^{k} a_i \alpha_i$

for some $\alpha_i \in S_1$ and $a_i \in F$.

Then, $\quad f(\alpha_i) = 0.$

Consequently, $\quad f(\alpha) = \sum a_i f(\alpha_i) = 0,$

so that $\quad f \in W^{\perp}.$

This gives $\quad S_1^{\perp} \subseteq W^{\perp}.$

Hence, $\quad W^{\perp} = S_1^{\perp}.$

Similarly, we can prove the following :

Lemma 13. Let V_F be any vector space. Then,

(*i*) For any subset A of V^*, $A^{\perp}$ is a subspace of V and $A \subseteq (A^{\perp})^{\perp}$

(*ii*) For $A_1 \subseteq A_2 \subseteq V^*$, $A_2^{\perp} \subseteq A_1^{\perp}$

(*iii*) If W is the subspace of V^* spanned by a subset A of V^*, then, $W^{\perp} = A^{\perp}$.

Example 13.a. Let $\quad V = \mathbf{R}^3$, $S = \{(1,\ 2,\ -1),\ (3,\ 0,\ 1)\}$.

Now, any $f \in V^*$ is given by

$$f(x_1, x_2, x_3) = a_1 x_1 + a_2 x_2 + a_3 x_3 \qquad \text{(Corollary 4).}$$

Then, $\quad f \in S^{\perp}$

iff $\quad f(1,\ 2,\ -1) = 0$

$\quad f(3,\ 0,\ 1) = 0$

iff $\quad a_1 + 2a_2 - a_3 = 0$

$\quad 3a_1 + 0a_2 + a_3 = 0.$

This is a system of two homogeneous linear equations in a_1, a_2, a_3 with coefficient matrix $A = \begin{bmatrix} 1 & 2 & -1 \\ 3 & 0 & 1 \end{bmatrix}$, whose rows are precisely the members of S. The general solution is

$$(a_1, a_2, a_3) = \left(-\frac{1}{2}r,\ r,\ \frac{3}{2}r\right) = (-s,\ 2s,\ 3s)$$

where $s = 2r$, is arbitrary, so that $S^{\perp}$ consists of those $f \in V^*$ which are given by the formula

$$f(x_1, x_2, x_3) = -sx_1 + 2sx_2 + 3sx_3 = s(-x_1 + 2x_2 + 3x_3)$$

For $s = 1$, $f = f_1$ is given by

$$f_1(x_1, x_2, x_3) = -x_1 + 2x_2 + 3x_3.$$

This f_1 spans $S^{\perp}$.

Hence, $$\dim S^{\perp} = 1.$$

Observe that S is a linearly independent subset of V, so that if W is the subspace of V spanned by S, then, $\dim W = 2$.

Also $$W^{\perp} = S^{\perp},$$

gives $$\dim W^{\perp} = 1,$$

so that $$\dim W + \dim W^{\perp} = 3 = \dim V$$

This last equation holds in general, as will be seen in the next theorem.

Theorem 14. Let W be any subspace of a finite dimensional vector space V_F. Then

(*i*) For any $\alpha \in V - W$, there exists a linear functional $f \in W^{\perp}$ such that $f(\alpha) \neq 0$

(*ii*) $$\dim W + \dim W^{\perp} = \dim V$$

Proof.

(*i*) Let $\dim V = n$ and $\dim W = m$. Let $B' = \{\alpha_1, \alpha_2, ..., \alpha_m\}$ be a basis of W.

Consider $$\alpha \in V - W.$$

As $\alpha \notin W$, by Lemma (2.2.14), the $m+1$ vectors $\alpha_1, \alpha_2, ..., \alpha_m, \alpha_{m+1} = (\alpha)$ are linearly independent. Therefrom we get a basis

$$B = \{\alpha_1, \alpha_2, ..., \alpha_n\}$$

of V.

Consider the dual basis B^* of B

$$B^* = \{\alpha_1^*, \alpha_2^*, ..., \alpha_n^*\}$$

Put $$f = \alpha_{m+1}^*.$$

For $1 \leq j \leq m$, as $$j \neq m+1,\ \alpha_{m+1}(\alpha_j) = 0,$$

i.e. $$f(\alpha_j) = 0.$$

So that $$f \in (B')^{\perp} = W^{\perp} \qquad \text{(Lemma 12)}$$

At the same time $$f(\alpha) = \alpha_{m+1}^*(\alpha_{m+1}) = 1 \neq 0.$$

This proves (*i*)

(*ii*) For $m+1 \leq k \leq n$ and $1 \leq j \leq m$, obviously $\alpha_k^*(\alpha_j) = 0$.

so that $$B_2 = \{\alpha_{m+1}^*, \alpha_{m+2}^*, ..., \alpha_n^*\} \subseteq W^{\perp}.$$

Now B_2 is a linearly independent subset with $n-m$ members. Consider any $f \in W^{\perp}$

As $f \in V^*$, $$f = \sum_{i=1}^{n} f(\alpha_i)\,\alpha_i^*.$$

For $1 \le j \le m$, $\alpha_j \in W$

gives $f(\alpha_j) = 0.$

So that $$f = \sum_{i=m+1}^{n} f(\alpha_i)\,\alpha_i^*$$

Consequently, B_2 spans $W^\perp$. Thus, B_2 is a basis of $W^\perp$.

This gives $\dim W^\perp = |B_2| = n - m = n - \dim W.$

Hence, $\dim W + \dim W^\perp = n = \dim V.$

Theorem 15. Let V_F be any finite dimensional vector space. Then, the following relations hold :

(*i*) For any subspace W_1 of V^*,

$$\dim W_1 + \dim W_1^\perp = \dim V$$

(*ii*) For any subspace W of V, $(W^\perp)^\perp = W$

(*iii*) For any subspace W_1 of V^*, $(W_1^\perp)^\perp = W_1$

Proof.

(*i*) Let $\dim V = n, \dim W_1 = m.$

Consider the canonical isomorphism

$$\sigma : V \to V^{**}.$$

Let W_1^0 be the annihilator of W_1 in $(V^*)^*$.

By Theorem 14, applied to V^*

$$\dim W_1^0 = n - m$$

Now, $\alpha \in W_1^\perp$ iff $f(\alpha) = 0$ for every $f \in W_1$.

However, $f(\alpha) = [\sigma(\alpha)](f),$

so that $\alpha \in W_1^\perp$ iff $[\sigma(\alpha)](f) = 0$ for every $f \in W_1$ iff $\sigma(\alpha) \in W_1^0$.

So that $W_1^\perp = \sigma^{-1}(W_1^0).$

This yields $\dim W_1^\perp = \dim W_1^0 = n - m.$

Hence, $\dim W_1^\perp + \dim W_1 = n = \dim V.$

(*ii*) Let $\dim W = k.$

By Theorem 14, $\dim W^\perp = n - k.$

By (*i*), $\dim (W^\perp)^\perp = n - (n - k) = k = \dim W.$

As $W \subseteq (W^\perp)^\perp,$

we get $$W = (W^{\perp})^{\perp}.$$

Similar arguments prove (*iii*).

Theorem 16. (Duality Theorem). Let V_F be any finite dimensional vector space, $L(V)$ and $L(V^*)$ be the lattices of subspaces of V and V^*, respectively.

Then, the mapping $$\lambda : L(V) \to L(V^*)$$

such that $\lambda(W) = W^{\perp}$, $W \in L(V)$, is one-to-one and onto.

It is order reversing in the sense that for any $W_1, W_2 \in L(V)$

$$W_1 \subseteq W_2$$

iff $$W_2^{\perp} \subseteq W_1^{\perp}$$

Proof. Consider two subspace W_1, W_2 of V.

Let $$\lambda(W_1) = \lambda(W_2).$$

Then, $$W_1^{\perp} = W_2^{\perp},$$

so that $$(W_1^{\perp})^{\perp} = (W_2^{\perp})^{\perp}.$$

By Theorem 15, $$W_1 = (W_1^{\perp})^{\perp}, W_2 = (W_2^{\perp})^{\perp}.$$

So that $$W_1 = W_2.$$

Hence, λ is one-to-one. Consider any subspace W' of V^*.

By Theorem 15 (*iii*), $$\lambda\{(W')^{\perp}\} = \{(W')^{\perp}\}^{\perp} = W'.$$

Hence, λ is also onto. The last part is an immediate consequence of Lemma 12 and 13, and Theorem 15.

Theorem 17. For any two subspace W_1, W_2 of a finite dimensional vector space V_F,

$$(W_1 + W_2)^{\perp} = W_1^{\perp} \cap W_2^{\perp}$$
$$(W_1 \cap W_2)^{\perp} = W_1^{\perp} + W_2^{\perp}$$

Proof. As $$W_1 \subseteq W_1 + W_2$$

and $$W_2 \subseteq W_1 + W_2,$$

by Lemma 12, $$(W_1 + W_2)^{\perp} \subseteq W_1^{\perp} \cap W_2^{\perp}.$$

Conversely, let $$f \in W_1^{\perp} \cap W_2^{\perp},$$

so that $$f \in W_1^{\perp}$$

and $$f \in W_2^{\perp}.$$

Consider $$\alpha \in W_1 + W_2.$$

Then, $$\alpha = \alpha_1 + \alpha_2$$

with $\alpha_1 \in W_1, \alpha_2 \in W_2$.

So $f(\alpha_1) = 0 = f(\alpha_2)$

and $f(\alpha) = f(\alpha_1) + f(\alpha_2) = 0$.

Consequently, $f \in (W_1 + W_2)^{\perp}$.

This proves that $W_1^{\perp} \cap W_2^{\perp} \subseteq (W_1 + W_2)^{\perp}$.

Hence, $(W_1 + W_2)^{\perp} = W_1^{\perp} \cap W_2^{\perp}$.

For any subspace W of V^*, let W^0 be the annihilator of W in V^{**}. If $\sigma : V \to V^{**}$ is the canonical isomorphism, as seen in the proof of Theorem 15,

$$W^{\perp} = \sigma^{-1}(W^0).$$

Put $U_1 = W_1^{\perp}, U_2 = W_2^{\perp}$.

As V^{**} is the dual of V^*, $(U_1 + U_2)^0 = U_1^0 \cap U_2^0$

This gives $\sigma^{-1}\left[(U_1 + U_2)^0\right] = \sigma^{-1}(U_1^0 \cap U_2^0)$.

As σ is one-to-one and onto,

$$\sigma^{-1}(U_1^0 \cap U_2^0) = \sigma^{-1}(U_1^0) \cap \sigma^{-1}(U_2^0).$$

so that $(U_1 + U_2)^{\perp} = U_1^{\perp} \cap U_2^{\perp}$

i.e., $(W_1^{\perp} + W_2^{\perp})^{\perp} = (W_1^{\perp})^{\perp} \cap (W_2^{\perp})^{\perp}$

i.e., $(W_1^{\perp} + W_2^{\perp})^{\perp} = W_1 \cap W_2$

Hence, $(W_1 \cap W_2)^{\perp} = \left[(W_1^{\perp} + W_2^{\perp})^{\perp}\right]^{\perp} = W_1^{\perp} + W_2^{\perp}$.

To give some applications of dual spaces, we first of all discuss a self duality in $V = F^n$.

For any subset S of F^n, define S^0 to be the set of those $\alpha = (x_1, x_2, \ldots, x_n) \in F^n$, such that for any $\beta = (y_1, y_2, \ldots, y_n) \in S$, the dot product

$$\alpha.\beta = x_1 y_1 + x_2 y_2 + \ldots + x_n y_n = 0.$$

It can be easily seen that S^0 is a subspace of F^n. As in Lemma 12, if W is the subspace spanned by a subset S of F^n, then, $W^0 = S^0$. Here, S^0 is called the *orthogonal* of S in F^n.

Theorem 18. For any subspace W of F^n

(i) $\dim W + \dim W^0 = n$

(ii) $(W^0)^0 = W$.

Proof. Let $V = F^n$.

For any $f \in V^*$, we have unique

$$\alpha_f = (a_1, a_2, \ldots, a_n) \in F^n$$

such that $f(x_1, x_2,..., x_n) = a_1x_1 + a_2x_2 +...+ a_nx_n$

and $$\eta : V^* \to F^n$$

given by $$\eta(f) = \alpha_f,\ f \in V^*$$

is an isomorphism (Corollary 4).

Consider a subspace W of F^n.

By definition $$\alpha = (a_1, a_2,..., a_n) \in W^0$$

iff $$x_1a_1 + x_2a_2 +...+ x_na_n = 0$$

for every $$\beta = (x_1, x_2,..., x_n) \in W$$

iff $$\eta^{-1}(\alpha) \in W^{\perp}.$$

Hence, $$\eta^{-1}(W^0) = W^{\perp}.$$

So that $$\dim W^0 = \dim W^{\perp}.$$

As $$\dim W + \dim W^{\perp} = n,$$

we get $$\dim W + \dim W^0 = n.$$

This proves (*i*).

Now, $$W \subseteq (W^0)^0.$$

Further, $$\dim W^0 + \dim (W^0)^0 = n = \dim W + \dim W^0.$$

This gives $$\dim W = \dim (W^0)^0$$

and, hence, $$W = (W^0)^0.$$

By using Theorem 16 and the theorem, given above, we get following, called *self-duality theorem.*

Theorem 19. The mapping $\lambda : L(F^n) \to L(F^n)$ such that for any subspace W of F^n, $\lambda(W) = W^0$, is one-to-one, onto and order reversing.

Let $A = \left[a_{ij}\right]$ be any $m \times n$-matrix over a field. Consider the system of homogeneous linear equations $AX = 0$. We had proved in chapter two that the dimension of the solution space W of this system is

$$n - \text{column-rank } (A)$$

and that $$\text{row-rank } (A) = \text{column-rank}(A).$$

We now give other proofs of these results.

Theorem 20. Let $A = \left[a_{ij}\right]$ be any $m \times n$-matrix over a field F.

(*i*) For W, the solution space of the system of homogeneous linear equations $AX = 0$, $\dim W = n - \text{row-rank } (A)$

(*ii*) $$\text{row-rank } (A) = \text{column-rank } (A).$$

Proof. Let W be the solution space of $AX = 0$, and W_1 be the row-space of A.

Consider the linear transformation $\sigma : F^{n\times 1} \to F^{m\times 1}$

such that $$\sigma(X) = AX,\ X \in F^{n\times 1}.$$

Clearly, $$W = \ker\sigma.$$

Now $$\alpha = (a_1, a_2, \ldots, a_n) \in W_1^0$$

iff for each $$A_i = (a_{i1}, a_{i2}, \ldots, a_{in}),\ A_i\alpha^T = 0,\ 1 \le i \le m$$

iff $$\alpha^T \in W.$$

Hence, $$W_1^0 \cong W.$$

Consequently,

$$\text{nullity}(\sigma) = \dim W = \dim W_1^0 = n - \dim W_1 = n - \text{row-rank}(A).$$

As seen in Example (3.1.6b),

$$\text{range}(\sigma) = \text{column-space}(A),$$

so that $$\text{column-rank}(A) = \text{rank}(\sigma).$$

But $$\text{nullity}(\sigma) = n - \text{rank}(\sigma).$$

Hence, $$n - \text{rank}(\sigma) = n - \text{row-rank}(A)$$

i.e., $$\text{rank}(\sigma) = \text{row-rank}(A)$$

i.e., $$\text{column-rank}(A) = \text{row-rank}(A).$$

Exercise 3.4

1. In $\mathbf{R}^3$, let $\alpha_1 = (1, 0, 1)$, $\alpha_2 = (0, 2, -1)$ and $\alpha_3 = (1, -1, 0)$.

   (*i*) Find the linear functional f on $V = \mathbf{R}^3$, such that
   $$f(\alpha_1) = 1,\ f(\alpha_2) = -3,\ f(\alpha_3) = -1$$

   (*ii*) Find a non-zero linear functional g on V such that
   $$g(\alpha_1) = 0 = g(\alpha_2).$$
   If $\alpha = (a, b, c) \in V$ is such that α is not in the subspace spanned by α_1, α_2, prove that $g(\alpha) \neq 0$.

   (*iii*) Find dual basis B^* of $B = \{\alpha_1, \alpha_2, \alpha_3\}$.

2. In $V = \mathbf{R}^3$, find the general formula for those linear functionals f on V such that $f \in W^{\perp}$, where W is the subspace of V spanned by
   $$\alpha_1 = (1, 0, 1),\ \alpha_2 = (1, -1, 0).$$
   Find $\dim W^{\perp}$.

3. Let $V = \mathbf{R}^4$ and W be the subspace spanned by

$$\alpha_1 = (2, -2, 3, 4), \alpha_2 = (3, -3, 2, 2)$$

(*i*) Find $\dim W$ **and** $\dim W^{\perp}$**, by explicitly giving their respective bases. verify that**

$$4 = \dim W + \dim W^{\perp}$$

(*ii*) Let $g \in V^*$ be given by $g(x_1, x_2, x_3, x_4) = x_1 + x_2 + x_3 - x_4$. Does $g \in W^{\perp}$?

(*iii*) Find a basis $\{\alpha_3, \alpha_4\}$ **of** W^0. **Prove that** $B = \{\alpha_1, \alpha_2, \alpha_3, \alpha_4\}$ **is a basis of** $\mathbf{R}^4$.

4. Let f_1, f_2, f_3 **be three linear functionals on** $V = \mathbf{R}^5$, **given by the formulae**

$$f_1(x_1, x_2, x_3, x_4, x_5) = 2x_1 + x_3$$
$$f_2(x_1, x_2, x_3, x_4, x_5) = x_1 - x_2 + x_3 - x_4 + x_5$$
$$f_3(x_1, x_2, x_3, x_4, x_5) = 3x_1 - x_2 + 2x_3 - x_4 + x_5$$

(*i*) Let W' **be the subspace of** V^* **spanned by** f_1, f_2, f_3. **Find** $\dim W'$.

(*ii*) Find a basis of the subspace W **of** V **such that** $W^{\perp} = W'$.

5. Let W **be the row-space of** $A = \begin{bmatrix} 1 & 2 & 2 & 1 \\ 1 & 2 & 2 & 2 \\ 2 & 4 & 3 & 3 \\ 0 & 0 & 1 & -1 \end{bmatrix}$

(*i*) Find the general formula for those $f \in (\mathbf{R}^4)^*$ **that belong to** $W^{\perp}$

(*ii*) Find the general formula for those $\alpha = (a, b, c, d) \in \mathbf{R}^4$, **which belong to** W^0.

6. Let $V = \mathbf{R}^2$. **Find the formulae for linear functionals** f **and** g **on** V **such that for a fixed** θ,

$$f(\cos\theta, \sin\theta) = 1, f(-\sin\theta, \cos\theta) = 2$$

$$g(\cos\theta, \sin\theta) = 2, g(-\sin\theta, \cos\theta) = 1$$

(*i*) Prove that f **and** g **are linearly independent**

(*ii*) Find an ordered basis $B = \{\alpha, \beta\}$ **of** V **such that** $B' = \{f, g\}$ **is the dual of** B.

7. Let $V = P_2(\mathbf{R})$. **Define** f_1, f_2, f_3 **in** V^* **such that for any** $p(x) = a_0 + a_1 x + a_2 x^2$,

$$f_1\{p(x)\} = \int_0^1 p(x)dx,$$

$$f_2\{p(x)\} = \int_0^2 p(x)dx$$

and

$$f_3\{p(x)\} = \int_0^{-1} p(x)dx.$$

Show that $B' = \{f_1, f_2, f_3\}$ is a basis of V^*, and find a basis B of V, of which B' is dual.

8. Let $V = P_3(\mathbf{R})$. Let $a_1 = 1$, $a_2 = 2$, $a_3 = 3$, $a_4 = -1$. Find four polynomials,

$$p_1(X),\ p_2(X),\ p_3(X),\ p_4(X)$$

in V such that $p_i(a_j) = \delta_{ij}$. Express, each of $1, X, X^2, X^3$, as linear combination of $p_1(X),\ p_2(X),\ p_3(X),\ p_4(X)$.

9. Let $V = P_3(\mathbf{R})$. Find the dual basis of $B = \{1,\ 1+X,\ (1+X)^2,\ (1+X)^3\}$

10. Let W be the subspace of $V = P_3(\mathbf{R})$, consisting of those $p(X) \in V$, such that

$$p(1) = 0,\ p(-1) = 0.$$

Find $\dim W$ and $\dim W^{\perp}$.

11. Let n and m be any two positive integers

(*i*) For any m linear functionals $f_1, f_2, \ldots, f_m$ on F^n, prove that $\sigma : F^n \to F^m$ given by $\sigma(\alpha) = (f_1(\alpha), f_2(\alpha), \ldots, f_m(\alpha))$ is a linear transformation.

(*ii*) Given any linear transformation $T : F^n \to F^m$, prove that there exist uniquely determined linear functionals $g_1, g_2, \ldots, g_m$ depending upon T such that

$$T(\alpha) = (g_1(\alpha),\ g_2(\alpha), \ldots,\ g_m(\alpha)).$$

12. Let V and W be two finite dimensional vector spaces over a field F. Let $B = \{\alpha_1, \alpha_2, \ldots, {}_n\}$ and $B' = \{\beta_1, \beta_2, \ldots, \beta_m\}$ be their respective ordered basis.

(*i*) For any $T \in \mathrm{Hom}_F(V, W)$, prove that $T^* : W^* \to V^*$ such that

$$[T^*(f)](\alpha) = f[T(\alpha)],\ f \in W^*,\ \alpha \in V,$$

is a linear transformation.

(*ii*) Prove that $f \in \ker(T^*)$ iff $f \in \{\mathrm{range}(T)\}^{\perp}$

(*iii*) By using (*ii*), prove that $P(T) = P(T^*)$

(*iv*) Let B^* and $(B')^*$ be the dual basis of B and B', respectively. If $A = [T]_{(B,\,B')}$, prove that the matrix of T^* relative to $(B')^*$, B^* is A^T. This T^* is called the *dual* of T.

(*v*) Given $T_1 : V_F \to W_F$, $T_2 : W_F \to U_F$, prove that $(T_2T_1)^* = T_1^*T_2^*$.

13. Let f and g be two linear functionals on a finite dimensional vector space V_F. Prove that $g = cf$ for some $c \in F$ iff $\ker(f) \subseteq \ker(g)$.

14. For any two $n \times n$-matrices A and B over a field F, prove that :

(*i*) $\mathrm{trace}(AB) = \mathrm{trace}(BA)$

(*ii*) If A and B are similar, then $\text{trace}(A) = \text{trace}(B)$. Further, prove that for no two matrices A and B in $\mathbf{R}^{2\times 2}$, $I_2 = AB - BA$.

15. Let W be the subspace of $V = \mathbf{R}^{2\times 2}$ consisting of those $A \in V$ such that $AB = BA$, $B = \begin{bmatrix} 1 & 3 \\ 2 & 6 \end{bmatrix}$. Find $\dim W$ and $\dim W^{\perp}$. Does there exist a non-zero $f \in V^*$ such that $f(I) = 0$, $f\left\{\begin{bmatrix} 0 & 0 \\ 0 & 1 \end{bmatrix}\right\} = 0$ and $f \in W^{\perp}$?

3.5 Linear Transformations on Direct Sums

Recall that a linear operator E on a vector space V_F is called an idempotent linear operator if $E^2 = E$; in that case, $V = W \oplus W'$, where $W = \text{range}(E)$ and $W' = \ker(E)$ {Lemma (3.3.15)}.

Definition 1. A collection of idempotent linear operators $E_1, E_2, \ldots, E_k$ on a vector space V_F is called an *orthogonal collection* if $E_i E_j = 0$ for $i \neq j$.

Example 1.a. Let $V = \mathbf{R}^2$.

Then, $E_1 : V \to V$, $E_2 : V \to V$

such that $E_1(x, y) = x$, $E_2(x, y) = y$

for any $\alpha = (x, y) \in V$, are idempotent linear operators and $E_1E_2 = E_2E_1 = 0$. So $\{E_1, E_2\}$ is an orthogonal set of idempotent linear operators.

We now give a characterisation of direct sum of subspaces in terms of orthogonal idempotent linear operators.

Theorem 2. Let $W_1, W_2, \ldots, W_k$ be subspaces of a vector space V_F.

Then, $$V = W_1 \oplus W_2 \oplus \ldots \oplus W_k$$

iff there exists an orthogonal set $B = \{E_1, E_2, \ldots, E_k\}$ of idempotent linear operators on V, such that

(*i*) $\text{range}(E_i) = W_i$, and

(*ii*) $I = E_1 + E_2 + \ldots + E_k$.

Proof. Let $$V = W_1 \oplus W_2 \oplus \ldots \oplus W_k.$$

Consider $\alpha \in V$. Then, $$\alpha = \alpha_1 + \alpha_2 + \ldots + \alpha_k$$

for some uniquely determined $\alpha_i \in W_i$.

For $1 \le i \le k$, define $E_i : V \to V$

such that $E_i(\alpha) = \alpha_i$.

For $\alpha_i \in W_i$, $$\delta_{ji}\alpha_i = 0 \in W_j$$

for $j \neq i$ and $\delta_{ii}\alpha_i = \alpha_i \in W_i$.

We can write $\qquad \alpha_i = \sum_{j=1}^{k} \delta_{ji}\alpha_j,\ \delta_{ji}\alpha_j \in W_j.$

So, by definition $\qquad E_i(\alpha_i) = \delta_{ii}\alpha_i = \alpha_i.$

Consequently, $\qquad E_i^2(\alpha) = E_i(\alpha_i) = \alpha_i.$

This gives $\qquad \text{range}(E_i) = W_i$

and

$$E_i^2 = E_i.$$

For $i \neq j$, $\qquad \alpha_j = \sum_{s=1}^{k} \delta_{sj}\alpha_s,\ \delta_{sj}\alpha_s \in W_s$

gives

$$E_i(\alpha_j) = \delta_{ij}\alpha_i = 0,$$

as

$$\delta_{ij} = 0.$$

Hence, $\qquad E_iE_j(\alpha) = E_i(\alpha_j) = 0 \qquad$ for $i \neq j$.

Consequently, $\{E_1, E_2, \ldots, E_k\}$ is an orthogonal set of idempotent linear operators on V.

Finally, $\qquad \alpha = \sum_{i=1}^{k} \alpha_i = \sum_{i=1}^{k} E_i(\alpha) = \left[\sum_{i=1}^{k} E_i\right](\alpha)$

gives

$$I = \sum_{i=1}^{k} E_i.$$

Conversely, let the given conditions be satisfied.

Consider $\alpha \in V$. Then, $\qquad \alpha = \sum_{i=1}^{k} E_i(\alpha),$

with

$$E_i(\alpha) \in W_i,$$

gives

$$V = \sum_{i=1}^{k} W_i$$

For some $\alpha_i \in W_i$, let $\qquad \alpha_1 + \alpha_2 + \ldots + \alpha_k = 0.$

However, $\qquad E_i(\alpha_i) = \alpha_i.$

This gives $\qquad \sum_{i=1}^{k} E_i(\alpha_i) = 0$

For $1 \leq j \leq k$, $\qquad E_j\left[\sum_{i=1}^{k} E_i(\alpha_i)\right] = E_j(0) = 0,$

i.e., $$\sum_{i=1}^{k} E_j E_i(\alpha_i) = 0$$

i.e., $$E_i(\alpha_i) = 0,$$

as $$E_j E_i = 0 \qquad \text{for } j \neq i$$

i.e., $$\alpha_i = 0.$$

Hence, $$V = W_1 \oplus W_2 \oplus \ldots \oplus W_k.$$

Given an idempotent linear operator E on V_F, if $W = \text{range}(E)$, then, E is called a *projection* of V onto W. If $W' = \ker(E)$, we say E is a *projection* of V onto W via W'. Given orthogonal set of projections

$$E_1, E_2, \ldots, E_k,$$

with $$V = W_1 \oplus W_2 \oplus \ldots \oplus W_k,\ W_i = \text{range}(E_i),$$

we say $$E_1, E_2, \ldots, E_k$$

determine the direct sum $$V = W_1 \oplus W_2 \oplus \ldots \oplus W_k.$$

Given a linear operator T on a non-zero finite dimensional vector space V_F, T admits different matrices relative to different ordered bases of V. Among these matrices, some are easily recognisable, and reflect different algebraic properties of T; these are known as matrices in canonical forms. These canonical forms will be discussed in details in the later chapters. To achieve this goal, we need the following concept :

Definition 3. Let T be a linear operator on a vector space V_F. A subspace W of V is said to be *invariant under T (or a T-invariant subspace)* if $T\alpha \in W$ for every $\alpha \in W$. [i.e., $T(W) \subseteq W$].

Example 3.a. Let T be the linear operator on $\mathbf{R}^5$ such that

$$T\epsilon_1 = \epsilon_2,\ T\epsilon_2 = \epsilon_3,\ T\epsilon_3 = 0,\ T\epsilon_4 = \epsilon_5,\ T\epsilon_5 = \epsilon_4.$$

Let W_1 be the subspace spanned by $\epsilon_1, \epsilon_2, \epsilon_3$.

Consider $$\alpha \in W_1.$$

Then, $$\alpha = a_1\epsilon_1 + a_2\epsilon_2 + a_3\epsilon_3,\ a_i \in \mathbf{R},$$

and $$T\alpha = a_1 T\epsilon_1 + a_2 T\epsilon_2 + a_3 T\epsilon_3 = a_1\epsilon_2 + a_2\epsilon_3 \in W_1,$$

so that W_1 is invariant under T.

Let W_2 be the subspace spanned by ϵ_4, ϵ_5.

Then, for any $\beta \in W_4$, $$\beta = b_1\epsilon_4 + b_2\epsilon_5,$$

gives $$T\beta = b_1\epsilon_5 + b_2\epsilon_4 \in W_2.$$

Hence, W_2 is also invariant under T.

Proposition 4. Let W be a subspace of a vector space V_F. Let Inv(W) be the set of those linear operator T on V for which W is T-invariant. Then,

(*i*) $I \in \text{Inv}(W)$, $0 \in \text{Inv}(W)$

(*ii*) Inv(W) is a subspace of $L(V)$

(*iii*) For any $T_1, T_2 \in \text{Inv}(W)$, $T_1T_2 \in \text{Inv}(W)$.

Proof. For any $\alpha \in W$, $I(\alpha) = \alpha \in W$

and $0(\alpha) = 0 \in W$,

give $I,\ 0 \in \text{Inv}(W)$.

Let $T_1, T_2 \in \text{Inv}(W)$ and $c \in F$. Consider $\alpha \in W$.

Then, $T_1\alpha \in W$, $T_2\alpha \in W$,

give $(T_1 + cT_2)(\alpha) = T_1\alpha + cT_2\alpha \in W$,

so that $T_1 + cT_2 \in \text{Inv}(W)$.

Hence, Inv(W) is a subspace of $L(V)$. Again $T_2\alpha \in W$ and W is T_1-invariant, yield

$$T_1(T_2\alpha) \in W,$$

i.e., $(T_1T_2)(\alpha) \in W$.

Hence, $T_1T_2 \in \text{Inv}(W)$.

Given $T \in \text{Inv}(W)$, the proposition, given above, shows that $T^k \in \text{Inv}(W)$ for every $k \geq 1$.

Also, $T^0 = I \in \text{Inv}(W)$,

so that as Inv(W) is a subspace of $L(V)$, $a_0I + a_1T + a_2T^2 + ... + a_sT^s \in \text{Inv}(W)$ for all $a_i \in F$, and any $s \geq 0$.

Define $T' : W \to W$ such that $T'(\alpha) = T(\alpha)$ for $\alpha \in W$. Then, T' is a linear operator on W, called the restriction of T to W.

Theorem 5. Let $V_F = W_1 \oplus W_2 \oplus ... \oplus W_k$ be determined by orthogonal projections E_1, $E_2, ..., E_k$ on V, and T be a linear operator on V. Then, every W_i is T-invariant iff $E_iT = TE_i$ for $1 \leq i \leq k$.

Proof. Let every W_i be T-invariant.

For any $\alpha_i \in W_i$, $E_i(\alpha_i) = \alpha_i$.

As $T\alpha_i \in W_i$

for an $\alpha_i \in W$, $E_i(T\alpha_i) = T\alpha_i$.

Consider $\alpha \in V$. Then, $E_i(\alpha) = \alpha_i \in W_i$,

yields $(TE_i)(\alpha) = T\alpha_i$

Again, $\alpha = \alpha_1 + \alpha_2 + ... + \alpha_k,\ \alpha_i \in W_i$, gives $$T\alpha = T\alpha_1 + T\alpha_2 + ... + T\alpha_k,\ T\alpha_i \in W_i.$$

This gives $$E_i(T\alpha) = T\alpha_i.$$

Hence, $$(TE_i)(\alpha) = (E_iT)(\alpha),$$

i.e., $$TE_i = E_iT.$$

Conversely, let $$E_iT = TE_i \quad \text{for } 1 \le i \le k.$$

Consider $\alpha_i \in W_i$. Then, $T\alpha_i = TE_i(\alpha_i) = E_i\{T(\alpha)\} \in W_i$.

This proves that W_i is T-invariant.

The following is trivial :

Lemma 6. Let T be a linear operator on V_F. If W_1 and W_2 are two T-invariant subspaces of V, then $W_1 + W_2$ and $W_1 \cap W_2$ are T-invariant subspaces.

Definition 7. Let $V_F = W_1 \oplus W_2 \oplus ... \oplus W_k$ and T be a linear operator on V such that each W_i is T-invariant. If T_i is the restriction of T to W_i, then, T is called a *direct sum* of $T_1, T_2, ..., T_k$ and we write

$$T = T_1 \oplus T_2 \oplus ... \oplus T_k.$$

Theorem 8. Let T be a linear operator on a finite dimensional vector space $V_F = W_1 \oplus W_2 \oplus ... \oplus W_k$ such that each W_i is T-invariant. Let T_i be the restriction of T to W_i. Then,

(*i*) For any $\alpha = \alpha_1 + \alpha_2 + ... + \alpha_k \in V$ with $\alpha_j \in W_j$, $1 \le j \le k$,

$$T(\alpha) = \sum_{j=1}^{k} T_j(\alpha_j)$$

(*ii*) $$\rho(T) = \sum_{j=1}^{k} \rho(T_j)$$

Proof. As $$\alpha = \sum_{j=1}^{k} \alpha_j,$$

$$T(\alpha) = \sum_{j=1}^{k} T(\alpha_j).$$

However, T_j being the restriction of T to W_j, gives $T(\alpha_j) = T_j(\alpha_j)$.

Hence, $$T(\alpha) = \sum_{j=1}^{k} T_j(\alpha_j).$$

This proves (*i*).

This also gives $$\text{range}(T) \subseteq \sum_{j=1}^{k} \text{range}(T_j).$$

However, $$T_j(W_j) = T(W_j),$$

gives $$\text{range}(T_j) \subseteq \text{range}(T).$$

Hence, $$\text{range}(T) = \sum_{j=1}^{k} \text{range}(T_j).$$

As $\text{range}(T_j) \subseteq W_j$ and the sum $W_1 + W_2 + \ldots + W_k$ is direct, we get

$$\text{range}(T) = \text{range}(T_1) \oplus \text{range}(T_2) \oplus \ldots \oplus \text{range}(T_k).$$

By Theorem (2.2.24) $\dim\{\text{range}(T)\} = \sum_{j=1}^{k} \dim\{\text{range}(T_j)\}$

i.e., $$\rho(T) = \sum_{j=1}^{k} \rho(T_j)$$

The decomposition of a linear operator T into a direct sum $T_1 \oplus T_2 \oplus \ldots \oplus T_k$, quite often helps us in studying various algebraic properties of T.

Example 8.a. Let $V = \mathbf{R}^5$. Consider the standard basis $\{\epsilon_1, \epsilon_2, \epsilon_3, \epsilon_4, \epsilon_5\}$ of V, and any two matrices A and B over $\mathbf{R}$, such that A is 3×3 and B is 2×2.

To be specific, let $$A = \begin{bmatrix} 0 & 0 & -1 \\ 1 & 0 & 3 \\ 0 & 1 & 1 \end{bmatrix},$$

$$B = \begin{bmatrix} 1 & -1 \\ 1 & 2 \end{bmatrix}$$

Let W_1 and W_2 be two subspaces of $\mathbf{R}^5$ having ordered bases $B_1 = \{\epsilon_1, \epsilon_2, \epsilon_3\}$ and $B_2 = \{\epsilon_4, \epsilon_5\}$, respectively. Let T_1 and T_2 be linear operators on W_1 and W_2, respectively, such that

$$[T_1]_{B_1} = A, [T_2]_{B_2} = B$$

We get linear operator T on V such that for $\alpha = \alpha_1 + \alpha_2$, $\alpha_i \in W_i$,

$$T(\alpha) = T_1(\alpha_1) + T_2(\alpha_2),$$

so that $$T = T_1 \oplus T_2.$$

By the theorem, given above,

$$\rho(T) = \rho(T_1) + \rho(T_2) = 3 + 2 = 5$$

Now, $$T_1 \epsilon_1 = \epsilon_2,$$

$$T_1 \in_2 = \in_3,$$
$$T_1 \in_3 = -\in_1 + 3\in_2 + \in_3,$$
$$T_2 \in_4 = \in_4 + \in_5,$$
$$T_2 \in_5 = -\in_4 + 2\in_5,$$

so that
$$T\in_1 = \in_2 = 0\in_1 + 1\in_2 + 0\in_3 + 0\in_4 + 0\in_5$$
$$T\in_2 = \in_3 = 0\in_1 + 0\in_2 + 1\in_3 + 0\in_4 + 0\in_5$$
$$T\in_3 = -\in_1 + 3\in_2 + \in_3 = -1\in_1 + 3\in_2 + 1\in_3 + 0\in_4 + 0\in_5$$
$$T\in_4 = \in_4 + \in_5 = 0\in_1 + 0\in_2 + 0\in_3 + 1\in_4 + 1\in_5$$
$$T\in_5 = -\in_4 + 2\in_5 = 0\in_1 + 0\in_2 + 0\in_3 - 1\in_4 + 2\in_5$$

So that for the standard ordered basis B of V

$$[T]_B = \begin{bmatrix} 0 & 0 & -1 & 0 & 0 \\ 1 & 0 & 3 & 0 & 0 \\ 0 & 1 & 1 & 0 & 0 \\ 0 & 0 & 0 & 1 & -1 \\ 0 & 0 & 0 & 1 & 2 \end{bmatrix} = \begin{bmatrix} A & 0 \\ 0 & B \end{bmatrix}.$$

The example, given above, motivates the following :

Definition 9. For $1 \le i \le k$, let A_i, be an $n_i \times n_i$-matrix over a field F. For $n = n_i + n_2 + \ldots + n_k$, the $n \times n$-matrix

$$A = \begin{bmatrix} A_i & 0 & 0 & \ldots & 0 & 0 \\ 0 & A_2 & 0 & \ldots & 0 & 0 \\ \vdots & \vdots & \vdots & \vdots & \vdots & \vdots \\ 0 & 0 & 0 & \ldots & 0 & A_k \end{bmatrix}$$

is called a *direct sum* of $A_1, A_2, \ldots, A_k$ and we write

$$A = A_1 \oplus A_2 \oplus \ldots \oplus A_k = \oplus \sum_{i=1}^{k} A_i .$$

Given
$$A = A_1 \oplus A_2 \oplus \ldots \oplus A_k, \qquad k \ge 2,$$

if
$$B = A_2 \oplus \ldots \oplus A_k = \begin{bmatrix} A_2 & 0 & 0 & \ldots & 0 & 0 \\ 0 & A_3 & 0 & \ldots & 0 & 0 \\ \vdots & \vdots & \vdots & \vdots & \vdots & \vdots \\ 0 & 0 & 0 & \ldots & 0 & A_k \end{bmatrix},$$

it is clear from the definition, that $A = A_1 \oplus B$.

So, we have :

Lemma 10. Let A_i for $1 \le i \le k$, be $n_i \times n_i$-matrices, with $k \ge 2$.

Then, $$A = A_1 \oplus A_2 \oplus ... \oplus A_k = A_1 \oplus (A_2 \oplus ... \oplus A_k)$$

We now prove the following useful theorem.

Theorem 11. Let $V_F = W_1 \oplus W_2 \oplus ... \oplus W_k$, be a finite dimensional vector space, with $W_i \ne 0$, $k \ge 1$. Let $T = T_1 \oplus T_2 \oplus ... \oplus T_k$, with each T_i a linear operator on W_i. Then, the following hold. For $1 \le i \le k$, let B_i be an ordered basis of W_i, and $A_i = [T]_{B_i}$.

Then for the ordered basis $$B = B_1 \cup B_2 \cup ... \cup B_k$$

of V, $$A = [T]_B = A_1 \oplus A_2 \oplus ... \oplus A_k.$$

Further, $$\text{rank}(A) = \sum_{i=1}^{k} \text{rank}(A_i).$$

Proof. For $$k = 1, V = W_1, T = T_1, B = B_1,$$

obviously $$[T]_B = [T_1]_{B_1} = A_1.$$

Hence, the result hold for $k = 1$.

Let $k = 2$. Then, $$V = W_1 \oplus W_2$$

We can write $$B_1 = \{\alpha_1, \alpha_2, ..., \alpha_{n_1}\}$$

$$B_2 = \{\alpha_{n_1+1}, \alpha_{n_1+2}, ..., \alpha_n\}$$

where $$n = n_1 + n_2,$$

so that $$B = \{\alpha_1, \alpha_2, ..., \alpha_{n_1}, ..., \alpha_n\}.$$

Here, $$\dim W_1 = n_1, \dim W_2 = n_2.$$

Let $$A_1 = [a_{ij}]_{n_1 \times n_1}$$

and $$A_2 = [b_{ij}]_{n_2 \times n_2}.$$

For $1 \le j \le n_1$ $$T\alpha_j = T_1\alpha_j$$

$$= \sum_{i=1}^{n_1} a_{ij}\alpha_i$$

$$= \sum_{i=1}^{n_1} a_{ij}\alpha_i + \sum_{i=n_1+1}^{n_1+n_2} 0\alpha_i.$$

This gives that the first n_1 columns of $[T]_B$ form the matrix

$$\begin{bmatrix} a_{11} & a_{12} & \dots & a_{1n_1} \\ a_{21} & a_{22} & \dots & a_{2n_1} \\ \vdots & \vdots & \dots & \vdots \\ \vdots & \vdots & \dots & \vdots \\ a_{n_1 1} & a_{n_1 2} & \dots & a_{n_1 n_1} \\ 0 & 0 & \dots & 0 \\ \vdots & \vdots & \dots & \vdots \\ \vdots & \vdots & \dots & \vdots \\ 0 & 0 & \dots & 0 \end{bmatrix} = \begin{bmatrix} A_1 \\ 0 \end{bmatrix}_{n \times n_1}.$$

Similarly, for $n_1 + 1 \le j \le n_1 + n_2 = n$

$$T\alpha_j = T_2\alpha_j$$

$$= \sum_{i=n_1+1}^{n_2} b_{ij}\alpha_i$$

$$= \sum_{i=1}^{n_1} 0\alpha_i + \sum_{i=n_1+1}^{n_1+n_2} b_{ij}\alpha_i.$$

This gives that the last n_2 columns of $[T]_B$ equal $\begin{bmatrix} 0 \\ A_2 \end{bmatrix}_{n \times n_2}$.

Consequently, $$[T]_B = \begin{bmatrix} A_1 & 0 \\ 0 & A_2 \end{bmatrix}.$$

Thus, the result hold for $k = 2$. Let $k > 2$, and to apply induction, let the result hold for direct sums of $k-1$ linear operators.

Now, $$T = T_1 \oplus (T_2 \oplus \dots \oplus T_k) = T_1 \oplus T'$$

where $T' = T_2 \oplus \dots \oplus T_k$ is a direct sum of $k-1$ linear operators.

Now, $B' = B_2 \cup B_3 \cup \dots \cup B_k$ is an ordered basis of $V' = W_2 \oplus \dots \oplus W_k$. So, by the induction hypothesis,

$$[T']_{B'} = B = \begin{bmatrix} A_2 & 0 & \dots & 0 \\ 0 & A_3 & \dots & 0 \\ \vdots & \vdots & \dots & \vdots \\ \vdots & \vdots & \dots & \vdots \\ 0 & 0 & \dots & A_k \end{bmatrix}$$

As $V = W_1 \oplus V'$, $B = B_1 \cup B'$, by the case for $k = 2$

$$[T]_B = [T]_{B_1} \oplus [T]_{B'}$$

$$= A_1 \oplus (A_2 \oplus \dots \oplus A_k)$$

$$= A_1 \oplus A_2 \oplus \dots \oplus A_k$$

Finally, for $A = [T]_B$, $\quad \text{rank}(A) = \text{rank}(T)$

and $\quad \text{rank}(T_i) = \text{rank}(A_i)$,

give $$\rho(A) = \sum_{i=1}^{k} \rho(A_i) \qquad \text{(Theorem 8)}$$

Proposition 4 and the remark after that give the following result :

Theorem 12. Let $V_F = W_1 \oplus W_2 \oplus \ldots \oplus W_k$. Let A be the set of those linear operators T on V for which every W_i is T-invariant. Let $E_1, E_2, \ldots, E_k$ be the orthogonal set of idempotent linear operators on V_F that determine the above decomposition. For any $T \in A$, let T_i denote its restriction to W_i. Then, the following hold :

(*i*) $I \in A$, $0 \in A$

(*ii*) A is a subalgebra of $L(V)$

(*iii*) For any $f(X) \in F[X]$, $T \in A$, $f(T)_i = f(T_i)$

Theorem 13. Let T be a linear operator on a finite dimensional vector space V_F. For some ordered basis $B = \{\alpha_1, \alpha_2, \ldots, \alpha_n\}$ of V, let $A = [T]_B = A_1 \oplus A_2$, where A_i is an $n_i \times n_i$-matrix. Then, $V = W_1 \oplus W_2$ for some T-invariant subspaces W_i of V, such that

(*i*) $B = B_1 \cup B_2$ for some ordered basis B_1 of W_1 and an ordered basis B_2 of W_2

(*ii*) If T_i is the restriction of T to W_i, then, $A_i = [T_i]_{B_i}$.

Proof. Now $$n = n_1 + n_2 = \dim V_F$$

and $$A = \begin{bmatrix} A_1 & 0 \\ 0 & A_2 \end{bmatrix} = [a_{ij}]$$

For $1 \le j \le n_1$, $a_{ij} = 0$, whenever $i > n_1$

This gives $$T\alpha_j = \sum_{i=1}^{n_1} a_{ij}\alpha_i, \; 1 \le j \le n_1 \qquad \ldots (1)$$

Thus, $$W_1 = \langle \alpha_1, \alpha_2, \ldots, \alpha_{n_1} \rangle$$

is T-invariant. For $n_1 < j \le n$, $a_{ij} = 0$, whenever $1 \le i \le n_1$.

This gives $$T\alpha_j = \sum_{i=n_1+1}^{n} a_{ij}\alpha_i, \; n_1 < j \le n \qquad \ldots (2)$$

so that $$W_2 = \langle \alpha_{n_1+1}, \alpha_{n_1+2}, \ldots, \alpha_n \rangle$$

is T-invariant.

Now, $$B_1 = \{\alpha_1, \alpha_2, \ldots, \alpha_{n_1}\}$$

and $$B_2 = \{\alpha_{n_1+1}, \alpha_{n_1+2}, \ldots, \alpha_n\}$$

are ordered bases of W_1 and W_2, respectively.

Let T_i be the restriction of T to W_i.

Then, (i) shows that $$A_1 = [T_1]_{B_1},$$

and (2) show that $$A_2 = [T_2]_{B_2}.$$

Obviously, $$B = B_1 \cup B_2.$$

14. External Direct Sum. Let $V_1, V_2, \ldots, V_n$ be any $n \geq 1$, vector spaces over the same field F. Consider the cartesian product

$$V = V_1 \times V_2 \times \ldots \times V_n$$

$$= \{(\alpha_1, \alpha_2, \ldots, \alpha_n) : \alpha_i \in V_i,\ 1 \leq i \leq n\}$$

For any $$\bar{\alpha} = (\alpha_1, \alpha_2, \ldots, \alpha_n)$$

$$\bar{\beta} = (\beta_1, \beta_2, \ldots, \beta_n)$$

in V and any scalar $a \in F$, define

$$\bar{\alpha} + \bar{\beta} = (\alpha_1 + \beta_1, \alpha_2 + \beta_2, \ldots, \alpha_n + \beta_n) \quad \ldots (1)$$

$$a\bar{\alpha} = (a\alpha_1, a\alpha_2, \ldots, a\alpha_n) \quad \ldots (2)$$

Here, the ith components $\alpha_i + \beta_i$, $a\alpha_i$ of $\bar{\alpha} + \bar{\beta}$ and $a\bar{\alpha}$, respectively are determined by the operations in V_i. The arguments similar to those used in Theorem (1.2.1), prove that V is a vector space over F. This space V_F is known as an *external direct sum* of V_1, $V_2, \ldots, V_n$. If 0_i is the zero vector in V_i, then, $\bar{0} = (0_1, 0_2, \ldots, 0_n)$ is the zero vector in V, for any $\bar{\alpha} = (\alpha_1, \alpha_2, \ldots, \alpha_n)$, its negative $-\bar{\alpha} = (-\alpha_1, -\alpha_2, \ldots, -\alpha_n)$. Let us denote V by $V_1 \times V_2 \times \ldots \times V_n$.

Example 14.a. Consider $V = F^n$. If we take $V_i = F$ for $1 \leq i \leq n_1$, it is immediate that $F^n = V_1 \times V_2 \times \ldots \times V_n$.

Example 14.b. Let $V_1 = F^2, V_2 = \mathbf{P}_2(F), V = V_1 \times V_2$.

Consider $\bar{\alpha} \in V$. By definition, $\bar{\alpha} = (\alpha_1, \alpha_2)$,

$$\alpha_1 \in V_1, \alpha_2 \in \mathbf{P}_2(F).$$

Thus, $$\alpha_1 = (x_1, x_2),\ x_1, x_2 \in F$$

$$\alpha_2 = a_0 + a_1X + a_2X^2,\ a_i \in F.$$

Thus, $$\bar{\alpha} = \{(x_1, x_2),\ a_0 + a_1X + a_2X^2\}.$$

This is the general form of a member of V.

Here, $$\overline{0} = \{(0, 0), 0\},$$

and $$-\overline{\alpha} = \{-(x_1, x_2), -(a_0 + a_1X + a_2X^2)\}$$
$$= \{(-x_1, -x_2), -a_0 - a_1X - a_2X^2\}.$$

Example 14.c. Let $V_1 = \mathbf{R}^{2\times1}$, $V_2 = \mathbf{R}^{3\times1}$ and $V_3 = \mathbf{R}^{2\times2}$. Then, any member of $V_1 \times V_2 \times V_3$ is of the form

$$\overline{\alpha} = \left\{ \begin{bmatrix} x_1 \\ x_2 \end{bmatrix}, \begin{bmatrix} y_1 \\ y_2 \\ y_3 \end{bmatrix}, \begin{bmatrix} a_{11} & a_{12} \\ a_{21} & a_{22} \end{bmatrix} \right\}$$

with x_i, y_j, a_{rs} all in **R**.

The zero vector is $$\overline{0} = \left\{ \begin{bmatrix} 0 \\ 0 \end{bmatrix}, \begin{bmatrix} 0 \\ 0 \\ 0 \end{bmatrix}, \begin{bmatrix} 0 & 0 \\ 0 & 0 \end{bmatrix} \right\}$$

Internal direct sum and external direct sums are closely linked concepts, as we see in the following. Once again, in any vector space V_i, we shall denote the zero vector 0_i simply by 0.

Theorem 15. Let $V_1, V_2, \ldots, V_n$ be any $n \geq 1$, vector spaces over the same field F, and $V = V_1 \times V_2 \times \ldots \times V_n$. Then,

(*i*) For $1 \leq j \leq n$, $W_j = \{\overline{\alpha} = (\alpha_1, \alpha_2, \ldots, \alpha_n) \in V : a_k = 0 \text{ for every } k \neq j\}$ is a subspace of V

(*ii*) $$V_j \cong W_j$$

(*iii*) $$V = W_1 \oplus W_2 \oplus \ldots \oplus W_n$$

Proof. Before we prove the result, observe that W_1 consists of those $\overline{\alpha} \in V$ such that all its components other than the first are zeros, i.e., any $\overline{\alpha} \in W_1$ is of the form

$$(\alpha_1, 0, 0 \ldots, 0).$$

Any vector in W_2 is of the form $(0, \alpha_2, 0, \ldots, 0)$,

and so on.

(*i*) Clearly, $$\overline{0} \in W_j.$$

Consider any $$\overline{\alpha}, \overline{\beta} \in W_j, c \in F.$$

Then, $$\overline{\alpha} = (\alpha_1, \alpha_2, \ldots, \alpha_n)$$
$$\overline{\beta} = (\beta_1, \beta_2, \ldots, \beta_n)$$

with $$\alpha_k = 0 = \beta_k \qquad \text{for every } k \neq j.$$

In $\quad \overline{\alpha}+c\overline{\beta}=(\alpha_1+c\beta_1,\ \alpha_2+c\beta_2,\ldots,\ \alpha_n+c_n\beta_n)$

$$\alpha_k+c\beta_k=0+c0=0 \qquad \text{for } k\neq j.$$

Thus $\quad \overline{\alpha}+c\overline{\beta}\in W_j.$

Hence, W_j is a subspace of V_j.

(*ii*) Define $\quad \sigma_j : V_j \to W_j,$

such that $\quad \sigma_j(\alpha_j)=(0,\ 0,\ldots,\ 0,\ \alpha_j^{j^{th}},\ 0,\ldots,\ 0),$

$$\alpha_j \in V_j.$$

It can be easily seen that σ_j is a vector space isomorphism.

Hence, $\quad V_j \cong W_j$

(*iii*) Consider $\quad \overline{\alpha}=(\alpha_1,\ \alpha_2,\ldots,\ \alpha_n)\in V.$

Then, $\quad \overline{\alpha}_1=(\alpha_1,\ 0,\ 0,\ldots,\ 0)\in W_1,$

$$\overline{\alpha}_2=(0,\ \alpha_2,\ 0,\ldots,\ 0)\in W_2,\ldots,\ \overline{\alpha}_n=(0,\ 0,\ldots,\ \alpha_n)\in W_n,$$

and $\quad \overline{\alpha}=\overline{\alpha}_1+\overline{\alpha}_2+\ldots+\overline{\alpha}_n$

Thus, $\quad V=W_1+W_2+\ldots+W_n$

Suppose for some $\quad \overline{\beta}_j\in W_j,\ 1\le j\le n,$

$$\overline{\beta}_1+\overline{\beta}_2+\ldots+\overline{\beta}_n=\overline{0}$$

As $\quad \overline{\beta}_j=(0,\ 0,\ldots,\ 0,\ \gamma_j^{j^{th}},\ 0,\ldots,\ 0)$

for some $\gamma_j\in V_j$, we get $\quad (\gamma_1,\ \gamma_2,\ldots,\ \gamma_n)=\overline{0},$

$$\gamma_j=0$$

and so $\quad \overline{\beta}_j=\overline{0}.$

By Theorem (2.2.25) $\quad V=W_1\oplus W_2\oplus\ldots\oplus W_n.$

Corollary 16. Let $V_1,\ V_2,\ldots,\ V_n$ be any finitely many, finite dimensional vector spaces over the same field F.

Then, for $\quad V=V_1\times V_2\times\ldots\times V_n,$

$$\dim V=\dim V_1+\dim V_2+\ldots+\dim V_n.$$

Proof. In the notations of the above theorem $V=W_1\oplus W_2\oplus\ldots\oplus W_n.$

So by Theorem (2.2.24), $\quad \dim V=\sum_{i=1}^{n}\dim W_i.$

As, $\quad W_i\cong V_i,\ \dim W_i=\dim V_i.$

Hence, $\quad \dim V=\sum_{i=1}^{n}\dim V_i.$

Theorem 17. If a vector space V_F is an internal direct sum of its subspaces W_1, $W_2, \ldots, W_n$, then, $V \cong W_1 \times W_2 \times \ldots \times W_n$.

Proof. Define $$T : W_1 \times W_2 \times \ldots \times W_n \to V$$

such that $$T(\alpha_1, \alpha_2, \ldots, \alpha_n) = \alpha_1 + \alpha_2 + \ldots + \alpha_n$$

For $$\overline{\alpha} = (\alpha_1, \alpha_2, \ldots, \alpha_n)$$

$$\overline{\beta} = (\beta_1, \beta_2, \ldots, \beta_n)$$

in $$W = W_1 \times W_2 \times \ldots \times W_n,$$

$$c \in F,$$

$$\begin{aligned} T(\overline{\alpha} + c\overline{\beta}) &= T(\alpha_1 + c\beta_1, \alpha_2 + c\beta_2, \ldots, \alpha_n + c\beta_n) \\ &= (\alpha_1 + c\beta_1) + (\alpha_2 + c\beta_2) + \ldots + (\alpha_n + c\beta_n) \\ &= (\alpha_1 + \alpha_2 + \ldots + \alpha_n) + c(\beta_1 + \beta_2 + \ldots + \beta_n) \\ &= T(\overline{\alpha}) + cT(\overline{\beta}), \end{aligned}$$

gives T is a linear transformation.

Now $$T(\overline{\alpha}) = 0,$$

gives $$\alpha_1 + \alpha_2 + \ldots + \alpha_n = 0,$$

so by Theorem (2.2.25), $$\alpha_i = 0, \qquad \text{for } 1 \le i \le n.$$

Hence, $\overline{\alpha} = \overline{0}$ and T is one-to-one.

Given $$\alpha' \in V, \alpha' = \alpha_1' + \alpha_2' + \ldots + \alpha_n'$$

for some $$\alpha_i' \in W_i.$$

Then for $$\overline{\alpha}' = (\alpha_1', \alpha_2', \ldots, \alpha_n') \in W,$$

$$T(\overline{\alpha}') = \alpha'.$$

This proves that T is an isomorphism.

Hence, $$V \cong W_1 \times W_2 \times \ldots \times W_n.$$

The two theorems, given above, show that there is very little difference (in algebraic sense) between internal direct sums and external direct sums of vector spaces. As a result, in most of situations, they are treated as identical concepts. So, we also denote the external direct $V_1 \times V_2 \times \ldots \times V_k$ by $V_1 \oplus V_2 \oplus \ldots \oplus V_k$, except, where there is a need to distinguish between the two concepts.

Exercise 3.5

1. Let $V = \mathbf{R}^2$, $W_1 = \{(a, a) : a \in \mathbf{R}\}$, $W_2 = \{(a, -a) : a \in \mathbf{R}\}$.

   Prove that $V = W_1 \oplus W_2$. Find orthogonal projections E_1, E_2 on V such that

   $$\text{range}(E_1) = W_1, \ \text{range}(E_2) = W_2.$$

2. Let $V = \mathbf{R}^2$ and T be the linear operator on V whose matrix relative to the standard ordered basis is $A = \begin{bmatrix} a & b \\ 0 & a \end{bmatrix}$ for some $a \neq 0, b \neq 0$. Prove the following

(*i*) $W_1 = \langle(1, 0)\rangle$ is T-invariant

(*ii*) $W_2 = \langle(0, 1)\rangle$ is not T-invariant

(*iii*) There exists no T-invariant subspace W_3 of V such that $V = W_1 \oplus W_3$.

3. Prove that the following are pairs of similar matrices over $\mathbf{R}$

(*i*) $\begin{bmatrix} 1 & 0 \\ 0 & 2 \end{bmatrix}$ and $\begin{bmatrix} 2 & 0 \\ 0 & 1 \end{bmatrix}$

(*ii*) $\begin{bmatrix} 1 & -1 & 0 \\ 0 & 1 & 0 \\ 0 & 0 & 2 \end{bmatrix}$ and $\begin{bmatrix} 2 & 0 & 0 \\ 0 & 1 & -1 \\ 0 & 0 & 1 \end{bmatrix}$

(*iii*) $\begin{bmatrix} 0 & 1 & 0 & 0 \\ -1 & 0 & 0 & 0 \\ 0 & 0 & 3 & 4 \\ 0 & 0 & 2 & -2 \end{bmatrix}$ and $\begin{bmatrix} 3 & 4 & 0 & 0 \\ 2 & -2 & 0 & 0 \\ 0 & 0 & 0 & 1 \\ 0 & 0 & -1 & 0 \end{bmatrix}$

[Hint : Use Theorem 11]

4. Give an example of a non-zero linear operator T on $\mathbf{R}^4$ such that T is not idempotent, but range$(T) \cap \ker(T) = \{0\}$. Prove that for any such

$$T, \mathbf{R}^4 = \text{range}(T) \oplus \ker(T).$$

5. Let A be any $n_1 \times n_1$-matrix and B be any $n_2 \times n_2$-matrix over a field F, and $D = \begin{bmatrix} A & 0 \\ 0 & B \end{bmatrix}$. Let $V = F^{n \times 1}$, $n = n_1 + n_2$ and W_1 and W_2 be the subspaces of $F^{n \times 1}$ spanned by sets of vectors $\{\in_1, \in_2, \ldots, \in_{n_1}\}$ and $\{\in_{n_1+1}, \in_{n_1+1}, \ldots, \in_n\}$, respectively. Let $T : V \to V$ be given by $T(X) = DX$. Prove the following

(*i*) W_1 and W_2 are T-invariant

(*ii*) If $B = \{\in_{n_1+1}, \in_{n_1+2}, \ldots, \in_n, \in_1, \in_2, \ldots, \in_{n_1}\}$. Then $[T]_B = \begin{bmatrix} B & 0 \\ 0 & A \end{bmatrix}$

(*iii*) $\begin{bmatrix} A & 0 \\ 0 & B \end{bmatrix}$ and $\begin{bmatrix} B & 0 \\ 0 & A \end{bmatrix}$ are similar matrices.

6. Let T be the linear operator on $\mathbf{R}^3$, whose matrix relative to the standard ordered basis is $A = \begin{bmatrix} 1 & 0 & 2 \\ 0 & 3 & 1 \\ 0 & 0 & 3 \end{bmatrix}$. Prove the following :

(*i*) Each of $W_1 = \langle \in_1 \rangle$, $W_2 = \langle \in_1, \ \in_2 \rangle$ is T-invariant

(*ii*) $W_3 = \langle \in_2 \rangle$ is T-invariant and there exists a T-invariant subspace W' of $\mathbf{R}^3$ such that $\mathbf{R}^3 = W_3 \oplus W'$

(*iii*) Prove that there exists no T-invariant subspace W' such that $\mathbf{R}^3 = W_2 \oplus W'$.

7. Let T be a linear operator on a finite dimensional vector space V_F, such that

$$\text{range}(T) \cap \ker(T) = \{0\}.$$

Prove the following

(*i*) $\rho(T) = \rho(T^2)$

(*ii*) If $\dim V = n$ and $\rho(T) = m$, then there exists an ordered basis B of V such that

$$[T]_B = \begin{bmatrix} A & 0 \\ 0 & 0 \end{bmatrix}_{n \times n}$$

where A is some non-singular $m \times m$-matrix.

8. Let A be any $n \times n$-matrix over a field F and T be the linear operator on $F^{n\times 1}$ given by $T(X) = AX$. Prove that A is similar to a diagonal matrix iff there exist one-dimensional T-invariant subspaces $W_1, W_2, \ldots, W_n$ of $F^{n\times 1}$ such that

$$F^{n\times 1} = W_1 \oplus W_2 \oplus \ldots \oplus W_n$$

9. For any two vector spaces V_1, V_2 over a field F, prove the following :

(*i*) $T : V_1 \oplus V_2 \to V_2 \oplus V_1$ such that $T(\alpha_1, \alpha_2) = (\alpha_2, \alpha_1)$ is an isomorphism

(*ii*) $T' : V_1 \oplus V_2 \to V_2$ such that $T'(\alpha_1, \alpha_2) = \alpha_2$ is an epimorphism. Find $\ker T'$.

10. Give an isomorphism of F^{10} onto $F^5 \oplus F^3 \oplus F^2$.

11. Let W_1 and W_2 be two subspaces of a vector space V_F. Prove that

$$T : W_1 \times W_2 \to W_1 + W_2$$

such that $\quad T(\alpha_1, \alpha_2) = \alpha_1 + \alpha_2$

is an epimorphism with

$$\ker T = \{(\alpha_1, -\alpha_1) : \alpha_1 \in W_1 \cap W_2\}$$

and $\quad \dim(\ker T) = \dim(W_1 \cap W_2)$.

Chapter 4

Determinants

4.1 Permutations

In this section, we review some properties of permutations. Any one-to-one mapping of a set A onto A itself is called a *permutation* of A. If A is a finite set with n members a_1, a_2, ..., a_n and f is a permutation of A, such that $f(a_i) = b_i$ for $i = 1, 2, \ldots, n$, we write

$$g = \left(\begin{vmatrix} 1 & 2 & 3 & 4 & 5 \\ 1 & 4 & 3 & 2 & 5 \end{vmatrix}\right)$$

This representation of f is known as two row notation of f.

For example,
$$g = \left(\begin{vmatrix} 1 & 2 & 3 & 4 \\ 3 & 2 & 4 & 1 \end{vmatrix}\right)$$

is the permutation of $\{1, 2, 3, 4\}$ such that $g(1) = 3$, $g(2) = 2$, $g(3) = 4$ and $g(4) = 1$. In such a situation, we say that under g, $1 \to 3$, $2 \to 2$, $3 \to 4$, $4 \to 1$.

For any set A, the set Sym(A) of all permutations of A is a group under the usual composition of mappings. For any positive integer n, S_n denotes the group of all permutations of $\{1, 2, \ldots, n\}$, called the *symmetric* group of degree n. The order of S_n is $n!$.

Example 0.a. S_3 is the symmetric group of all permutations of $\{1, 2, 3\}$. It consists of the following six permutations :

$$I = \left(\begin{vmatrix} 1 & 2 & 3 \\ 1 & 2 & 3 \end{vmatrix}\right),$$

$$\sigma_1 = \left(\begin{vmatrix} 1 & 2 & 3 \\ 2 & 1 & 3 \end{vmatrix}\right),$$

$$\sigma_2 = \left(\begin{vmatrix} 1 & 2 & 3 \\ 3 & 2 & 1 \end{vmatrix}\right),$$

Chapter 4

Determinants

4.1 Permutations

In this section, we review some properties of permutations. Any one-to-one mapping of a set A onto A itself is called a *permutation* of A. If A is a finite set with n members a_1, a_2, ..., a_n and f is a permutation of A, such that $f(a_i) = b_i$ for $i = 1, 2, ..., n$, we write

$$f = \begin{pmatrix} a_1 & a_2 & \dots & a_n \\ b_1 & b_2 & \dots & b_n \end{pmatrix}$$

This representation of f is known as two row notation of f.

For example,

$$g = \begin{pmatrix} 1 & 2 & 3 & 4 \\ 3 & 2 & 4 & 1 \end{pmatrix}$$

is the permutation of $\{1, 2, 3, 4\}$ such that $g(1) = 3$, $g(2) = 2$, $g(3) = 4$ and $g(4) = 1$. In such a situation, we say that under g, $1 \to 3$, $2 \to 2$, $3 \to 4$, $4 \to 1$.

For any set A, the set Sym(A) of all permutations of A is a group under the usual composition of mappings. For any positive integer n, S_n denotes the group of all permutations of $\{1, 2, ..., n\}$, called the *symmetric* group of degree n. The order of S_n is $n!$.

Example 0.a. S_3 is the symmetric group of all permutations of $\{1, 2, 3\}$. It consists of the following six permutations :

$$I = \begin{pmatrix} 1 & 2 & 3 \\ 1 & 2 & 3 \end{pmatrix},$$

$$\sigma_1 = \begin{pmatrix} 1 & 2 & 3 \\ 2 & 1 & 3 \end{pmatrix},$$

$$\sigma_2 = \begin{pmatrix} 1 & 2 & 3 \\ 3 & 2 & 1 \end{pmatrix},$$

(*i*) If $f(x) \neq x$, then, $g(x) = x$

(*ii*) If $g(x) \neq x$, then, $f(x) = x$

Any two disjoint permutations commute.

Example 2.a. $f = (1\ 5\ 3)$, $g = (2\ 4)$ in S_5 are disjoint cyclic permutations. Observe that $f(1) = 5 \neq 1$ but $g(1) = 1$, $g(2) \neq 2$ but $f(2) = 2$, etc.

The following is a useful result:

Theorem 3. Any non-identity permutation of a finite set is a product of pairwise disjoint cyclic permutations each of length, at least two.

Example 3.a. Consider $$f = \left(\begin{vmatrix} 1 & 2 & 3 & 4 & 5 \\ 3 & 4 & 5 & 2 & 1 \end{vmatrix}\right),$$

under f, $1 \to 3 \to 5 \to 1,\ 2 \to 4 \to 2,$

so that $f = (1\ 3\ 5).(2\ 4)$

a product of two disjoint cycles, one of length 3 and other of length two.

As remarked before, any cyclic permutation is a product of transpositions. As any permutation is a product of pairwise disjoint cycles, any permutation f of a finite set is expressible as a product of transpositions; identity permutation is treated as a product of zero number of transpositions. A permutation f may be expressible as a product of transposition, in several ways. However, if an expression of f as a product of transpositions has even number of terms, then, there must always be even number of terms in any expression of f as a product of transpositions.

Definition 4. A permutation f of finite set X is called an *even permutation* if it is a product of even number of transpositions. A permutation that is not even is called an *odd permutation.*

In particular, identity permutation is an even permutation. Any cycle of length $k \geq 1$, being a product of $k - 1$ transpositions is even or odd according as k odd or even. Thus, the cycles (1 2), (1 4 3 5), (2 3 4 5 7 1), being of even lengths, are odd permutations. The cycle (1 3 2), being of odd length, is an even permutation.

For $n \geq 2$, the set A_n of all even permutations in S_n is a subgroup of S_n with $0(A_n) = \frac{1}{2}0(S_n) = \frac{n!}{2}$. Thus, in S_n, $n \geq 2$, the number of odd and even permutations is the same. In S_3, $A_3 = \{I,\ (1\ 2\ 3),\ (1\ 3\ 2)\}$. The following lemma is obvious :

Lemma 5. For any two permutations σ and η of a finite set X, the following hold :

(*i*) If σ and η are even, so is $\sigma\eta$

(*ii*) If σ and η are odd, then $\sigma\eta$ is even

(*iii*) If one of σ and η is odd and the other is even, the product $\sigma\eta$ is odd.

Definition 6. For any permutation σ of a finite set X, the *sign* of σ, denoted by $(-1)^{\sigma}$ is equal to 1, if σ is even, and is equal to -1, if σ is odd.

The following is an immediate consequence of Lemma 5 :

Lemma 7. For any two permutations σ and η of a finite set X, $(-1)^{\sigma\eta} = (-1)^{\sigma}(-1)^{\eta}$.

Exercise 4.1

1. Write each of the following permutations as a product of disjoint cycles

   (*i*) $\begin{pmatrix} 1 & 2 & 3 & 4 \\ 3 & 1 & 2 & 4 \end{pmatrix}$,

   (*ii*) $\begin{pmatrix} 1 & 2 & 3 & 4 & 5 \\ 3 & 2 & 1 & 4 & 5 \end{pmatrix}$

   (*iii*) $\begin{pmatrix} 1 & 2 & 3 & 4 & 5 & 6 \\ 3 & 4 & 5 & 6 & 1 & 2 \end{pmatrix}$

   (*iv*) $\begin{pmatrix} 1 & 2 & 3 \\ 2 & 1 & 3 \end{pmatrix}$

2. Write each of the following permutations as a product of disjoint cycles. Further, find which of these permutations are even and which are odd

   (*i*) (1 2 3) (3 2) (1 4 2 3)

   (*ii*) (1 3 2) (1 4 2) (2 4 5)

   (*iii*) (4 5) (1 2 3) (2 3 1) (ϵ 2)

   (*iv*) $(a\ b\ c)\ (a\ c)\ (a\ b\ d)$

3. If two permutations f and g of a set X are disjoint, prove that f^k and g^s are disjoint for any two non-negative integers k and s.

4. For any cyclic permutation $f = (a_1 a_2 \dots a_k)$, prove that $f^{-1} = (a_1 a_k a_{k-1} \dots a_2)$

5. For $f = (1\ 2\ 3)(3\ 2)$ in S_4, answer the following :

   (*i*) What is the order of f?

   (*ii*) Write down all the members of the subgroup H of S_4 generated by f

   (*iii*) Find every $g \in S_4$ such that $fg = gf$.

6. For each of the following cyclic permutations in S_5, verify that its order equals the length of that cyclic permutation.

   (*i*) (1 3 5), (*ii*) (1 4 2 3) (*iii*) (1 5), (*iv*) (1 5 3 2 4)

7. Write down all the members of A_4. Further,

(*i*) Prove that $H = \{I, (1\ 2)(3\ 4),\ (1\ 4)(2\ 3),\ (1\ 3)\ (2\ 4)\}$ is a normal subgroup of A_4.

(*ii*) Prove that $K = \{I,\ (1\ 2)(3\ 4)\}$ is a normal subgroup of *H*, but not of A_4.

(*iii*) Prove that A_4 has no subgroup of order 6.

8. Prove that any permutation *f* in $S_n\ (n > 2)$ is expressible as a product of the transpositions (1 2), (2 3), (3 4), ..., (*n*–1 *n*)

9. For $n \geq 3$, prove that any $f \in A_n$ is expressible as a product of 3-cycles (1 2 3), (1 2 4),..., (1 2 *n*).

4.2 Determinants

Throughout, *R* is a commutative ring with identity $1 \neq 0$ and *F* is a field. For any $\sigma \in S_n$, recall that $(-1)^\sigma = 1$ if σ is even, and $(-1)^\sigma = -1$ if σ is odd.

Definition 1. For any $n \times n$-matrix $A = [a_{ij}]$ over a commutative ring *R*, the *determinant* of *A*, denoted by $\det(A)$, is defined by

$$\det(A) = \sum_{\sigma} (-1)^\sigma a_{1\sigma(1)}\ a_{2\sigma(2)} \ldots a_{n\sigma(n)}$$

where σ runs over all permutations in S_n.

The $\det(A)$ is also denoted by $|A|$.

As there are *n*! permutations in S_n, the expression of $\det(A)$, involves sum of *n*! terms, with each term a product of *n* entries of *A*, and a $(-1)^\sigma$. The $\det(A)$ will also be denoted by

$$\begin{vmatrix} a_{11} & a_{12} & \ldots & a_{1n} \\ a_{21} & a_{22} & \ldots & a_{2n} \\ \ldots & \ldots & \ldots & \ldots \\ \ldots & \ldots & \ldots & \ldots \\ a_{n1} & a_{n2} & \ldots & a_{nn} \end{vmatrix}$$

Example 1.a. For $n = 1$, $A = [a_{11}]$

has $\det(A) = a_{11}$.

For $n = 2$, $$A = \begin{bmatrix} a_{11} & a_{12} \\ a_{21} & a_{22} \end{bmatrix}$$

$$S_2 = \{I,\ (1\ 2)\}.$$

Now, $(-1)^I = 1$

and $(-1)^\sigma = -1,$

where $\sigma = (1\ 2)$.

By definition, $\det(A) = (-1)^I a_{1I(1)}\, a_{2I(2)} + (-1)^\sigma a_{1\sigma(1)}\, a_{2\sigma(2)}$
$= a_{11}a_{22} - a_{12}a_{21}$.

Thus, $$\det\begin{bmatrix} a & b \\ c & d \end{bmatrix} = ad - bc.$$

In particular, $$\det\begin{bmatrix} 1 & 5 \\ 4 & 3 \end{bmatrix} = 1\times 3 - 5\times 4 = -17.$$

Example 1.b. In S_3, there are six permutations. Among them I, $\sigma_1 = (1\ 2\ 3)$ and $\sigma_2 = (1\ 3\ 2)$ are even, and $\eta_1 = (1\ 2)$, $\eta_2 = (1\ 3)$, $\eta_3 = (2\ 3)$ are odd.

So, for $$A = \begin{bmatrix} a_{11} & a_{12} & a_{13} \\ a_{21} & a_{22} & a_{23} \\ a_{31} & a_{32} & a_{33} \end{bmatrix}$$

over a ring R, $\det(A)$ is the sum of the following six terms

$$(-1)^I a_{1I(1)}\, a_{2I(2)}\, a_{3I(3)} = a_{11}\, a_{22}\, a_{33}$$

$$(-1)^{\sigma_1} a_{1\sigma_1(1)}\, a_{2\sigma_1(2)}\, a_{3\sigma_1(3)} = a_{12}\, a_{23}\, a_{31}$$

$$(-1)^{\sigma_2} a_{1\sigma_2(1)}\, a_{2\sigma_2(2)}\, a_{3\sigma_2(3)} = a_{13}\, a_{21}\, a_{32}$$

$$(-1)^{\eta_1} a_{1\eta_1(1)}\, a_{2\eta_1(2)}\, a_{3\eta_1(3)} = -a_{12}\, a_{21}\, a_{33}$$

$$(-1)^{\eta_2} a_{1\eta_2(1)}\, a_{2\eta_2(2)}\, a_{3\eta_2(3)} = -a_{13}\, a_{22}\, a_{31}$$

$$(-1)^{\eta_3} a_{1\eta_3(1)}\, a_{2\eta_2(2)}\, a_{3\eta_2(3)} = -a_{11}\, a_{23}\, a_{32}$$

so that $\det(A) = a_{11}a_{22}a_{33} + a_{12}a_{23}a_{31} + a_{13}a_{21}a_{32} - a_{12}a_{21}a_{33} - a_{13}a_{22}a_{31} - a_{11}a_{23}a_{32}$

This expression can be represented diagrammatically as :

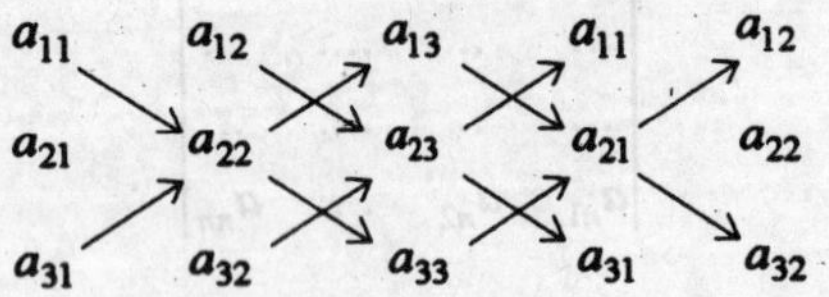

In the diagram, the three terms along the arrows in a line are multiplied together; before each such product, sign + or − is put according as these arrows are directed downward or upward. The $\det(A)$ is, then, the sum of these six terms. Warning : Similar diagrammatic representation of the determinant of any $n \times n$-matrix with $n \neq 3$ does not hold.

Lemma 2. For any $n \geq 1$, the determinant of the identity matrix I_n is 1.

Proof. Now, $I_n = \left[\delta_{ij}\right]$ where $\delta_{ij} = 0$ if $i \neq j$ and $\delta_{ij} = 1$ for $i = j$, so that

$$\det(I_n) = \sum (-1)^\sigma \delta_{1\sigma(1)}\, \delta_{2\sigma(2)} \ldots \delta_{n\sigma(n)}.$$

For $\sigma \neq I$, there exists i such that $\sigma(i) \neq i$, so that $\delta_{i\sigma(i)} = 0$.

Consequently, for $\sigma \neq I$, $\quad \delta_{1\sigma(1)}\, \delta_{2\sigma(2)} \ldots \delta_{n\sigma(n)} = 0$.

Hence, $\quad \det(I_n) = (-1)^I\, \delta_{11}\delta_{22} \ldots \delta_{nn} = 1$.

This completes the proof.

Theorem 3. For any $n \times n$-matrix $A = \left[a_{ij}\right]$ over a commutative ring R,

$$\det(A) = \det(A^T)$$

Proof. By definition, $\det(A) = \sum_{\sigma \in S_n} (-1)^\sigma a_{1\sigma(1)}\, a_{2\sigma(2)} \ldots a_{n\sigma(n)}$. ... (1)

Now,
$$A^T = \left[b_{ij}\right]$$
with
$$b_{ij} = a_{ji}.$$

This gives
$$\det(A^T) = \sum_{\sigma \in S_n} (-1)^\sigma b_{1\sigma(1)}\, b_{2\sigma(2)} \ldots b_{n\sigma(n)}$$
$$= \sum_{\sigma \in S_n} (-1)^\sigma a_{\sigma(1)1}\, a_{\sigma(2)2} \ldots a_{\sigma(n)n} \quad \text{... (2)}$$

$\sigma(1), \sigma(2), \ldots, \sigma(n)$ take all the values $1, 2, \ldots, n$.

Thus, if $\sigma(i) = j$, then $\quad a_{\sigma(i)i} = a_{\sigma(i)\sigma^{-1}\sigma(i)} = a_{j\sigma^{-1}(j)}$.

Consequently, by rearranging terms, (2) becomes
$$\det(A^T) = \sum_{\sigma \in S_n} (-1)^\sigma a_{1\sigma^{-1}(1)}\, a_{2\sigma^{-1}(2)} \ldots a_{n\sigma^{-1}(n)}$$
$$= \sum_{\sigma \in S_n} (-1)^{\sigma^{-1}} a_{1\sigma^{-1}(1)}\, a_{2\sigma^{-1}(2)} \ldots a_{n\sigma^{-1}(n)},$$
as
$$(-1)^{\sigma^{-1}} = (-1)^\sigma.$$

As σ ranges over S_n, σ^{-1} ranges over S_n and conversely. So, (3) becomes
$$\det(A^T) = \sum_{\eta \in S_n} (-1)^\eta a_{1\eta(1)}\, a_{2\eta(2)} \ldots a_{n\eta(n)} = \det(A).$$

Theorem 4. Let $A = \left[a_{ij}\right]$ be any $n \times n$-matrix over R with $n \geq 2$, such that for some $k \neq l$, the kth row and the lth row of A are identical. Then, $\det(A) = 0$.

Proof. Consider the transposition $\lambda = (kl) \in S_n$.

Then,
$$S_n = A_n \cup A_n\lambda,$$

so that $\det(A)=\sum_{\sigma\in A_n}\left[(-1)^{\sigma}a_{1\sigma(1)}\,a_{2\sigma(2)}\ldots a_{n\sigma(n)}+(-1)^{\sigma\lambda}a_{1\sigma\lambda(1)}\ldots a_{n\sigma\lambda(n)}\right].$

As $\sigma\in A_n$ is even, λ is odd, $\sigma\lambda$ is odd. So $(-1)^{\sigma\lambda}=-1$.

Thus, $$\det(A)=\sum_{\sigma\in A_n}\left[a_{1\sigma(1)}\,a_{2\sigma(2)}\ldots a_{n\sigma(n)}-a_{1\sigma\lambda(1)}\,a_{2\sigma\lambda(2)}\ldots a_{n\sigma\lambda(n)}\right] \quad \ldots (1)$$

for $i\neq k,\ l\ \lambda(i)=i$, gives $(\sigma\lambda)(i)=\sigma(i)$

and, hence, $a_{i\sigma(i)}=a_{i(\sigma\lambda)(i)}$.

Further, by hypothesis $a_{kj}=a_{lj}$ for $1\le j\le n$

so that $a_{k(\sigma\lambda)(k)}=a_{k\sigma(l)}=a_{l\sigma(l)}$

and $a_{l(\sigma\lambda)(l)}=a_{k\sigma(k)}$.

Consequently, $a_{1\sigma\lambda(1)}\,a_{2\sigma\lambda(2)}\ldots a_{n\sigma\lambda(n)}=a_{1\sigma(1)}\,a_{2\sigma(2)}\ldots a_{n\sigma(n)}$

Consequently, (1) gives $\det(A)=0$.

Lemma 5. Let $A=\left[a_{ij}\right]$ be any $n\times n$-matrix over R with $n\ge 2$.

For some k, let $a_{kj}=b_{kj}+c_{kj}$ for $1\le j\le n$.

Then, $\det(A)=\det(B)+\det(C)$

where $B=\left[b_{ij}\right],\ C=\left[c_{ij}\right]$

are such that for $i\neq k,\ b_{ij}=a_{ij}=c_{ij}$.

Proof. $$\det(A)=\sum_{\sigma\in S_n}(-1)^{\sigma}a_{1\sigma(1)}\,a_{2\sigma(2)}\ldots a_{n\sigma(n)}$$
$$=\sum_{\sigma\in S_n}(-1)^{\sigma}a_{1\sigma(1)}\ldots a_{k-1\sigma(k-1)}\left\{b_{k\sigma(k)}+c_{k\sigma(k)}\right\}\ldots a_{n\sigma(n)}$$

As $b_{ij}=a_{ij}=c_{ij}$ for $i\neq k$, this gives

$$\det(A)=\sum_{\sigma\in S_n}(-1)^{\sigma}b_{1\sigma(1)}\,b_{2\sigma(2)}\ldots b_{n\sigma(n)}+\sum_{\sigma\in S_n}(-1)^{\sigma}c_{1\sigma(1)}\,c_{2\sigma(2)}\ldots c_{n\sigma(n)}$$
$$=\det(B)+\det(C).$$

Theorem 6. Let A be any $n\times n$-matrix over R with $n\ge 2$. Let B be obtained by interchanging two distinct rows (columns). Then, $\det(B)=-\det(A)$.

Proof. Let $A=\left[a_{ij}\right]$, $B=\left[b_{ij}\right]$ be obtained from A by interchanging k^{th} and l^{th} rows of A with $k\neq l$. Consider $A'=\left[a'_{ij}\right]_{n\times n}$ such that for $i\neq k,\ l$, the ith row of A' is the same as that of A, but the kth and the lth rows of A' are both equal to

$$[a_{k1}+a_{l1},\ a_{k2}+a_{l2},\dots,\ a_{kn}+a_{ln}] \quad (1)$$

By Theorem 4, $\det(A') = 0.$

By Lemma 5, $0 = \det(A') = \det(A_1) + \det(A_2)$ ... (2)

where A_1 and A_2 have same i^{th} rows as that of A' for $i \neq k$; the kth rows of A_1 and A_2 are

$$[a_{k1},\ a_{k2},\dots,\ a_{kn}] \quad \text{... (3)}$$

and

$$[a_{l1},\ a_{l2},\dots,\ a_{ln}], \quad \text{... (4)}$$

respectively.

Thus, for $i \neq l$, i^{th} row of A_1 is the same as that of A, and the ith row of A_2 is the same as that of B. Further, the lth rows of A_1 and A_2 are given by (1). By applying Lemma 5 once again, we get

$$\det(A_1) = \det(A) + \det(A_3)$$

where the lth row of A_3 is given by (3) and all the other rows of A_3 are the same as those of A_1.

Consequently, k^{th} row of A_3 is also given by (3).

So, by Theorem 4, $\det(A_3) = 0.$

Hence, $\det(A_1) = \det(A).$

Similarly, $\det(A_2) = \det(B).$

Consequently, by (2) $0 = \det(A) + \det(B)$

i.e., $\det(B) = -\det(A).$

Let now B be obtained from A by interchanging k^{th} and l^{th} columns with $k \neq l$.

Then, $\det(B^T) = -\det(A^T).$

By Theorem 3, $\det(B) = \det(B^T) = -\det(A).$

Remark 7. Let σ be any permutation of 1, 2,..., n, i.e., $\sigma \in S_n$. For any $n \times n$-matrix A over R, let $\sigma(A)$ be the $n \times n$-matrix whose i^{th} column is the $\sigma(i)^{\text{th}}$ column of A for $1 \leq i \leq n$.

For example, if $n = 3$, $\sigma = (1\ 3\ 2)$.

$$\sigma(A) = \begin{bmatrix} a_{13} & a_{11} & a_{12} \\ a_{23} & a_{21} & a_{22} \\ a_{33} & a_{31} & a_{32} \end{bmatrix}$$

as $\sigma(1) = 3,\ \sigma(2) = 1,\ \sigma(3) = 2.$

Given any transposition $\lambda = (kl) \in S_n$, $\lambda(A)$ is the matrix whose kth column is the lth column of A and the lth column is the kth column of A; other columns of $\lambda(A)$ are the same as the corresponding column of A.

So, by Theorem 6, $\det\{\lambda(A)\} = -\det(A)$.

By the repeated application of this result, we get

$$\det\{\lambda_1\lambda_2\ldots\lambda_t(A)\} = (-1)^t \det(A)$$

for any transpositions $\lambda_i \in S_n$.

For $n \geq 2$, given $\sigma \in S_n$, $\sigma = \mu_1\mu_2\ldots\mu_s$

for some transpositions $\mu_i \in S_n$ with $(-1)^\sigma = (-1)^s$,

so that $\det\{\sigma(A)\} = (-1)^s \det(A) = (-1)^\sigma \det(A)$

Theorem 8. Let $A = [a_{ij}]$ be any $n \times n$-matrix over R. For any $1 \leq k \leq n$ and $c \in R$, if $B = [b_{ij}]$ is the $n \times n$-matrix obtained by multiplying the kth row of A by c, then,

$$\det(B) = c\det(A).$$

Proof. By the hypothesis $b_{ij} = a_{ij}$ for $i \neq k$,

and $b_{kj} = ca_{kj}$.

Consequently,

$$\begin{aligned}\det(B) &= \sum_{\sigma\in S_n}(-1)^\sigma b_{1\sigma(1)}\ldots b_{k\sigma(k)}\ldots b_{n\sigma(n)} \\ &= \sum_{\sigma\in S_n}(-1)^\sigma a_{1\sigma(1)}\, a_{2\sigma(2)}\ldots ca_{k\sigma(k)}\ldots a_{n\sigma(n)} \\ &= c\sum_{\sigma\in S_n}(-1)^\sigma a_{1\sigma(1)}\, a_{2\sigma(2)}\ldots a_{k\sigma(k)}\ldots a_{n\sigma(n)} \\ &= c\det(A)\end{aligned}$$

The theorem, given above, and Theorem 3 yield the following :

Theorem 9. For any $n \times n$-matrix A over R and $c \in R$, if an $n \times n$-matrix B is obtained from A by multiplying a column of A by c, then, $\det(B) = c\det(A)$.

Remark 10. By using Theorem 3, Lemma 5 can be re-interpreted.

Consider $A = [a_{ij}] = [C_1, C_2, \ldots, C_n]$

where C_j is the jth column of A.

If for some k, $C_k = C'_k + C''_k$ for some $C'_k, C''_k \in R^{n\times 1}$, then,

$$\det(A) = \det(A') + \det(A'')$$

where A' and A'' are $n \times n$-matrices, whose j^{th} columns for $j \neq k$ equal C_j, but their k^{th} columns are C'_k and C''_k, respectively.

Determinant defines a function $f: R^{n\times n} \to R$

$$f((C_1, C_2, \ldots, C_n)) = \det[C_1, C_2, \ldots, C_n]$$

This remark and Theorem 9 show that this function is linear in each variable C_k. Thus, if $C_1 = \sum_{j=1}^{t} a_j C'_{j1}$ for some $c_j \in R$ and $C'_{j1} \in R^{n\times 1}$, then,

$$f\left(\left[\sum_{j=1}^{t} a_j C'_{j1}, C_2, \ldots, C_n\right]\right) = \sum_{j=1}^{t} a_j f\left(\left[C'_{j1}, C_2, \ldots, C_n\right]\right)$$

If a row of A consists of zeroes only, say k^{th} row, then

$$a_{kj} = 0 = 0a_{kj}$$

yields $$\det(A) = 0\det(A) = 0.$$

Similarly, if some column of A consists of zeroes only, then, $\det(A) = 0$.

Theorem 11. Let A and B be any two $n \times n$-matrices over a commutative ring R. Then, $\det(AB) = \det(A)\det(B)$.

Proof. We write $$A = [C_1, C_2, \ldots, C_n], B = [b_{ij}].$$

Then, $$AB = [C'_1, C'_2, \ldots, C'_n]$$

with $$C'_j = \sum_{i=1}^{n} C_i b_{ij},$$

so that
$$\det(AB) = \det[C'_1, C'_2, \ldots, C'_n]$$
$$= \det\left[\sum_{i=1}^{n} C_i b_{i1}, \sum_{i=1}^{n} C_i b_{i2}, \ldots, \sum_{i=1}^{n} C_i b_{in}\right].$$

By using Remark 10, we get

$$\det(AB) = \sum \det\left[C_{i_1}, C_{i_2}, \ldots, C_{i_n}\right]\left(b_{i_1 1} b_{i_2 2} \ldots b_{i_n n}\right)$$

where summation is taken over all n-tuples $(i_1, i_2, \ldots, i_n)$ of integer i_j satisfying $1 \le i_j \le n$.

If for some such n-tuple $(i_1, i_2, \ldots, i_n)$, $i_k = i_l$ for some $k \neq l$, by Theorem 4,

$$\det\left[C_{i_1}, C_{i_2}, \ldots, C_{i_n}\right] = 0,$$

so that $$\det(AB) = \sum \det\left[C_{i_1}, C_{i_2}, \ldots, C_{i_n}\right]\left(b_{i_1 1} b_{i_2 2} \ldots b_{i_n n}\right)$$

where summation runs over all those n-tuples $(i_1, i_2, \ldots, i_n)$ of integers i_j, $1 \le i_j \le n$, such that no two of their components are equal.

These n-tuples can be written as

$$(\sigma(1), \sigma(2), \ldots, \sigma(n)),$$

$$\sigma \in S_n,$$

so that $$\det(AB) = \sum_{\sigma \in S_n} \det\left[C_{\sigma(1)}, C_{\sigma(2)}, \ldots, C_{\sigma(n)}\right] b_{\sigma(1)} b_{\sigma(2)2} \ldots b_{\sigma(n)n}$$

By Remark 7, $$\det\left[C_{\sigma(1)}, C_{\sigma(2)}, \ldots, C_{\sigma(n)}\right] = (-1)^{\sigma} \det(A)$$

Hence, $$\det(AB) = \sum_{\sigma \in S_n} (-1)^{\sigma} \det(A)\, b_{\sigma(1)1} b_{\sigma(2)2} \ldots b_{\sigma(n)n}$$

$$= \det(A)\det(B^T)$$

$$= \det(A)\det(B)$$

This theorem shows that for any $n \times n$-matrices $A_1, A_2, \ldots, A_k$ over R

$$\det(A_1 A_2 \ldots A_k) = \det(A_1)\det(A_2)\ldots\det(A_k).$$

In particular, for any $n \times n$-matrix A and any positive integer k, by putting $A_i = A$, we get

$$\det(A^k) = [\det(A)]^k.$$

If A has inverse, then, $$AA^{-1} = I_n,$$

so that $$1 = \det(I_n) = \det(A)\det(A^{-1}).$$

This gives $$\det(A^{-1}) = \{\det(A)\}^{-1}.$$

For any $k \ge 1$ $$(A^k)^{-1} = (A^{-1})^k = A^{-k}.$$

Hence, $$\det(A^{-k}) = \{\det(A^{-1})\}^k = \{\det(A)\}^{-k}.$$

We collect these results below.

Theorem 12. Let A be any $n \times n$ invertible matrix over a commutative ring R.

Then, $$\det(A^k) = \{\det(A)\}^k \quad \text{for any integer } k.$$

Further, $$\det(A) \ne 0.$$

Theorem 13. An $n \times n$-matrix A over a field F is singular iff $\det(A) = 0$.

Proof. If A is non-singular, by Theorem 12, $\det(A) \ne 0$, so that if $\det(A) = 0$, then, A is singular. Let A be singular. There exists a non-singular $n \times n$-matrix P such that $B = PA$ is in reduced row echelon form. Then, B is singular. By Theorem (2.3.8), rank$(B) < n$. By Proposition (2.3.2), the number of non-zero rows of B being equal to rank(B), is less than n, so that B has a row consisting entirely of zeros.

Hence, $$\det(B) = 0.$$

As $$A = P^{-1}B,$$

we get $$\det(A) = \det(P^{-1})\det(B) = 0.$$

Theorem 14. Let $A = \left[a_{ij}\right]$ be any $n \times n$-matrix over R. with $n \geq 2$. For some $k \neq l$ and $c \in R$, let $B = \left[b_{ij}\right]$ be obtained from A by adding c times the kth row (column) of A to the lth row (column) of A, then, $\det(B) = \det(A)$.

Proof. Let B be obtained by adding c times the kth column of A to the lth column of A.

Thus, if $$A = \left[C_1, C_2, \ldots, C_n\right]$$
$$B = \left[C_1', C_2', \ldots, C_n'\right]$$

then, $$C_i' = C_i$$

for $i \neq l$, but $$C_l' = C_l + cC_k.$$

Consequently, $$\det(B) = \det\left[C_1, C_2, \ldots, C_n\right] + \det(A'),$$
$$= \det(A) + \det(A'),$$

where $$A' = \left[C_1, C_2, \ldots, cC_k^{l^{th}}, \ldots, C_n\right],$$

so that $$\det(A') = c\det\left[C_1, C_2, \ldots, C_k^{l^{th}}, \ldots, C_n\right] = 0,$$

as the matrix on the right has k^{th} and l^{th} columns equal.

Hence, $$\det(B) = \det(A).$$

Similarly, for the row case.

Lemma 15. If for a $\sigma \in S_n$, $i \leq \sigma(i)$ for $1 \leq i \leq n$, then, $\sigma = I$.

Proof. Now, $$n \leq \sigma(n).$$

As $$\sigma(n) \in \{1, 2, \ldots, n\},$$

we get $$n = \sigma(n).$$

Let k be the smallest positive integer such that $\sigma(i) = i$ for $k \leq i \leq n$. If $k = 1$, we get $\sigma = I$. Let $k > 1$.

By the choice $$\sigma(k-1) \neq k-1$$

and for $i > k$, $$\sigma(k-1) \neq \sigma(i) = i.$$

Consequently, $$\sigma(k-1) \leq k-1.$$

By the hypothesis, $$k-1 \leq \sigma(k-1).$$

Hence, $$\sigma(k-1) = k-1.$$

This contradicts the minimality of k. Hence, $k=1$. This proves the result.

Theorem 16. For any $n \times n$-triangular matrix $A=\left[a_{ij}\right]$, $\det(A)=a_{11}a_{22}\ldots a_{nn}$, the product of the diagonal entries of A.

Proof. Let A be an upper triangular matrix. Then, $a_{ij}=0$ for $i>j$. Given $\sigma \neq I$, in S_n, this lemma gives an i such that $i>\sigma(i)$, so that $a_{i\sigma(i)}=0$.

Consequently, $$a_{1\sigma(1)}\,a_{2\sigma(2)}\ldots a_{n\sigma(n)}=0.$$

Hence,
$$\begin{aligned}\det(A) &= \sum_{\sigma\in S_n}(-1)^{\sigma}a_{1\sigma(1)}\,a_{2\sigma(2)}\ldots a_{n\sigma(n)}\\ &= (-1)^{I}\,a_{11}a_{22}\ldots a_{nn}\\ &= a_{11}a_{22}\ldots a_{nn}\end{aligned}$$

Let A be a lower triangular matrix. Then, $A^T=\left[a'_{ij}\right]$ is an upper triangular matrix with $a'_{ij}=a_{ji}$, gives $a'_{ii}=a_{ii}$.

Hence, $$\det(A)=\det(A^T)=a'_{11}a'_{22}\ldots a'_{nn}=a_{11}a_{22}\ldots a_{nn}$$

17. Practical method of finding determinant. Let A be any $n\times n$-matrix over a field F. A is row equivalent to a matrix in reduced row echelon form. In prticular, A is row equivalent to an upper triangular matrix A'. By Theorem 16, det(A') is the product of the diagonal entries of A'. For an elementary row operation θ, $\det\{\theta(A)\}$ is known from det(A) and Theorems 6, 8 and 14. Now, A' is obtainable from A by a finite sequence of elementary row operations, $\theta_1, \theta_2, \ldots, \theta_t$, so that det($A'$) is known from det($A$). However, det($A'$) is also given by Theorem 16, and it helps to find det(A). To illustrate the point, we consider the following matrix.

$$A=\begin{bmatrix}0 & 2 & 1 & 2\\ 6 & 3 & 3 & 3\\ 3 & 1 & 2 & 0\\ -3 & -1 & 1 & 5\end{bmatrix}.$$

By the operation $r_1 \leftrightarrow r_3$, we get

$$A'=\begin{bmatrix}3 & 1 & 2 & 0\\ 6 & 3 & 3 & 3\\ 0 & 2 & 1 & 2\\ -3 & -1 & 1 & 5\end{bmatrix}.$$

By Theorem 6, $$\det(A')=-\det(A),$$

so that
$$\det(A) = -\det(A') = -\begin{vmatrix} 3 & 1 & 2 & 0 \\ 6 & 3 & 3 & 3 \\ 0 & 2 & 1 & 2 \\ -3 & -1 & 1 & 5 \end{vmatrix}.$$

To A', apply $r_2 \to r_2 - 2r_1$, $r_4 \to r_4 + r_1$ in succession. By Theorem 14, these operation do not change the value of the determinant, so that

$$\det(A) = -\begin{vmatrix} 3 & 1 & 2 & 0 \\ 0 & 1 & -1 & 3 \\ 0 & 2 & 1 & 2 \\ 0 & 0 & 3 & 5 \end{vmatrix}.$$

Apply $r_3 \to r_3 - 2r_2$.

By Theorem 14, we get
$$\det(A) = -\begin{vmatrix} 3 & 1 & 2 & 0 \\ 0 & 1 & -1 & 3 \\ 0 & 0 & 3 & -4 \\ 0 & 0 & 3 & 5 \end{vmatrix}.$$

Apply $r_4 \to r_4 - r_3$, we get

$$\det(A) = -\begin{vmatrix} 3 & 1 & 2 & 0 \\ 0 & 1 & -1 & 3 \\ 0 & 0 & 3 & -4 \\ 0 & 0 & 0 & 9 \end{vmatrix}.$$

By Theorem 16, we have $\det(A) = -(3 \times 1 \times 3 \times 9) = -81$.

Exercise 4.2

1. **Find each of the following determinants by directly using the definition**

$$\begin{vmatrix} 5 & 1 \\ 4 & 3 \end{vmatrix}, \begin{vmatrix} 0 & 1 \\ 1 & 0 \end{vmatrix}, \begin{vmatrix} 1 & 2 & 3 \\ 0 & 5 & -1 \\ 1 & 0 & 1 \end{vmatrix}$$

2. Use the diagrammatic expression of the determinant of a 3×3-matrix, given in Example 1.b to find the following determinants :

$$\begin{vmatrix} 3 & 4 & 2 \\ 1 & 5 & 1 \\ 0 & 3 & 2 \end{vmatrix}, \begin{vmatrix} -3 & 1 & 4 \\ 1 & 2 & 5 \\ 3 & 1 & 1 \end{vmatrix}$$

3. **Find each of the following determinants by using Theorem 16 :**

$$\begin{vmatrix} -1 & 5 \\ 0 & 1 \end{vmatrix}, \begin{vmatrix} 5 & 0 \\ -3 & 4 \end{vmatrix}, \begin{vmatrix} 8 & 1 & 9 \\ 0 & -1 & 4 \\ 0 & 0 & 3 \end{vmatrix}, \begin{vmatrix} 5 & 1 & 0 & -1 \\ 0 & 7 & -8 & 9 \\ 0 & 0 & 3 & 4 \\ 0 & 0 & 0 & 2 \end{vmatrix}, \begin{vmatrix} 3 & 0 & 0 \\ 0 & -4 & 0 \\ 0 & 0 & 9 \end{vmatrix}$$

4. For any two $n \times n$-matrices A, B over a commutative ring R, prove the following relations :

(*i*) $\det(AB) = \det(BA) = \det(A^T B) = \det(AB^T)$

(*ii*) If A is invertible, then, $\det(A^{-1}BA) = \det(B)$

[Hint : $\det(AB) = \det(A)\det(B)$]

5. By finding an upper triangular matrix row equivalent to each of the following matrices, find their determinants

(*i*) $\begin{bmatrix} 2 & 4 & 5 \\ 0 & 1 & 3 \\ 1 & 4 & 9 \end{bmatrix}$

(*ii*) $\begin{bmatrix} 1 & 3 & 2 \\ 2 & 1 & 4 \\ 1 & -7 & 2 \end{bmatrix}$

(*iii*) $\begin{bmatrix} 1 & 3 & -1 & 2 \\ 2 & -6 & 4 & 1 \\ 3 & 5 & -1 & 3 \\ 4 & -6 & 5 & 2 \end{bmatrix}$

(*iv*) $\begin{bmatrix} 0 & 1 & 1 \\ 1 & 2 & 3 \\ 2 & 0 & 4 \end{bmatrix}$

(*v*) $\begin{bmatrix} 1 & a & a^2 \\ 1 & b & b^2 \\ 1 & c & c^2 \end{bmatrix}$ **a, b, c any three real numbers.**

By using Theorem 13, find which of the above matrices are non-singular.

6. If $\begin{vmatrix} a_1 & b_1 & c_1 \\ a_2 & b_2 & c_2 \\ a_3 & b_3 & c_3 \end{vmatrix} = 5$ find

(*i*) $\begin{vmatrix} a_1 + 2b_1 & b_1 & 3c_1 \\ a_2 + 2b_2 & b_2 & 3c_2 \\ a_3 + 2b_3 & b_3 & 3c_3 \end{vmatrix}$

(*ii*) $\begin{vmatrix} c_3 & b_3 & a_3 \\ c_2 & b_2 & a_2 \\ c_1 & b_1 & a_1 \end{vmatrix}$.

7. Let $A = \begin{bmatrix} 2 & 1 & 2 \\ 1 & 1 & 1 \\ a & 2 & 3 \end{bmatrix}$. **Without expanding the determinant, find the value of a for which** $\det(A) = 0$.

8. **Find the values of** x **for which** $\begin{vmatrix} x-1 & 2 \\ 2 & x+1 \end{vmatrix} = 0$

9. **For any four real numbers** a_1, a_2, a_3, a_4, **expand**

$$\begin{vmatrix} a_1 & a_2 & a_3 & a_4 \\ a_2 & a_3 & a_4 & a_1 \\ a_3 & a_4 & a_1 & a_2 \\ a_4 & a_1 & a_2 & a_3 \end{vmatrix}$$

10. **Find the values of** x **for which** $\begin{vmatrix} x-1 & 5 & -10 \\ 0 & x-4 & 6 \\ 0 & 0 & x+7 \end{vmatrix} = 0$

11. **If** w **is the cube root of** -1 **other than** -1 **in C, prove that**

$$\begin{vmatrix} 1 & w & w^2 & w^3 \\ w & w^2 & w^3 & 1 \\ w^2 & w^3 & 1 & w \\ w^3 & 1 & w & w^2 \end{vmatrix} = \pm 3i\sqrt{3}$$

[Hint : $w^3 = -1$**]**

12. **For any** $n \times n$**-matrix** A **over a commutative ring** R, **and any** $a \in R$, **prove that** $|aA| = a^n |A|$

[Hint : If $A = [a_{ij}]$, **then,** $\quad aA = [b_{ij}]$

with $\quad b_{ij} = aa_{ij}$,

gives $\quad |aA| = \sum_{\sigma \in S_n} (-1)^\sigma b_{1\sigma(1)}\, b_{2\sigma(2)} \cdots b_{n\sigma(n)} = a^n |A|$**]**

13. **Let** G **be any group of** $n \times n$**-non-singular real matrices, under matrix multiplication. Prove the following relations :**

(*i*) $H = \{A \in G : \det(A) = 1\}$ **is a normal subgroup of** G.

(*ii*) $K = \{A \in G : \det(A) = 1 \text{ or } -1\}$ **is a normal subgroup of G.**

4.3 Cofactor Expansion

Throughout R is a commutative ring with identity $1 \neq 0$ and F is a field.

Definition 1. Let $A = [a_{ij}]$ be any $n \times n$-matrix over R with $n \geq 2$. Let X_{ij} be the $(n-1) \times (n-1)$ submatrix of A obtained by deleting the ith row and j^{th} column of A. Then, $\det(X_{ij}) = M_{ij}$ is called the $(i, j)^{\text{th}}$ *minor* of A. Further, $A_{ij} = (-1)^{i+j} M_{ij}$ is called the $(i, j)^{\text{th}}$ *cofactor* of A; A_{ij} is also known as the cofactor of a_{ij}.

Example 1.a. Let
$$A = \begin{bmatrix} a_{11} & a_{12} \\ a_{21} & a_{22} \end{bmatrix}.$$

Then,
$$X_{11} = [a_{22}],\ A_{11} = (-1)^{1+1} \det(X_{11}) = a_{22}$$
$$X_{12} = [a_{21}],\ A_{12} = (-1)^{1+2} |X_{12}| = -a_{21}$$
$$X_{21} = [a_{12}],\ A_{21} = (-1)^{2+1} |X_{21}| = -a_{12}$$
$$X_{22} = [a_{11}],\ A_{22} = (-1)^{22} |X_{22}| = a_{11}$$

Consider
$$A' = \begin{bmatrix} A_{11} & A_{21} \\ A_{12} & A_{22} \end{bmatrix} = \begin{bmatrix} a_{22} & -a_{12} \\ -a_{21} & a_{11} \end{bmatrix}.$$

Clearly,
$$AA' = A'A = \begin{bmatrix} a_{11}a_{22} - a_{12}a_{21} & 0 \\ 0 & a_{11}a_{22} - a_{12}a_{21} \end{bmatrix} = |A|.I_2.$$

Observe the following
$$|A| = a_{11}A_{11} + a_{12}A_{12} = a_{21}A_{21} + a_{22}A_{22}.$$

Example 1.b. Let
$$A = \begin{bmatrix} a_{11} & a_{12} & a_{13} \\ a_{21} & a_{22} & a_{23} \\ a_{31} & a_{32} & a_{33} \end{bmatrix}.$$

To get the (1, 1)th submatrix X_{11} we delete the first row and the first column of A, so that
$$X_{11} = \begin{bmatrix} a_{22} & a_{23} \\ a_{32} & a_{33} \end{bmatrix}.$$

The minor
$$M_{11} = |X_{11}| = a_{22}a_{33} - a_{23}a_{32}.$$

The cofactor
$$A_{11} = (-1)^{1+1} M_{11} = M_{11}.$$

Again, X_{12} is obtained by deleting the 1st-row and the 2nd column of A.

So, $$X_{12} = \begin{bmatrix} a_{21} & a_{23} \\ a_{31} & a_{33} \end{bmatrix}.$$

Consequently, $$M_{12} = |X_{12}| = a_{21}a_{33} - a_{23}a_{31}$$

and $$A_{12} = (-1)^{1+2} M_{12} = M_{12} = a_{23}a_{31} - a_{21}a_{33}$$

Similarly, $$A_{13} = (a_{21}a_{32} - a_{22}a_{31}).$$

As in Example (4.2.1b)

$$\begin{aligned} |A| &= a_{11}a_{22}a_{33} + a_{12}a_{23}a_{31} + a_{13}a_{21}a_{32} - a_{12}a_{21}a_{33} - a_{13}a_{22}a_{31} - a_{11}a_{23}a_{32} \\ &= a_{11}(a_{22}a_{33} - a_{23}a_{32}) + a_{12}(a_{23}a_{31} - a_{21}a_{33}) + a_{13}(a_{21}a_{32} - a_{22}a_{31}) \\ &= a_{11}A_{11} + a_{12}A_{12} + a_{13}A_{13}. \end{aligned}$$

Similarly, $|A| = a_{21}A_{21} + a_{22}A_{22} + a_{23}A_{23} = a_{31}A_{31} + a_{32}A_{32} + a_{33}A_{33}$.

For any $n \times n$-matrix $A = [a_{ij}]$ with $n \geq 2$, the $(n-1) \times (n-1)$ submatrix X_{ij} obtained by deleting the ith row and the jth column of A will be called $(i, j)^{\text{th}}$ *submatrix* of A.

Lemma 2. Let $A = [a_{ij}]$ be any $n \times n$-matrix over a commutative ring R, with $n \geq 2$. For any fixed i, j with $1 \leq i,\ j \leq n$, let B_{ij} be the sum of those terms in the expression

$$\det(A) = \sum_{\sigma \in S_n} (-1)^{\sigma} a_{1\sigma(1)}\, a_{2\sigma(2)} \cdots a_{n\sigma(n)}$$

for which $\sigma(i) = j$.

Then, $$B_{ij} = (-1)^{i+j} a_{ij} M_{ij} = a_{ij} A_{ij}$$

where M_{ij} and A_{ij} are the $(i, j)^{\text{th}}$ minor and the cofactor respectively of A.

Proof. Let X_{ij} be the $(n-1) \times (n-1)$-submatrix of A obtained by deleting the ith row and the jth column of A.

Its rows are indexed by $1, 2, 3, \ldots, i-1,$
$i+1, i+2, \ldots, n$

and the columns by $1, 2, 3, \ldots, j-1,$
$j+1, j+2, \ldots, n$

Define $\lambda \in S_n$ as follows.

$\lambda = I$ if $i = j$, $\lambda = (j, j+1, \ldots, i-1, i)$, if $i > j$

and $\lambda = (j, j-1, \ldots, i+1, i)$ if $j > i$

In respective cases, λ is of length 1, $i-j+1$ and $j-i+1$. In each of these cases $(-1)^{\lambda} = (-1)^{i+j}$. We find that the columns of X_{ij} are indexed by

$$\lambda(1),\ \lambda(2),\ldots,\ \lambda(i-1),\ \lambda(i+1),\ldots,\ \lambda(n).$$

Now, we can write $\quad X_{ij} = [a_{kl}]_{(n-1)\times(n-1)}$,

with $1 \le k, l \le n, k \ne i$, and $l \ne j$.

Let

$E_i = \{1, 2, \ldots, i-1, i+1, \ldots, n\}$. By using the permutation λ, we can write $X_{ij} = [b_{kl}]$ with $k, l \in E_i$ and $b_{kl} = a_{k\lambda(l)}$.

Thus, by the definition of determinant

$$M_{ij} = |X_{ij}| = \sum_{\eta} (-1)^{\eta} \left[\prod_{k \in E_i} b_{k\eta(k)}\right]$$

where η runs over all the permutations of E_i.

Such an η can be taken as a member of S_n with $\eta(i) = i$.

Let $\quad H_i = \{\eta \in S_n : \eta(i) = i\}$,

so that

$$M_{ij} = \sum_{\eta \in H_i} (-1)^{\eta} \left[\prod_{k \in E_i} b_{k\eta(k)}\right]$$

$$= \sum_{\eta \in H_i} (-1)^{\eta} \left[\prod_{k \in E_i} a_{k\lambda\eta(k)}\right].$$

Observe that $\quad \eta(i) = i,$

gives $\quad \lambda\eta(i) = \lambda(i) = j,\ a_{ij} = a_{i\lambda\eta(i)}.$

Consequently,

$$a_{ij} M_{ij} = a_{i\lambda\eta(i)} \sum_{\eta \in H_i} (-1)^{\eta} \left[\prod_{k \in E_i} a_{k\lambda\eta(k)}\right]$$

$$= \sum_{\eta \in H_i} (-1)^{\eta} \left[\prod_{k=1}^{n} a_{k\lambda\eta(k)}\right].$$

By definition,

$$B_{ij} = \sum_{\sigma(i)=j} (-1)^{\sigma} \left[\prod_{k=1}^{n} a_{k\sigma(k)}\right].$$

For any $\sigma \in S_n$ with $\sigma(i) = j$, $\sigma = \lambda(\lambda^{-1}\sigma) = \lambda\eta$,

$$\eta \in H_i,$$

so that

$$B_{ij} = \sum_{\eta \in H_i} (-1)^{\lambda\eta} \left[\prod_{k=1}^{n} a_{k\lambda\eta(k)}\right]$$

$$= (-1)^{\lambda} \sum_{\eta \in H_i} (-1)^{\eta} \left[\prod_{k=1}^{n} a_{k\lambda\eta(k)} \right]$$

$$= (-1)^{i+j} a_{ij} \acute{M}_{ij}.$$

As $$A_{ij} = (-1)^{i+j} M_{ij}, \; B_{ij} = a_{ij} A_{ij}.$$

Theorem 3. Let $A = \left[a_{ij}\right]$ be any $n \times n$-matrix over a commutative ring R. Then, for any fixed i, $1 \le i \le n$

$$\det(A) = a_{i1}A_{i1} + a_{i2}A_{i2} + \ldots + a_{in}A_{in}.$$

Proof. For $1 \le j \le n$, let $K_j = \{\sigma \in S_n : \sigma(i) = j\}$.

Then, $$S_n = \bigcup_{j=1}^{n} K_j$$

and these K_j's are pairwise disjoint.

By Lemma 2, $$a_{ij}A_{ij} = \sum_{\sigma \in K_j} (-1)^{\sigma} a_{1\sigma(1)}\, a_{2\sigma(2)} \ldots a_{n\sigma(n)}.$$

However, $$\det(A) = \sum_{j=1}^{n} \sum_{\sigma \in K_j} (-1)^{\sigma} a_{1\sigma(1)}\, a_{2\sigma(2)} \ldots a_{n\sigma(n)}.$$

As $$S_n = \bigcup_{j=1}^{n} K_j,$$

this gives $$\det(A) = \sum_{j=1}^{n} a_{ij}A_{ij}.$$

Remark 4. The expression $|A| = \det(A) = \sum_{j=1}^{n} a_{ij}A_{ij}$ is called the *cofactor expansion* of $\det(A)$ by the ith row.

By considering A^T, we also get $|A| = \det(A^T) = \sum_{i=1}^{n} a_{ij}A_{ij}$, $\quad 1 \le j \le n$.

This is called the *cofactor expansion* of $\det(A)$ by the jth column.

Definition 5. For any $n \times n$-matrix $A = \left[a_{ij}\right]$ over a commutative ring R, with $n \ge 2$, the *adjoint* of A is the $n \times n$-matrix $\left[a'_{ij}\right]$ such that $a'_{ij} = A_{ji}$, the cofactor of a_{ji}. For a 1×1-matrix $A = [a]$, its adjoint is the 1×1-matrix $[1]$. Further, adj(A) denotes the adjoint of A.

Example 5.a. For $$A=\begin{bmatrix} a_{11} & a_{12} \\ a_{21} & a_{22} \end{bmatrix},$$

$$A_{11}=a_{22},\ A_{12}=-a_{21},\ A_{21}=-a_{12},\ A_{22}=a_{11},$$

so that $$\operatorname{adj}(A)=\begin{bmatrix} a'_{11} & a'_{12} \\ a'_{21} & a'_{22} \end{bmatrix}$$

with $$a'_{ij}=A_{ji}$$

gives $$\operatorname{adj}(A)=\begin{bmatrix} A_{11} & A_{21} \\ A_{12} & A_{22} \end{bmatrix}=\begin{bmatrix} a_{22} & -a_{12} \\ -a_{21} & a_{11} \end{bmatrix}.$$

For $$B=\begin{bmatrix} b_{11} & b_{12} & b_{13} \\ b_{21} & b_{22} & b_{23} \\ b_{31} & b_{32} & b_{33} \end{bmatrix}$$

$$\operatorname{adj}(B)=\begin{bmatrix} B_{11} & B_{21} & B_{31} \\ B_{12} & B_{22} & B_{32} \\ B_{13} & B_{23} & B_{33} \end{bmatrix}$$

where B_{ij} is the cofactor of b_{ij}.

Theorem 6. For any $n \times n$-matrix A over a commutative ring R

$$A.\operatorname{adj}(A)=|A|I_n=\operatorname{adj}(A)I_n.$$

Proof. Let $$A=\left[a_{ij}\right],\ \operatorname{adj}(A)=\left[a'_{ij}\right].$$

For $n = 1$, $$\operatorname{adj}(A)=[1],$$

so that $$A.\operatorname{adj}(A)=A=a_{11}[1]=|A|I_1.$$

This proves the result for $n = 1$. So, let $n \geq 2$.

Let $$A.\operatorname{adj}(A)=\left[b_{ij}\right] \quad \text{... (1)}$$

By definition, $$b_{ij}=\sum_{k=1}^{n} a_{ik}a'_{kj}=\sum_{k=1}^{n} a_{ik}A_{jk}.$$

In particular, $$b_{ii}=\sum_{k=1}^{n} a_{ik}A_{ik}=\det(A)=|A| \quad \text{by Theorem 3.}$$

Let $i \neq j$. For a fixed such pair of i and j, let $C=\left[c_{rs}\right]$ be such that for any $l \neq j$, the lth row of C is the same as the lth row of A, but the jth row of C is the ith row of A. It is, then, clear that in C, the cofactor of any c_{jk} is A_{jk}.

Now $$c_{jk}=a_{ik},$$

so that $$\det(C) = \sum_{k=1}^{n} c_{jk} A_{jk} = \sum_{k=1}^{n} a_{ik} A_{jk} = b_{ij}.$$

At the same time, it is clear that the ith and j^{th} rows of C are the same.

So, $$\det(C) = 0.$$

Thus, $$b_{ij} = 0 \qquad \text{for } i \neq j,$$

and $$b_{ii} = \det(A).$$

Hence, $$A.\text{adj}(A) = \det(A) I_n.$$

Similarly, one gets $$\text{adj}(A)A = |A| I_n.$$

Corollary 7. For any $n \times n$-non-singular matrix A over a field F,

$$A^{-1} = \frac{1}{|A|}\text{adj}(A)$$

Proof. As A is non-singular $$|A| \neq 0 \qquad \{\text{Theorem}(4.2.13)\}$$

By this theorem, $$A.\text{adj}(A) = |A| I_n = \text{adj}(A).A.$$

This gives $$A\left[\frac{1}{|A|}\text{adj}(A)\right] = I_n = \left[\frac{1}{|A|}\text{adj}(A)\right].A.$$

Hence, $$A^{-1} = \frac{1}{|A|}\text{adj}(A).$$

This corollary gives another method of finding the inverse of a non-singular matrix.

Example 7.a. Consider $$A = \begin{bmatrix} 1 & 2 & 1 \\ 0 & 1 & 2 \\ -1 & 3 & 2 \end{bmatrix}.$$

Hence, $$A_{11} = \begin{vmatrix} 1 & 2 \\ 3 & 2 \end{vmatrix} = -4,$$

$$A_{12} = -\begin{vmatrix} 0 & 2 \\ -1 & 2 \end{vmatrix} = -2$$

$$A_{13} = \begin{vmatrix} 0 & 1 \\ -1 & 3 \end{vmatrix} = 1.$$

So that $|A| = a_{11}A_{11} + a_{12}A_{12} + a_{13}A_{13} = 1.(-4) + 2.(-2) + 1.(-1) = -7$.

Thus, A is non-singular.

Further, $$A_{21} = -\begin{vmatrix} 2 & 1 \\ 3 & 2 \end{vmatrix} = -1, \quad A_{22} = \begin{vmatrix} 1 & 1 \\ -1 & 2 \end{vmatrix} = 3$$

$$A_{23} = -\begin{vmatrix} 1 & 2 \\ -1 & 3 \end{vmatrix} = -5, \; A_{31} = \begin{vmatrix} 2 & 1 \\ 1 & 2 \end{vmatrix} = 3$$

$$A_{32} = -\begin{vmatrix} 1 & 1 \\ 0 & 2 \end{vmatrix} = -2, \; A_{33} = \begin{vmatrix} 1 & 2 \\ 0 & 1 \end{vmatrix} = 1,$$

so that
$$\operatorname{adj}(A) = \begin{bmatrix} A_{11} & A_{21} & A_{31} \\ A_{12} & A_{22} & A_{32} \\ A_{13} & A_{23} & A_{33} \end{bmatrix} = \begin{bmatrix} -4 & -1 & 3 \\ -2 & 3 & -2 \\ 1 & -5 & 1 \end{bmatrix}.$$

Obviously,
$$A.\operatorname{adj}(A) = \begin{bmatrix} -7 & 0 & 0 \\ 0 & -7 & 0 \\ 0 & 0 & -7 \end{bmatrix} = -7I_3 = \operatorname{adj}(A).A$$

$$A^{-1} = \frac{1}{-7}\operatorname{adj}(A) = \begin{bmatrix} \frac{4}{7} & \frac{1}{7} & \frac{-3}{7} \\ \frac{2}{7} & \frac{-3}{7} & \frac{2}{7} \\ \frac{-1}{7} & \frac{5}{7} & \frac{-1}{7} \end{bmatrix}.$$

We now give an application of the adjoint of a matrix in the solution of certain special types of systems of linear equations.

Theorem 8. (Cramer's rule). Let $AX = B$ be a system of a linear equations in n variables over a field F, such that the coefficient matrix A is non-singular. Let A_j be the $n \times n$-matrix, obtained from A by replacing the jth column of A by B.

Then, in $X = \begin{bmatrix} x_1 \\ x_2 \\ \vdots \\ \vdots \\ x_n \end{bmatrix}$ $\quad x_j = \dfrac{|A_j|}{|A|}.$

Proof. Let
$$A = [a_{ij}], \; \operatorname{adj}(A) = [a'_{ij}]$$

with
$$a'_{ij} = A_{ji}$$

and
$$B = \begin{bmatrix} b_1 \\ b_2 \\ \vdots \\ \vdots \\ b_n \end{bmatrix}.$$

As all the entries of A_j other than in the jth column are the same as the corresponding of A, the cofactor of the $(i, j)^{\text{th}}$ entry b_i of A_j is the same as the $(i, j)^{\text{th}}$ cofactor A_{ij} of A. Consequently, the cofactor expansion of $|A_j|$ by the jth column gives

$$\left|A_j\right| = \sum_{i=1}^{n} b_i A_{ij}.$$

Now, $$AX = B$$

gives $$\text{adj}(A).AX = \text{adj}(A)B$$

i.e., $$|A|X = \text{adj}(A)B.$$

Consequently, by comparing the jth components on both sides, we get

$$|A|x_j = \sum_{i=1}^{n} a'_{ji}b_i = \sum_{i=1}^{n} b_i A_{ij} = \left|A_j\right|.$$

Hence, $$x_j = \frac{\left|A_j\right|}{|A|}.$$

Example 8.a. Consider the following system of linear equations :

$$\begin{aligned} 2x + y + z &= 6 \\ 3x + 2y - 2z &= -2. \\ x + y + 2z &= 4 \end{aligned}$$

Here the coefficient matrix is $A = \begin{bmatrix} 2 & 1 & 1 \\ 3 & 2 & -2 \\ 1 & 1 & 2 \end{bmatrix}$.

By the cofactor expansion of $|A|$, by the first row,

$$|A| = 2\begin{vmatrix} 2 & -2 \\ 1 & 2 \end{vmatrix} - \begin{vmatrix} 3 & -2 \\ 1 & 2 \end{vmatrix} + \begin{vmatrix} 3 & 2 \\ 1 & 1 \end{vmatrix} = 5.$$

Hence, $$\left|A_1\right| = \begin{vmatrix} 6 & 1 & 1 \\ -2 & 2 & -2 \\ 4 & 1 & 2 \end{vmatrix} = 22$$

$$\left|A_2\right| = \begin{vmatrix} 2 & 6 & 1 \\ 3 & -2 & -2 \\ 1 & 4 & 2 \end{vmatrix} = -26$$

$$\left|A_3\right| = \begin{vmatrix} 2 & 1 & 6 \\ 3 & 2 & -2 \\ 1 & 1 & 4 \end{vmatrix} = 12,$$

so that $$x = \frac{\left|A_1\right|}{|A|} = \frac{22}{5},$$

$$y = \frac{|A_2|}{|A|} = \frac{-26}{5},$$

$$z = \frac{12}{5}.$$

Theorem 9. For any square matrix C of the form $\begin{bmatrix} A & 0 \\ D & B \end{bmatrix}$ over a commutative ring R, with A and B square matrices, $\det(C) = \det(A)\det(B)$.

Proof. Let $$A = [a_{ij}]_{k \times k}$$

and $$B = [b_{ij}]_{s \times s}.$$

Then, $$C = [c_{ij}]_{n \times n}$$

with $$n = k + l,$$

and $$D = [d_{ij}]_{s \times k}.$$

Observe that for $1 \le j \le k$, the jth column of C is of the form

$$\begin{bmatrix} a_{1j} \\ \vdots \\ a_{kj} \\ d_{1j} \\ \vdots \\ d_{sj} \end{bmatrix}$$

So the removal of such a column leaves B unaffected. The removal of first row and j^{th} column of C, with $1 \le j \le k$, does not affect B and it gives $(1, j)^{th}$ submatrix C_{1j} of C such that

$$C_{1j} = \begin{bmatrix} A_{1j} & 0 \\ D'_{1j} & B \end{bmatrix} \qquad \text{... (1)}$$

where A_{1j} is the $(1, j)^{th}$ submatrix of A, and D'_{1j} is obtained from D by deleting its jth column.

Now, $$c_{1j} = a_{1j} \qquad \text{for } 1 \le j \le k$$

and $$c_{1j} = 0 \qquad \text{for } j > k.$$

This gives $$\det(C) = \sum_{j=1}^{n} (-1)^{1+j} c_{1j} \det(C_{1j}) = \sum_{j=1}^{k} (-1)^{1+j} a_{1j} \det(C_{1j}) \qquad \text{....(2)}$$

We now apply induction on k.

For $k = 1$, $$\det(C) = a_{11} \det(C_{11}).$$

In this case, $C_{11} = B$

and $\det(A) = a_{11}$.

So that $\det(C) = \det(A)\det(B)$.

Hence, the result holds for $k = 1$. To apply induction, let $k > 1$ and the result hold for $k-1$. In (1), A_{1j} is a $(k-1)\times(k-1)$-matrix.

So, by the induction hypothesis $\det(C_{1j}) = \det(A_{1j})\det(B)$.

Then, (2) yields
$$\det(C) = \sum_{j=1}^{k}(-1)^{1+j}a_{1j}\det(A_{1j})\det(B)$$
$$= \det(A)\det(B).$$

Given any square matrices $A_1, A_2, ..., A_k$

over R recall that their direct sum $A_1 \oplus A_2 \oplus ... \oplus A_k$

is the block matrix
$$A = \begin{bmatrix} A_1 & 0 & .... & 0 \\ 0 & A_2 & .... & 0 \\ \vdots & \vdots & .... & \vdots \\ \vdots & \vdots & .... & \vdots \\ 0 & 0 & .... & A_k \end{bmatrix}.$$

Theorem 10. Let $A_1, A_2, ..., A_k$ be any $k \geq 1$ square matrices over a commutative ring R, and $A = A_1 \oplus A_2 \oplus ... \oplus A_k$.

Then, $\det(A) = \det(A_1)\det(A_2)...\det(A_k)$.

Proof. We prove the result by induction on k. For $k = 1$, $A = A_1$ and hence $\det(A) = \det(A_1)$. Thus, the result holds for $k = 1$. To apply induction, let $k > 1$ and the result hold for $k-1$.

We can write $A = B \oplus A_k = \begin{bmatrix} B & 0 \\ 0 & A_k \end{bmatrix}$

where $B = A_1 \oplus A_2 \oplus ... \oplus A_{k-1}$.

By the induction hypothesis, $\det(B) = \det(A_1)...\det(A_{k-1})$.

But by Theorem 9, $\det(A) = \det(B)\det(A_k)$.

Hence, $\det(A) = \det(A_1)\det(A_2)...\det(A_k)$.

Exercise 4.3

1. Let
$$A = \begin{bmatrix} 2 & 2 & 1 & -3 \\ 0 & 1 & 2 & 1 \\ -1 & 1 & 0 & 3 \\ 4 & 0 & 1 & -5 \end{bmatrix}$$

(*i*) Find the cofactors A_{11}, A_{23}, A_{41}.

(*ii*) Find $|A|$ by the cofactor expansion by the first row.

2. (*a*) Find the adjoint of each of the following matrices :

(*i*) $[-5]$,

(*ii*) $\begin{bmatrix} 3 & 0 \\ 4 & -1 \end{bmatrix}$,

(*iii*) $\begin{bmatrix} 0 & 1 & -2 \\ -1 & 3 & 1 \\ 2 & -2 & 3 \end{bmatrix}$

(*iv*) $\begin{bmatrix} 2 & 1 & -3 \\ 1 & 0 & 2 \\ 2 & -4 & 1 \end{bmatrix}$

(*v*) $\begin{bmatrix} 3 & -3 & 0 \\ 0 & 1 & -5 \\ 0 & 0 & 6 \end{bmatrix}$

(*b*) By using cofactor expansion, find the determinant of each of the matrices in (*a*).

(*c*) By using Theorem (4.2.13), find which of the matrices in (*a*) are non-singular.

(*d*) By using Corollary 7, find the inverse of each of the non-singular matrices in (*a*).

3. By using Theorem 10, find the determinants of the matrices

(*i*)
$$\begin{bmatrix} 3 & 4 & -1 & 0 & 0 & 0 & 0 & 0 \\ -1 & 2 & 0 & 0 & 0 & 0 & 0 & 0 \\ 7 & 9 & -1 & 0 & 0 & 0 & 0 & 0 \\ 0 & 0 & 0 & 0 & 5 & 0 & 0 & 0 \\ 0 & 0 & 0 & -7 & 0 & 0 & 0 & 0 \\ 0 & 0 & 0 & 0 & 0 & 3 & 5 & -1 \\ 0 & 0 & 0 & 0 & 0 & -8 & 1 & 0 \\ 0 & 0 & 0 & 0 & 0 & 0 & 0 & 1 \end{bmatrix}$$

(ii) $$\begin{bmatrix} 4 & 0 & 0 & 0 & 0 & 0 \\ -5 & 7 & 0 & 0 & 0 & 0 \\ 9 & 0 & 8 & 0 & 0 & 0 \\ 0 & 0 & 0 & -1 & 5 & 10 \\ 0 & 0 & 0 & 0 & 11 & 0 \\ 0 & 0 & 0 & 0 & 0 & 6 \end{bmatrix}$$

4. By using Theorem 9, find the determinant of

$$\begin{bmatrix} 8 & 9 & 0 & 0 & 0 \\ 1 & 2 & 0 & 0 & 0 \\ 30 & 40 & 1 & 2 & 3 \\ -91 & 45 & -1 & 1 & 3 \\ -89 & 93 & 0 & 0 & 5 \end{bmatrix}.$$

5. For any two square matrices A and B over a commutative ring R, prove that

$$\text{adj}(A \oplus B) = \text{adj}(A) \oplus \text{adj}(B).$$

Further, if A and B are non-singular, prove that

$$(A \oplus B)^{-1} = A^{-1} \oplus B^{-1}.$$

6. By using Cramer's rule, solve each of the following systems of linear equations :

(a) $$\begin{aligned} 3x+4y &= 8 \\ 2x+3y &= 7 \end{aligned}$$

(b) $$\begin{aligned} 2x+4y+6z &= 2 \\ x \quad\quad +2z &= 0 \\ 2x+3y-\ z &= -5 \end{aligned}$$

(c) $$\begin{aligned} x+y+\ z &= 7 \\ 2x+3y+4z &= 1 \\ 4x+9y+16z &= -1 \end{aligned}$$

(d) $$\begin{aligned} x+2y \quad\quad + w &= 1 \\ x+2y+3z \quad\quad &= 1 \\ z+2w &= 0 \\ y+2z-w &= 0 \end{aligned}$$

7. For any $n \times n$-matrix A over a field F, prove that the system of homogeneous linear equations $AX = 0$ over F has non-trivial solution iff $\det(A) = 0$.

8. By using the result in Exercise 7, find which of the following systems of linear homogeneous equations have non-trivial solutions :

(a) $$\begin{aligned} 2x+y &= 0 \\ x+2y &= 0 \end{aligned}$$

(*b*)
$$\begin{aligned} 3x+2y&=0\\ 6x+4y&=0 \end{aligned}$$

(*c*)
$$\begin{aligned} x-2y+\ z&=0\\ 2x+3y+\ z&=0\\ 3x+\ y+2z&=0 \end{aligned}$$

(*d*)
$$\begin{aligned} x+\ y+\ z+\ w&=0\\ -2x+2y+2z\qquad&=0\\ y+\ z+2w&=0\\ x+4y+4z+3w&=0 \end{aligned}$$

(*e*)
$$\begin{aligned} y+2z&=0\\ x+2y\qquad&=0\\ x+3y+4z&=0 \end{aligned}$$

9. Find the value of a for which $\begin{vmatrix} 3 & a & 5 \\ -1 & 2 & 3 \\ 1 & -2 & 3 \end{vmatrix} = 0$.

4.4 Characteristic Values

Throughout, V is a non-zero finite dimensional vector space over a field F.

Let
$$B = \{\alpha_1, \alpha_2, ..., \alpha_n\}$$
and
$$B' = \{\beta_1, \beta_2, ..., \beta_n\}$$
be two ordered bases of V.

Let T be a linear operator on V. Then, $A = [T]_B$, $A' = [T]_{B'}$ are two $n \times n$-matrices over F. Further, $A' = P^{-1}AP$, where P is the matrix of B' relative to B {Corollary(3.3.7)}, so that $|A'| = |P|^{-1}|A||P| = |A|$. Hence, irrespective of the choice of an ordered basis of V, the determinant $|[T]_B|$ is uniquely determined by T. This determinant is called the *determinant* of T and is denoted by $\det(T)$ or by $|T|$.

Theorem 1. Let T_1 and T_2 be two linear operators on V_F. Then,

(*i*)
$$|T_2T_1| = |T_2||T_1|$$

(*ii*) T_1 is invertible iff $|T_1| \neq 0$

(*iii*) If T_1 is invertible, then $\quad |T_1^{-1}| = |T_1|^{-1}$

(*iv*) For any $a \in F$, $\quad |aT_1| = a^n|T_1|$,
$$n = \dim V_F.$$

Proof. For any ordered basis B of V,

$$|T_2T_1|_B = |T_2|_B\,|T_1|_B$$

gives $$|T_2T_1| = \left|[T_2]_B[T_1]_B\right| = \left|[T_2]_B\right|\left|[T_1]_B\right| = |T_2||T_1|$$

by using Theorem (4.2.11).

This proves (*i*).

By Theorem (3.3.5), $$[aT_1]_B = a[T_1]_B.$$

This gives $$|aT_1| = |aA|,$$

where $$A = [T_1]_B.$$

Then, $$|aT_1| = a^n|A| \qquad \{\text{see Exercise (4.2.12)}\},$$

i.e., $$|aT_1| = a^n|T_1|.$$

This proves (*iv*). Now T_1 is invertible iff $[T_1]_B$ is invertible. However, by Theorem (4.2.13), $[T_1]_B$ is invertible iff its determinant $|T_1| \neq 0$. This proves (*ii*). Finally, let T_1 be invertible.

Then, $$T_1^{-1}T_1 = I$$

gives $$|T_1^{-1}||T_1| = |I| = 1.$$

Consequently, $$|T_1^{-1}| = |T_1|^{-1}.$$

Definition 2.

(*i*) Let A be any $n \times n$-matrix over a field F. Then, an element $c \in F$ is called a *characteristic value* (or an eigen value) of A if $AB = cB$ for some non-zero $B \in F^{n\times 1}$; any such B is called a *characteristic vector* (or an eigen vector) of A relative to c.

(*ii*) For any linear operator T on V_F, an element $c \in F$ is called a *characteristic value* (or an eigen value) of T if $T\alpha = c\alpha$ for some non-zero $\alpha \in V$; any such α is called a *characteristic vector* (or an eigen vector) of T relative to c.

Example 2.a. Let T be the linear operator on $\mathbf{R}^{3\times 1}$ whose matrix relative to the standard ordered basis is

$$A = \begin{bmatrix} 2 & 0 & 0 \\ 1 & 2 & 0 \\ 0 & 0 & -1 \end{bmatrix}.$$

Here $$T\in_1 = 2\in_1 + \in_2,$$

$$T\in_2 = 2\in_2,$$

$$T\in_3 = -\in_3 = (-1)\in_3.$$

By definition, 2 and -1 are characteristic values of T. As $A\in_2 = 2\in_2$, $A\in_3 = (-1)\in_3$, 2 and -1 are characteristic values of A. Here, $\in_2$ is a characteristic vector of T as well as A, relative to the characteristic value 2.

Also, $$|T| = |A| = -4.$$

Observe that $$|2I_3 - A| = \begin{vmatrix} 0 & 0 & 0 \\ -1 & 0 & 0 \\ 0 & 0 & 3 \end{vmatrix} = 0$$

and $$|(-1)I_3 - A| = \begin{vmatrix} -1 & 0 & 0 \\ -1 & -1 & 0 \\ 0 & 0 & 0 \end{vmatrix} = 0.$$

We shall see that the characteristic values play a very important role in studying linear operators.

Theorem 3. Let T be a linear operator on V_F, $n = \dim V$ and $A = [T]_B$ where B is an ordered basis of V. Then, for any $c \in F$, the following statements are equivalent :

(*i*) c is a characteristic value of T

(*ii*) c is a characteristic value of A

(*iii*) $\det(cI_n - A) = 0$.

Proof. Consider $$\alpha \in V.$$

Then, the co-ordinate vector $$[\alpha]_B \in F^{n\times 1}$$

and by Theorem (3.3.11), $$[T\alpha]_B = A[\alpha]_B,$$

so that $$T\alpha = c\alpha$$

iff $$c[\alpha]_B = A[\alpha]_B$$

iff $$cI_n[\alpha]_B = A[\alpha]_B$$

iff $$(cI_n - A)[\alpha]_B = 0.$$

Thus, if c is a characteristic value of T, then for some $\alpha \neq 0$ in V, $T\alpha = c\alpha$. Then $[\alpha]_\beta \neq 0$ and $A[\alpha]_\beta = c[\alpha]_\beta$ yields that c is a characteristic value of A. Conversely, le c be a characteristic value of A. Then, for some $B \neq 0$ in $F^{n\times 1}$, $AB = cB$. As $V \cong F^{n\times 1}$ we get $\beta \neq 0$ in V such that $B = [\beta]_B$ {Theorem (2.4.5)}. Consequently, $T\beta = c\beta$. This gives that c is a characteristic value of T. Hence, (*i*) and (*ii*) are equivalent. This also shows that $c \in F$ is a characteristic value of A iff $(cI_n - A)B = 0$ for some $B \neq 0$ in $F^{n\times 1}$ iff the system of homogeneous linear equations $(cI_n - A)X = 0$ has a non-trivial solution iff $cI_n - A$ is singular iff $\det(cI_n - A) = 0$ {Theorem (4.2.13)}. Hence (*ii*) and (*iii*) are equivalent.

Definition 4. For any $n \times n$-matrix $A = [a_{ij}]$ over a field F,

$$|XI_n - A| = \begin{vmatrix} X - a_{11} & -a_{12} & -a_{13} & \dots & -a_{1n} \\ -a_{21} & X - a_{22} & -a_{23} & \dots & -a_{2n} \\ -a_{31} & -a_{23} & X - a_{33} & \dots & -a_{3n} \\ \dots & \dots & \dots & \dots & \dots \\ \dots & \dots & \dots & \dots & \dots \\ -a_{n1} & -a_{n2} & -a_{n3} & \dots & X - a_{nn} \end{vmatrix}$$

$$= X^n + a_1 X^{n-1} + \dots + a_n$$

where X is an indeterminate, is called the *characteristic polynomial* of A.

The equation $|XI_n - A| = 0$ is called the *characteristic equation* of A.

Theorem 3 shows that given $A \in F^{n \times n}$, any $c \in F$ is a characteristic value of A iff c is a root of the characteristic polynomial of A. As the characteristic polynomial of A is of degree n, it can have, at most, n distinct root in F, so that an $A \in F^{n \times n}$ cannot have more than n characteristic values.

Example 4.a. Consider
$$A = \begin{bmatrix} 2 & -2 & 3 \\ 0 & 3 & -2 \\ 0 & -1 & 2 \end{bmatrix}.$$

The characteristic polynomial of A is

$$|xI_3 - A| = \begin{bmatrix} x-2 & 2 & -3 \\ 0 & x-3 & 2 \\ 0 & 1 & x-2 \end{bmatrix}.$$

The cofactor expansion by first column gives

$$|xI_3 - A| = (x-2)[(x-3)(x-2)-2)] = (x-2)(x-1)(x-4).$$

The equation $|xI_3 - A| = 0$, gives $x = 1$, 2 or 4, so that A has three characteristic values namely 1, 2, 4. We now determine characteristic vector relative to each of the characteristic values.

For $x = 1$,
$$xI_3 - A = \begin{bmatrix} -1 & 2 & -3 \\ 0 & -2 & 2 \\ 0 & 1 & -1 \end{bmatrix} = A'.$$

A' is row equivalent to the matrix in reduced row echelon form,

$$A'' = \begin{bmatrix} 1 & 0 & 1 \\ 0 & 1 & -1 \\ 0 & 0 & 0 \end{bmatrix},$$

so that
$$A'X = 0$$

has general solution $$X = \begin{bmatrix} -r \\ r \\ r \end{bmatrix}, \ r \in \mathbf{R}.$$

It satisfies $$AX = 1X.$$

For any value of $r \neq 0$, say for $r = 1$, we get $\begin{bmatrix} -1 \\ 1 \\ 1 \end{bmatrix}$ is a characteristic vector of A relative to the characteristic value 1. Indeed, every non-zero member of the solution space of $A'X = 0$ is a characteristic vector of A relative to 1. Similarly, by considering the homogeneous linear equations

$$(2I_3 - A)X = 0$$

$$(4I_3 - A)X = 0$$

we get $$X_2 = \begin{bmatrix} 1 \\ 0 \\ 0 \end{bmatrix}$$

and $$X_3 = \begin{bmatrix} 7 \\ -4 \\ 2 \end{bmatrix}$$

are characteristic vectors of A relative to 2 and 4, respectively.

Consider $$P = \begin{bmatrix} -1 & 1 & 7 \\ 1 & 0 & -4 \\ 1 & 0 & 2 \end{bmatrix}.$$

Then, $$AP = \begin{bmatrix} -1 & 2 & 28 \\ 1 & 0 & -16 \\ 1 & 0 & 8 \end{bmatrix} = P\begin{bmatrix} 1 & 0 & 0 \\ 0 & 2 & 0 \\ 0 & 0 & 4 \end{bmatrix},$$

so that $$P^{-1}AP = \text{diag}(1,\ 2,\ 4).$$

Thus, A is similar to a diagonal matrix. One asks the question : under what conditions a square matrix over a field F is similar to a diagonal matrix ? This question will be answered after a while.

Lemma 5. Any two similar matrices over a field F have the same characteristic polynomials and the same characteristic values.

Proof. Let A and B be two similar matrices in $F^{n \times n}$.

Then, $$B = P^{-1}AP,$$

for some non-singular $P \in F^{n \times n}$.

Consequently,
$$|XI_n - B| = |XP^{-1}I_nP - P^{-1}AP|$$
$$= |P^{-1}(XI_n - A)P|$$
$$= |XI_n - A|.$$

The second part is obvious.

Given any linear operator T on V_F, as $[T]_B$ and $[T]_{B'}$ are similar; they have the same characteristic polynomials. So given $A = [T]_B$, we define the characteristic polynomial of T to be $|XI_n - A|$.

For any characteristic value c of T, the set $W_c = \{\alpha \in V : T\alpha = c\alpha\}$ is a subspace of V. To see this, notice that $0 \in W_c$.

Consider
$$\alpha, \beta \in W_c$$
and
$$a \in F.$$
As
$$T\alpha = c\alpha$$
and
$$T\beta = c\beta,$$
$$T(\alpha + a\beta) = T\alpha + aT\beta = c\alpha + ca\beta = c(\alpha + a\beta),$$
we get
$$\alpha + a\beta \in W_c.$$

Hence, W_c is a subspace of V. Over that, as c is a characteristic value of T, there exists an $\alpha_0 \neq 0$ in V such that $T\alpha_0 = c\alpha_0$. Hence, W_c is a non-zero subspace of V. This subspace W_c is called the *characteristic space* of T (or the eigen space of T) relative to c. Similarly, we define the characteristic space of a matrix in $F^{n\times n}$.

Theorem 6. Let T be a linear operator on V_F and $A = [T]_B$. Let c be a characteristic value of T. Then, the following hold

(*i*) For any $\alpha \in V$
$$T\alpha = c\alpha$$
iff
$$A[\alpha]_B = c[\alpha]_B$$

(*ii*) For any polynomials $p(X) = a_0 + a_1X + ... + a_kX^k$

and any $\alpha \in W_c$,
$$p(T)\alpha = p(c)\alpha.$$

Proof. (*i*). Was seen in the proof of Theorem 3.

Now,
$$T\alpha = c\alpha,$$
gives
$$T^2\alpha = T(T\alpha) = T(c\alpha) = cT\alpha = c^2T\alpha.$$

Repeated operation by T gives $\quad T^i\alpha = c^i\alpha \quad$ for every $i \geq 1$.

At the same time
$$T^0 = I,$$
gives
$$T^0\alpha = I\alpha = \alpha = c^0\alpha.$$

Thus,
$$\begin{aligned} p(T)\alpha &= a_0 I\alpha + a_1 T\alpha + a_2 T^2\alpha + \ldots + a_k T^k \alpha \\ &= a_0\alpha + a_1 c\alpha + a_2 c^2 \alpha + \ldots + a_k c^k \alpha \\ &= (a_0 + a_1 c + \ldots + a_k c^k)\alpha \\ &= p(c)\alpha. \end{aligned}$$

Theorem 7. Let T be a linear operator on V_F and $c_1, c_2, \ldots, c_k$ be $k \geq 1$ distinct characteristic values of T.

Then,
$$W_{c_1} + W_{c_2} + \ldots + W_{c_k} = W_{c_1} \oplus W_{c_2} \oplus \ldots \oplus W_{c_k}.$$

Proof. The result is obvious for $k = 1$. Let $k > 1$. By Theorem (3.4.6), we get k polynomials $p_1(X), p_2(X), \ldots, p_k(X)$ each of degree $k-1$ such that $p_i(c_j) = \delta_{ij}$.

Let for some $\alpha_i \in W_{c_i}$, $$\alpha_1 + \alpha_2 + \ldots + \alpha_k = 0 \qquad \ldots (1)$$

Then, for any $1 \leq i \leq k$, $$p_i(T)\left[\sum_{j=1}^{k} \alpha_j\right] = p_i(T)0 = 0 \qquad \ldots (2)$$

i.e., $$\sum_{j=1}^{k} p_i(T)\alpha_j = 0.$$

As $$T\alpha_j = c_j \alpha_j.$$

By Theorem 6, $$p_i(T)\alpha_j = p_i(c_j)\alpha_j = \delta_{ij}\alpha_j.$$

For $i \neq j$, $$\delta_{ij} = 0,$$

gives $$p_i(T)\alpha_j = 0.$$

Hence, (2) gives $$p_i(T)\alpha_i = 0$$

i.e., $$p_i(c_i)\alpha_i = 0$$

i.e., $$\alpha_i = 0,$$

as $$p_i(c_i) = 1.$$

Hence, in (1) every $\alpha_i = 0$. This proves the result.

Corollary 8. Let T be a linear operator on a finite dimensional vector space V_F with $\dim V_F = n \geq 1$. Let $c_1, c_2, \ldots, c_k$ be $k \geq 1$ distinct characteristic values of T Let for $1 \leq i \leq k$, W_i be the characteristic space of T relative to c_i. Then,

(*i*) $$\dim(W_1 + W_2 + \ldots + W_k) = \sum_{i=1}^{k} \dim W_i$$

(*ii*) $$k \leq n$$

(*iii*) If $k = n$, then $$\dim W_i = 1 \qquad \text{for } 1 \leq i \leq n.$$

Proof.

(*i*) As $W = W_1 + W_2 + \ldots + W_k = W_1 \oplus W_2 \oplus \ldots \oplus W_k$, by Theorem (2.2.24),

$$\dim W = \sum_{i=1}^{k} \dim W_i$$

(*ii*) As $\dim W_i \geq 1$,

and $\dim W \leq \dim V = n$,

we get $k \leq n$.

(*iii*) For $k = n$, $$n \geq \sum_{i=1}^{n} \dim W_i \geq n,$$

gives $$n = \sum_{i=1}^{n} \dim W_i .$$

This immediately gives $\dim W_i = 1$.

The following result is immediately clear from this Corollary and Theorem 7.

Corollary 9. Let A be any $n \times n$-matrix over a field F and let $c_1, c_2, \ldots, c_k$ be any $k \geq 1$ distinct characteristic values of A. If for $1 \leq i \leq k$, X_i is a characteristic vector of A in $F^{n \times 1}$ relative to c_i, then, $X_1, X_2, \ldots, X_k$ are linearly independent over F.

In Example 4.b, the matrix A is not a diagonal matrix, but is similar to a diagonal matrix. This motivates the following inference :

Definition 10.

(*i*) A matrix A in $F^{n \times n}$ is said to be *diagonalizable* if for some non-singular P in $F^{n \times}$ $P^{-1}AP$ is a diagonal matrix.

(*ii*) A linear operator T on a finite dimensional vector space V_F is said to be a *diagonalizable* operator if for some ordered basis B, the matrix $[T]_B$ is a diagonal matrix.

Lemma 11. If a matrix A in $F^{n \times n}$ is diagonalizable, then, the characteristic polynomial of A is of the form $(X - c_1)(X - c_2)\ldots(X - c_n)$ for some $c_i \in F$.

Proof. As A is diagonalizable, it is similar to a diagonal matrix

$$A' = \text{diag}(c_1, c_2, \ldots, c_n).$$

Then, $$XI_n - A' = \text{diag}(X - c_1, X - c_2, \ldots, X - c_n),$$

so that $$|XI_n - A'| = (X - c_1)(X - c_2)\ldots(X - c_n).$$

By Lemma 5, $$|XI_n - A| = |XI_n - A'|.$$

Hence, $$|XI_n - A| = (X - c_1)(X - c_2)\ldots(X - c_n).$$

This lemma shows that if a square matrix A in $F^{n\times n}$ is such that its characteristic polynomial is not factorizable into degree one factors over F, then, A is not diagonalizable.

For example, $$A = \begin{bmatrix} 0 & -1 \\ 1 & 0 \end{bmatrix}$$

over **R** has $$|XI_2 - A| = X^2 + 1,$$

which cannot be factored into degree one factors over **R**.

So, A is not similar to a diagonal matrix.

However, over **C**, $X^2 + 1 = (X + i)(X - i)$, $i = \sqrt{-1}$.

Thus, i and $-i$ are two characteristic values of A in **C**,

$$X_1 = \begin{bmatrix} i \\ 1 \end{bmatrix}$$

and $$X_2 = \begin{bmatrix} -i \\ 1 \end{bmatrix}$$

are characteristic vectors of A relative to i and $-i$, respectively.

As X_1, X_2 are linearly independent over **C** (Corollary 9), $P = [X_1 X_2] = \begin{bmatrix} i & -i \\ 1 & 1 \end{bmatrix}$ is a non-singular matrix, $P^{-1}AP = \begin{bmatrix} i & 0 \\ 0 & -i \end{bmatrix}$, so that A is diagonalizable over **C**.

The converse of Lemma 11 is not true as is seen in the following example :

Example 11.a. Consider $$A = \begin{bmatrix} 1 & 1 \\ 0 & 1 \end{bmatrix}$$

over **R**.

Then, $$|XI_2 - A| = (x - 1)^2.$$

If A is similar to a diagonal matrix $A' = \begin{bmatrix} c_1 & 0 \\ 0 & c_2 \end{bmatrix}$

then $$(X - 1)(X - 1) = |XI_2 - A'| = (X - c_1)(X - c_2).$$

This gives $$c_1 = c_2 = 1.$$

So that $$A' = I_2.$$

Now, $$A' = P^{-1}AP$$

for some non-singular matrix P, i.e., $I_2 = P^{-1}AP$.

This gives $$A = PI_2P^{-1} = I_2.$$

This is a contradiction. Hence, A is not similar to a diagonal matrix.

In a diagonal matrix $A = \text{diag}(a_1, a_2,..., a_n)$, the diagonal entries a_i need not be distinct. Let $c_1, c_2,..., c_k$ the totality of distinct entries among $a_1, a_2,..., a_n$. Let c_i occur λ_i times.

Then,
$$\begin{aligned} |XI_2 - A| &= (X - a_1)(X - a_2)...(X - a_n) \\ &= (X - c_1)^{\lambda_1}(X - c_2)^{\lambda_2}...(X - c_k)^{\lambda_k} \end{aligned}$$

and $c_1, c_2,..., c_k$ are the distinct characteristic values of A.

If $\in_1, \in_2,..., \in_n$ constitute the standard basis of $F^{n\times 1}$, then, $A\in_i = a_i \in_i$, so that $\in_i$ is a characteristic vector of A relative to a_i. For $1 \le j \le k$, λ_j of a_i's equal c_j. Thus, W_j, the characteristic space of A relative to c_j, contains, at least, λ_j, $\in_i$'s.

So,
$$\dim W_j \ge \lambda_j,$$

$$\sum_{j=1}^{k} \dim W_j \le n = \sum_{j=1}^{k} \lambda_j \le \sum_{j=1}^{k} \dim W_j$$

gives
$$n = \sum_{j=1}^{k} \dim W_j.$$

This, in turn, yields
$$\dim W_j = \lambda_j.$$

Let now T be a diagonalizable operator on V_F. By definition, V_F has an ordered basis $B' = \{\beta_1, \beta_2,..., \beta_n\}$ such that $A = [T]_{B'}$ is a diagonal matrix, say $\text{diag}(a_1, a_2,..., a_n)$. Here, $T\beta_i = a_i\beta_i$. If $c_1, c_2,..., c_k$ are distinct characteristic values of T, and for $1 \le j \le k$, λ_j of a_i's equal to c_j, we also get that W_j the characteristic space of T relative to c_j has dimension λ_j; further, the set B_j of those β_i for which $a_i = c_j$, is a basis of W_j.

Theorem 12. Let T be a linear operator on a finite dimensional vector space V_F. Let $c_1, c_2,..., c_k$ be all the distinct characteristic values of T in F, with corresponding characteristic spaces $W_1, W_2,..., W_k$. Then, the following, are equivalent :

(*i*) T is diagonalizable

(*ii*) V_F has a basis B whose every member is a characteristic vector of T

(*iii*) The characteristic polynomial $f(X)$ of T is

$$f(X) = (X - c_1)^{\lambda_1}(X - c_2)^{\lambda_2}...(X - c_k)^{\lambda_k}$$

with
$$\dim W_i = \lambda_i,\ 1 \le i \le k$$

(*iv*)
$$V = W_1 \oplus W_2 \oplus ... \oplus W_k$$

Proof. (*i*) implies (*ii*). Let T be diagonalizable.

So, we have an ordered basis $B=\{\alpha_1, \alpha_2, \ldots, \alpha_n\}$ of V such that $A=[T]_B$ is a diagonal matrix.

Let $$A=\text{diag}(a_1, a_2, \ldots, a_n).$$

Then, $$T\alpha_i = a_i\alpha_i.$$

This shows that every member of B is a characteristic vector of T.

(*ii*) implies (*iii*). Let (*ii*)hold.

We get an ordered basis $$B=\{\alpha_1, \alpha_2, \ldots, \alpha_n\}$$

such that $$T\alpha_i = a_i\alpha_i \quad \text{for some } a_i \in F.$$

As explained above, this theorem,

$$f(X)=(X-c_1)^{\lambda_1}(X-c_2)^{\lambda_2}\ldots(X-c_k)^{\lambda_k}$$

with $$\lambda_i = \dim W_i.$$

(*iii*) implies (*iv*). Let (*iii*) hold.

Then, $$f(X)=(X-c_1)^{\lambda_1}(X-c_2)^{\lambda_2}\ldots(X-c_k)^{\lambda_k}$$

with $$\lambda_i = \dim W_i$$

gives $$n=\deg f(X)=\sum_{i=1}^{k}\dim W_i.$$

This gives $$\dim V=\sum_{i=1}^{k}\dim W_i=\dim\left[\sum_{i=1}^{k}W_i\right].$$

Consequently, $$V=\sum_{i=1}^{n}W_i=\oplus\sum_{i=1}^{k}W_i \quad \text{(Theorem 7)}$$

(*iv*) implies (*i*). Let $$V=W_1\oplus W_2\oplus\ldots\oplus W_k.$$

Let B_i be an ordered basis of W_i.

For any $\alpha\in B_i$, $$T\alpha=c_i\alpha.$$

Then, $$B=B_1\cup B_2\cup\ldots\cup B_k$$

is an ordered basis of V.

Let $$B=\{\alpha_1, \alpha_2, \ldots, \alpha_n\}.$$

For $1\le i\le n$, $$\alpha_i\in B_j$$

for some j, so that $$T\alpha_i = c_j\alpha_i.$$

Consequently, $[T]_B$ is a diagonal matrix. Hence, T is diagonalizable.

Corollary 13. Let T be a linear operator on a finite dimensional vector space V_F, such that T has $n = \dim V$, the number of distinct characteristic values. Then, T is diagonalizable.

Proof. Let $c_1, c_2, \ldots, c_n$ be the n distinct characteristic values of T and $W_1, W_2, \ldots, W_n$ be the corresponding characteristic spaces.

Then, $$n \le \sum_{i=1}^{k} \dim W_i \le n,$$

with $$\dim W_i \ge 1,$$

gives $$\sum_{i=1}^{n} \dim W_i = \dim V,$$

so that $$V = \oplus \sum_{i=1}^{n} W_i.$$

Hence, by this theorem, T is diagonalizable.

Corollary 14. Let A be any $n \times n$-matrix over a field F such that A has n distinct characteristic values in F. Then, A is diagonalizable.

Proof. Let T be the linear operator on $F^{n\times 1}$ whose matrix relative to the standard ordered basis is A. For any $X \in F^{n\times 1}$, $T(X) = AX$, gives T and A have same characteristic values, so that T has n distinct characteristic values in F. Hence, by Corollary 13, there exists an ordered basis B such that $A' = [T]_B$ is diagonal. Then, A is similar to the diagonal matrix A'.

Theorem 15. Let $A = [a_{ij}]$ be any triangular matrix in $F^{n\times n}$. Then, the characteristic polynomial of A is

$$(X - a_{11})(X - a_{22})\ldots(X - a_{nn}),$$

and $$a_{11}, a_{22}, \ldots, a_{nn}$$

are the characteristic values of A.

Proof. As XI_n is a diagonal matrix, $XI_n - A$ is a triangular matrix with diagonal entries $X - a_{11}, X - a_{22}, \ldots, X - a_{nn}$.

By Theorem (4.2.16), $|XI_n - A| = (X - a_{11})(X - a_{22})\ldots(X - a_{nn})$.

The last part is obvious.

Theorem 16. (Caylay-Hamilton Theorem). Let A be any $n \times n$-matrix over a field F and $f(X) = X^n + a_{n-1}X^{n-1} + \ldots + a_n$ be its characteristic polynomial.

Then, $$f(A) = A^n + a_{n-1}A^{n-1} + \ldots + a_n I_n = 0.$$

Proof. Now, $(XI_n - A)\ \text{adj}(XI_n - A) = |XI_n - A| I_n = f(X)I_n$. Observe that each entry of $XI_n - A$ is a polynomial of degree less than or equal to 1. So, (i, j)th cofactor of $XI_n - A$, is of degree at most $n-1$; thus, we can write

$$\text{adj}(XI_n - A) = B_{n-1}X^{n-1} + B_{n-2}X^{n-2} + ... + B_1X + B_0$$

for some matrices B_i in $F^{n\times n}$.

Then, $(XI_n - A)(B_{n-1}X^{n-1} + B_{n-2}X^{n-2} + ... + B_0) = (X^n + a_{n-1}X^{n-1} + ... + a_0)I_n$

By comparing the coefficients of $X^0, X^1, X^2, ..., X^n$, we get

$$\begin{aligned} -AB_0 &= a_0I_n \\ B_0 - AB_1 &= a_1I_n \\ B_1 - AB_2 &= a_2I_n \\ &\vdots \\ B_{n-2} - AB_{n-1} &= a_{n-1}I_n \\ B_{n-1} &= I_n \end{aligned}$$

These are $n+1$ equations. Starting from top, multiply them on the left by, I, $A, A^2, ..., A^n$, respectively and then, add the results, this gives

$$0 = a_0I_n + a_1A + a_2A^2 + ... + a_{n-1}A^{n-1} + A^n = f(A).$$

Theorem 17. **(Caylay-Hamilton Theorem for Linear Operators).** Let T be a linear operator on a finite dimensional vector space V_F with $f(X)$ as its characteristic polynomial. Then, $f(T) = 0$.

Proof. Let $A = [T]_B$, where B is an ordered basis of V_F.

Then $\qquad f(X) = |XI_n - A|.$

By Theorem 16, $\qquad f(A) = 0.$

By the isomorphism between $\text{Hom}_F(V, V)$ and $F^{n\times n}$, we get $f(T) = 0$.

Definition 18. Given any polynomial $f(X) = \sum_{i=0}^{k} a_iX^i$ of degree $k \geq 0$ over a commutative ring R, it is called a *monic polynomial* if its leading coefficient $a_k = 1$.

Thus, $X^2 + 2$ is monic, but $2X^2 + 1$ is not monic.

Definition 19. Let $f(X) = X^n - a_{n-1}X^{n-1} - ... - a_0$ be a monic polynomial of degree $n \geq 1$ over a field F. The companion matrix $\text{Com}[f(X)]$ of $f(X)$ is the $n\times n$-matrix, given as follows.

For $n = 1$, $\qquad \text{Com}[f(X)] = [a_0].$

For $n \geq 2$, $\quad \text{Com}[f(X)] = \begin{bmatrix} 0 & 0 & 0 & \dots & 0 & a_0 \\ 1 & 0 & 0 & \dots & 0 & a_1 \\ 0 & 1 & 0 & \dots & 0 & a_2 \\ \dots & \dots & \dots & \dots & 0 & \dots \\ \dots & \dots & \dots & \dots & 0 & \dots \\ 0 & 0 & 0 & \dots & 1 & a_{n-1} \end{bmatrix}.$

For example, the companion matrices of $X+2$ and $X^2+1 = X^2+0X+1$ are $[-2]$ and $\begin{bmatrix} 0 & -1 \\ 1 & 0 \end{bmatrix}$, respectively. The following theorem helps in constructing matrices with pre-assigned monic polynomials as their characteristic polynomials.

Theorem 20. For any monic polynomial $f(X)$ of positive degree, over a field F, the characteristic polynomial of the companion matrix of $f(X)$, is $f(X)$.

Proof. Let $f(X) = X^n - a_{n-1}X^{n-1} - \dots - a_0$. We prove the result by induction on n.

For $n = 1$,
$$f(X) = X - a_0,$$
$$\text{Com}[f(X)] = [a_0],$$
so that
$$|XI_1 - \text{Com}[f(X)]| = |[X - a_0]| = X - a_0 = f(X).$$

Thus, the result hold for $n = 1$. Let $n \geq 2$ and let the result hold for monic polynomials of degree $n-1$.

Let
$$A = \text{Com}[f(X)].$$

Then,
$$XI_n - A = \begin{bmatrix} X & 0 & 0 & \dots & 0 & -a_0 \\ -1 & X & 0 & \dots & 0 & -a_1 \\ 0 & -1 & X & \dots & 0 & -a_2 \\ \dots & \dots & \dots & \dots & \dots & \dots \\ \dots & \dots & \dots & \dots & \dots & \dots \\ 0 & 0 & 0 & \dots & -1 & X - a_{n-1} \end{bmatrix}.$$

Expansion of $|XI_n - A|$ by the first row gives
$$|XI_n - A| = XA_{11} + (-a_0)A_{1n} \qquad \dots (1)$$
where the cofactors A_{11} and A_{1n} are given below :
$$A_{11} = \begin{vmatrix} X & 0 & 0 & \dots & 0 & -a_1 \\ -1 & X & 0 & \dots & 0 & -a_2 \\ 0 & -1 & X & \dots & 0 & -a_3 \\ \dots & \dots & \dots & \dots & \dots & \dots \\ \dots & \dots & \dots & \dots & \dots & \dots \\ 0 & 0 & 0 & \dots & -1 & X - a_{n-1} \end{vmatrix}.$$

This is the characteristic polynomial of the companion matrix of the polynomial

$$g(X) = X^{n-1} - a_{n-1}X^{n-2} - ... - a_1.$$

By the induction hypothesis, $A_{11} = g(X)$.

$$\text{Further,} \quad A_{1n} = (-1)^{n+1}\begin{vmatrix} -1 & 0 & 0 & .... & 0 \\ 0 & -1 & 0 & .... & 0 \\ .. & .. & .. & .... & .. \\ .. & .. & .. & .... & .. \\ 0 & 0 & 0 & .... & -1 \end{vmatrix} = (-1)^{n+1}(-1)^{n-1} = 1$$

$$\text{Hence, (1) gives} \quad |XI_n - A| = X(X^{n-1} - a_{n-1}X^{n-2} - ... - a_1) - a_0$$
$$= X^n - a_{n-1}X^{n-1} - ... - a_1X - a_0$$
$$= f(X).$$

Another useful result, to construct matrices with pre-assigned characteristic polynomials, is the following :

Theorem 21. Let $A_1, A_2, ..., A_k$ be any square matrices over a field F, and $A = A_1 \oplus A_2 \oplus ... \oplus A_k$. Then, the characteristic polynomial of A is the product of the characteristic polynomials of $A_1, A_2, ..., A_k$.

Proof. Let A_i be $n_i \times n_i$-matrix. Then, A is an $n \times n$-matrix with $n = n_1 + n_2 + ... + n_k$.

We can write $$I_n = I_{n_1} \oplus I_{n_2} \oplus ... \oplus I_{n_k},$$

so that $$XI_n - A = (XI_{n_1} \oplus XI_{n_2} \oplus ... \oplus XI_{n_k}) - (A_1 \oplus ... \oplus A_k)$$
$$= (XI_{n_1} - A_1) \oplus (XI_{n_2} - A) \oplus ... \oplus (XI_{n_k} - A_k).$$

This immediately gives $|XI_n - A| = |XI_{n_1} - A_1||XI_{n_2} - A|...|XI_{n_k} - A_k|$.

Example 21.a. Consider $f(X) = X^4 + X^3 - X - 1$

$$A = \text{Com}[f(X)] = \begin{bmatrix} 0 & 0 & 0 & 1 \\ 1 & 0 & 0 & 1 \\ 0 & 1 & 0 & 0 \\ 0 & 0 & 1 & -1 \end{bmatrix}$$

has characteristic polynomial $f(X)$, by Theorem 20.

However, $$f(X) = (X^2 - 1)(X^2 + X + 1).$$

The companion matrices of $X^2 - 1$ and $X^2 + X + 1$ are

$$A_1 = \begin{bmatrix} 0 & 1 \\ 1 & 0 \end{bmatrix}, A_2 = \begin{bmatrix} 0 & -1 \\ 1 & -1 \end{bmatrix},$$

respectively.

Their respective characteristic polynomials are X^2-1 and X^2+X+1. So, by Theorem 21, the characteristic polynomial of

$$A' = A_1 \oplus A_2 = \begin{bmatrix} 0 & 1 & 0 & 0 \\ 1 & 0 & 0 & 0 \\ 0 & 0 & 0 & -1 \\ 0 & 0 & 1 & -1 \end{bmatrix}$$

is also $f(X)$.

Clearly, $$A \neq A'.$$

Exercise 4.4

1. Find the characteristic polynomial of each of the following matrices over **R** :

(*i*) $\begin{bmatrix} 2 & 1 \\ 1 & 2 \end{bmatrix}$

(*ii*) $\begin{bmatrix} 0 & 1 \\ -1 & 0 \end{bmatrix}$

(*iii*) $\begin{bmatrix} 0 & 1 \\ -1 & 2 \end{bmatrix}$

(*iv*) $\begin{bmatrix} 1 & 1 \\ 2 & 2 \end{bmatrix}$

(*v*) $\begin{bmatrix} 0 & -2 \\ 1 & 0 \end{bmatrix}$.

2. For each of the matrices in Exercise 1, find its characteristic values in **R** and in **C**. Using the characteristic values, determine which of these matrices are diagonalizable over **R**.

3. Prove that $\begin{bmatrix} 0 & -1 \\ 1 & 2 \end{bmatrix}$ is not diagonalizable over **R**.

4. For each of the following matrices, find its characteristic values in **R**. For each characteristic value, find a basis of the characteristics space. By using this information, find which of these matrices are similar over **R** to diagonal matrices

(*i*) $\begin{bmatrix} 1 & 1 \\ 1 & 1 \end{bmatrix}$

(*ii*) $\begin{bmatrix} 0 & 1 \\ 0 & 0 \end{bmatrix}$

(*iii*) $\begin{bmatrix} 2 & 0 & 5 \\ 0 & -1 & 3 \\ 0 & 0 & 4 \end{bmatrix}$

(iv) $\begin{bmatrix} 0 & 1 & 5 \\ 0 & 0 & 3 \\ 0 & 0 & 0 \end{bmatrix}$

(v) $\begin{bmatrix} 2 & 1 & 0 & 0 \\ 1 & 2 & 0 & 0 \\ 0 & 0 & 0 & 1 \\ 0 & 0 & 1 & 0 \end{bmatrix}$

(vi) $\begin{bmatrix} 1 & 0 & 0 \\ 0 & 5 & 0 \\ 1 & 3 & -1 \end{bmatrix}$.

5. For each matrix A in Exercise 4 that is similar to a diagonal matrix over R, find a non-singular matrix P such that $P^{-1}AP$ is a diagonal matrix.

6. Verify Caylay-Hamilton Theorem for each matrix in Exercise 4.

7. For $A = \begin{bmatrix} 2 & 0 & 5 \\ 0 & -1 & 3 \\ 0 & 0 & 4 \end{bmatrix}$, prove that for no non-zero polynomial $g(X)$ of degree less than 3 over R, $g(A) = 0$.

8. Let $T : \mathbf{R}^3 \to \mathbf{R}^3$ be the linear operator such that

$$T(\epsilon_1) = 6\,\epsilon_1 + 4\,\epsilon_2 + 10\,\epsilon_3,$$

$$T(\epsilon_2) = -3\,\epsilon_1 - 1\,\epsilon_2 - 5\,\epsilon_3,$$

$$T(\epsilon_3) = -2\,\epsilon_1 - 2\,\epsilon_2 - 2\,\epsilon_3.$$

(*a*) Find the characteristic polynomial of *T*.

(*b*) Find its characteristic values.

(*c*) For each characteristic value, find the corresponding characteristic space.

(*d*) Is *T* diagonalizable ?

9. Let A and B be any two $n \times n$-matrices over a field F. If $I_n - AB$ is invertible, prove that $I_n - BA$ is invertible and $(I_n - BA)^{-1} = I_n + B(I_n - AB)^{-1}A$. Use this result to prove that AB and BA have the same characteristic values in F.

10. Write down the companion matrix of each of the following polynomials :

$$X,\ X^3 - 3X^2 + 3X + 1,\ X^4 + 2X,\ X^3 + X + 1$$

and verify Theorem 20 for each of these polynomials.

11. Let $T : \mathbf{R}^2 \to \mathbf{R}^2$ be the linear operator such that

$$T(\epsilon_1 + \epsilon_2) = 3\,\epsilon_1 + 3\,\epsilon_2,\ T(\epsilon_1 - \epsilon_2) = \epsilon_1 - \epsilon_2.$$

Find the characteristic polynomial and the characteristic values of T. If possible, find an ordered basis B of $\mathbf{R}^2$ such that $[T]_B$ is a diagonal matrix.

12. Prove that any 2×2-matrix A over $\mathbf{R}$ satisfying $A^2 = A$ is similar over $\mathbf{R}$ to a diagonal matrix.

13. Let N be any 2×2-matrix over any field F such that $N^2 = 0$. Prove that either $N = 0$ or N is similar to $\begin{bmatrix} 0 & 0 \\ 1 & 0 \end{bmatrix}$.

14. Let $A = \begin{bmatrix} a & b \\ c & d \end{bmatrix}$ be any 2×2-matrix over $\mathbf{R}$. Prove that A diagonalizable over $\mathbf{R}$ iff either $(a-d)^2 + 4bc > 0$ or $A = \left[\dfrac{a+d}{2}\right] I_2$.

15. If an $A \in F^{n \times n}$ is invertible, prove that A^{-1} is expressible as a polynomial in A over F.

16. If $A \in F^{n \times n}$ is non-zero and non-invertible, prove that there exists $B \in F^{n \times n}$ such that $AB = 0$ but $BA \neq 0$.

Chapter 5

Single Linear Operator

5.1 Polynomial Rings

Throughout, F is a field and $R = F[X]$, the ring of polynomials over F. Here, we shall review some results on R that we need in subsequent sections. We shall be proving a few of the results and the others we shall only state. We start with the following :

Theorem 1. Let $f(X)$, $g(X) \in F[X]$ with $g(X) \neq 0$. Then, there exist unique $q(X)$, $r(X) \in F[X]$ such that

$$f(X) = q(X)g(X) + r(X)$$

and either $$r(X) = 0$$

or $$\deg r(X) < \deg g(X).$$

Proof. Let A be the set of all polynomials in $F[X]$ of the type

$$f(X) - h(X)g(X)$$

where $$h(X) \in F[X].$$

$$A \neq \phi,$$

as $$f(X) = f(X) - 0g(X) \in A$$

so that $0 \in A$ or A has a non-zero member of smallest degree. If $0 \in A$, put $r(X) = 0$. If $0 \notin A$, let $r(X) \in A$ be one of smallest degree among members of A.

In any case, $$r(X) = f(X) - q(X)g(X) \quad \text{... (1)}$$

for some $$q(X) \in F[X].$$

If $r(X) \neq 0$, we prove that $\deg r(X) < \deg g(X)$. Let deg $r(X) = s$ and deg $g(X) = m$.

Then, $$\left.\begin{aligned} r(X) &= c_0 + c_1X + \ldots + c_sX^s \\ g(X) &= b_0 + b_1X + \ldots + b_mX^m \end{aligned}\right\} \quad \text{... (2)}$$

with $$c_s \neq 0,\ b_m \neq 0.$$

Let $$m \le s.$$

Then, $$c_s b_m^{-1} X^{s-m} g(X) = c_s b_m^{-1} b_0 X^{s-m} + \ldots + c_s b_m^{-1} b_m X^{s-m} X^m$$

$$= d_{s-m} X^{s-m} + \ldots + c_s X^s \qquad \ldots (3)$$

(2) and (3) give $$r_1(X) = r(X) - c_s b_m^{-1} X^{s-m} g(X)$$

is either 0 or of degree less than $s = \deg r(X)$.

However, $$r_1(X) = f(X) - q(X) - c_s b_m^{-1} X^{s-m} g(X)$$

$$= f(X) - \left[q(X) + c_s b_m^{-1} X^{s-m}\right] g(X) \in A.$$

This contradicts the choice of $r(X)$ in A.

Hence, $$\deg r(X) < \deg g(X).$$

Again, (1) gives $$f(X) = q(X)g(X) + r(X) \qquad \ldots (4)$$

with $$r(X) = 0$$

or $$\deg r(X) < \deg g(X).$$

We now prove the uniqueness of $q(X)$ and $r(X)$.

Let also $$f(X) = q'(X)g(X) + r'(X) \qquad \ldots (5)$$

with either $$r'(X) = 0$$

or $$\deg r'(X) < \deg g(X).$$

Suppose $$r(X) \ne r'(X).$$

Then, $$[r(X) - r'(X)] < \deg g(X).$$

However, (4) and (5) give

$$r(X) - r'(X) = [q'(X) - q(X)]g(X) \qquad \ldots (6)$$

This gives $$\deg[r(X) - r'(X)] \ge \deg g(X).$$

This is a contradiction.

Hence, $$r(X) = r'(X).$$

Then, (4) and (5) give $$q(X)g(X) = q'(X)g(X).$$

As $g(X) \ne 0$, we get $$q(X) = q'(X).$$

The result, given in the above theorem, is called *Division Algorithm* in $F[X]$. In the equation $f(X) = q(X)g(X) + r(X)$, $f(X)$ is called the *divident*, $g(X)$ is called the *divisor*, $q(X)$ is called the *quotient* and $r(X)$ is called the *remainder*.

Given a non-zero polynomial

$$f(X) = a_0 + a_1 X + \ldots + a_k X^k$$

in $F[X]$ with $a_k \ne 0$, we have $f(X) = a_k g(X)$,

where, $$g(X) = a_k^{-1}a_0 + a_k^{-1}a_1X + ... + a_k^{-1}a_{k-1}X^{k-1} + X^k$$

is a monic polynomial.

Before we prove more results on polynomial rings, we state the following general result :

Lemma 2. A non-empty subset A of a commutative ring S is an *ideal* of S iff

(*i*) $a - b \in A$ for $a, b \in A$, and

(*ii*) $ar \in A$ for all $a \in A$, $r \in S$.

Consider a commutative ring S with identity $1 \neq 0$ and $a \in S$.

Let $$A = aS = \{ar : r \in S\}.$$

Clearly, $0 = a0 \in A$, so $A \neq \phi$.

Let $$x, y \in A, r \in S.$$

Then, $$x = ar_1, y = ar_2$$

for some $$r_i \in S, x - y = a(r_1 - r_2) \in A, xr = a(r_1 r) \in A.$$

Hence, A is an ideal of S.

Further, $$a = a1 \in A.$$

This ideal aS is called a *principal ideal* of S and a is called a generator of A. If every ideal of an integral domain D' is principal, then, D' is called a *principal ideal domain* (in short a *PID*).

Theorem 3. Let F be any field. For any non-zero ideal A of $F[X]$, there exists a unique monic polynomial $g(X) \in A$ such that $g(X)$ generates A ; $F[X]$ is a principal ideal domain.

Proof. As $A \neq \{0\}$, it has some non-zero polynomials. Among the non-zero polynomials in A, let $h(X)$ be of the smallest degree. As remarked just before this theorem, we can write $h(X) = ag(X), a \neq 0$ in F and $g(X)$ monic. Clearly, $g(X) = a^{-1}h(X) \in A$ and deg $g(X) =$ deg $h(X)$. Hence, the monic polynomial $g(X) \in A$ is of smallest degree among all the non-zero members of A. Consider any $f(X) \in A$. By the division algorithm,

$$f(X) = q(X)g(X) + r(X),\ q(X),\ r(X) \in F[X]$$

such that either $$r(X) = 0$$

or $$\deg r(X) < \deg g(X).$$

As $$g(X) \in A,\ q(X)g(X) \in A.$$

However, $$f(X) \in A.$$

So, we get $$f(X) - q(X)g(X) \in A.$$

The minimality of $\deg g(X)$ among the degrees of non-zero members of A, yields $r(X) = 0$.

Hence, $$f(X) = q(X)g(X),$$
so that $g(X)$ generates A.

Let $g_1(X)$ be another monic generator of A.

Then, $$g(X) = f_1(X)g_1(X)$$
for some $$f_1(X) \in F[X]$$
so that $$g_1(X) \leq \deg g(X).$$

As the degree of $g(X)$ is the smallest,

we get $$\deg g_1(X) = \deg g(X).$$

Consequently, $$f_1(X) = c \in F.$$

Then, $$g_1(X) = cg(X),$$
by comparing the leading coefficients on both sides, we get $c = 1$.

Hence, $$g_1(X) = g(X).$$

This proves the uniqueness of $g(X)$. As $\{0\}$ is also a principal ideal, every ideal in $F[X]$ is principal. Hence, $F[X]$ is a *PID*.

Remark 4. For any $f(X), g(X) \in F[X]$, $g(X)$ is called a *factor* of $f(X)$ [i.e., $g(X)$ divides $f(X)$, (in symbol, $g(X) | f(X)$) if $f(X) = g(X)h(X)$ for some $h(X) \in F[X]$.

Now, $$(X^2 - 1) = (X-1)(X+1),$$
gives $$(X-1) \big| (X^2 - 1).$$

Given a non-zero ideal A of $F[X]$, Theorem 3 shows that there exists a unique monic polynomial $g(X) \in A$ such that any $f(X)$ in $F[X]$ is in A iff $g(X) \big| f(X)$.

The following is immediately clear :

Theorem 5. Let $f(X), g(X), h(X) \in F[X]$.

(*i*) If $f(X) \big| g(X)$ and $f(X) \big| h(X)$,

then, $$f(X) \big| [q(X)g(X) + s(X)h(X)]$$
for every $$q(X), s(X) \in F[X].$$

(*ii*) If $$f(X) \big| g(X)$$
and $$g(X) \big| h(X),$$

then $f(X)|h(X)$.

(*iii*) If $f(X) \neq 0 \neq g(X)$,

then, $f(X)|g(X)$

and $g(X)|f(X)$

iff $f(X) = ag(X)$

for some $0 \neq a \in F$;

in particular, if $f(X)$ and $g(X)$ are nonic, then, $f(X) = g(X)$.

Definition 6. Let $f(X)$, $g(X)$ be any two non-zero members of $F[X]$. Then, a monic polynomial $d(X) \in F[X]$ is called the *greatest common divisor* (in short *gcd*) of $f(X)$ and $g(X)$, if

(*i*) $d(X)|f(X)$ and $d(X)|g(X)$, and

(*ii*) for any $h(X) \in F[X]$, if $h(X)|f(X)$ and $h(X)|g(X)$, then $h(X)|d(X)$.

Definition 7. Let $f(X)$ and $g(X)$ be any two non-zero members of $F[X]$. Then, a monic polynomial $m(X) \in F[X]$ is called *least common multiple* (in short, 1cm) of $f(X)$ and $g(X)$ if

(*i*) $f(X)|m(X)$ and $g(X)|m(X)$, and

(*ii*) given any $k(X) \in F[X]$ such that $f(X)|k(X)$ and $g(X)|k(X)$, then, $m(X)|k(X)$.

These two definitions show that the *gcd* of $f(X)$ and $g(X)$ is a largest degree monic common factor of $f(X)$ and $g(X)$; the 1cm of $f(X)$ and $g(X)$ is smallest degree monic polynomial of which $f(X)$ and $g(X)$ both are factors. Before we prove the existence of *gcd* and 1cm of two non-zero polynomials in $F[X]$, we give the following Lemma :

Lemma 8. Let *S* be any commutative ring, *A* and *B* be two ideals of *S*. then,

(*i*) $A \cap B$ is an ideal of *S*

(*ii*) $A + B = \{a + b : a \in A,\ b \in B\}$ is an ideal of *S* containing $A \cup B$.

Proof.

(*i*) Now, $0 \in A$ and $0 \in B$, gives $0 \in A \cap B$. So, $A \cap B \neq \phi$. Let $x,\ y \in A \cap B$ and $r \in S$. Then, $x,\ y \in A$, yield $x - y \in A$, $xr \in A$, by Lemma 2. Similarly, $x - y \in B$, $xr \in B$. This gives $x - y \in A \cap B$, $xr \in A \cap B$. Hence, by Lemma 2, $A \cap B$ is an ideal of *S*.

(*ii*) For any $a \in A$, $a = a + 0$, with $a \in A$, $0 \in B$, gives $a \in A + B$. Thus, $A \subseteq A + B$. Similarity, $B \subseteq A + B$. Hence, $A \cup B \subseteq A + B$. As $A \neq \phi$, we get $A + B \neq \phi$. Let $x,\ y \in A + B$, $r \in S$.

Then, $x = a_1 + b_1,$

$$y = a_2 + b_2,$$

$$x - y = (a_1 - a_2) + (b_1 - b_2) \in A + B,$$

$$xr = a_1 r + b_1 r \in A + B,$$

yield $A+B$ is an ideal of *S*.

($A+B$ is called the *sum* of the ideals *A* and *B*).

In particular, for any $a, b \in S$, $aS + bS$ is an ideal of *S*. The ideal $aS+bS$ is said to be generated by $\{a, b\}$.

Theorem 9. Let $f(X)$ and $g(X)$ be any two non-zero members of $F[X]$. Then,

(*i*) $f(X)$ and $g(X)$ have unique *gcd* $d(X)$.

Further, $d(X) = f(X)h(X) + g(X)u(X)$

for some $h(X),\ u(X) \in F[X],$

and $d(X)F[X] = f(X)F[X] + g(X)F[X]$

(*ii*) $f(X)$ and $g(X)$ have unique 1cm $m(X)$.

Further, $f(X)F[X] \cap g(X)F[X] = m(X)F[X].$

Proof. $A = f(X)F[X] + g(X)F[X]$

being sum of two non-zero ideals of $F[X]$ is a non-zero ideal of $F[X]$.

By Theorem 3, there exists unique monic polynomial $d(X)$ that generates this ideal,

i.e., $$A = f(X)F[X] + g(X)F[X] = d(X)F[X]. \quad \text{... (1)}$$

As $d(X) \in A$, $$d(X) = f(X)k(X) + g(X)u(X) \quad \text{... (2)}$$

for some $k(X),\ g(X) \in F[X].$

We now prove that $d(X)$ is the *gcd* of $f(X)$ and $g(X)$ to complete the proof of (*i*).

As $f(X) = f(X).1 + g(X).0 \in A,$

we get $f(X) = d(X) f_1(X)$

for some $f_1(X) \in F[X]$

i.e. $d(X) | f(X).$

Similarly, $d(X) | g(X).$

Consider any $d'(X) \in F[X]$

such that $d'(X) | f(X)$

and $d'(X) | g(X).$

Then, (2) and Theorem 5 (*i*) give $d'(X)|d(X)$.

Hence, $d(X)$ is a *gcd*. Let $d_1(X)$ be another *gcd* of $f(X)$ and $g(X)$. Then, by the condition (*ii*) of Definition 6, $d(X)| d_1(X)$ and $d_1(X)| d(X)$. So by Theorem 5 (*iii*) $d_1(X) = ad(X)$ for some $a \in F$. As $d(X)$ and $d_1(X)$ are both monic, by comparing the leading coefficients on both sides of $d_1(X) = ad(X)$, we get $a = 1$. Hence, $d(X)$ is the only gcd of $f(X)$ and $g(X)$. This proves (*i*).

As $$0 \neq f(X)g(X) \in f(X)F[X] \cap g(X)F[X] = A',$$

A' is a non-zero ideal of $F[X]$.

By Theorem 3, there exists a unique monic polynomial $m(X)$ such that $A' = m(X)F[X]$.

Then, $$m(X) \in f(X)F[X]$$

and $$m(X) \in g(X)F[X],$$

give $$f(X)|m(X)$$

and $$g(X)|m(X).$$

Let $$m'(X) \in F[X]$$

such that $$f(X)|m'(X)$$

and $$g(X)|m'(X).$$

Then, $$m'(X) \in f(X)F[X] \cap g(X)F[X] = m(X)F[X],$$

yields $$m(X)|m'(X).$$

Hence, $m(X)$ is an lcm of $f(X)$ and $g(X)$. The uniqueness of the lcm follows on the same line as that of the *gcd*.

Definition 10. Two non-zero polynomials $f(X)$, $g(X)$ are said to be *co-prime* to each other, if their *gcd* is 1.

The following is immediately inferred from Theorem 7.

Theorem 11. For any two non-zero polynomials $f(X)$, $g(X)$ in $F[X]$, the following are equivalent :

(*i*) $f(X)$ and $g(X)$ are co-prime to each other.

(*ii*) $1 = f(X)h(X) + g(X)u(X)$ for some $h(X), u(X) \in F[X]$.

(*iii*) $F[X] = f(X)F[X] + g(X)F[X]$.

Example 11.a. Consider $f(X) = X + 2$, $g(X) = X + 1$ in $\mathbf{R}[X]$.

As $1 = (X + 2)1 + (X + 1)(-1)$, $f(X)$ and $g(X)$ are co-prime to each other.

Theorem 12. Let $f(X)$, $g(X)$ and $h(X)$ be any three non-zero members of $F[X]$ such that $h(X)|f(X)g(X)$ and $\gcd[f(X), h(X)] = 1$.

Then,
$$h(X)|g(X).$$

Proof. As
$$1 = \gcd[f(X), h(X)]$$
by Theorem 9,
$$1 = f(X)u(X) + h(X)v(X)$$
for some
$$u(X),\ v(X) \in F[X].$$
This yields
$$g(X) = f(X)g(X)u(X) + h(X)g(X)v(X).$$
As
$$h(X)|f(X)g(X)$$
and
$$h(X)|h(X),$$
this equation yields
$$h(X)|g(X) \qquad \text{(Theorem 5).}$$

Example 12.a. Consider
$$f(X) = (X+1)(X+2),$$
$$g(X) = (X+1)(X+3)$$
in **R**[*X*].

Let
$$d(X) = \gcd[f(X), g(X)].$$
As $X+1$ is a common factor of $f(X)$ and $g(X)$,
$$(X+1)|d(X)|(X+1)(X+2),$$
if
$$d(X) \neq (X+1),$$
we have
$$d(X) = (X+1)(X+2).$$
Then,
$$(X+2)|d(X),$$
gives
$$(X+2)|g(X) = (X+1)(X+3).$$
But $X+2$ and $X+3$ are coprime ; this gives $(X+2)|(X+1)$.

This is not possible, as $X+2$ and $X+1$ are coprime. Hence, we get a contradiction.

Hence,
$$d(X) = X+1.$$
Let
$$m(X) = l\text{cm}[f(X), g(X)].$$
As
$$f(X)|m(X), m(X) = (X+1)(X+2)m_1(X)$$
for some
$$m_1(X) \in F[X].$$
As
$$g(X)|m(X),$$
we get
$$(X+3)|(X+1)(X+2)m_1(X).$$
However, $X+3$ is co-prime to $X+1$ and $X+2$.

By Theorem 12, $(X+3)|m_1(X)$

Consequently, $m_1(X)=(X+3)h(X)$,

for some $h(X)\in F[X]$.

This yields $m(X)=(X+1)(X+2)(X+3)h(X)$.

But $f(X)$ and $g(X)$ divide $(X+1)(X+2)(X+3)$.

Consequently, $m(X)|(X+1)(X+2)(X+3)$.

This yields $m(X)=(X+1)(X+2)(X+3)$.

Theorem 13. Let $f(X)$ and $g(X)$ be two non-zero polynomials in $F[X]$ such that they are co-prime to each other. If for some polynomial $m(X)\in F[X]$, both $f(X)$ and $g(X)$ divide $m(X)$, then, $f(X)$, $g(X)$ divides $m(X)$.

Proof. As $f(X)|m(X)$,

$$m(X)=f(X)m_1(X)$$

for some $m_1(X)\in F[X]$.

As $g(X)|f(X)m_1(X)$

and $\gcd[f(X), g(X)]=1$,

by Theorem 12, $g(X)|m_1(X)$

so that $m_1(X)=g(X)h(X)$

for some $h(X)\in F[X]$.

Consequently, $m(X)=f(X)g(X)h(X)$;

hence, $f(X)g(X)|m(X)$.

Theorem 14. Let $f(X)$ and $g(X)$ be any two monic polynomials in $F[X]$ with $d(X)$ and $m(X)$ as their respective gcd and 1cm.

Then, $f(X)g(X)=m(X)d(X)$.

Proof. We write $f(X)=f_1(X)d(X)$,

$$g(X)=g_1(X)d(X).$$

Let $d_1(X)=\gcd[f_1(X), g_1(X)]$.

Then, $d_1(X)d(X)|f_1(X)d(X)=f(X)$.

Similarly, $d_1(X)d(X)|g(X)$.

Hence, $d(X)=d_1(X)d(X)$

and $$d_1(X) = 1.$$

Now, $$m_1(X) = f_1(X)g_1(X)d(X) = f(X)g_1(X) = g(X)f_1(X).$$

Thus, $$m(X)\big|m_1(X).$$

Again, $$f(X)\big|m(X),$$

yields $$m(X) = f(X)h(X) = f_1(X)d(X)h(X),$$

for some $$h(X) \in F[X].$$

As $$g(X) = g_1(X)d(X)\big|f_1(X)d(X)h(X)$$

we get $$g_1(X)\big|f_1(X)h(X).$$

However, $g_1(X)$ and $f_1(X)$ are co-prime to each other.

So, $$g_1(X)\big|h(X).$$

Consequently, $$h(X) = g_1(X)h_1(X),$$

$$h_1(X) \in F[X].$$

Hence, $$m(X) = f_1(X)d(X)g_1(X)h_1(X) = m_1(X)h_1(X),$$

and $$m_1(X)\big|m(X).$$

Also, $$m(X)\big|m_1(X).$$

This gives $$m(X) = m_1(X).$$

Hence, $$m(X)d(X) = f_1(X)g_1(X)d(X)d(X) = f(X)g(X).$$

Corollary 15. If two monic polynomials $f(X)$, $g(X)$ in $F[X]$ are coprime to each other, then, their 1cm is $f(X)g(X)$.

Proof. Here, $$gcd\big(f(X), g(X)\big) = 1.$$

So, by Theorem 14, $$m(X) = m(X)d(X) = f(X)g(X).$$

Remark 16. Let $f(X)$, $g(X)$ be any two non-zero member of $F[X]$.

Write $$f(X) = af_1(X),$$

$$g(X) = bg_1(X),$$

with $a, b \in F$, $f_1(X)$ and $g_1(X)$ monic; if this gcd is $d(X)$, by Theorem 14,

we also get $$m(X) = \frac{[f_1(X)g_1(X)]}{d(X)}$$

as the 1cm of $f(X)$ and $g(X)$.

If deg $f(X) > 0$, we can find a non-constant factor $p(X)$ of $f(X)$ of the smallest degree. Then, $p(X)$ cannot have a non-constant factor $q(X)$ of degree less than deg $p(X)$. This leads to the following :

Definition 17. A polynomial $p(X)$ in $F[X]$ of positive degree is said to be *irreducible* in $F[X]$, if there exists no polynomial $q(X) \in F[X]$ such that $0 < \deg q(X) < \deg p(X)$ and $q(X)$ divides $f(X)$; if in addition, $p(X)$ is monic, it is called a *prime* element.

Any polynomial of degree one in $F[X]$ is irreducible.

As $X^2 + 1 \in \mathbf{R}[X]$ has no factor of degree one in $\mathbf{R}[X]$, it is irreducible (indeed prime) in $\mathbf{R}[X]$.

However, in $\mathbf{C}[X]$, $$X^2 + 1 = (X + i)(X - i),$$
gives that $X^2 + 1$ is not irreducible in $\mathbf{C}[X]$.

Theorem 18. Let $0 \neq f(X) \in F(X)$ and let $p(X)$, $q(X)$ be two distinct primes in $F(X)$. Then,

(*i*) Either $p(X) | f(X)$ or $[p(X), f(X)] = 1$

(*ii*) $\gcd[p(X), q(X)] = 1$

Proof.

(*i*) Let $$d(X) = \gcd[p(X), f(X)].$$

As $d(X)|p(X)$, by definition, either deg $d(X) = 0$ or deg $d(X) = p(X)$. In the former case, $d(X) = 1$ and, in the latter case, $d(X)|p(X)$, gives $d(X) = p(X)$, as $d(X)|f(X)$, we get $p(X)|f(X)$.

(*ii*) Let $\gcd[p(X), q(X)] \neq 1$; by (*i*) $p(X)|q(X)$ and $q(X)|p(X)$.

So, $p(X) = q(X)$ by Theorem 5 ; This is a contradiction.

Hence, $$\gcd[p(X), q(X)] = 1.$$

Theorem 19. Let $f(X)$, $g_1(X)$, $g_2(X)$ be any three non-zero members of $F[X]$, such that $\gcd[g_1(X), g_2(X)] = 1$.

(*i*) If $$f(X)|g_1(X)g_2(X)$$

then, $$f(X) = f_1(X)f_2(X)$$

for some $$f_1(X)|g_1(X)$$

and $$f_2(X)|g_2(X).$$

(*ii*) If $f(X)$ is coprime to $g_1(X)$ as well as to $g_2(X)$, then, $f(X)$ is coprime to $g_1(X)\ g_2(X)$.

Proof.

(*i*) Let $$f_1(X) = \gcd[f(X),\ g_1(X)]$$

Then, $$f(X) = f_1(X) f_2(X),$$
$$g_1(X) = f_1(X)h(X),$$

and $$\gcd[f_2(X),\ h(X)] = 1.$$

Now, $$f(X) \mid g_1(X)g_2(X),$$

i.e., $$f_1(X)f_2(X) \mid f_1(X)h(X)g_2(X).$$

This yields $$f_2(X) \mid h(X)g_2(X).$$

By Theorem 12, $$f_2(X) \mid g_2(X).$$

(*ii*) Let $$\gcd[f(X),\ g_i(X)] = 1,$$

for $$i = 1, 2,$$

and $$d(X) = \gcd[f(X),\ g_1(X)g_2(X)].$$

Then, $$d(X) \mid g_1(X)g_2(X),$$

gives $$d(X) = d_1(X)d_2(X),$$

with $$d_i(X) \mid g_i(X).$$

As $$d_i(X) \mid f(X)$$

as well, we get $$d_i(X) = a_i \in F.$$

So $d(X)$ has degree zero.

Consequently, $$d(X) = 1.$$

Remark 20.

(*i*) Let $f_1(X),\ f_2(X),\ \ldots,\ f_m(X)$ and $g_1(X),\ g_2(X), \ldots,\ g_n(X)$ be any non-zero polynomials in $F[X]$ such that $[f_i(X),\ g_j(X)] = 1$ for $1 \le i \le m,\ 1 \le j \le n$.

Then, repeated application of the theorem, given above, gives that the products
$$f_1(X)\, f_2(X) \ldots.. f_n(X)$$
and $$g_1(X)\, g_2(X) \ldots.. g_m(X)$$
are coprimes.

In particular, given any two distinct primes $p(X),\ q(X)$ in $F[X]$, $p(X)^m$ and $q(X)^n$ are coprimes for every $m \ge 0$ and $n \ge 0$.

(*ii*) Let $g_1(X)$, $g_2(X)$,..., $g_n(X)$ in $F[X]$ be non-zero and pairwise co-primes. If some $d(X)$ in $F[X]$ divides $\prod_{i=1}^{n} g_i(X)$, by Theorem 19,

$$d(X) = \prod_{i=1}^{n} d_i(X)$$

for some factors $d_i(X)$ of $g_i(X)$.

Theorem 21 (Fundamental Theorem on factorization). Any non-constant monic polynomial $f(X) \in F[X]$ can be factored as a product of primes in $F[X]$. The factorization is unique in the sense that if

$$\begin{aligned} f(X) &= p_1(X)p_2(X)....p_t(X) \\ &= q_1(X)q_2(X)....q_u(X) \end{aligned}$$

for some primes $p_i(X)$, $q_j(X)$, then, $t = u$ and for some permutation σ of 1, 2, ..., t, $q_i(X) = p_{\sigma(i)}(X)$.

Proof. If $f(X)$ is irreducible itself, it has only one factorization into prime, $f(X) = p_1(X)$; so, the result holds for this case. In particular, it holds for deg $f(X) = 1$. We apply induction on $f(X) = n$. We suppose that $n > 1$, and that the result holds for all polynomials of degree less than n. If $f(X)$ is prime, there is nothing to prove. Let $f(X)$ be not prime.

Then, $$f(X) = f_1(X)f_2(X)$$
for some monic $$f_i(X) \in F[X]$$
each with $$\deg f_i(X) < n.$$

By the induction hypothesis, each $f_i(X)$ is a product of primes. This, in turn, gives $f(X)$ is a product of primes. We now prove the uniqueness.

Let $$f(X) = p_1(X)p_2(X)....p_t(X) \quad ...(1)$$
$$= q_1(X)q_2(X)....q_u(X) \quad ...(2)$$
for some primes $$p_i(X),\ q_j(X).$$

As $p_1(X)$ divides $q_1(X)q_2(X)...q_u(X)$, if $p_1(X) \neq q_j(X)$ for all j, Theorems 18 and 19 give that $p_1(X)$ is coprime to $\prod_{j=1}^{u} q_j(X) = f(X)$. This is a contradiction.

Hence, $$p_1(X) = q_{j_1}(X)$$
for some j_1.

By re-indexing $q_j(X)$'s, we take $j_1 = 1$.

Then, in (1) as $$p_1(X) = q_1(X),$$

we get $$g(X) = p_2(X)p_3(X)....p_t(X) \quad ... (3)$$

$$= q_2(X)q_3(X)....q_u(X). \quad ... (4)$$

As $\deg g(X) < \deg f(X)$, the induction hypothesis shows the numbers of primes in (3) and (4) are the same,

i.e., $$t-1 = u-1,\ t = u.$$

Further, by re-indexing of suffixes, we can get

$$p_i(X) = q_i(X)$$

for $$2 \le i \le t.$$

This proves the theorem.

Remark. 22

(*i*) In the factorization $f(X) = p_1(X)p_2(X)....p_t(X)$,

given in this theorem, $p_i(X)$ need not be distinct.

By re-indexing, we can take $p_1(X),...,p_s(X)$, as distinct primes, and that any other $p_j(X)$ equals some, $p_i(X)$, $1 \le i \le s$. For $1 \le i \le s$, let $p_i(X)$ equal $p_j(X)$ for r_i values of j.

Then, $$f(X) = p_1(X)^{r_1} p_2(X)^{r_2} ... p_s(X)^{r_s};$$

we say *$p_i(X)$ is a factor of $f(X)$ of multiplicity r_i*; if $r_i = 1$, we say that $p_i(X)$ is a *simple factor of $f(X)$* and if $r_i > 1$, we say that $p_i(X)$ is a *multiple factor of* $f(X)$. In particular, if $p_i(X) = (X - a_i)$ for some $a_i \in F$, we say a_i is a *simple* (*multiple*) root of $f(X)$ according as $r_i = 1(r_i > 1)$.

Further, observe that

$$p_1(X)^{r_1}, p_2(X)^{r_2}, ..., p_s(X)^{r_s},$$

are pairwise coprimes.

For example, in

$$f(X) = (X^2+1)^2(X-2)(X^3+2) \in \mathbf{R}[X],$$

$$p_1(X) = X^2+1,$$

$$p_2(X) = X-2,$$

$$p_3(X) = X^3+2$$

are the distinct prime factors.

Here, $$r_1 = 2,\ r_2 = 1,\ r_3 = 1$$

so that $p_1(X)$ is a multiple factor, while $p_2(X)$, $p_3(X)$ are simple factors.

As $p_2(X) = X-2$, 2 is a simple root of $f(X)$.

In $$g(X) = (X-1)^3(X-3)(X+5)^7,$$

1 and -5 are multiple roots, while 3 is a simple root.

(*ii*) Let $f(X)$ be any non-constant polynomial in $F[X]$. Then, $f(X) = af_1(X)$, where a is the leading coefficient of $f(X)$, and $f_1(X)$ is monic.

By using Theorem 19, we get

$$f(X) = ap_1(X)^{r_1}\, p_2(X)^{r_2} \ldots\ldots p_k(X)^{r_k}.$$

where $p_i(X)$, $1 \le i \le s$, are the distinct prime factors of $f(X)$.

This factorization of $f(X)$ is called *prime factorization*. If a prime $p(X)$ does not divide $f(X)$, then, as $p(X)^0 = 1$ divides $f(X)$, we can say that $p(X)$ is a factor of $f(X)$ of multiplicity zero, and we can write

$$f(X) = ap(X)^0\, p_1(X)^{r_1}\, p_2(X)^{r_2} \ldots\ldots p_k(X)^{r_k}.$$

Consider $$f(X) = (X+1)(X^2+1)^2,$$
$$g(X) = (X+2)^2(X+1)(X+3)^3,$$

we can write $$f(X) = (X+1)(X^2+1)^2(X+2)^0(X+3)^0$$
$$g(X) = (X+1)(X^2+1)^0(X+2)^2(X+3)^3.$$

By keeping this example in view, the following theorem gives a method of finding gcd and lcm of two polynomials. We leave its proof as an exercise.

Theorem 23. Let $f(X)$ and $g(X)$ be any two non zero polynomials in $F[X]$ such that

$$f(X) = ap_1(X)^{r_1}\, p_2(X)^{r_2} \ldots\ldots p_k(X)^{r_k}.$$
$$g(X) = bp_1(X)^{s_1}\, p_2(X)^{s_2} \ldots p_k(X)^{s_k}$$

for some $a, b \in F$, distinct primes $p_i(X)$, $1 \le i \le k$, and $r_i \ge 0$.

Then, $$\gcd[f(X), g(X)] = p_1(X)^{u_1}\, p_2(X)^{u_2} \ldots\ldots p_k(X)^{u_k}$$
$$\operatorname{lcm}[f(X), g(X)] = p_1(X)^{v_1}\, p_2(X)^{v_2} \ldots\ldots p_k(X)^{v_k}$$

where $$u_i = \min(r_i, s_i)$$
and $$v_i = \max(r_i, s_i) \qquad \text{for } 1 \le i \le k.$$

Remark 24. The concept of gcd and lcm of a pair of polynomials can be analogously defined for any finite number of polynomials. Given any $k \ge 1$, non-zero polynomials $g_1(X)$, $g_2(X)$,..., $g_k(X)$, if $d(X)$ and $m(X)$ are their gcd and lcm, respectively, then, as in Theorem 9, we have

$$d(X)F[X] = \sum_{i=1}^{k} g_i(X)F[X]$$

$$m(X)F[X]=\bigcap_{i=1}^{k} g_i(X)F[X].$$

One can easily observe that the proof of Theorem 1, giving the division algorithm, depends mainly on the fact that $g(X)$ has its leading coefficient a unit (i.e., has inverse). Without altering the proof, we have the following more general version.

Theorem 25. (Division Algorithm). Let R be any commutative ring with identity $1 \neq 0$. Let $f(X)$, $g(X) \in R[X]$ with $g(X) \neq 0$ and that the leading co-efficient of $g(X)$ is a unit. Then, there exist unique $q(X)$, $g(X) \in R[X]$, such that

$$f(X)=q(X)g(X)+r(X)$$

and either $$r(X)=0$$

or $$\deg r(X) < \deg g(X).$$

In particular, $g(X)$ could be a monic polynomial, in the theorem given above.

Exercises 5.1

Note : Throughout, F is a field.

1. By using the following division algorithm in Z, prove that Z is a PID : For any integer n, $m \in Z$ with $m > 0$, there exist unique integers q, r such that $n = mq + r$ and $0 \leq r \leq m-1$.

2. For any n, $m \in Z$ with $m \neq 0$, prove that there exist q, $r \in Z$ such that $n = mq + r$ and $|r| \leq |m/2|$.

3. Consider the ring $R = Z[i] = \{a+bi : a, b \in Z\}, i = \sqrt{-1}$. For any $z = a+bi \in R$, the integer a^2+b^2 is called the *norm* of z, and is denoted by $N(z)$.

   (*a*) Prove the following :

   (*i*) $N(z)=0$ iff $z=0$

   (*ii*) $N(z_1z_2)=N(z_1)N(z_2)$, $z_1, z_2 \in R$

   (*iii*) $N(z)=z\bar{z}$

   (*b*) Let $x, y \in R$ with $y \neq 0$. Let $x\bar{y}=a+bi$ with $a, b \in Z$ and let $N(y)=m$.

   Write $a=mq_1+r_1$, $b=mq_2+r_2$ with q_i, $r_i \in Z$ and $|r_i| \leq m/2$. Prove the following

   (*i*) $x\bar{y}=m(q_1+iq_2)+r_1+ir_2$

   (*ii*) $r_1+ir_2=u\bar{y}$ for some $u \in R$

   (*iii*) $N(r_1+ir_2)<N(y)^2$

   (*iv*) $x=y(q_1+iq_2)+u$, with $N(u)<N(y)$

   (*v*) Use the result in (*b*) (*iv*), to conclude that $Z[i]$ is a PID.

4. In the ring $Z\left[\sqrt{-5}\right]=\left\{a+b\sqrt{-5} : a,b \in Z\right\}$, prove that the ideal generated by $2+\sqrt{-5}$ and $2-\sqrt{-5}$ is not principal.

5. Let $R=Z\left[\sqrt{2}\right]=\left\{a+b\sqrt{2} : a,b \in Z\right\}$. For , define $N(x)=a^2-2b^2$. By similar arguments, as in Exercise 3, prove that R is a PID.

6. Let D be any integral domain and $a \in D$ be non-zero and non-unit. Prove that the ideal A of $D[X]$ generated by X and a is not principal. Prove that $Z[x]$ is not a PID.

7. Given $f(X)$, $g(X)$ over a field F, as under, find $q(X)$, $r(X) \in F[X]$, such that

$$f(X)=q(X)g(X)+r(X)$$

with $r(X)=0$ or $\deg r(X)<\deg g(X)$.

(*i*) $f(X)=X^5+3X^2+1$, $g(X)=X^2+1$, $F=\mathbf{Q}$

(*ii*) $f(X)=X^4+\sqrt{2}X+3$, $g(X)=X-\sqrt{2}$,

$F=\mathbf{Q}\left[\sqrt{2}\right]=\left\{a+b\sqrt{2} : a,\ b \in Q\right\}$

(*iii*) $f(X)=X^3+X+1$, $g(X)=X^2+\sqrt{-1}$, $F=\mathbf{C}$

(*iv*) $f(X)=X^3+X^2+1$, $g(X)=X^2+X+1$, $F=Z_2$

(*v*) $f(X)=X^7+1$, $g(X)=X^4+X+1$, $F=\mathbf{R}$

(*vi*) $f(X)=X^7+X+1$, $g(X)=X^3+1$, $F=Z_5$

8. Let $f(X)=\sum_{i=0}^{n} a_i X^i \in Z[X]$, $a_n \neq 0$. Let $u=a/b \in Q$ with $a \neq 0$, $b \neq 0$, $\gcd(a,b)=1$. If $f(a/b)=0$, prove that $b \mid a_n$.

9. By using Exercise 8, prove that any monic polynomial over Z, has no rational roots other than integers.

10. Let F be a field of characteristic $p>0$. for any $f(X)$, $g(X) \in F[X]$, prove that

$$[f(X)+g(X)]^p=f(X)^p+g(X)^p.$$

11. For any $f(X) \in F[X]$, $X-a \in F[X]$,

prove that $f(X)=(X-a)q(X)+f(a)$

for some $q(X) \in F[X]$.

12. Prove that any polynomial $f(X) \in F[X]$ of degree 2 or 3 is irreducible iff it has no root in $F[X]$.

13. Which of the following polynomials are irreducible in $\mathbf{Q}[X]$?

$$X^2+2X+1, X^2+2X+2, X^3-2, X^4+4X^2+4$$

14. **If a polynomial $f(X) \in F[X]$ is irreducible, prove that for any $a \in F$, $f(X+a)$ is irreducible in $F[X]$.**

15. **For any irreducible polynomial $p(X) \in F[X]$, $A = p(X)F[X]$, prove that $K = F[X]/A$ is a field.**

    [Hint : given $f(X) \notin p(X)F[X]$,

    $$1 = f(X)u(X) + p(X)v(X)$$

    for some $u(X), v(X) \in F[X]$**]**

16. **For each of the following pairs of polynomials, find their gcd and lcm.**

    (*i*) X^4+2X^3+4X+2, X^3+3X+3 in $\mathbf{Q}[X]$

    (*ii*) $(X+1)(X^2+X+1), (X+1)(X^3+1)$ in $\mathbf{Q}[X]$

    (*iii*) $X^2+2, (X^3+2)(X-i\sqrt{2})$ in $\mathbf{C}[X]$

    (*iv*) $X^2+2\pi X+\pi^2, X^3+1$ in $\mathbf{R}[X]$.

17. (Eisens'ein Criterion) Let $f(X) = a_0 + a_1X + ... + a_nX^n$ be a polynomial in $Z[X]$ with $a_n \neq 0$. If there exits a prime number p such that $p | a_i$ for $0 \le i \le n-1$, $p \nmid a_n$ and $p^2 \nmid a_0$, prove that f is an irreducible polynomial over Q.

18. **Use Eisenstein's criterion to conclude that each of the following polynomials is irreducible in $\mathbf{Q}[X]$, X^4-2, $3X^4+6X^3+2X^2+2X+6$, $X^5+9X+15$, X^n-p (p any prime number).**

19. **For any prime number p, prove that $f(X) = X^{p-1}+X^{p-2}+...+X+1$ is irreducible in $\mathbf{Q}[X]$.**

    [Hint : Consider $g(X) = f(X+1)$, and apply Eisenstein's criterion to $g(X)$]

20. **By knowing that one root of X^3+2X^2-5X-6 is a rational number, factorize this polynomial into primes in $\mathbf{Q}[X]$.**

21. **Write down all the polynomials of degree 2 over Z_2. Prove that X^2+X+1 is the only irreducible polynomial of degree 2 in $Z_2[X]$. Find the order of the field $K = \dfrac{Z_2[X]}{A}$, $A = (X^2+X+1)Z_2[X]$.**

22. **Prove that each of the following polynomials is prime in $\mathbf{Q}[X]$: X^3+3X^2+X+8, X^3+X+1.**

23. **Find the prime factors of the following in $\mathbf{Q}[X]$:**

(*i*) $(X^2+1)^3(X-2)(X^3+X+1)^3$

(*ii*) $(X^4+X^3+X^2+X+1)(X^2+2X+1)$

(*iii*) $X^3+4X^2+5X+X+2$

(*iv*) $(X^4+4X^2+4)(X^5+7)^3(X^2+X+1)$

Find the prime factors in $\mathbf{C}[X]$ as well.

24. Let $f(X) \in F[X]$ be of degree $n \geq 2$. Prove that $f(X)$ is irreducible iff it has no prime factor of degree $\leq n/2$.

25. For $f(X)$, $g(X)$, any two non-zero polynomials in $F[X]$,

let $$d(X) = \gcd[f(X), g(X)] = f(X)u(X) + g(X)v(X).$$

Prove that the only monic polynomial in $F[X]$, that can divide $u(X)$ as well as $v(X)$ is 1.

26. (*a*) Find all primes of degree 3 in $Z_2[X]$.

(*b*) Factorize each of the followings into prime factors in $Z_2[X]$:

(*i*) X^4+X^2+X+1

(*ii*) X^6+1

(*iii*) X^3+X^2+X+1

(*iv*) X^7+1

5.2 Modules and Minimal Polynomials

Definition 1. Let **R** be any ring with identity $1 \neq 0$. An abelian group $\langle M, + \rangle$ together with a function for from $R \times M$ to M which for each $(a, \alpha) \in R \times M$, determines $a\alpha \in M$ is called a left *R-module* if it satisfies the following axioms. For any $\alpha, \beta \in M$ and $a, b \in R$

(*i*) $a(\alpha+\beta) = a\alpha + a\beta$

(*ii*) $(a+b)\alpha = a\alpha + b\alpha$

(*iii*) $(ab)\alpha = a(b\alpha)$

(*iv*) $1\alpha = \alpha$.

Thus, any vector space *V* over a field *F* is a left *F*-module. $_RM$ will denote that *M* is a left *R-module*. For $a \in R$, $\alpha \in M$, $a\alpha$ is known as a scalar multiple of α by *a*. Analogous definition of a right *R-module* N_R, by taking scalar multiplication αa, $\alpha \in N$, $a \in R$, can be given. Consider any left ideal *A* of *R*. As *A* is an additive abelian group, and for any $\alpha \in A$, $a \in R$, we have $a\alpha \in A$, this gives that *A* is a left *R*-module. Similarly, any right ideal of *R* is a right *R*-module. The concepts of submodules, internal direct sum of submodules of an $_RM$ can be analogously defined, as in a vector space $_FV$.

Theorem 2. Let N be a submodule of a module ${}_RM$, and $\overline{M} = M/N = \{\alpha + N : \alpha \in M\}$. The $\overline{M}$ is left R-module under the addition and scalar multiplication operations given by

$$(\alpha + N) + (\beta + N) = (\alpha + \beta) + N \quad \text{... (1)}$$

$$a(\alpha + N) = a\alpha + N \quad \text{... (2)}$$

for any $\alpha, \beta \in M$, $a \in R$.

Proof. As $\langle M, + \rangle$ is an abelian group, and N is a subgroup of $\langle M, + \rangle$, $\overline{M} = M/N$ is an abelian group under the operation given by (1).

Consider any $\quad \overline{\alpha} = \alpha + N,\ \overline{\beta} = \beta + N \in \overline{M}$.

We know that $\overline{\alpha} = \overline{\beta}$ iff $\alpha - \beta \in N$. The scalar multiplication given by (2) can be stated as $a\overline{\alpha} = \overline{a\alpha}$. First of all, we show its well definedness.

Let $\quad \overline{\alpha} = \overline{\beta}$.

Then, $\quad \alpha - \beta \in N$.

As N is a submodule, $\quad a(\alpha - \beta) \in N$,

i.e., $\quad a\alpha - a\beta \in N$.

This gives $\quad \overline{a\alpha} = \overline{a\beta}$,

i.e., $\quad a\overline{\alpha} = a\overline{\beta}$.

Hence, the scalar multiplication is well defined.

Observe that $\quad \overline{\alpha} + \overline{\beta} = \overline{\alpha + \beta}$.

Consider any $a, b \in R$. Then,

(*i*)
$$\begin{aligned} a(\overline{\alpha} + \overline{\beta}) &= a(\overline{\alpha + \beta}) \\ &= \overline{a(\alpha + \beta)} \\ &= \overline{a\alpha + a\beta} \\ &= \overline{a\alpha} + \overline{a\beta} \\ &= a\overline{\alpha} + a\overline{\beta} \end{aligned}$$

(*ii*)
$$\begin{aligned} (a + b)\overline{\alpha} &= \overline{(a + b)\alpha} \\ &= \overline{a\alpha + b\alpha} \\ &= \overline{a\alpha} + \overline{b\alpha} \\ &= a\overline{\alpha} + b\overline{\alpha} \end{aligned}$$

(*iii*)
$$\begin{aligned} (ab)\overline{\alpha} &= \overline{(ab)\alpha} \\ &= \overline{a(b\alpha)} \\ &= a\overline{(b\alpha)} \end{aligned}$$

$$= a(b\overline{\alpha})$$

(*iv*) $1\overline{\alpha} = \overline{1\alpha} = \overline{\alpha}$.

Hence, $\overline{M}$ is a left R-module.

The module $\overline{M}$ is known as a *quotient module* of M. Consider a vector space V over a field F. By definition, V is a left F-module. If for any $\alpha \in V$ and $a \in F$, we define $\alpha a = a\alpha \in V$, it can be easily checked that V also becomes a right F-module. So, we shall treat, wherever needed, V as a left as well as a right F-module. Consider a subspace W of V. We get quotient F-module V/W, i.e., a *Quotient space* of V by W. The following theorem gives the dimension of V/W.

Theorem 3. For any subspace W of a finite dimensional vector space V_F, $\dim(V/W) = \dim V - \dim W$.

Proof. Let W' be a complement of W in V.

Then, $\dim W' = \dim V - \dim W$ [Theorem (2.2.19)]

Let $\dim W' = k$.

If $W = V$, then $V/W = \{0\}$;

obviously then, $\dim(V/W) = 0 = \dim V - \dim W$.

Let $W \neq V$ and $B = [\alpha_1, \alpha_2, ..., \alpha_k]$ be a basis of W'.

Consider $\overline{\alpha}_i = \alpha_i + W \in \overline{V} = V/W$.

Let $\alpha \in V$.

As $V = W + W'$, $\alpha = \beta + \beta'$,

$\beta \in W$, $\beta' \in W'$.

Now, $\beta + W = W$.

So, $\overline{\alpha} = (\beta' + \beta) + W = \beta' + W$.

Now, $$\beta' = \sum_{i=1}^{k} a_i\alpha_i,$$

yields $$\overline{\alpha} = \overline{\beta'} = \sum_{i=1}^{k} \overline{a_i\alpha_i} = \sum_{i=1}^{k} a_i\overline{\alpha}_i,$$

so that $\overline{B} = [\overline{\alpha}_1, \overline{\alpha}_2, ..., \overline{\alpha}_k]$ spans $\overline{V}$.

Suppose for some $b_i \in F$, $$\sum_{i=1}^{k} b_i\overline{\alpha}_i = \overline{0}.$$

This yields $$\sum_{i=1}^{k} b_i\alpha_i + W = W.$$

Hence, $$\sum_{i=1}^{k} b_i\alpha_i \in W \cap W' = \{0\},$$

$$\sum_{i=1}^{k} b_i\alpha_i = 0.$$

This yields $$b_i = 0;\ 1 \le i \le k.$$

Hence, $\overline{B}$ is a basis of $\overline{V}$.

Consequently, $$\dim\overline{V} = \dim V - \dim W.$$

Analogous to the concept of linear transformation, we have the following :

Definition 4. Let M and M' be two left modules over a ring R.

Then a mapping $T : M \to M'$ is called an *R-homomorphism* if it satisfies the following conditions :

(*i*) $T(\alpha+\beta) = T(\alpha) + T(\beta)$, and

(*ii*) $T(a\alpha) = aT(\alpha)$

for all $\alpha,\ \beta \in M$, $a \in R$; further the set $\{\alpha \in M : T(\alpha) = 0\}$ is called the *kernel of T.*

For any R-homomorphism $T{:}_R M \to {}_R M'$ the kernel of T, denoted by ker(T) is a submodule of M. T is one-to-one iff $\ker(T) = \{0\}$. T is called an *R-isomorphism* if T is one-to-one and onto. Two left R-modules M and M' are said to be *isomorphic* if there exists an R-isomorphism of M onto M'.

Theorem 5. (Fundamental Theorem of Homomorphisms). Let T be an R-homomorphism of a module ${}_R M$ onto a module ${}_R M'$, with $K = \ker(T)$.

Then, the mapping $$\overline{T} :\ M/K \to M'$$

such that $$\overline{T}(\alpha + K) = T(\alpha)$$

for $\alpha \in M$, is an R-isomorphism.

Proof. Consider any two members $\overline{\alpha} = \alpha + K$,

$$\overline{\beta} = \beta + K \in \overline{M} = M/K.$$

If $\overline{\alpha} = \overline{\beta}$, then, $$\alpha - \beta \in K.$$

So, by definition, $$T(\alpha - \beta) = 0,$$

i.e., $$T(\alpha) - T(\beta) = 0.$$

This gives $$\overline{T}(\overline{\alpha}) = T(\alpha) = T(\beta) = \overline{T}(\overline{\beta}),$$

so that $\overline{T}$ is a well defined mapping.

Now, $$\overline{T}(\overline{\alpha} + \overline{\beta}) = \overline{T}(\overline{\alpha + \beta})$$

$$= T(\alpha+\beta)$$
$$= T(\alpha)+(\beta)$$
$$\overline{T}(\overline{\alpha})+\overline{T}(\overline{\beta}),$$

and for $a \in R$, $\quad \overline{T}(a\overline{\alpha}) = \overline{T}(\overline{a\alpha}) = T(a\alpha) = aT(\alpha) = a\overline{T}(\overline{\alpha}),$

give $\overline{T}$ is an R-homomorphism.

Now, $\alpha \in \ker(\overline{T})$, yields $\overline{T}(\overline{\alpha}) = 0$, i.e., $T(\alpha) = 0$, so that $\alpha \in K$ and, hence, $\overline{\alpha} = \overline{0}$.

Consequently, $\quad \ker(\overline{T}) = \{\overline{0}\}.$

This proves that $\overline{T}$ is one-to-one. Finally, consider $\alpha' \in M'$. As $T : M \to M'$ is onto, there exists an $\alpha \in M$ such that $T(\alpha) = \alpha'$

Then, $\quad \overline{T}(\overline{\alpha}) = \alpha'$

Hence, $\overline{T}$ is also onto. This all proves that $\overline{T}$ is an R-isomorphism.

This theorem shows that given any left R-module M, any homomorphic image of M is isomorphic to some quotient module of M. The structure theory of modules has been extensively developed. But in the present book, we do not intend to go into this development. we confine ourselves to those parts of the theory, that are relevant to the study of vector spaces.

A module $_RM$ is said to be *finitely generated* if there exist finitely many elements $\alpha_1, \alpha_2, \ldots, \alpha_n$ in M, such that any $\alpha \in M$ is expressible as

$$\alpha = a_1\alpha_1 + a_2\alpha_2 + \ldots,\ a_n\alpha_n,\ a_i \in R.$$

In this case, the set $\{\alpha_1, \alpha_2, \ldots, \alpha_n\}$ is called a *generating set*.

If for $\alpha \in M$, $\quad R\alpha = \{a\alpha : a \in R\},$

then, $R\alpha$ is a submodule of M, generated by one element α : $R\alpha$ is called a cyclic submodule; in particular, if $M = R\alpha$ for some $\alpha \in M$, M itself is called a *cyclic module* with α as its *generator*.

For example, $R = R.1$, gives that $_RR$ is a cyclic module. Any one-dimensional vector space $_FV$ (or V_F) is a cyclic F-module. Any infinite dimensional vector space $_FV'$, is not a finitely generated F-module.

Consider any finite dimensional vector space V_F and a linear operator T on V_F. Given $f(X) \in F[X]$, we know that $f(T)$ is a linear operator on V (see Remark (3.2.13)). Thus, for any $a \in V$, $f(T)\alpha$, the image of α under $f(T)$ is a well-defined member of V.

Define $\quad f(X)\alpha = f(T)\alpha.$

By using Remark (3.2.13), it follows that V becomes a left $F[X]$-module. Here, we say that V becomes *a left $F[X]$-module via T*. Let $\alpha_1, \alpha_2, \ldots, \alpha_n$ generate V_F. As

$F \subseteq F[X]$, given $\alpha \in V$, $\alpha = \sum_{i=1}^{n} a_i\alpha_i$, $\alpha_i \in F$ yields $\alpha_1, \alpha_2, ..., \alpha_n$ generate V as a left $F[X]$-module, so that V is a finitely generated left $F[X]$-module. Let $\dim V = n$. Consider $a \in V$.

Then, the $n + 1$ vectors $\quad \alpha, T\alpha, T^2\alpha, ..., T^n\alpha$

in V are linearly dependent.

So, for some a_i in F, not all zeros,

$$a_0\alpha + a_1T\alpha + ..., + a_nT^n\alpha = 0$$

i.e.,
$$f(T)\alpha = 0, f(X)\alpha = 0,$$

where
$$f(X) = \sum_{i=0}^{n} a_i X^i$$

is a non-zero member of $F[X]$.

This motivates the following :

Definition 6. Let R be an integral domain. A module $_RM$ is called a *torsion* module if for each $\alpha \in M$, there exists $a \neq 0$ in R such that $a\alpha = 0$.

Thus, V as a left $F[X]$-module is a torsion module.

Definition 7. Let M be a left module over a ring R. For any subset S of M, the *annihilator* of S in R, denoted by $\text{ann}_R(S)$, is the set of those $x \in R$ such that $x\alpha = 0$ for all $\alpha \in S$. for any subset J of R, its *annihilator* in M, denoted by $\text{ann}_M(J)$, is the set of those $\alpha \in M$ such that $a\alpha = 0$ for all $a \in J$.

In case there is no confusion concerning the underlying ring R, $\text{ann}_R(S)$ will be simply denoted by $\text{ann}(S)$. If $S = \{\alpha\}$, then $\text{ann}_R(S)$ is denoted by $\text{ann}_R(\alpha)$.

Example 7.a. Consider the linear operator T on $V = \mathbf{R}^2$, whose matrix relative to the standard ordered basis is $A = \begin{bmatrix} 0 & 1 \\ 1 & 0 \end{bmatrix}$.

So, for the standard basis

$$B = \{\epsilon_1, \ \epsilon_2\}, T\epsilon_1 = \epsilon_2, T\epsilon_2 = \epsilon_1$$

This gives
$$T^2\epsilon_1 = T\epsilon_2 = \epsilon_1,$$

so that
$$(T^2 - I)\epsilon_1 = 0,$$

Consider V as a left $\mathbf{R}[X]$-module via T.

We get
$$(X^2 - 1)\epsilon_1 = 0.$$

Thus, $$X^2 - 1 \in \text{ann}_{\mathbf{R}[X]}(\epsilon_1).$$

Let $$f(X) \in \mathbf{R}[X].$$

By the division algorithm $f(X) = g(X)(X^2 - 1) + r(X)$

with $$r(X) = 0$$

or $$\deg r(X) < \deg(X^2 - 1).$$

So that $$r(X) = aX + b,\ a,\ b \in \mathbf{R}.$$

Consequently,

$$f(X)\epsilon_1 = g(T)(T^2 - I)\epsilon_1 + aT\epsilon_1 + bI\epsilon_1 = a\epsilon_2 + b\epsilon_1.$$

Thus, $$f(X)\epsilon_1 = 0$$

iff $$a\epsilon_2 + b\epsilon_1 = 0$$

iff $$a = 0 = b$$

iff $$(X^2 - 1) \mid f(X).$$

Consequently, ann(ϵ_1) is the ideal

$$(X^2 - 1)\mathbf{R}[X].$$

Consider $$\epsilon = \epsilon_1 + \epsilon_2.$$

Then, $$\epsilon \neq 0$$

and $$T\epsilon = T\epsilon_1 + T\epsilon_2 = \epsilon_2 + \epsilon_1 = \epsilon.$$

This gives $$(T - 1)\epsilon = 0,$$

i.e., $$(X - 1)\epsilon = 0.$$

For $f(X) \in \mathbf{R}[X]$, $\quad f(X) = (X - 1)q(X) + a,\ a \in \mathbf{R},$

gives $$f(X)\epsilon = a\epsilon.$$

Consequently, $$f(X)\epsilon = 0$$

iff $$a\epsilon = 0$$

iff $$a = 0$$

iff $$(X - 1) \mid f(X)$$

iff $$f(X) \in (X - 1)\mathbf{R}[X].$$

Hence, $$\text{ann}(\epsilon) = (X - 1)\mathbf{R}[X].$$

Finally, $$(T^2 - I)\epsilon_1 = 0 = (T^2 - I)\epsilon_2,$$

yields for any $$\alpha = a_1\epsilon_1 + a_2\epsilon_2 \in V,$$

$$(T^2 - I)\alpha = 0.$$

This, in turn, yields, $\text{ann}_{\mathbf{R}[X]}(V) = (X^2 - 1)\mathbf{R}[X]$.

Theorem 8. Let S and S' be any two subsets of a module ${}_RM$. Then,

(*i*) $\text{Ann}_R(S)$ is a left ideal of R.

(*ii*) If $S \subseteq S'$, then, $\text{ann}_R(S') \subseteq \text{ann}_R(S)$.

(*iii*) $\text{Ann}_R(S \cup S') = \text{ann}_R(S) \cap \text{ann}_R(S')$

(*iv*) $\text{Ann}_R(0) = R = \text{ann}_R(\phi)$.

Proof.

(*i*) For any $\alpha \in S$, $0\alpha = 0$, gives $0 \in \text{ann}_R(S)$. Consider $a, b \in \text{ann}_R(S)$ and $r \in R$.

Then, for $\alpha \in S$, $a\alpha = 0 = b\alpha$,

yields $(a-b)\alpha = a\alpha - b\alpha = 0 - 0 = 0$,

and, $(ra)\alpha = r(a\alpha) = r0 = 0$.

Consequently, $a - b \in \text{ann}_R(S)$

and $ra \in \text{ann}_R(S)$.

Hence, $\text{ann}_R(S)$ is a left ideal of R.

(*ii*) Let $S \subseteq S'$. consider $a \in \text{ann}_R(S')$. Let $\alpha \in S$. Then, $\alpha \in S'$, gives $a\alpha = 0$, i.e. $a \in \text{ann}_R(S)$.

Hence, $\text{ann}_R(S') \subseteq \text{ann}_R(S)$

(*iii*) As $S \subseteq S \cup S'$, by (*ii*) $\text{ann}_R(S \cup S') \subseteq \text{ann}_R(S)$.

Similarly, $\text{ann}_R(S \cup S') \subseteq \text{ann}_R(S')$.

Consequently, $\text{ann}_R(S \cup S') \subseteq \text{ann}_R(S) \cap \text{ann}_R(S')$.

Let $a \in \text{ann}_R(S) \cap \text{ann}_R(S')$

and $\alpha \in S \cup S'$

As $\alpha \in S$ or $\alpha \in S'$ and $a \in \text{ann}_R(S)$, as well as $a \in \text{ann}_R(S')$, we get $a\alpha = 0$, i.e. $a \in \text{ann}_R(S \cup S')$.

This gives $\text{ann}_R(S) \cap \text{ann}_R(S') \subseteq \text{ann}_R(S \cup S')$.

Hence, $\text{ann}_R(S \cup S') = \text{ann}_R(S) \cap \text{ann}_R(S')$.

(*iv*) By definition $\text{ann}_R(0) \subseteq R$.

Conversely, given $a \in R$, $a0 = 0$, gives $a \in \text{ann}_R(0)$.

Hence, $R = \text{ann}_R(0)$

Finally, $\phi \subset \{0\}$,

yields $R = \text{ann}_R\{0\} \subseteq \text{ann}_R(\phi) \subseteq R$.

So that $R = \text{ann}_R(\phi)$.

Let S be any subset of a module M. The smallest submodule of M that contains S, that we denote by RS, is given by the following :

(*i*) If $S = \phi$, $RS = \{0\}$.

(*ii*) If $S \neq \phi$, then, any $\alpha \in RS$ iff

$$\alpha = a_1\alpha_1 + a_2\alpha_2 + ..., + a_k\alpha_k$$

for some $\alpha_i \in S$ and $a_i \in R$.

The submodule RS is said to be generated by S.

In case $$S = \{\alpha_1, \alpha_2, ..., \alpha_k\},$$
$$RS = R\alpha_1 + R\alpha_2 + ... + R\alpha_k.$$

Theorem 9. Let M be a module over a commutative ring R and S be a subset of M. Then,

(*i*) $\text{Ann}_R(S)$ is ideal of R.

(*ii*) If A is the submodule generated by S,

then, $$\text{ann}_R(S) = \text{ann}_R(A).$$

Proof. By Theorem 8, $\text{ann}_R(S)$ is a left ideal of R. As R is commutative, it follows that $\text{ann}_R(S)$ is an ideal of R.

As $S \subseteq A$, $$\text{ann}_R(A) \subseteq \text{ann}_R(S), \quad \text{by Theorem 8}(ii)$$

Consider $a \in \text{ann}_R(S)$, and $\alpha \in A$. If $\alpha = 0$, obviously, $a\alpha = 0$. Let $\alpha \neq 0$.

Then, $$\alpha = a_1\alpha_1 + a_2\alpha_2 + ..., + a_k\alpha_k,$$
$$\alpha_i \in S,\ a_i \in R.$$

Then, $$a\alpha_i = 0,\ 1 \le i \le k,$$

gives $$a\alpha = \sum_{i=1}^{k} aa_i\alpha_i = \sum_{i=1}^{k} a_i a\alpha_i = 0.$$

Consequently, $$a \in \text{ann}_R(A).$$

Thus, $$\text{ann}_R(S) \subseteq \text{ann}_R(A).$$

Hence, $$\text{ann}_R(A) = \text{ann}_R(S).$$

Theorem 10. Let M be a left module over a ring R, and A_1, A_2 be two subsets of R. Then,

(*i*) $\text{ann}_M(A_1)$ is a subgroup of $\langle M, + \rangle$

(*ii*) If $A_1 \subseteq A_2$, then $\text{ann}_M(A_2) \subseteq \text{ann}_M(A_1)$.

(*iii*) If R is commutative, then $\text{ann}_M(A_1)$ is a submodule of M, and $\text{ann}_M(A_1) = \text{ann}_M(A)$, where A is the ideal of R generated by A_1.

We now return to the study of linear operators. Throughout, V_F is a non-zero, finite dimensional vector space of dimension n.

Theorem 11. Let T be a linear operator on a finite dimensional vector space V_F. Then, there exists a unique monic polynomial $f(X) \in F[X]$ such that

(*i*) $f(T) = 0$

(*ii*) for any $g(X) \in F[X]$, $g(T) = 0$ iff $f(X) | g(X)$.

Proof. Let $\dim V_F = n$ and V be made left $F[X]$-module via T.

As $$\dim[\text{Hom}_F(V, V)] = n^2,$$

the $n^2 + 1$ linear operators $$I, T, T^2, ..., T^{n^2}$$

are linearly dependent.

Consequently, for some $$a_i \in F,$$

not all zeroes, $$0 \le i \le n^2, \sum_{i=0}^{n^2} a_i T^i = 0.$$

Thus, $$h(X) = \sum_{i=0}^{n^2} a_1 X^i \ne 0$$

And $$h(X)\alpha = h(T)\alpha = 0$$

for all $$a \in V.$$

Consequently, $$A = \text{ann}_{F[X]}(V)$$

is a non-zero ideal of $F[X]$.

By Theorem (5.1.3), A has a unique monic generator $f(X)$, so that $f(T) = 0$.

For any $$g(X) \in F[X],$$
$$g(T) = 0$$

iff $$g(X)\alpha = g(T)\alpha = 0$$

for every $$\alpha \in V$$

iff $$g(X) \in \text{ann}_{F[X]}(V) = A$$

iff $$f(X) | g(X).$$

Observe that in this theorem $f(X)$ is a non-zero polynomial in $F[X]$ of smallest degree, satisfying $f(T) = 0$. This motivates the following :

Definition 12. Let T be a linear operator on a finite dimensional space V_F. The monic polynomial $f(X) \in F[X]$ of smallest degree satisfying $f(T) = 0$, is called the *minimal polynomial* of T.

Example 12.a. Let T be the linear operator on $\mathbf{R}^3$, whose matrix relative to the standard ordered basis is $A = \begin{bmatrix} 0 & 0 & -1 \\ 1 & 0 & 2 \\ 0 & 1 & 1 \end{bmatrix}$.

Then, $$T \in_1 = \in_2,\ T \in_2 = \in_3,$$
$$T \in_3 = - \in_1 + 2 \in_2 + \in_3.$$

Consider any $$0 \neq g(x) = A + bX + cX^2 \in \mathbf{R}[X]$$
of degree ≤ 2.

Then, $$g(T) \in_1 = a \in_1 + bT \in_1 + cT^2 \in_1 = a \in_1 + b \in_2 + c \in_3 \neq 0.$$

That gives $g(T) \neq 0$. Thus, the minimal polynomial $f(X)$ of T has degree at least three.

Now, $$h(X) = X^3 - X^2 - 2X + 1$$
is a monic polynomial of degree 3.

As $$T^2 \in_1 = \in_3,$$
$$T^3 \in_1 = T \in_3 = - \in_1 + 2 \in_2 + \in_3 = - \in_1 + 2T \in_1 + T^2 \in_1,$$
we get $$(T^3 - T^2 - 2T + I) \in_1 = 0,$$
i.e., $$h(T) \in_1 = 0.$$

This further gives $$T[h(T) \in_1] = 0,$$
i.e., $$h(T)T \in_1 = 0,$$
i.e., $$h(T) \in_2 = 0.$$

Further, $$T[h(T) \in_2] = 0,$$
gives $$h(T)T \in_2 = 0,$$
i.e., $$h(T) \in_3 = 0.$$

Hence, $$h(T) = 0.$$

Hence $h(X)$ is the minimal polynomial of T.

Theorem 13. Let T be a linear operator on a finite dimensional vector space V_F. Then, the minimal polynomial of T is a factor of the characteristic polynomial of T and its degree is less than or equal to dimension of V.

Proof. Let $f(X)$ be the characteristic polynomial of T and $m(X)$ be the minimal polynomial of T in $F[X]$. By Theorem (4.4.17), $f(T) = 0$. By Theorem 11, $m(X) | f(X)$. As $\dim V = n = \deg f(X)$, it is obvious that $\deg m(X) \leq n$.

In Example 12.a, the degree of the minimal polynomial of T is 3, and the dimension of the vector space is also three. so, by this theorem, the minimal polynomial of T and the characteristic polynomial of T are the same.

Consider $$A = \begin{bmatrix} 2 & 0 & 0 \\ 0 & 2 & 0 \\ 0 & 0 & 3 \end{bmatrix}.$$

Let T' be the linear operator on $V = \mathbf{R}^3$, whose matrix relative to the standard ordered basis of V is A. Then, the characteristic polynomial of T' is $f(X) = (X-2)^2(X-3)$.

However, as $$(A-2I)(A-3I) = \begin{bmatrix} 0 & 0 & 0 \\ 0 & 0 & 0 \\ 0 & 0 & 1 \end{bmatrix}\begin{bmatrix} -1 & 0 & 0 \\ 0 & -1 & 0 \\ 0 & 0 & 0 \end{bmatrix} = 0,$$

T' satisfies $$m(T') = 0,$$

where $$m(X) = (X-2)(X-3).$$

As A is not a scalar matrix, for any

$$g(X) = aX + b,\ a \neq 0,\ g(T') \neq 0.$$

Hence, $m(X)$ is the minimal polynomial of T'.

Clearly $$m(X) \neq f(X).$$

However, in certain cases minimal polynomial and the characteristic polynomial are the same.

Theorem 14. Let $f(X) = X^n - a_{n-1}X^{n-1} - \ldots - a_1X - a_0$ be any monic polynomial of degree $n \geq 1$ in $F[X]$ and A be the companion matrix of $f(X)$. If T is a linear operator on an n-dimensional vector space V_F such that relative to some ordered basis $B = \{\alpha_1, \alpha_2, \ldots, \alpha_n\}$, $A = [T]_B$, then, the minimal polynomial and the characteristic polynomial of T both equal $f(X)$.

Proof. By Theorem (4.4.20), the characteristic polynomial of T is $f(X)$. Let $m(X)$ be the minimal polynomial of T.

Now, $$A = \begin{bmatrix} 0 & 0 & 0 & .. & .... & 0 & a_0 \\ 1 & 0 & 0 & .. & ... & 0 & a_1 \\ 0 & 1 & 0 & .. & ... & 0 & a_2 \\ .. & .. & .. & .. & ... & .. & .. \\ .. & .. & .. & .. & ... & .. & .. \\ 0 & 0 & 0 & .. & ... & 1 & a_{n-1} \end{bmatrix}$$

so that $$T\alpha_1 = \alpha_2,\ T\alpha_2 = \alpha_3, \ldots,\ T\alpha_{n-1} = \alpha_n,$$

$$T\alpha_n = a_0\alpha_1 + a_1\alpha_2 + \ldots + a_{n-1}\alpha_n.$$

Then, $$T^2\alpha_1 = T\alpha_2 = \alpha_3.$$

and so on.

In general, $$T^i\alpha_1 = \alpha_{i+1}$$
for $$0 \le i \le n-1.$$

Consider any non-zero polynomial
$$g(X) = b_0 + b_1X + \dots + b_{n-1}X^{n-1}$$
with $$\deg g(X) \le n-1.$$

Here, some $b_i \ne 0$ so that
$$\begin{aligned} g(T)\alpha_1 &= b_0\alpha_1 + b_1T\alpha_1 + b_2T^2\alpha_1 + \dots + b_{n-1}T^{n-1}\alpha_1 \\ &= b_0\alpha_1 + b_1\alpha_2 + b_2\alpha_3 + \dots + b_{n-1}\alpha_n \\ &\ne 0, \end{aligned}$$
as α_i are linearly independent.

Hence, T does not satisfy any non-zero polynomial of degree less than n, so that $n \le \deg m(X) \le \deg f(X) = n$.

As $$m(X) \mid f(X),$$
we get $$m(X) = f(X).$$

Given an $n \times n$-matrix A over F, we define the minimal polynomial of A on similar lines as in Definition 12. If T is the linear operator on V_F with $\dim V_F = n$, such that $A = [T]_B$ for some ordered basis B of V, the minimal polynomials of A and T are the same.

Consider $$A = \begin{bmatrix} A_1 & 0 & \dots & 0 \\ 0 & A_2 & \dots & 0 \\ .. & .. & \dots & .. \\ .. & .. & \dots & .. \\ 0 & 0 & \dots & A_k \end{bmatrix} = A_1 \oplus A_2 \oplus A_k.$$

Then, for any polynomial $f(X) = a_0 + a_1X + \dots + a_sX^s$,
$$f(A) = a_0I + a_1A + \dots + a_sA^s,$$
it is clear that $$f(A) = f(A_1) \oplus f(A_2) \oplus \dots \oplus f(A_k)$$
so that $$f(A) = 0$$
iff $$f(A_i) = 0$$
for $$1 \le i \le k.$$

Let $m(X)$, $m_i(X)$ be the minimal polynomials of A, A_i.

We get from $$f(A) = 0$$
iff $$f(A_i) = 0,\ 1 \le i \le k$$
that $$m(X) \mid f(X)$$

iff $$m_i(X) \mid f(X),\ 1 \le i \le k$$

iff $$m'(X) = \text{lcm}[m_1(X), m_2(X), \dots, m_k(X)] \mid f(X).$$

Hence, $$m(X) = m'(X).$$

This gives the following :

Theorem 15. Let A be any $n \times n$-matrix over F, and $A = A_1 \oplus A_2 \oplus \dots \oplus A_k$. Then, the minimal polynomial $m(X)$ of A is the lcm of the minimal polynomials of A_i, $1 \le i \le k$. Further, let T be a linear operator on V_F, such that

$$V = W_1 \oplus W_2 \oplus \dots \oplus W_k$$

for some T-invariant subspaces W_i of V with the property that for the restriction T_i of T to W_i, $[T_i]_{B_i} = A_i$ for some ordered basis B_i of W_i, then, the minimal polynomial of T is the lcm of the minimal polynomials of T_i's (i.e. of A_i's).

Example 15.a. $A = \begin{bmatrix} 0 & 1 & 0 & 0 \\ 1 & -1 & 0 & 0 \\ 0 & 0 & 3 & 0 \\ 0 & 0 & 0 & 3 \end{bmatrix} = A_1 \oplus A_2 \oplus A_3$ where $A_1 = \begin{bmatrix} 0 & 1 \\ 1 & -1 \end{bmatrix}$ being the companion matrix of $X^2 + X - 1$, has minima polynomial $m_1(X) = X^2 + X - 1$.

As $$A_2 = A_3 = [3],$$

their minimal polynomials are $m_2(X) = m_3(X) = X - 3$,

so that the minimal polynomial $m(X)$ of A being the lcm of

$$X^2 + X - 1,\ X - 3,$$

equals $(X^2 + X - 1)(X - 3) \ne$ the characteristic polynomial of A.

For any positive integer n, the companion matrix of X^n is the $n \times n$-matrix

$$A_n = \begin{bmatrix} 0 & 0 & 0 & \dots & 0 & 0 \\ 1 & 0 & 0 & \dots & 0 & 0 \\ 0 & 1 & 0 & \dots & 0 & 0 \\ .. & .. & .. & \dots & .. & .. \\ .. & .. & .. & \dots & .. & .. \\ 0 & 0 & 0 & \dots & 1 & 0 \end{bmatrix}.$$

By Theorem 14, X^n is the minimal polynomial of A_n. That means $A_n^n = 0$, but $A_n^k \ne 0$, $0 \le k \le n-1$. Thus, A_n is a nilpotent matrix of index n. For any $n \times n$-matrix A we say that T is the *linear operator on* F^n (or $F^{n \times 1}$) is *determined by* A, if A is the matrix of T relative to the standard ordered basis. Consider the linear operator on $\mathbf{R}^4$ given by the A in Example 15.a.

Then, $$T \in_1 = \in_2,\ T \in_2 = \in_1 - \in_2,$$

$$T \in_3 = 3 \in_3,\ T \in_4 = 3 \in_3$$

so that $$T^2 \in_1 = T \in_2 = \in_1 - \in_2 = \in_1 - T \in_1.$$

This gives $$(T^2 + T - I) \in_1 = 0.$$

For any non-zero $$g(X) = a + bX,\ a,\ b \in \mathbf{R},$$

$$g(T) \in_1 = a \in_1 + bT \in_2 = (a+b) \in_1 - b \in_2$$

so that $X^2 + X - 1$ is the smallest degree monic polynomial $h(X)$, satisfying $h(T) \in_1 = 0$.

Clearly, $$u_1(X) = X^2 + X - 1$$

divides $m(X)$.

Again, $$T \in_3 = 3 \in_3,$$

gives $$(T - 3I) \in_3 = 0.$$

Thus, $$u_2(X) = X - 3$$

is the smallest degree monic polynomial such that $u_2(T) \in_2 = 0$.

Consider $$\in = \in_1 + \in_3.$$

As $$m(T) = 0,$$

obviously, $$m(T) \in = 0.$$

Here, $$\deg m(T) = 3.$$

Consider any $$g(X) = a + bX + cX^2$$

of degree less than 3, but $$g(X) \neq 0.$$

Now, $$T \in = T \in_1 + T \in_3 = \in_2 + 3 \in_3,$$

$$T^2 \in = T(\in_2 + 3 \in_3) = T \in_2 + 3T \in_3 = \in_1 - \in_2 + 9 \in_3.$$

This gives $$g(T) \in = a(\in_1 + \in_3) + b(\in_2 + 3 \in_3) + c(\in_1 - \in_2 + 9 \in_3)$$

$$= (a+c) \in_1 + (b-c) \in_2 + (a + 3b + 9c) \in_3.$$

Let $$g(T) \in = 0$$

The linear independence of $\in_1,\ \in_2,\ \in_3$ yield

$$a + c = 0,\ b - c = 0,\ a + 3b + 9c = 0.$$

These three equations yield $a = 0,\ b = 0,\ c = 0.$

This is a contradiction.

Hence, $$g(T) \in \neq 0.$$

Thus, $m(X)$, itself is the smallest degree monic polynomial such that $m(T) \in = 0$

These examples motivate the following theorem :

Theorem 16. Let T be a linear operator on a finite dimensional vector space V_F. Given $\alpha \in V$, there exists a unique monic polynomial $g(X) \in F[X]$ such that

(*i*) $g(T)\alpha = 0$

(*ii*) For any $f(X) \in F[X]$, $f(T)\alpha = 0$ iff $g(X)|f(X)$. Further, $g(X)$ is a factor of the minimal polynomial of T.

Proof. Let V be made a left $F[X]$-module via T.

Let $$A = \{f(X) \in F[X] : f(X)\alpha = 0\}.$$

This A is an ideal of $F[X]$. Let $m(X)$ be the minimal polynomial of T. Then $m(X) \neq 0$, and $m(T) = 0$, gives $m(X)\alpha = 0$ so that $m(X) \in A$. Consequently, $A \neq \{0\}$. Hence, A has unique monic generator $g(X)$.

By definition $$g(T)\alpha = g(X)\alpha = 0.$$

For any $f(X) \in F[X]$, $$f(T)\alpha = 0$$

iff $$f(X)\alpha = 0$$

iff $$f(X) \in A$$

iff $$g(X)|f(X).$$

Finally, as $m(X) \in A$, $$g(X)|m(X).$$

This theorem motivates the following definition :

Definition 17. Let T be a linear operator on a finite dimensional vector space V_F. For any $\alpha \in V$, the smallest degree monic polynomial $m_\alpha(X) \in F[X]$ such that $m_\alpha(T) = 0$ is called the *annihilator polynomial* of α relative to T (or the T-annihilator polynomial of α).

For the linear operator T on $\mathbf{R}^3$, given by the matrix A in Example 15.a, the annihilator polynomials of $\in_1$ and $\in_3$ relative to T are $X^2 + X - 1$ and $X - 3$, respectively.

Theorem 18. Let T be linear operator on V_F and let $\alpha, \beta \in V$ have $m_1(X)$, $m_2(X)$ as their respective annihilator polynomials relative to T. Then,

(*i*) For any monic factor $h(X)$ of $m_1(X)$, the annihilator polynomial of $h(T)\alpha$ relative to T is $m_1(X)/h(X)$.

(*ii*) Let $0 \neq g(X) \in F[X]$ and $d(X) = \gcd[m_1(X), g(X)]$. Then, the annihilator polynomial of $g(T)\alpha$ relative to T is $m_1(X)/d(X)$.

(*ii*) If $m'(X)$ is the annihilator polynomial of $\alpha + \beta$ relative to T, then $m'(X)$ is a factor of $\operatorname{lcm}[m_1(X), m_2(X)]$.

Further, if $$\gcd[m_1(X), m_2(X)] = 1,$$

then, $$m'(X) = \operatorname{lcm}[m_1(X), m_2(X)] = m_1(X)\, m_2(X).$$

Proof.

(*i*) Let $m_1(X) = h(X)g(X)$

and $\alpha' = h(T)\alpha$.

As $m_1(T)\alpha = 0$,

we get $g(T)\alpha' = g(T)h(T)\alpha = m_1(T)\alpha = 0$.

Consequently, the *T*-annihilator polynomial $m'(X)$ of α' divides $g(X)$.

Then, $m'(X)\alpha' = 0$,

yield $m'(X)h(X)\alpha = 0$

so that $\deg m_1(X) \le \deg m'(X)h(X)$,

i.e., $\deg g(X)h(X) \le \deg m'(X)h(X)$.

Thus, $\deg g(X) \le \deg m'(X) \le \deg g(X)$.

Hence, $g(X) = m'(X)$.

This proves that the *T*-annihilator polynomial of $h(T)\alpha$ is $g(X) = m_1(X)/h(X)$.

(*ii*) Now, $d(X) = m_1(X)u(X) + g(X)v(X)$,

$u(X),\ v(X) \in F[X]$.

Thus, $d(T)\alpha = v(T)g(T)\alpha$.

By (*i*), the *T*-annihilator of $d(T)\alpha$ is $k(X) = m_1(X)/d(X)$. Let $k'(X)$ be the *T*-annihilator of $g(T)\alpha$.

Then, $k'(X)d(T)\alpha = k'(T)v(T)g(T)\alpha = v(T)k'(T)[g(T)\alpha] = 0$.

Consequently, $k(X)|k'(X)$.

Again $g(X) = g_1(X)d(X)$,

$g_1(X) \in F[X]$,

yields $k(T)[g(T)\alpha] = g_1(T)k(T)[d(T)\alpha] = g_1(T)m_1(T)\alpha = 0$,

$k'(X)|k(X)$.

Hence, $k'(X) = k(X)$.

This proves (*ii*).

(*iii*) Let $l(X) = \text{lcm}[m_1(X),\ m_2(X)]$.

As $m_1(X)|l(X)$,

we get $l(X)\alpha = 0$.

Similarly, $l(X)\beta = 0$.

Consequently, $l(X)(\alpha + \beta) = 0$.

Hence, the *T*-annihilator $m(X)$ of $\alpha + \beta$ is a factor of $l(X)$.

Finally, let $\gcd[m_1(X),\ m_2(X)] = 1$.

By Theorem (5.1.19)

$$m(X) = h_1(X)h_2(X), h_i(X) \mid m_i(X).$$

Now, $\quad m(X) \mid m_1(X)h_2(X)$

yields, $\quad 0 = m_1(T)h_2(T)(\alpha + \beta) = m_1(T)h_2(T)\beta.$

So $\quad m_2(X) \mid m_1(X)h_2(X).$

As $\quad \gcd[m_1(X), m_2(X)] = 1,$

we get $\quad m_2(X) \mid h_2(X).$

Consequently, $\quad m_2(X) = h_2(X).$

Similarly, $\quad m_1(X) = h_1(X).$

Hence, $\quad m(X) = m_1(X)m_2(X).$

By applying mathematical induction to the result in the above theorem we get the following corollary :

Corollary 19. Let T be a linear operator on a finite dimensional vector space V_F. Let α_i, $1 \le i \le k$ be any finitely many members of V with respective T-annihilator polynomials $m_i(X)$, and $m(X)$ be the T-annihilator polynomial of $\alpha = \sum_{i=1}^{k} \alpha_i$. Then

(*i*) $m(X)$ divides the lcm of $m_i(X)$, $1 \le i \le k$.

(*ii*) If $\gcd[m_i(X), m_j(X)] = 1$ for $i \ne j$, then

$$m(X) = m_1(X)m_2(X)...m_k(X).$$

Theorem 20. (Primary Decomposition Theorem). Let T be a non-zero linear operator on a finite dimensional vector space V_F, with minimal polynomial

$$m(X) = p_1(X)^{r_1} p_2(X)^{r_2} ... p_k(X)^{r_k},$$

where $p_i(X)$ are distinct primes in $F[X]$, and $r_i \ge 1$, $1 \le i \le k$.

Then, V is uniquely expressible as $V = V_1 \oplus V_2 \oplus ... \oplus V_k$ such that V_i is a T-invariant subspace of V and the minimal polynomial of the restriction T_i of T to V_i is $p_i(X)^{r_i}$.

Proof. For $1 \le i \le k$, consider $V_i = \text{ann}_V(p_i(X)^{r_i})$; here we are regarding V, a left $F[X]$-module via T. As V_i is a submodule of V, it is T-invariant.

For $k = 1$ $\quad m(X) = p_1(X)^{r_1},$

gives $\quad p_1(T)^{r_1} = 0$

and, hence, $\quad V = V_1.$

Hence the result holds for $k = 1$.

Let $k \geq 2$. Define $$q_i(X) = \frac{m(X)}{p_i(X)^{r_i}}.$$

As $q_i(X)$ is a product of prime powers $p_j(X)^{r_j}$, $j \neq i$, $q_i(X)$ and $p_i(X)^{r_i}$ are coprimes. This, in turn, yields that $q_i(X)$, $1 \leq i \leq k$, have gcd 1.

So $$1 = \sum_{i=1}^{k} \lambda_i(X) q_i(X),$$
$$\lambda_i(X) \in F[X]$$

Thus, $$I = \sum_{i=1}^{k} \lambda_i(T) q_i(T).$$

Consider $\alpha \in V$. Then, $$\alpha = \sum_{i=1}^{k} \lambda_i(T) q_i(T)\alpha = \sum_{i=1}^{k} \alpha_i$$

where $$\alpha_i = \lambda_i(T) q_i(T)\alpha.$$

As $$m(X) = q_i(X) p_i(X)^{r_i},$$
$$p_i(T)^{r_i} \alpha_i = \lambda_i(T) m(T)\alpha = 0.$$

So that $$\alpha_i \in V_i.$$

This proves that $$V = V_1 + V_2 + \ldots + V_k \qquad \ldots (1)$$

We claim that this sum is direct.

Let for some $\beta_j \in V_j$, $$\sum_{j=1}^{k} \beta_j = 0 \qquad \ldots (2)$$

For any i, $1 \leq i \leq k$, $$p_j(X)^{r_j} \,\Big|\, q_i(X)$$

for $j \neq i$, gives $$q_i(X)\beta_j = 0.$$

Consequently, (2) gives $$\sum_{j=1}^{k} q_i(T)\beta_j = 0,$$

i.e., $$q_i(T)\beta_i = 0.$$

Also, $$p_i(T)^{r_i} \beta_i = 0.$$

Consequently, the T-annihilator polynomial $h_i(X)$ of β_i divides $q_i(X)$ as well as $p_i(X)^{r_i}$. As these are coprimes, we get $h_i(X) = 1$.

Thus, $$0 = I\beta_i = \beta_i.$$

Hence, $$V = V_1 \oplus V_2 \oplus \ldots \oplus V_k \qquad \ldots (3)$$

For any $\alpha \in V_i$, $p_i(T)^{r_i}\alpha = 0$,

yields that the minimal polynomial of the restriction T_i of T to V_i is a factor of $p_i(X)^{r_i}$, and, hence, it equals $p_i(X)^{s_i}$ for some $s_i \le r_i$.

Then by Theorem 15, $m(X) = p_1(X)^{s_1} p_2(X)^{s_2} .. p_k(X)^{s_k}$.

This immediately gives $r_i = s_i$.

Let also $$V = W_1 \oplus W_2 \oplus ... \oplus W_k \quad ... (4)$$

where each W_i is a T-invariant subspace of V such that the restriction T_i' of T to W_i has minimal polynomial $p_i(X)^{r_i}$.

Consequently, for any $\alpha \in W_i$, $p_i(X)^{r_i}\alpha = 0$.

This yields $W_i \subseteq V_i$.

Now, (3) and (4) give $V = V_i \oplus V_i' = W_i \oplus W_i'$

where $$V_i' = \sum_{j+1} V_j, \; W_i' = \sum_{j \ne 1} W_j.$$

Obviously, $W_i' \subseteq V_i'$,

so that $V = W_i + V_i'$.

Then, $V_i = V \cap V_i = W_i + (V_i \cap V_i') = W_i$,

as $V_i \cap V_i' = \{0\}$.

This proves the uniqueness.

This theorem leads to the following definition :

Definition 21. Let T be a non-zero linear operator on a finite dimensional vector space V_F, with minimal polynomial
$$m(X) = p_1(X)^{r_1} p_2(X)^{r_2} .. p_k(X)^{r_k}$$
where $p_i(X)$ are distinct prime factors of $m(X)$, and $r_i \ge 1$.

Then, the expression $V = V_1 \oplus V_2 \oplus ... \oplus V_k$

where each $$V_i = \{\alpha \in V : p_i(T)^{r_i}\alpha = 0\},$$

is called the *primary decomposition* of V relative to T, and each V_i is called the $p_i(X)$-*primary* (or simply a *primary*) *component* of V.

Proposition 22. Let T be a linear operator on a finite dimensional vector space V_F. For any prime factor $p_i(X)$ of the minimal polynomial $m(X)$ of T, the $p_i(X)$-primary component of V relative to T is given by $V_i = \{\alpha \in V : p_i(X)^{s_i}\alpha = 0 \text{ for some } s_i \ge 0\}$.

Proof. Let $$W_i = \{\alpha \in V : p_i(T)^{s_i}\alpha = 0 \text{ for some } s_i \ge 0\}.$$

Let $r_i \geq 1$ be the largest positive integer such that $p_i(X)^{r_i} | m(X)$.

By definition $\alpha \in V_i$ gives $\quad p_i(X)^{r_i}\alpha = 0.$

So that $\quad V_i \subseteq W_i.$

Consider $\quad \beta \in W_i.$

Then, $\quad p_i(T)^{s_i}\beta = 0$

for some $\quad s_i \geq 0.$

Thus, the T-annihilator polynomial of β is $p_i(T)^{u_i}$ for some u_i.

Now, $\quad p_i(X)^{u_i} \Big| p_i(X)^{r_i}$

yields $\quad p_i(T)^{r_i}\beta = 0,$

so that $\quad W_i \subseteq V_i.$

Hence, $\quad W_i = V_i.$

For any prime $p(X) \in F[X]$, that may or may not divide the minimal polynomial of a linear operator T, the T-invariant subspace $W_{p(X)} = \{\alpha \in V : p(T)^s\alpha = 0 \text{ for some } s \geq 0\}$ is called the $p(X)$-*primary component of* V relative to T. Here, if $p(X)$ does not divide the minimal polynomial of T, $W_{p(X)} = \{0\}$.

We now give a characterization of a diagonalizable operator. Observe that given a diagonal matrix $A = \text{diag}(a_1, a_2, ..., a_n)$, the minimal polynomial being lcm of the polynomials $X - a_i$, $1 \leq i \leq n$, equals $(X - b_1)(X - b_2)...(X - b_k)$, where $b_1, b_2, ..., b_k$ are all the distinct entries on the diagonal of A ; $b_1, b_2, ..., b_k$ are precisely the distinct characteristic values of A.

Theorem 23. Let T be a linear operator on a finite dimensional vector space V_F. Then T is diagonalizable iff the minimal polynomial $m(X)$ of T is given by $m(X) = (X - b_1)(X - b_2)...(X - b_k)$ for some distinct b_i, $1 \leq i \leq k$ in F.

Proof. Let T be diagonalizable. Then, for some ordered basis B of V, $A = [T]_B$ is a diagonal matrix. so, as remarked above this theorem,

$$m(X) = (X - b_1)(X - b_2)...(X - b_k)$$

where $b_1, b_2, ..., b_k$ are distinct characteristic values of A (or of T).

Conversely, let $m(X)$ be given as above.

By the primary decomposition theorem

$$V = V_1 \oplus V_2 \oplus ... \oplus V_k$$

such that $\quad V_i = \{\alpha \in V : (T - b_i)\alpha = 0\}.$

Thus, each $\alpha \in V_i$ satisfies $\quad T\alpha = b_i\alpha.$

So if B_i is a basis of V_i, then, each member of B_i is a characteristic vector of T.

Then,
$$B = \bigcup_{i=1}^{k} B_i$$
is a basis of V, whose every member is a characteristic vector of T.

By Theorem (4.4.12), T is diagonalizable.

Theorem 24. Let $f(X)$ and $m(X)$ be the characteristic polynomial and the minimal polynomial of a linear operator T on a finite dimensional vector space V_F. Then, $f(X)$ and $m(X)$ have same roots in F.

Proof. As $m(X)$ divides $f(X)$, obviously any root of $m(X)$ is a root of $f(X)$. conversely, let some $c \in F$ be a root of $f(X)$. Then, c is a characteristic value of T [Theorem (4.4.3)]. So, for some $\alpha \neq 0$ in V, $T\alpha = c\alpha$ i.e. $(T - cI)\alpha = 0$. The T-annihilator polynomial $m'(X)$ of α divides $X - c$.

As $\deg m'(X) \geq 1$, $m'(X) = X - c$.

As $m'(X) \mid m(X)$,

we get c is a root of $m(X)$.

Exercise 5.2

1. Find the minimal polynomials of the following matrices over **R**

$$\begin{bmatrix} 1 & 2 \\ 2 & 1 \end{bmatrix}, \begin{bmatrix} -1 & 0 \\ 0 & -1 \end{bmatrix}, \begin{bmatrix} 5 & 0 & 0 \\ 0 & 5 & 0 \\ 0 & 0 & 3 \end{bmatrix}, \begin{bmatrix} 0 & 0 & 0 \\ 1 & 0 & 1 \\ 0 & 1 & 0 \end{bmatrix}$$

$$\begin{bmatrix} 0 & 1 & 3 \\ 0 & 0 & 4 \\ 0 & 0 & 0 \end{bmatrix}, \begin{bmatrix} 0 & 1 & 0 & 0 \\ 1 & 0 & 0 & 0 \\ 0 & 0 & 1 & 2 \\ 0 & 0 & 2 & 1 \end{bmatrix}, \begin{bmatrix} 0 & 0 & 0 & 5 \\ 1 & 0 & 0 & -6 \\ 0 & 1 & 0 & 7 \\ 0 & 0 & 0 & 9 \end{bmatrix}$$

2. Let T be the linear operator on $\mathbf{R}^3$ such that
$$T \epsilon_1 = \epsilon_1 + \epsilon_2,\ T \epsilon_2 = 2\,\epsilon_1 \text{ and } T \epsilon_3 = 0.$$
   (*i*) Find the minimal polynomial and the characteristic polynomial of T.
   (*ii*) Find the T-annihilator polynomials of ϵ_1, ϵ_2 and ϵ_3. Use Theorem 18, to find the T-annihilator polynomial of $\epsilon_1 + \epsilon_3$.
   (*iii*) Give the T-primary decomposition of $\mathbf{R}^3$.

3. Let T be the linear operator on $\mathbf{C}^3$ such that
$$T \epsilon_1 = \epsilon_1 + \epsilon_2,\ T \epsilon_2 = 2\,\epsilon_1 + \epsilon_2 \text{ and } T \epsilon_3 = 0.$$
   Find the number of T-primary components of $\mathbf{C}^3$.

4. Let $T : \mathbf{R}^{3\times 1} \to \mathbf{R}^{3\times 1}$ be given by

$$T\begin{bmatrix} x \\ y \\ z \end{bmatrix} = \begin{bmatrix} -6x+4y-12z \\ 4x-4y+8z \\ 6x-4y+12z \end{bmatrix}.$$

(*i*) Show that $T^3 - 2T^2 - 8T = 0$

(*ii*) Find the minimal polynomial of T

(*iii*) Find the dimension of each T-primary component of T

(*iv*) Find a vector $\alpha \in \mathbf{R}^3$, such that the T- annihilator polynomial of α is the same as the minimal polynomial of T.

5. Let T be a linear operator on a finite dimensional vector space V_F, with minimal polynomial $m(X) = X^k + a_{k-1}X^{k-1} + ... + a_1 X + a_0$. Prove that T is invertible iff $a_0 \neq 0$.

6. Let $A = \begin{bmatrix} 1 & 0 & 0 & 0 \\ 0 & 0 & 1 & 0 \\ 0 & -1 & 0 & 0 \\ 0 & 0 & 0 & -1 \end{bmatrix}$. Find the minimal polynomial of A and the prime factors of this minimal polynomial in $\mathbf{Q}[X]$. Find the prime factors of the minimal polynomial of A in $\mathbf{C}[X]$. If T is the linear operator on $\mathbf{C}^4$ given by A, find the T-annihilators of ϵ_1, ϵ_2, ϵ_3, ϵ_4. Is A diagonalizable over $\mathbf{R}$?

7. Let S and T be two linear operators on a finite dimensional vector space V_F, such that S is invertible. Prove that T and $S^{-1}TS$ have the same minimal polynomials.

8. Give an example of two linear operators S and T on $\mathbf{R}^2$ such that ST and TS have different minimal polynomials.

9. Find all 2×2-matrices A over $\mathbf{R}$ such that the minimal polynomial of A is X^2.

10. For any two invertible linear operators S and T on a finite dimensional vector space V_F, prove that ST and TS have the same minimal polynomials.

11. For any idempotent linear operator T on V_F, what are the possible minimal polynomials of T?

12. Let T be a linear operator on a finite dimensional vector space V_F. Let $V = V_1 \oplus V_2 \oplus ... \oplus V_k$ be the T-primary decomposition of V. Let W be any T-invariant subspace of V, and T' be the restriction of T to W. For $W_i = W \cap V_i$, prove that $W = W_1 \oplus W_2 \oplus ... \oplus W_k$ give the T'-primary decomposition of W.

13. By using Theorem 23, find which of the matrices in Exercise 1 are diagonalizable?

5.3 Triangularization

A square matrix A over a field F is said to be *Triangularizable* over F if A is similar over F to a triangular matrix. The matrix $\begin{bmatrix} 1 & 1 \\ 0 & 1 \end{bmatrix}$ is not diagonalizable, but is obviously triangulizable. So, a triangulizable matrix need not be diagonalizable. Obviously, every diagonalizable matrix is triangulization. Given a triangular matrix $A = \left[a_{ij}\right]_{n \times n}$ we know that its characteristic polynomial is $f(X) = \prod_{i=1}^{n} (X - a_{ii})$. Consequently, the characteristic polynomial of a triangulizable matrix is a product of linear factors; so is its minimal polynomial. We shall prove the converse. A linear operator T on a finite dimensional, non-zero vector space V_F is said to be *triangularizable*, if for some ordered basis B of V, $[T]_B$ is a triangular matrix. Thus, the characteristic polynomial of a triangularizable linear operator is a product of linear factors.

Consider a linear operator T on V_F. Let V be made a left $F[X]$-module via T. Let W be a T-invariant subspace of V_F. Then, for any $f(X) \in F[X]$, W is $f(T)$-invariant [see the remark following Proposition (3.5.4)], so that for any $\alpha \in W$, $f(X)\alpha = f(T)\alpha \in W$. This gives that W is a submodule of V. Conversely, if W is a submodule of V, then, for any $\alpha \in W$, $T\alpha = X\alpha \in W$, yields W is a T-invariant subspace of V. So, we get the following Lemma :

Lemma 1. Let T be a linear operator on V_F and V be made a left $F[X]$-module via T. Then, a subspace W of V is T-invariant iff W is an $F[X]$-submodule of V.

Lemma 2. Let T be a linear operator on V_F and W be a T-invariant subspace of V. Then,

(*i*) The mapping $$\overline{T} : \frac{V}{W} \to \frac{V}{W}$$

such that $$\overline{T}(\overline{\alpha}) = \overline{T(\alpha)},\ \overline{\alpha} = \alpha + W \in \overline{V},$$

is a linear operator on $\overline{V} = \frac{V}{W}$. ($\overline{T}$ is said to be induced by T).

(*ii*) The minimal polynomial $\overline{m}(X)$ of $\overline{T}$ is a factor of the minimal polynomial $m(X)$ of T.

(*iii*) Let $\alpha \in V$ and $\overline{\alpha} = \alpha + W$. Then the $\overline{T}$-annihilator polynomial $\overline{q}(X)$ of $\overline{\alpha}$ is a factor of the T-annihilator polynomial $g(X)$ of α

Proof. By Theorem (5.2.2), $$\overline{V} = \frac{V}{W}$$

is a left $F[X]$-module such that for any $f(X) \in F[X]$,

$$\overline{\alpha} \in \overline{V},\ f(X)\overline{\alpha} = \overline{f(X)\alpha} = \overline{f(T)\alpha}$$

In particular, $$X\overline{\alpha} = \overline{T\alpha}$$

is well defined.

Then, $\overline{T}(\overline{\alpha}+c\overline{\beta}) = \overline{T}(\overline{\alpha+c\beta}) = \overline{T\alpha+cT\beta} = \overline{T}\overline{\alpha}+c\overline{T}\overline{\beta}$,

gives $\overline{T}$ is a linear operation on $\overline{V}$.

It follows by induction that $\overline{T}^i\ \overline{\alpha} = \overline{T^i\alpha}$ for every $i \geq 0$. This, in turn, yields that for any polynomial $f(X) \in F[X]$,

if $$f(X) = a_0 + a_1X + ... + a_kX^k$$

then, $$f(\overline{T})\overline{\alpha} = a_0\overline{\alpha} + a_1\overline{T}\overline{\alpha} + ... + a_k\overline{T}^k\overline{\alpha} = \overline{f(T)\alpha}.$$

Thus, in particular, as $m(T) = 0$, $m(\overline{T})\overline{\alpha} = \overline{m(T)\alpha} = \overline{0}$.

Hence, $\overline{m}(X)$ divides $m(X)$ by Theorem (5.2.11).

This proves (*ii*). For (*iii*), as $q(T)\alpha = 0$, we have $q(\overline{T})\overline{\alpha} = \overline{q(T)\alpha} = \overline{0}$.

Hence, by Theorem (5.2.16) $\overline{q}(X)$ divides $q(X)$.

Proposition 3. Let T be a linear operator on V_F and W be a T-invariant subspace of V such that $W \neq V$. Then, there exists a prime factor $p(X)$ of the minimal polynomial $m(X)$ of T and an $\alpha \in V - W$ such that $p(X)\alpha \in W$ and for any $f(X) \in F[X]$, $f(X)\alpha \in W$ iff $p(X)|f(X)$.

Proof. Consider the linear operator $\overline{T}$ on $\overline{V} = V/W$ induced by T. Let $\overline{m}(X)$ be the minimal polynomial of $\overline{T}$.

As $$\overline{V} \neq \{\overline{0}\},\ \deg \overline{m}(X) > 0.$$

So, $\overline{m}(X)$ has a prime factor say $p(X)$.

As $$\overline{m}(X)|m(X),$$

we get $$p(X)|m(X).$$

Write $$\overline{m}(X) = p(X)q(X).$$

Then, $$q(\overline{T}) \neq 0.$$

So for some $$\overline{\beta} = \beta + W \in \overline{V},$$

$$q(T)\overline{\beta} \neq \overline{0},$$

i.e., $$\overline{q(T)\beta} \neq \overline{0}.$$

Consequently, $$\alpha = q(T)\beta \in V - W.$$

Then, $$p(\overline{T})\overline{\alpha} = m(\overline{T})\overline{\beta} = \overline{0}$$

yields $$p(T)\alpha \in W.$$

Let $m'(X)$ be the $\overline{T}$-annihilator polynomial of $\overline{\alpha}$.

As $\quad \overline{\alpha} \neq \overline{0},\ m'(X) \neq 1.$

As $\quad m'(X) \mid p(X),$

we get $\quad m'(X) = p(X).$

For $\quad f(X) \in F[X],\ f(X)\alpha \in W$

iff $\quad f(\overline{T})\overline{\alpha} = \overline{0}$

iff by Theorem (5.2.16) $\quad p(X) \mid f(X).$

In this theorem if $m(X)$ is a product of linear factors, then, the prime factors of $m(X)$ are of the type $X - c,\ c \in F$, so that $p(X) = X - c$ for some $c \in F$. Clearly, c being a root of $m(X)$, is a characteristic value of *T*. This gives us a corollary.

Corollary 4. Let *T* be a linear operator on V_F such that its minimal polynomial factors into linear factors over *F*. Let *W* be a *T*-invariant subspace *V* such that $W \neq V$. Then, there exists a factor $X - c$ of $m(X)$ and an $\alpha \in V - W$ such that $T\alpha - c\alpha \in W$. Further, for any $f(X) \in F[X]$ if $f(T)\alpha \in W$, then, $f(c) = 0$.

Any non-constant polynomial over **C** factors into linear factors over **C**. So, we get the following.

Corollary 5. Let *T* be a linear operator on $V_{\mathbf{C}}$ and *W* be a *T*-invariant subspace of *V* such that $W \neq V$. Then, there exists an $\alpha \in V - W$ and a characteristic value c of *T* such that $T\alpha - c\alpha \in W$.

Theorem 6. Let *T* be a linear operator on a finite dimensional vector space V_F. Then, *T* is triangularizable iff the minimal polynomial of *T* is a product of linear factors.

Proof. "Only if" has already been discussed in the introduction to this section. Conversely, let the minimal polynomial $m(X)$ of *T* be given by

$$m(X) = \prod_{i=1}^{s} (X - d_i),\ d_i \in F.$$

Let $\dim V = n$. We construct an ordered basis $B = \{\alpha_1, \alpha_2, \ldots, \alpha_n\}$ of *V* such that for $1 \leq i \leq n$, if W_i is the subspace of *V* spanned by $\alpha_1, \alpha_2, \ldots, \alpha_i$, then W_i is *T*-invariant, and for $i < n$, there exists a characteristic value c_{i+1} of *T* such that $T\alpha_{i+1} - c_{i+1}\alpha_{i+1} \in W_i$. Let c_1 be any root of $m(X)$. Then, $c_1 = d_j$ for some *j* and it is a characteristic value of *T*. Choose α_1 any characteristic vector of *T* relative to c_1.

Consider $\quad W_1 = \langle \alpha_1 \rangle.$

Given $\quad \alpha \in W_1,\ \alpha = a\alpha_1,$

yields $\quad T\alpha = aT\alpha_1 = ac_1\alpha_1 \in W_1,$

so that W_1 is *T*-invariant.

Suppose for some, i, $1 < i < n$, we have constructed i linearly independent vectors α_1, $\alpha_2, \ldots, \alpha_i$ such that the subspace W_i spanned by them is T-invariant. By Corollary 4, there exists a characteristic value c_{i+1} [indeed a root of $m(X)$] and $\alpha_{i+1} \in V - W_i$ such that $T\alpha_{i+1} - c_{i+1}\alpha_{i+1} \in W_i$.

Then, $\alpha_1, \alpha_2, \ldots, \alpha_{i+1}$ are linearly independent.

Let W_{i+1} be the subspace spanned by these $i+1$ vectors.

Then, $$W_i \underset{\neq}{\subset} W_{i+1}.$$

Consider $$\alpha \in W_{i+1}.$$

Then, $$\alpha = \beta + c\alpha_{i+1}, \beta \in W_i, \; c \in F,$$

$$T\alpha = cT\alpha_{i+1} + T\beta = [T\beta + c(T\alpha_{i+1} - c_{i+1}\alpha_{i+1})] + cc_{i+1}\alpha_{i+1}.$$

As $\beta \in W_i$, $$T\beta \in W_i \subset W_{i+1}.$$

Also. by the construction $$T\alpha_{i+1} - c_{i+1}\alpha_{i+1} \in W_i.$$

Consequently, $$T\alpha \in W_{i+1},$$

hence W_{i+1} is T-invariant.

Now, $$T\alpha_1 = c_1\alpha_1 = c_{11}\alpha_1, \; c_{11} = c_1. \quad \ldots (1)$$

For $1 \le i \le n$, $$T\alpha_i \in W_i,$$

yields $$T\alpha_i = \sum_{j=1}^{i} c_{ji}\,\alpha_j \; . \quad \ldots (2)$$

(1) and (2) yield $$[T]_B = \begin{bmatrix} c_{11} & c_{12} & c_{13} & \ldots & c_{1n} \\ 0 & c_{22} & c_{23} & \ldots & c_{2n} \\ 0 & 0 & c_{33} & \ldots & c_{3n} \\ .. & .. & .. & \ldots & .. \\ .. & .. & .. & \ldots & .. \\ 0 & 0 & 0 & \ldots & c_{nn} \end{bmatrix}$$

an upper triangular matrix.

Since any polynomial over **C** is a product of linear factors, we get the following corollary :

Corollary 7. Let V be a finite dimensional vector space over **C**. Then, any linear operator on V is triangularizable. Any square matrix A over **C**, is similar over **C**, to an upper triangular matrix.

In most of the applications, one studies vector spaces over **C**. Let T be a linear operator on V_C such that the minimal polynomial of T is $m(X) = \prod_{i=1}^{k} (X - c_i)^{r_i}$ ...(1)

for some distinct complex numbers $c_1, c_2, \ldots, c_k$, and $r_i \geq 1$.

We have the primary decomposition of V relative to T.

$$V = V_1 \oplus V_2 \oplus \ldots \oplus V_k \quad \ldots (2)$$

where the restriction T_i of T to V_i has minimal polynomial $(X - c_i)^{r_i}$

Thus, if $N_i = (T - c_i I)$, its restriction K_i to V_i satisfies $K_i^{r_i} = 0$. Let $E_1, E_2, \ldots, E_k$ be the collection of idempotent linear operators that determine (2).

By Theorem (3.5.5), $\quad TE_i = E_i T$

such that $\quad N_i E_i = TE_i - c_i IE_i = E_i (T - c_i I) = E_i N_i$.

Consider $\quad N = N_1 E_1 + N_2 E_2 + \ldots + N_k E_k \quad \ldots (3$

As $\quad E_i E_j = 0$

for $i \neq j$ and $E_i^2 = E_i$, (3) gives $\quad N^t = N_1^t E_1 + N_2^t E_2 + \ldots + N_k^t E_k$

for every $t \geq 1$.

For $\quad t = \max(r_1, r_2, \ldots, r_k)$,

the fact that the restriction of N_i^t to V_i is zero and that $N_i^t E_i$ is zero on all V_j with $j \neq i$,

yields $\quad N^t = 0 \quad \ldots (4)$

so that N is nilpotent.

$$D = c_1 E_1 + c_2 E_2 + \ldots + c_k E_k \quad \ldots (5)$$

is diagonalizable on V.

It can be easily seen that $ND = DN$.

Now, $\quad T = TE_1 + TE_2 + \ldots + TE_k$

$$= (TE_1 - cE_1) + \ldots + (T - cE_2) + cE_1 + \ldots + cE_k$$

$$= N + D.$$

Infact, more generally, we get the following theorem :

Theorem 8. Let T be a triangularizable linear operator on V_F (T could be any linear operator in case $F = \mathbf{C}$). Then, $T = D + N$ for some diagonalizable linear operator D and some nilpotent linear operator N on V_F, such that $ND = DN$.

Example 8.a. Consider $\quad A = \begin{bmatrix} 0 & 1 & 0 \\ 2 & -2 & 2 \\ 2 & -3 & 2 \end{bmatrix}$.

Then $\quad \det(XI_3 - A) = X^3$.

By Theorem 6, A is triangularizable. Its only characteristic value is 0.

Now $\alpha_1 = \begin{bmatrix} 1 \\ 0 \\ -1 \end{bmatrix}$ is a characteristic vector of A.

Let T be the linear operator on $\mathbf{R}^{3\times 1}$ determined by A. Then, $W_1 = \langle \alpha_1 \rangle$ is T-invariant.

Now, $$\epsilon_1 \notin W_1,\ T\epsilon_1 = \begin{bmatrix} 0 \\ 2 \\ 2 \end{bmatrix} \notin W_1,$$

but $$T^2\epsilon_1 = \begin{bmatrix} 2 \\ 0 \\ -2 \end{bmatrix} \in W_1.$$

Let $$\alpha_2 = \begin{bmatrix} 0 \\ 2 \\ 2 \end{bmatrix}.$$

Clearly, $$T\alpha_2 - 0\alpha_2 \in W_1.$$

Again, $$\epsilon_1 \notin W_2 = \langle \alpha_1, \alpha_2 \rangle,$$

but $$T\epsilon_1 \in W_2.$$

Choose $$\alpha_3 = \epsilon_1.$$

Consider the ordered basis $B = \{\alpha_1, \alpha_2, \alpha_3\}$ of $\mathbf{R}^{3\times 1}$.

Then, $$T\alpha_1 = 0 = 0\alpha_1 + 0\alpha_2 + 0\alpha_3$$
$$T\alpha_2 = 2\alpha_1 = 2\alpha_1 + 0\alpha_2 + 0\alpha_3$$
$$T\alpha_3 = \alpha_2 = 0\alpha_1 + 1\alpha_2 + 0\alpha_3$$

yields $[T]_B = \begin{bmatrix} 0 & 2 & 0 \\ 0 & 0 & 1 \\ 0 & 0 & 0 \end{bmatrix}$ an upper triangular matrix.

As X^3 has repeated roots and is the minimal polynomial of A, the matrix A is not diagonalizable. Here, as T itself is nilpotent, $T = N + D$ is satisfied by $N = T$ and $D = 0$.

Example 8b. Consider $$A = \begin{bmatrix} 0 & -1 \\ 1 & -2 \end{bmatrix}.$$

This being the companion matrix of $m(X) = (X+1)^2$ is triangularizable. Its only characteristic value is -1, $\alpha_1 = \begin{bmatrix} 1 \\ 1 \end{bmatrix}$ is a characteristic vector, $\epsilon_1 \notin W_1 = \langle \alpha_1 \rangle$ in $\mathbf{R}^{2\times 1}$ but $A\epsilon_1 + \epsilon_1 \in W_1$.

Choose $\alpha_2 = \epsilon_1$.

Let $$P = \begin{bmatrix} 1 & 1 \\ 1 & 0 \end{bmatrix}.$$

Then, $$P^{-1}AP = \begin{bmatrix} -1 & 1 \\ 0 & -1 \end{bmatrix}$$

an upper triangular matrix.

$$P^{-1}AP = D' + N',$$

where $$D' = \begin{bmatrix} -1 & 0 \\ 0 & -1 \end{bmatrix},\ N' = \begin{bmatrix} 0 & 1 \\ 0 & 0 \end{bmatrix}.$$

Clearly, N' **is nilpotent,** D' **is a diagonal matrix and** $D'N' = N'D'$.

Then, $$A = D + N$$

where $$D = PD'P^{-1} = D',$$

$$N = PN'P^{-1} = \begin{bmatrix} 1 & -1 \\ 1 & -1 \end{bmatrix}.$$

Exercise 5.3

1. **Which of the following matrices are triangularizable over R? If any one is triangularizable, find an upper triangular form similar to that matrix.**

$$\begin{bmatrix} 1 & 4 \\ 1 & -2 \end{bmatrix}, \begin{bmatrix} 0 & 2 \\ -2 & 2 \end{bmatrix}, \begin{bmatrix} 0 & 2 \\ -2 & 4 \end{bmatrix}$$

$$\begin{bmatrix} 1 & 1 & 0 \\ 0 & 1 & 2 \\ 1 & 0 & 2 \end{bmatrix}, \begin{bmatrix} 0 & 1 & 0 & 1 \\ 1 & 0 & 1 & 0 \\ 0 & 1 & 0 & 1 \\ 1 & 0 & 1 & 1 \end{bmatrix}.$$

2. **If a** 2×2**-matrix over R is not triangularizable over R, prove that it is diagonalizable over C.**

3. If a matrix over a field F is triangularizable, prove that A^T, $f(A)$ for any $f(X) \in F[X]$, are triangularizable over F.

4. **Prove that any lower triangular matrix over a field** F **is similar to an upper triangular matrix over** F.

5. **Give an example of** 2×2**-matrices** A **and** B **over R, such that** $A + B$ **is not triangularizable ; however,** A **and** B **are triangularizable.**

6. **If two** $n \times n$**-matrices** A **and** B **over a field** F **are triangularizable and** $AB = BA$, **prove that** AB **is triangularizable.**

7. Find an invertible matrix P such

$$P^{-1}\begin{bmatrix} 0 & 0 & 0 \\ 1 & 0 & 0 \\ 0 & 1 & 0 \end{bmatrix} P = \begin{bmatrix} 0 & 1 & 0 \\ 0 & 0 & 1 \\ 0 & 0 & 0 \end{bmatrix}.$$

8. If any matrix A in Exercise 1 is triangularizable over **C**, find corresponding non-singular matrix P over **C** such that $P^{-1}AP$ is upper triangular.

9. Given example of 2×2-matrices A and B which are triangularizable over **R**, but AB is not triangularizable over **R**.

[Hint : $A = \begin{bmatrix} 2 & 0 \\ 1 & 3 \end{bmatrix}, B = \begin{bmatrix} 1 & -10 \\ 0 & 3 \end{bmatrix}$**]**

10 Give example of 2×2-matrices A and B over **R** that are triangularizable over **R**, such that AB is triangularizable, but $AB \neq BA$.

5.4 Canonical Forms

Throughout, V_F is a non-zero finite dimensional vector space over a field F. Given an $n \times n$-matrix A over F, can we find some $n \times n$-matrix B similar to A, which can easily describe the various properties A, viz., its characteristic polynomial, minimal polynomial, primary decomposition of F^n given by A? There are various such forms of matrices, known as *canonical forms*. Let T be a linear operator on V_F. Consider V a left $F[X]$- module via T. For any $\alpha \neq 0$, $F[X]\alpha = F(T)\alpha$ is the cyclic submodule of V generated by α. By Lemma (5.3.1), $W = F[X]\alpha$ is a T-invariant subspace of V. W consists of all vectors of the form $f(T)\alpha, f(X) \in F[X]$. Such a W is called a *T-cyclic subspace* of V. W is the smallest T-invariant subspace of V that contains α. If for some $\alpha \in V, V = F[X]\alpha$, V is called a *T-cyclic space* and any such α is called a *T-generator* of V.

Example 0.a. Let T be the linear operator on $\mathbf{R}^3$, given by

$$A = \begin{bmatrix} 0 & 0 & 0 \\ 1 & 0 & 0 \\ 0 & 1 & 0 \end{bmatrix}.$$

Consider $\alpha = \epsilon_1,\ W = F[T]_\alpha$.

As $T\alpha = \epsilon_2$

and $T^2\alpha = T\epsilon_2 = \epsilon_3$, W

contains the standard basis $\{\epsilon_1, \epsilon_2, \epsilon_3\}$ of $\mathbf{R}^3$, so that $\mathbf{R}^3 = \mathbf{R}[X]\epsilon_1$.

Hence, $\mathbf{R}^3$ is T-cyclic with ϵ_1 as a T-generator.

Consider $\beta = \epsilon_2$.

Then, $$T\beta = \epsilon_3$$

and $$T\epsilon_3 = 0.$$

So that $F[T]\beta$ contains ϵ_2 and ϵ_3.

Thus, $$W' = \langle \epsilon_2, \epsilon_3 \rangle \subseteq F[T]\beta.$$

At the same time for $\alpha' \in W'$, $\alpha' = a_2 \epsilon_2 + a_3 \epsilon_3$,

$$T\alpha' = a_2 \epsilon_3 \in W',$$

yields W' is T-invariant.

As $F[T]\beta$ is the smallest T-invariant subspace that contains $\beta = \epsilon_2$, we get $W' = F[T]\beta$. Obviously $W' \neq \mathbf{R}^3$, so that β is not a T-generator of $\mathbf{R}^3$. Still we say, β is a T-generator of W'.

Lemma 1. Let T be a linear operator on V_F, with minimal polynomial $m(X)$. Let $\{\alpha_1, \alpha_2, ..., \alpha_k\}$ be any subset of V that generates V_F. Then,

(*i*) $m(X)$ is the 1cm of T-annihilator polynomials of α_i, $1 \leq i \leq k$

(*ii*) Given a prime factor $p(X)$ of $m(X)$ of multiplicity $r \geq 1$, it is a prime factor of T-annihilator polynomial of some α_i, of multiplicity r.

Proof.

(*i*) Let $m_i(X)$ be the T-annihilator polynomial of α_i. For any $c \in F$ with $c \neq 0$, and for $g(X) \in F[X]$, $g(T)(c\alpha_i) = cg(T)\alpha_i = 0$ iff $g(T)\alpha_i = 0$. Hence, $m_i(X)$ is the T-annihilator polynomial of $c\alpha_i$. In case $c = 0$, $c\alpha_i = 0$, has 1 as its T-annihilator polynomial, so that in any case the T-annihilator polynomial of a $c_i\alpha_i$, $c_i \in F$ is a factor of $m_i(X)$.

Consider $$\alpha \in V.$$

Then, $$\alpha = \sum_{i=1}^{k} c_i\alpha_i, \ c_i \in F.$$

By Corollary (5.2.19), the T-annihilator polynomial of α is a factor of the 1cm of the T-annihilator polynomials $n_i(X)$ of $c_i\alpha_i$. As $n_i(X)|m_i(X)$, we get the T-annihilator polynomial of α is a factor of $m'(X)$, the 1cm of $m_i(X)$'s.

Consequently, $$m'(T)\alpha = 0 \quad \text{for every } \alpha \in V.$$

So that $$m'(T) = 0$$

and $$m(X)|m'(X).$$

On the other hand, as every $m_i(X)|m(X)$, we get

$$m'(X)|m(X).$$

Hence, $$m(X) = m'(X)$$

(*ii*) Let $p_1(X), p_2(X), ..., p_s(X)$ be the distinct prime factors of $m(X)$.

Then, $m(X) = p_1(X)^{r_1} p_2(X)^{r_2} \dots p_s(X)^{r_s}$

$$m_i(X) = p_1(X)^{r_{i1}} p_2(X)^{r_{i2}} \dots p_s(X)^{r_{is}}$$

such that $r_j \geq 1$

and $0 \leq r_{ij} \leq r_j$ for $1 \leq i < k$.

Now, $p(X) = p_i(X)$ for some i, say $i = 1$.

Then, $p(X) = p_1(X)$

and $r = r_1$.

Now, $m(X) = \text{lcm}[m_1(X), m_2(X), \dots, m_k(X)]$,

$$r_j = \max(r_{1j}, r_{2j}, \dots, r_{kj}).$$

In particular, $r = r_1 = \max(r_{11}, r_{21}, \dots, r_{k1})$.

Consequently, for some i, say for $i = i'$, $r = r_{i1}$. Hence $p(X)$ is a factor of $m_i(X)$, of multiplicity r.

Proposition 2. Let T be a linear operator on a finite dimensional vector space V_F. Then, there exists an $\alpha \in V$ such that the T-annihilator polynomial of α is the same as the minimal polynomial of T.

Proof. Let $m(X)$ be the minimal polynomial of T. As the T-annihilator polynomial of every member of V is a factor of $m(X)$ [Theorem (5.2.16)], we get an $\alpha \in V$ having largest degree T-annihilator polynomial $m'(X)$. Suppose that $m'(X) \neq m(X)$. However, $m'(X) | m(X)$. So, we can find a prime $p(X) \in F[X]$ such that the multiplicity of $p(X)$, as a factor of $m(X)$ is some $k \geq 1$, but $p(X)$ is a factor of $m'(X)$, of multiplicity $l < k$.

Write $m'(X) = m''(X)p(X)^l$.

Then, $\gcd[m''(X), p(X)] = 1$.

By Theorem (5.2.18), the T-annihilator polynomial $m_1(X)$ of $\alpha' = p(T)^l\alpha$, equals $m''(X)$. By Lemma 1, there exists $\beta \in V$ such that $p(X)$ is a factor of the T-annihilator polynomial $m_2(X)$ of β, of multiplicity k.

Then $m_2(X) = g(X)p(X)^k$

with $[p(X), g(X)] = 1$.

Consequently, $\beta' = g(T)\beta$ has $p(X)^k$ as its T-annihilator polynomial. By Theorem (5.2.18) (*iii*) $\alpha_1 = \alpha' + \beta'$ has T-annihilator polynomial $h(X) = m''(X)p^k(X)$ of degree greater than deg $m'(X)$. This contradicts the choice of α.

Hence, $m'(X) = m(X)$.

Lemma 3. Let T be a linear operator on V_F, and $\alpha \in V$. Then, the dimension of the T-cyclic subspace $F[T]\alpha$ equals the degree of the T-annihilator polynomial of α.

Proof. Let $m(X)$ be the T-annihilator polynomial of α, and $\deg m(X) = s$. If $\alpha = 0$, $m(X) = 1$, then, $[F(T)\alpha] = \dim[\{0\}] = 0 = \deg[m(X)]$.

Let $\alpha \neq 0$.

Then, $\dim[F(T)\alpha] \geq 1$.

Consider the s vectors $\alpha, T\alpha, \ldots, T^{s-1}\alpha$

in $W = F[T]\alpha$.

If for some $a_0, a_1, \ldots, a_{s-1} \in F$,

not all zeros, $a_0\alpha + a_1 T\alpha + \ldots + a_{s-1}T^{s-1}\alpha = 0$,

then
$$m'(X) = \sum_{i=0}^{s-1} a_i X^i$$
is a non-zero polynomial of degree less than s,

such that $m'(T)\alpha = 0$.

This is a contradiction, as $m(X) \mid m'(X)$,

gives $s \leq \deg m'(X)$.

Hence, $B = \{\alpha, T\alpha, \ldots, T^{s-1}\alpha\}$

is a linearly independent subset of W.

Consider any $\beta \in W$.

Then $\beta = f(T)\alpha$,

$f(X) \in F[X]$.

By the division algorithm $f(X) = q(X)m(X) + r(X)$,

$q(X), r(X) \in F[X]$

with $r(X) = 0$

or $\deg r(X) < s$.

Consequently, $r(X) = b_0 + b_1(X) + \ldots + b_{s-1}X^{s-1}$,

$b_i \in F$

and
$$\begin{aligned}\beta &= f(T)\alpha \\ &= q(T)m(T)\alpha + r(T)\alpha \\ &= r(T)\alpha, \qquad \text{as } m(T)\alpha = 0 \\ &= b_0\alpha + b_1 T\alpha + \ldots + b_{s-1}T^{s-1}\alpha.\end{aligned}$$

This proves that B spans W. Hence, B is a basis of W.

This proves that $\dim[F[T]\alpha] = s = \deg m(X)$

The proof of this Lemma gives :

Lemma 4. Let T be a linear operator on V_F. For any $\alpha \neq 0$ in V, if the degree of T-annihilator polynomial of α is s, then, $B = \{\alpha, T\alpha, T^2\alpha, ..., T^{s-1}\alpha\}$ is a basis of $F[T]\alpha$. Further, each $\beta \in F[T]\alpha$ is expressible as $f(T)\alpha$, for a uniquely determined $f(X)$ of degree less than s.

Theorem 5. For any linear operator T on V_F with $f(X)$ and $m(X)$ as its characteristic polynomial and minimal polynomial respectively, the following are equivalent :

(*i*) $f(X) = m(X)$

(*ii*) V is T-cyclic

(*iii*) There exists an ordered basis B of V such that $[T]_B$ is the companion matrix of $f(X)$.

Proof. Let $\dim V = n$.

Then $\deg f(X) = n$.

Let $f(X) = m(X)$.

By Proposition 2, there exists $\alpha \in V$, whose T-annihilator polynomial is $f(X)$.

By Lemma 3, $\dim[F[T]\alpha] = \deg[m(X)] = n = \dim V$.

Hence, $V = F[T]\alpha$, a T-cyclic space. Thus, (*i*) implies (*ii*). Let $V = F[T]\alpha$. Then, the degree of the T-annihilator polynomial $m'(X)$ of α is n.

As $$m'(X) | m(X) | f(X)$$

and $$\deg f(X) = n,$$

we get $$f(X) = m(X) = m'(X).$$

By Lemma 4, $$B = \{\alpha, T\alpha, ..., T^{n-1}\alpha\}$$

is an ordered basis of V.

Write $$\alpha_i = T^{i-1}\alpha,$$

$$1 \le i \le n,$$

$$f(X) = X^n + a_{n-1}X^{n-1} + ... + a_0.$$

Then, $$T\alpha_i = \alpha_{i+1},\ 1 \le i \le n-1 \quad ... (1)$$

$$f(T)\alpha = 0,$$

yields $$T\alpha_n = T^n\alpha$$

$$= -a_0\alpha - a_1T\alpha - a_2T^2\alpha - ... - a_{n-1}T^{n-1}\alpha$$

$$= -a_0\alpha_1 - a_1\alpha_2 - a_2\alpha_3 - \ldots - a_{n-1}\alpha_n.$$

This gives $$[T]_B = \begin{bmatrix} 0 & 0 & 0 & \ldots & & -a_0 \\ 1 & 0 & 0 & \ldots & 0 & -a_1 \\ 0 & 1 & 0 & \ldots & 0 & -a_2 \\ \ldots & \ldots & \ldots & \ldots & & \ldots \\ \ldots & \ldots & \ldots & \ldots & & \ldots \\ 0 & 0 & 0 & \ldots & 1 & -a_{n-1} \end{bmatrix} = \text{com}[f(X)]$$

This proves, (*ii*) implies (*iii*). Finally, to prove, (*iii*) implies (*i*), let V have an ordered basis B such that $[T]_B = \text{com}[f(X)]$. By Theorem (5.2.14), the minimal polynomial of $[T]_B$ is $f(X)$.

Hence, $$f(X) = m(X).$$

As the companion matrix of X^n is

$$\begin{bmatrix} 0 & 0 & 0 & \ldots & 0 & 0 \\ 1 & 0 & 0 & \ldots & 0 & 0 \\ 0 & 1 & 0 & \ldots & 0 & 0 \\ \ldots & \ldots & \ldots & \ldots & \ldots & \ldots \\ \ldots & \ldots & \ldots & \ldots & \ldots & \ldots \\ 0 & 0 & 0 & \ldots & 1 & 0 \end{bmatrix}$$

We get the following result :

Corollary 6. If a linear operator T on V_F is such that

$$T^n = 0,\ T^{n-1} \neq 0,\ n = \dim V,$$

then, there exists an ordered basis B of V such that $[T]_B$ is the $n \times n$-matrix

$$C_n = \begin{bmatrix} 0 & 0 & 0 & \ldots & 0 & 0 \\ 1 & 0 & 0 & \ldots & 0 & 0 \\ 0 & 1 & 0 & \ldots & 0 & 0 \\ \ldots & \ldots & \ldots & \ldots & \ldots & \ldots \\ \ldots & \ldots & \ldots & \ldots & \ldots & \ldots \\ 0 & 0 & 0 & \ldots & 1 & 0 \end{bmatrix}.$$

Any $n \times n$-matrix A over F satisfying $A^n = 0, A^{n-1} \neq 0$ is similar to the matrix C_n. Conversely, any $n \times n$-matrix A similar to C_n satisfies $A^n = 0, A \neq 0$.

Theorem 7. Let T be a linear operator on V_F and $\dim V_F = n$. Then, the minimal polynomial of T is $(X - c)^n$ for some $c \in F$ iff there exists an ordered basis B of V such that

$$[T]_B = \begin{bmatrix} c & 0 & 0 & \dots & 0 & 0 \\ 1 & c & 0 & \dots & 0 & 0 \\ 0 & 1 & c & \dots & 0 & 0 \\ \dots & \dots & \dots & \dots & \dots & \dots \\ \dots & \dots & \dots & \dots & \dots & \dots \\ 0 & 0 & 0 & \dots & 1 & c \end{bmatrix} \quad \dots (1)$$

Proof. Let $T' = T - cI$.

Let $(X-c)^n$ be the minimal polynomial of T.

Then, $(T-cI)^n = 0$

but $(T-cI)^{n-1} \neq 0$

i.e., $(T')^n = 0$,

but $(T')^{n-1} \neq 0$.

By Corollary 6, there exists an ordered basis B such that $[T']_B = C_n$.

As $T = T' + cI$

$$[T]_B = [T']_B + [cI]_B = C_n = cI_n = \begin{bmatrix} c & 0 & 0 & \dots & 0 & 0 \\ 1 & c & 0 & \dots & 0 & 0 \\ 0 & 1 & c & \dots & 0 & 0 \\ \dots & \dots & \dots & \dots & \dots & \dots \\ \dots & \dots & \dots & \dots & \dots & \dots \\ 0 & 0 & 0 & \dots & 1 & c \end{bmatrix}.$$

Conversely, if $[T]_B$ is given by (1),

then, for $T' = T - cI$, $[T']_B = C_n$.

By Corollary 6, $(T')^n = 0$

and $(T')^{n-1} \neq 0$.

Hence, $(X-c)^n$ is the minimal polynomial of T.

Consider $A = \begin{bmatrix} 0 & -1 \\ 1 & 2 \end{bmatrix}$.

Being the companion matrix of $(X-1)^2$, its minimal polynomial is $m(X) = (X-1)^2$.

By Theorem 7, A is similar to $A' = \begin{bmatrix} 1 & 0 \\ 1 & 1 \end{bmatrix}$.

Indeed, $A' = P^{-1}AP$,

$$P = \begin{bmatrix} 1 & -1 \\ 0 & 1 \end{bmatrix}.$$

Consider the linear operator T on $\mathbf{R}^2$ such that

$$T\in_1=\frac{1}{2}(\in_1-\in_2)=T\in_2.$$

Then, $$T\neq 0,$$

but $$T^2=0.$$

Let $$\alpha_1=\frac{1}{2}(\in_1+\in_2),$$

$$\alpha_2=\frac{1}{2}(\in_1-\in_2).$$

Then, $$T\alpha_1=\alpha_2$$

and $$T\alpha_2=0.$$

Clearly, the matrix of T relative to the ordered basis $B\{\alpha_1,\alpha_2\}$ is $\begin{bmatrix}0&0\\1&0\end{bmatrix}$.

For $$n=3,\ c=-2,$$

the matrix in Theorem 7 is $$\begin{bmatrix}-2&0&0\\1&-2&0\\0&1&-2\end{bmatrix}.$$

Its minimal polynomial is $(X+2)^3$.

We are going to prove that any complex $n\times n$-matrix is a direct sum of the types of matrices given by Theorem 7. The theorem, given above and the results on direct sums of matrices proved before, immediately determine the minimal polynomial and the characteristic polynomial of such a direct sum.

Definition 8. for any $n\geq 1$, and any $c\in F$, the $n\times n$-matrix

$$J_n(c)=\begin{bmatrix}c&0&0&\cdots&0&0\\1&c&0&\cdots&0&0\\0&1&c&\cdots&0&0\\\vdots&\vdots&\vdots&\cdots&\vdots&\vdots\\\vdots&\vdots&\vdots&\cdots&\vdots&\vdots\\0&0&0&\cdots&1&c\end{bmatrix}$$

is called an *elementary Jordan-Matrix* with characteristic value c.

Here, $$J_1(c)=[c],\ J_2(c)=\begin{bmatrix}c&0\\1&c\end{bmatrix}$$

$$J_4(i)=\begin{bmatrix}i&0&0&0\\1&i&0&0\\0&1&i&0\\0&0&1&i\end{bmatrix},\ i=\sqrt{-1}.$$

Example 8.a. Consider $A = \begin{bmatrix} 0 & 0 & 0 & -1 \\ 1 & 0 & 0 & 0 \\ 0 & 1 & 0 & -2 \\ 0 & 0 & 1 & 0 \end{bmatrix}$.

It being a companion matrix of $(X^2+1)^2$, has characteristic polynomial $f(X) =$ minimal polynomial $m(X) = (X^2+1)^2 = (X+i)^2(X-i)^2$ over **C**. By the primary decomposition theorem over **C**, A is similar to $A_1 \oplus A_2$ with minimal polynomials of A_1, A_2 given, respectively, by

$$m_1(X) = (X+i)^2,$$

$$m_2(X) = (X-i)^2.$$

By Theorem 7, A_1 is similar to $\begin{bmatrix} -i & 0 \\ 1 & -i \end{bmatrix}$ and A_2 is similar to $\begin{bmatrix} i & 0 \\ 1 & i \end{bmatrix}$.

Hence, A is similar over **C**, to

$$\begin{bmatrix} -i & 0 \\ 1 & -i \end{bmatrix} \oplus \begin{bmatrix} i & 0 \\ 1 & i \end{bmatrix} = \begin{bmatrix} -i & 0 & 0 & 0 \\ 1 & -i & 0 & 0 \\ 0 & 0 & i & 0 \\ 0 & 0 & 1 & i \end{bmatrix}.$$

Example 8.b. Consider matrix

$$A = \begin{bmatrix} 3 & 0 \\ 1 & 3 \end{bmatrix} \oplus \begin{bmatrix} 3 & 0 \\ 1 & 3 \end{bmatrix} \oplus \begin{bmatrix} -5 & 0 & 0 \\ 1 & -5 & 0 \\ 0 & 1 & -5 \end{bmatrix}.$$

It is a direct sum of three elementary Jordan matrices $J_2(3)$, $J_2(3)$, $J_3(-5)$, with respective minimal (as well as characteristic polynomials) $(X-3)^2$, $(X-3)^2$ and $(X+5)^3$; their respective characteristic values are 3, 3 and –5, so that A has two distinct characteristic values.

The characteristic polynomial of A is

$$f(X) = (X-3)^2(X-3)^2(X+5)^3 = (X-3)^4(X+5).$$

By Theorem (5.2.15), the minimal polynomial of A is

$$m(X) = \text{lcm} = \left[(X-3)^2,\ (X-3)^2,\ (X+5)^3\right] = (X-3)^2(X+5)^3$$

so that

$$f(X) \neq m(X).$$

Consequently, $\mathbf{R}^7$ is not a cyclic T-space, where T is the linear operator on $\mathbf{R}^7$ given by A.

Theorem 9. Let T be a linear operator on a finite dimensional vector space V_F. Let $\alpha \in V$ be such that the T-annihilator polynomial of α is same as the minimal polynomial of T. Then, there exists a T-invariant subspace W of V such that $V = F[T]\alpha \oplus W$.

Proof. Now, $W' = \{0\}$ is a T-invariant subspace of V satisfying $W' \cap F[T]\alpha = \{0\}$. As $\dim V$ is finite, we can find largest dimensional T-invariant subspace W of V satisfying $F[T]\alpha \cap W = \{0\}$. We prove that $V = F[T]\alpha \oplus W$. Suppose the contrary. Then, for $V' = F[T]\alpha + W$, $V' \neq V$. Let $m(X)$ be the minimal polynomial of T. By Proposition (5.3.3), there exists $\beta \in V - V'$ and a prime factor $p(X)$ of $m(X)$ such that $p(T)\beta \in V'$.

Then, $$p(T)\beta = q(T)\alpha + \beta' \qquad \text{... (1)}$$

for some $$q(X) \in F[X],\ \beta' \in W$$

Write $$m(X) = p(X)m_1(X).$$

Then, (1) gives $$0 = m(T)\beta = m_1(T)p(T)\beta = m_1(T)q(T)\alpha + m_1(T)\beta'$$

Consequently, $$m_1(T)q(T)\alpha = -m_1(T)\beta' \in F[T]\alpha \cap W = \{0\}.$$

Hence, $$m_1(T)q(T)\alpha = 0.$$

By the hypothesis on α, we get $m(X) | m_1(X)q(X)$,

i.e., $$p(X)m_1(X) | m_1(X)q(X).$$

This gives $$p(X) | q(X).$$

Write $$q(X) = p(X)q_1(X).$$

Then, (1) gives $$p(T)[\beta - q_1(T)\alpha] = \beta'$$

i.e., $$p(T)\alpha_1 = \beta',\ \alpha_1 = \beta - q_1(T)\alpha \qquad \text{... (2)}$$

As $$\beta \notin V'$$

and $$q_1(T)\alpha \in V',$$

we get $$\alpha_1 \notin V', \text{ with } p(T)\alpha_1 = \beta' \in V'.$$

By Proposition (5.3.3) if $$f(T)\alpha_1 \in V'$$

for some $f(X) \in F[X]$, then $$p(X) | f(X).$$

Consider $$W' = W + F[T]\alpha_1 \qquad \text{... (3)}$$

Clearly, W' is T-invariant and $W < W'$.

Suppose $$\beta'' \in F[T]\alpha \cap W'.$$

Then, $$\beta'' = g(T)\alpha = \beta_2 + h(T)\alpha_1,\ g(X), h(X) \in F[X],$$

and $$\beta_2 \in W.$$

Consequently, $$h(T)\alpha_1 = -\beta_2 + g(T)\alpha \in V'.$$

As remarked above, $$p(X) | h(X).$$

Write $$h(X) = h_1(X)p(X).$$

Then, $$h(T)\alpha_1 = h_1(T)\beta'$$

by (2), so that $h(T)\alpha_1 \in W$.

Consequently, $\beta'' \in F(T)\alpha \cap W = \{0\}$.

Hence, $F(T)\alpha \cap W' = \{0\}$.

As $W < W'$ and W is a largest dimensional T-invariant subspace satisfying $F(T)\alpha \cap W = \{0\}$,

we should have $F(T)\alpha \cap W' \neq \{0\}$.

This is a contradiction.

Hence, $V' = V$,

i.e., $V = F(T)\alpha \oplus W$.

We now discuss one of the most important theorems on linear operators.

Theorem 10. (Cyclic Decomposition Theorem). Let T be a linear operator on a finite dimensional vector space V_F.

Then, $$V = F[T]\alpha_1 \oplus F[T]\alpha_2 \oplus .... \oplus F[T]\alpha_k \quad \text{... (1)}$$

for some non-zero vectors $\alpha_i \in V$ with T-annihilator polynomials $m_i(X)$, $1 \le i \le k$, satisfying the condition :

$$m_{i+1}(X) \mid m_i(X),\ 1 \le i \le k \quad \text{... (2)}$$

Further, the positive integer k and $m_i(X)$ are uniquely determined by T, in the sense that if also

$$V = F[T]\beta_1 \oplus F[T]\beta_2 \oplus ... \oplus F[T]\beta_s$$

for some non-zero $\beta_j \in V$,

with T-annihilator polynomials $m'_j(X)$ such that $m'_{j+1}(X) \mid m'_j(X)$ for $1 \le j < s$, then, $k = s$ and $m_i(X) = m'_i(X)$ for $1 \le i \le k$.

Proof. First of all we prove the existence of the decomposition (1) satisfying condition (2).

Let $\dim V_F = n$.

For $n = 1$, choose any $\alpha_1 \neq 0$ in V, then $V = F[T]\alpha_1$.

So, the result holds for $n = 1$. Let $n > 1$. To apply induction, let the result hold for all vector spaces of dimensions less than n.

Let $m(X)$ be the minimal polynomial of T. By Proposition 2, there exists $\alpha_1 \in V$ whose T-annihilator polynomial $m_1(X)$ equals $m(X)$.

As $\deg m(X) > 0$, $\alpha_1 \neq 0$.

By Theorem 9, $$V = F[T]\alpha_1 \oplus W \quad \text{... (3)}$$

for some T-invariant subspace W of V. Let T' be the restriction of T to W.

Then, for any $\beta \in W$, $$F[T]\beta = F[T']\beta$$

and the T'-annihilator polynomial of β is same as its T-annihilator polynomial.

If $W = \{0\}$, (3) becomes $$V = F[T]\alpha_1,$$
which is like (1).

Let $$W \neq \{0\}.$$

Then, $$\dim W < \dim V.$$

By the induction hypothesis $$W = F[T']\alpha_2 \oplus F[T']\alpha_2 \oplus \ldots \oplus F[T']\alpha_k$$
$$= F[T]\alpha_2 \oplus F[T]\alpha_2 \oplus \ldots \oplus F[T]\alpha_k$$

for some non-zero $\alpha_i \in W$, with T-annihilator polynomials $m_i(X)$, $2 \le i \le k$, satisfying $m_{j+1}(X) \mid m_j(X)$ for $2 \le j < k$.

Consequently, (3) gives $$V = F[T]\alpha_1 \oplus F[T]\alpha_2 \oplus \ldots \oplus F[T]\alpha_k. \quad \ldots (4)$$

This decomposition satisfies (2).

Let also $$V = F[T]\beta_1 \oplus F[T]\beta_2 \oplus \ldots \oplus F[T]\beta_s \quad \ldots (5)$$

for some non-zero $\beta_j \in V$ with T-annihilator polynomials $m'_j(X)$ satisfying $m'_{j+1}(X) \mid m'_j(X)$ for $1 \le j < s$.

Let $$\alpha \in V$$

then, $$\alpha = f_1(T)\beta_1 + f_2(T)\beta_2 + \ldots + f_s(T)\beta_s$$

for some $$f_i(X) \in F[T].$$

As $$m'_i(X) \mid m'_1(X),\ m'_1(X)\beta_i = 0$$

for $$1 \le i \le s.$$

Consequently, $$m'_1(T)\alpha = \sum_{i=1}^{s} f_i(T) m'_1(T)\beta_i = 0.$$

Thus, $$m'_1(T) = 0.$$

Consequently, $$m_1(X) \mid m'_1(X).$$

However, $$m'_1(X) \mid m_1(X).$$

Hence, $$m_1(X) = m'_1(X).$$

This further gives $$\dim F[T]\alpha_1 = \dim F[T]\beta_1 = \deg m_1(X). \quad \ldots (6)$$

If $k = 1$, (6), proves the result. Let $k > 1$. Suppose for some i, $1 \le i < k$, we have already proved that $m_j(X) = m'_j(X)$ for $1 \le j \le i$.

Then, given any $g(X) \neq 0$ in $F[X]$, for $1 \leq j \leq s$, the T-annihilator polynomial of $g(T)\alpha_j$ is $h_j(X) = \dfrac{m_j(X)}{d(X)}$,

where $$d(X) = \gcd\left[m_j(X), g(X)\right] \quad \text{(Theorem 5.2.18)}$$

Similarly, the T-annihilator polynomial of $g(T)\beta_j$ is

$$h_j'(X) = \frac{m_j'(X)}{d'(X)},$$

$$d'(X) = \gcd\left[m_j'(X), g(X)\right].$$

For $1 \leq j \leq i$, $$m_j'(X) = m_j(X),$$

gives $$h_j(X) = h_j'(X).$$

Hence $$\dim F[T]g(T)\alpha_j = \dim F[T]g(T)\beta_j \quad \text{... (7)}$$

for $$1 \leq j \leq i,$$

and $$0 \neq g(X) \in F[X].$$

Further, $$\dim\left[\sum_{j=1}^{i} F[T]\alpha_j\right] = \dim\left[\sum_{j=1}^{i} F[T]\beta_j\right] \quad \text{... (8)}$$

Thus, $$\dim\left[\sum_{j=i+1}^{k} F[T]\alpha_j\right] = \dim\left[\sum_{j=i+1}^{s} F[T]\beta_j\right]. \quad \text{... (9)}$$

Thus, $$F[T]\alpha_{i+1} \neq \{0\},$$

we get $$F[T]\beta_{i+1} \neq \{0\}.$$

For $l \geq i+1$, $$m_l(X) \mid m_{i+1}(X),$$

yields $$m_{i+1}(T)\alpha_l = 0.$$

Consequently, by (4) $$m_{i+1}(T)V = \oplus\sum_{j=1}^{i} F[T]m_{i+1}(T)\alpha_j. \quad \text{... (10)}$$

By (5) $$m_{i+1}(T)V = \left[\sum_{j=1}^{i} F[T]m_{i+1}(T)\beta_j\right] \oplus \left[\sum_{j=i+1}^{s} F[T]m_{i+1}(T)\beta_j\right]. \quad \text{... (11)}$$

By (7) $$\dim F[T]m_{i+1}(T)\alpha_j = \dim F[T]m_{i+1}(T)\beta_j$$

for $$1 \leq j \leq i.$$

So, by comparing (10) and (11), we get

$$\left[\sum_{j=i+1}^{k} F[T]m_{i+1}(T)\beta_j\right] = 0.$$

So that $$m_{i+1}(T)\beta_{i+1}=0.$$

Hence, $$m'_{i+1}(X)\big|m_{i+1}(X).$$

Similarly, $$m_{i+1}(X)\big|m'_{i+1}(X).$$

Hence, $$m_{i+1}(X)=m'_{i+1}(X).$$

This proves that $k=s$ and that $m_i(X)=m'_i(X)$ for $1\le i\le k$.

We now give a matrix version of this theorem :

Theorem 11. **Let A be any $n\times n$-matrix over a field F. Then, there exists a uniquely determined finite sequence $m_1(X), m_2(X), ..., m_k(X)$ of monic polynomials in $F[X]$ such that $m_{i+1}(X)|m_i(X)$ for $1\le i\le k$ and A is similar to**

$$A'=\text{Com}[m_1(X)]\oplus\text{Com}[m_2(X)]\oplus...\oplus\text{Com}[m_k(X)].$$

Moreover, $m_1(X)$ is the minimal polynomial of A. The matrix A' is uniquely determined by A. The characteristic polynomial $f(X)$ of A divides $m_1(X)^n$.

Proof. **Let T be the linear operator on $V=F^n$ given by A. Then, the minimal polynomial of A and T are the same, say $m(X)$.**

By the cyclic decomposition theorem

$$V=F[T]\alpha_i\oplus F[T]\alpha_2\oplus...F[T]\alpha_k$$

for some non-zero $\alpha_i\in V$, with T-annihilator polynomial $m_i(X)$, $1\le i\le k$, such that $m_{j+1}(X)\big|m_j(X)$ for $1\le j<k$.

These, $m_i(X)$ are uniquely determined by T and, hence, by A.

Let T_i be the restriction of T to $F[T]\alpha_i$. By Theorem 5, there exists an ordered basis B_i of $F[T]\alpha_i$ such that $A_i=[T_i]_{B_i}=\text{Com}[m_i(X)]$.

Then, $$B'=\bigcup_{i=1}^{k}B_i$$

is an ordered basis of V such that

$$[T]_{B'}=\text{Com}[m_1(X)]\oplus\text{Com}[m_2(X)]\oplus...\oplus\text{Com}[m_k(X)].$$

As $$A=[T]_B,$$

where B is the standard ordered basis of V, A is similar to $[T]_{B'}$.

Clearly, $$m(X)=m_1(X).$$

The uniqueness of A' follows from Theorem 10.

Finally, as $$A' = \oplus \sum_{i=1}^{k} \text{Com}[m_i(X)],$$

the characteristic polynomial $f(X)$ of A equal $\prod_{i=1}^{k} m_i(X)$.

Clearly, $$k \le n.$$

As $$m_i(X) \mid m_1(X),$$

we get $f(X)$ divides $m_1(X)^n$.

This theorem motivates the following *canonical form* of a matrix :

Definition 12. Any square matrix A over a field F, which is of the form

$$A = \begin{bmatrix} A_1 & 0 & 0 & \dots & 0 \\ 0 & A_2 & 0 & \dots & 0 \\ 0 & 0 & A_3 & \dots & 0 \\ \vdots & \vdots & \vdots & \dots & \\ \vdots & \vdots & \vdots & & \\ 0 & 0 & 0 & \dots & A_k \end{bmatrix}$$

where each $A_i = \text{com}[m_i(X)]$ for some monic polynomial $m_i(X)$ of positive degree, and $m_{j+1}(X) \mid m_j(X)$ for $1 \le j < k$, is called a matrix in *rational form* ;

$$m_1(X),\ m_2(X), \dots, m_k(X)$$

are called the *invariant factors* of A.

Given any square matrix A over F, Theorem 11 tells that there exists only one matrix B in rational form that is similar to A. The *invariant factors* of A. Observe that the characteristic polynomial of A is the product of its invariant factors.

Example 12.a. The characteristic polynomial of a 4 × 4 nilpotent matrix is X^4, so that following are the possibilities for the invariants factors of a 4 × 4-nilpotent matrix in rational form.

(*i*) X^4 (*ii*) X^3, X

(*iii*) X^2, X^2 (*iv*) X^2, X, X

(*v*) X, X, X, X.

We get the following five nilpotent 4 × 4 matrices in rational form.

$$A_1 = \begin{bmatrix} 0 & 0 & 0 & 0 \\ 1 & 0 & 0 & 0 \\ 0 & 1 & 0 & 0 \\ 0 & 0 & 1 & 0 \end{bmatrix}$$

$$A_2 = \begin{bmatrix} 0 & 0 & 0 & 0 \\ 1 & 0 & 0 & 0 \\ 0 & 1 & 0 & 0 \\ 0 & 0 & 0 & 0 \end{bmatrix}$$

$$A_3 = \begin{bmatrix} 0 & 0 & 0 & 0 \\ 1 & 0 & 0 & 0 \\ 0 & 0 & 0 & 0 \\ 0 & 0 & 1 & 0 \end{bmatrix}$$

$$A_4 = \begin{bmatrix} 0 & 0 & 0 & 0 \\ 1 & 0 & 0 & 0 \\ 0 & 0 & 0 & 0 \\ 0 & 0 & 0 & 0 \end{bmatrix}$$

$$A_2 = \begin{bmatrix} 0 & 0 & 0 & 0 \\ 0 & 0 & 0 & 0 \\ 0 & 0 & 0 & 0 \\ 0 & 0 & 0 & 0 \end{bmatrix}.$$

Any 4×4-matrix is similar to one of these five matrices.

Example 12.b. Consider $A = \begin{bmatrix} 1 & 0 & 0 \\ 0 & 2 & 0 \\ 0 & 0 & 1 \end{bmatrix}.$

Its minimal polynomial is $m_1(X) = (X-1)(X-2)$.

Let T be the linear operator on $\mathbf{R}^3$ given by A. As $T \in_1 = \in_1$ and $T \in_2 = 2 \in_2$, the T-annihilator polynomials of $\in_1$ and $\in_2$ are $X-1$ and $X-2$, respectively, so that $\alpha_1 = \in_1 + \in_2$ has T-annihilator polynomial $(X-1)(X-2) = m_1(X)$.

Then, $$\dim[R(T)\alpha_1] = 2,$$

and $$\{\alpha_1, T\alpha_1\} = \{\in_1 + \in_2, \ \in_1 + 2\in_2\}$$

is a basis of $F[T]\alpha_1$.

Clearly, $$\in_3 \notin R[T]\alpha_1.$$

As $$T \in_3 = \in_3,$$

the T-annihilator polynomial of $\in_3$ is $m_2(X) = X-1$,

$$\dim[F(T) \in_3] = 1.$$

Obviously, $$\mathbf{R}^3 = F[T]\alpha_1 \oplus F[t]\alpha_2, \ \alpha_2 = \in_3$$

is a T-cyclic decomposition of $\mathbf{R}^3$.

For the ordered basis $B = \{\epsilon_1 + \epsilon_2,\ \epsilon_1 + 2\epsilon_2,\ \epsilon_3\}$,

$$[T]_B = \begin{bmatrix} 0 & -2 & 0 \\ 1 & 3 & 0 \\ 0 & 0 & 1 \end{bmatrix} = A'$$

is in rational form.

A is similar to A'. The invariant factors of A are $(X-1)(X-2)$, $X-1$.

Theorem 13. Let T be a linear operator on a finite dimensional vector space V_F, such that the minimal polynomial of T is $(X-c)^k$ for some $c \in F$ and $k \geq 1$. Then, there exists a uniquely determined sequence of positive integers $k_1 \geq k_2 \geq \ldots \geq k_r$ with $k = k_1$, such that for some ordered basis B of V, $[T]_B$ is the direct sum

$$J_{k_1}(c) \oplus J_{k_2}(c) \oplus \ldots \oplus J_{k_r}(c).$$

Proof. By the cyclic decomposition theorem

$$V = F[T]\alpha_1 \oplus F[T]\alpha_2 \oplus \ldots \oplus F[T]\alpha_r \qquad \ldots (1)$$

such that each $\alpha_i \neq 0$ and the T-annihilator polynomials $m_i(X)$ of α_i, satisfy

$$m_{i+1}(X) \mid m_i(X).$$

Further, $m_1(X)$ is the minimal polynomial of T, so that $m_1(X) = (X-c)^k$.

As $\quad m_i(X) \mid (X-c)^k,\ m_i(X) = (X-c)^{k_i},$

with $\quad 0 < k_i \leq k.$

As $\quad m_{i+1}(X) \mid m_i(X),$

we get $\quad k_i \geq k_{i+1}.$

Consequently, $\quad k = k_1 \geq k_2 \geq \ldots \geq k_r \qquad \ldots (2)$

Let T_i be the restriction of T to $W_i = F[T]\alpha_i$. Then, W_i is a T_i-cyclic space and the minimal polynomial of T_i is $(X-c)^{k_i}$. By Theorem 7, W_i has an ordered basis B_i such that $[T_i]_{B_i} = J_{k_i}(c)$.

Then, $\quad B = B_1 \cup B_2 \cup \ldots \cup B_k$

is an ordered basis of V

and $\quad [T]_B = [T_1]_{B_1} \oplus \ldots \oplus [T]_B = J_{k_1}(c) \oplus \ldots \oplus J_{k_r}(c)$

To prove uniqueness, let for some ordered basis B' of V,

$$[T]_{B'} = J_{s_1}(c) + \ldots + J_{s_t}(c)$$

for some $\quad s_1 \geq s_2 \geq \ldots \geq s_t > 0.$

By Theorem (3.5.13), $V = V_1 \oplus V_2 \oplus ... \oplus V_t$

where each V_i is T-invariant, and for some ordered basis B_i' of V_i, the restriction T_i' of T to V_i satisfies $J_{s_i}(c) = [T']_{B_i'}$, so that $s_i = \dim V_i$ and the minimal polynomial of T_i' is $(X-c)^{s_i}$, by Theorem 7. Consequently, V_i is T_i'-cyclic, i.e., T-cyclic, so that $V_i = F[T]\beta_i$.

We get $$V = F[T]\beta_1 \oplus ... \oplus F[T]\beta_t \qquad ... (3)$$

(1) and (3) are two T-cyclic decompositions of V.

By the cyclic decomposition theorem,

$$r = t, \ s_i = k_i, \ 1 \le i \le r.$$

This proves the theorem.

Remark 14. Let T be a linear operator on V_F such that the minimal polynomial $m(X)$ of T is given by

$$m(X) = (X-c_1)^{s_1}(X-c_2)^{s_2}...(X-c_t)^{s_t}$$

for some distinct members $c_1, c_2, ..., c_t$ of F.

By the primary decomposition theorem,

$$V = V_1 \oplus V_2 \oplus ... \oplus V_t$$

where each V_i is T-invariant and the minimal polynomial of the restriction T_i of T to V_i is $(X-c_i)^{s_i}$.

By Theorem 13, there exists an ordered basis B_i of V_i such that

$$[T_i]_{B_i} = A_i = J_{s_{i1}}(c_i) \oplus J_{s_{i2}}(c_i) \oplus ... \oplus J_{s_{it_i}}(c_i)$$

with $$s_i = s_{i1} \ge s_{i2} \ge ... \ge s_{it_i} > 0.$$

So, we have an ordered basis B of V such that

$$[T]_B = A_1 \oplus A_2 \oplus ... \oplus A_t = \oplus \sum_{i=1}^{t} \sum_{j=1}^{t_i} J_{s_{ij}}(c_i),$$

a direct sum of elementary Jordan matrices.

This remark leads to the following definition :

Definition 15. Any $n \times n$-matrix A over a field F is said to be in *Jordan form* if

$$A = \begin{bmatrix} A_1 & 0 & 0 & .... & 0 \\ 0 & A_2 & 0 & .... & 0 \\ .. & .. & .. & .... & .. \\ .. & .. & .. & .... & .. \\ 0 & 0 & 0 & .... & A_k \end{bmatrix}$$

such that for some distinct $c_1, c_2, \ldots, c_k$,

$$A_i = J_{s_{i1}}(c_i) \oplus J_{s_{i2}}(c_i) \oplus \ldots \oplus J_{s_{it_i}}(c_i)$$

with $s_{i1} \geq s_{i2} \geq \ldots \geq s_{it_i}$.

Remark 14 shows that if a linear operator T on V_F has minimal polynomial

$$m(X) = (X - c_1)^{s_1}(X - c_2)^{s_2} \ldots (X - c_t)^{s_t},$$

then there exists an ordered basis B such that $[T]_B$ is in Jordan form.

In the case $F = \mathbf{C}$, any monic polynomial over $\mathbf{C}$ is of the form $\prod_i (X - c_i)^{s_i}$. So, given any linear operator T on a finite dimensional vector space $V_{\mathbf{C}}$, there exists an ordered basis of V, such that $[T]_B$ is in Jordan form. Consequently, any square matrix A over $\mathbf{C}$ is similar over $\mathbf{C}$ to a matrix in Jordan form.

We now generalize Theorem 7. For this we prove some additional results.

Theorem 16. Let $m(X)$ be any non-constant polynomial of degree k over a field F. Given any positive integer n, any polynomial $f(X) \in F[X]$ of degree $\leq kn - 1$ is expressible as

$$f(X) = \sum_{i=0}^{n-1} m(X)^i q_i(X)$$

for some $q_i(X)$, $0 \leq i \leq n-1$ in $F[X]$, each of degree less than k and uniquely determined by $f(x)$.

Proof. First of all we prove the existence of $q_i(X)$ by induction on n.

For $n = 1$, $\deg f(X) \leq k - 1$.

Then, $f(X) = m(X)q_1(X) + m(X)^0 q_0(X)$

with $q_1(X) = 0$

and $q_0(X) = f(X)$.

Thus, the result holds for $n = 1$. To apply induction, let $n > 1$ and the result hold for $n - 1$.

By the division algorithm, $f(X) = m(X)q(X) + r(X)$ ... (1)

for some $q(X), r(X) \in F[X]$

with $\deg r(X) < k$.

As $\deg m(X) = k$, $\deg f(X) \leq kn - 1$,

(1) gives $\deg m(X)q(X) \leq kn - 1$

so that $\deg q(X) \leq kn - 1 - \deg m(X) = k(n-1) - 1$.

By the induction hypothesis, $q(X) = \sum_{i=0}^{n-2} m(X)^i s_i(X)$, ... (2)

for some $s_i(X) \in F[X]$, satisfying $\deg s_i(X) < k$.

(1) and (2) give $$f(X) = \sum_{i=0}^{n-2} m(X)^{i+1} s_i(X) + m(X)^0 r(X)$$

$$= \sum_{i=0}^{n-1} m(X)^i q_i(X) \qquad \text{... (3)}$$

where $q_0(X) = r(X),\ q_i(X) = s_{i-1}(X)$ for $1 \le i \le n-1$.

Let also $$f(X) = \sum_{i=0}^{n-1} m(X)^i q_i'(X) \qquad \text{... (4)}$$

with $\deg q_i'(X) < k$.

Once again we apply induction on n. For $n = 1$, (3) and (4) give

$$f(X) = m(X)^0 q_0(X) = q_0(X) = m(X)^0 q_0'(X) = q_0'(X).$$

Hence, $q_0(X) = q_0'(X)$.

Thus, the result hold for $n = 1$. Let $n > 1$, and the uniqueness hold for $n-1$.

Again (3) and (4), give $f(X) = m(X)q(X) + r(X)$ ... (5)

$$= m(X)q'(X) + r'(X) \qquad \text{... (6)}$$

where $$q(X) = \sum_{i=1}^{n-1} m(X)^{i-1} q_i(X),$$

$$r(X) = q_0(X),$$

$$q'(X) = \sum_{i=1}^{n-1} m(X)^{i-1} q_i'(X),$$

$$r'(X) = q_0'(X).$$

By the division algorithm, $r(X) = r'(X)$, so $q_0(X) = q_0'(X)$. Also $q(X) = q'(X)$. Hence

$$q(X) = \sum_{i=0}^{n-2} m(X)^i q_{i+1}(X) \qquad \text{... (7)}$$

$$= \sum_{i=0}^{n-2} m(X)^i q_{i+1}'(X) \qquad \text{... (8)}$$

Observe that $\deg q(X) \le k(n-1) - 1$.

By the induction hypothesis on uniqueness, we get

$$q_{i+1}(X) = q'_{i+1}(X), \qquad 0 \le i \le n-2$$

i.e.,

$$q_i(X) = q'_i(X), \qquad 1 \le i \le n-1.$$

This proves the uniqueness.

Proposition 17. Let T be a linear operator on a finite dimensional vector space V_F such that for some monic polynomial $m(X) \in F[X]$ of degree $k \ge 1$, the minimal polynomial of T is $g(X) = m(X)^n$ and that V is T-cyclic.

Further, let

$$V = F[T]\alpha$$

and

$$m(X) = X^k - c_{k-1}X^{k-1} - \ldots - c_0.$$

Then,

(*I*) $\dim V = nk$

(*II*) V has an ordered basis $B = \{\beta_0, \beta_1, \ldots, \beta_{nk-1}\}$

such that given $1 \le s \le n$, $(s-1)k \le i < sk$

(*a*) $\beta_i = m(T)^{s-1} T^j \alpha$ with $j = i - (s-1)k$

(*b*) $T\beta_i = \beta_{i+1}$ for $(s-1)k \le i < sk - 1$

(*c*) $T\beta_{sk-1} = \beta_{sk} + \sum_{j=0}^{k-1} c_j \beta_{(s-1)k+j}$, $1 \le s \le n-1$

(*d*) $T\beta_{nk-1} = \sum_{j=0}^{k-1} c_j \beta_{(n-1)k+j}$.

Proof. As V is T-cyclic, $\dim V = \deg m(X)^n = nk$.

This proves (*I*). Consider $\beta \in V$.

By Lemma 4,

$$\beta = f(T)\alpha \qquad \ldots (1)$$

for some unique $f(X) \in F[X]$ of degree $\le nk - 1$.

By Theorem 15,

$$f(X) = \sum_{i=0}^{n-1} m(X)^i q_i(X)$$

for some uniquely determined polynomials $q_i(X) \in F[X]$, each of degree $\le k-1$, so that

$$\beta = f(T)\alpha = \sum_{i=0}^{n-1} m(T)^i q_i(T)\alpha \qquad \ldots (2)$$

Consider $V_i = \{m(T)^i q(T)\alpha : q(X) \in F[X],\ \deg q(X) < k\}$.

V_i is a subspace of V.

Given

$$q(X) \in F[X]$$

with

$$\deg q(X) < k,$$

$$q(X) = \sum_{j=0}^{k-1} a_j X^j,$$

gives
$$m(T)^i q(T)\alpha = \sum_{j=0}^{k-1} a_j \left[m(T)^i T^j \alpha\right],$$

so that
$$B_i = \left\{m(T)^i \alpha,\ m(T)^i T\alpha, \ldots,\ m(T)^i T^{k-1}\alpha\right\}$$

spans V_i.

Consequently,
$$\dim V_i \le k.$$

But (2) gives
$$V = \sum_{i=0}^{n-1} V_i$$

so that
$$nk \le \sum_{i=0}^{n-1} \dim V_i.$$

Clearly, if for some i, $\dim V_i < k$, then,
$$\sum_{i=0}^{n-1} \dim V_i < nk.$$

This will lead to a contradiction.

Hence,
$$\dim V_i = k.$$

This, in turn, gives that B_i is a basis of V_i.

Then,
$$B = \bigcup_{i=0}^{n-1} B_i = \left\{m(T)^i T^j \alpha : 0 \le i \le n-1,\ 0 \le j \le k-1\right\}$$

is a basis of V.

We write
$$B = \left\{\beta_0,\ \beta_1, \ldots,\ \beta_{nk-1}\right\}$$

such that given
$$1 \le s \le n, (s-1)k \le i < sk,$$

then,
$$\beta_i = m(T)^{(s-1)} . T^{i-(s-1)k} \alpha$$

Clearly, if
$$i < sk - 1,$$

then
$$T\beta_i = m(T)^{(s-1)} T^{i+1-(s-1)k} = \beta_{i+1}.$$

Now,
$$X^k = m(X) + \sum_{j=0}^{k-1} c_j X^j,$$

for
$$m(X) = X^k - c_{k-1} X^{k-1} - \ldots . c_0$$

yields
$$T^k = m(T) + \sum_{j=0}^{k-1} c_j T^j.$$

Thus, for $\quad 1 \le s \le n-1$

$$\begin{aligned} T\beta_{sk-1} &= Tm(T)^{(s-1)}T^{k-1}\alpha \\ &= m(T)^{(s-1)}T^{k}\alpha \\ &= m(T)^{s-1}\left[m(T)\alpha + \sum_{j=0}^{k-1} c_j T^j\alpha\right] \\ &= m(T)^{s}\alpha + \sum_{j=0}^{k-1} c_j m(T)^{s-1} T^j\alpha \\ &= \beta_{sk} + \sum_{j=0}^{k-1} c_j \beta_{(s-1)k+j}. \end{aligned}$$

As $m(T)^n = 0$, similar arguments give (d).

Remark 18. Consider $\quad J_n(c) = \begin{bmatrix} c & 0 & 0 & \cdots & 0 & 0 \\ 1 & c & 0 & \cdots & 0 & 0 \\ 0 & 1 & c & \cdots & \cdots & \cdots \\ \cdots & \cdots & \cdots & \cdots & \cdots & \cdots \\ \cdots & \cdots & \cdots & \cdots & \cdots & \cdots \\ 0 & 0 & 0 & \cdots & 1 & c \end{bmatrix}.$

For $m(X) = X - c \quad \text{com}(X-c) = [c]$.

Let $\quad E_1 = [1]$.

Then, $\quad J_n(c) = \begin{bmatrix} C_1 & 0 & 0 & \cdots & 0 & 0 \\ E_1 & C_1 & 0 & \cdots & 0 & 0 \\ 0 & 1 & c & \cdots & \cdots & \cdots \\ \cdots & \cdots & \cdots & \cdots & \cdots & \cdots \\ \cdots & \cdots & \cdots & \cdots & \cdots & \cdots \\ 0 & 0 & 0 & \cdots & E_1 & C_1 \end{bmatrix}$

where C_1 is the companion matrix of the degree one polynomial $X - c$.

Keeping this in mind, one sees that the theorem, given below, generalizes Theorem 7.

Theorem 19. Let T be linear operator on a finite dimensional vector space V_F such that V is T-cyclic. Let $m(X)$ be a monic polynomial in $F[X]$ of degree $k \ge 1$. Then, the minimal polynomial of T is $m(X)^n$ for some $n \ge 1$ iff $\dim V = nk$ and for some ordered basis B of V,

$$[T]_B = \begin{bmatrix} C & 0 & 0 & \cdots & 0 & 0 \\ E & C & 0 & \cdots & 0 & 0 \\ 0 & E & C & \cdots & 0 & 0 \\ \cdots & \cdots & \cdots & \cdots & \cdots & \cdots \\ \cdots & \cdots & \cdots & \cdots & \cdots & \cdots \\ 0 & 0 & 0 & \cdots & E & C \end{bmatrix}$$

where C is the companion matrix of $m(X)$ and E is the $k \times k$-matrix

$$\begin{bmatrix} 0 & 0 & \dots & 0 & 1 \\ 0 & 0 & \dots & 0 & 0 \\ 0 & 0 & \dots & 0 & 0 \\ .. & .. & \dots & .. & .. \\ .. & .. & \dots & 0 & 0 \\ 0 & 0 & \dots & 0 & 0 \end{bmatrix}.$$

Proof. Let $$m(X) = X^k - c_{k-1}X^{k-1} - \dots - c_0$$

and $$g(X) = [m(X)]^n$$

be the minimal polynomial of T.

By Proposition 17, $\dim V = nk$

and there exists an ordered basis $B = \{\beta_0, \beta_1, \dots, \beta_{nk-1}\}$

of V such that for $1 \le s \le n,\ 0 \le j \le k$

$$\beta_{(s-1)k+j} = m(T)^{s-1}\, T^j \alpha \qquad \dots (1)$$

where $0 \ne \alpha \in V$ is such that $V = F[T]\alpha$.

Further, if $j < k-1$, then, $T\beta_{(s-1)k+j} = \beta_{(s-1)k+j+1}$.

so that $$\left[T\beta_{(s-1)k+j}\right]_B = [0\ 0 \dots 0\ \overset{(s-1)k+j+1}{1}\ 0 \dots 0]^T.$$

Also, $$T\beta_{sk-1} = \sum_{j=0}^{k-1} c_j \beta_{(s-1)k+j} + \beta_{sk}, \qquad \text{if } s < n$$

gives $$[T\beta_{sk-1}]_B = [0\ 0 \dots 0 \dots 0\ \overset{(s-1)k}{c_0}\ c_1 \dots c_{k-1}\ 1\ 0\ 0 \dots 0]^T$$

so that in $[T]_B$, the column numbered from $(s-1)k$ to $sk-1$ for $s \le n-1$ form the block matrix

$$\begin{bmatrix} 0 \\ 0 \\ \vdots \\ \vdots \\ C \\ E \\ 0 \\ \vdots \\ \vdots \\ 0 \end{bmatrix} \begin{matrix} \\ \\ \\ \\ \dots s\text{-th block} \\ \dots (s+1)\text{-th block} \\ \\ \\ \\ \\ \end{matrix}$$

here each 0 is a $k \times k$-matrix, $C = \text{Com}[m(X)]$ and E is the $k \times k$-matrix whose (1, k)th entry is 1 and the other entries are zeros.

For $s=n$, by using *II*(*b*) and *II*(*d*) in Proposition 17, we get that the columns numbered from $(n-1)k$ to $nk-1$ from the block matrix

$$\begin{bmatrix} 0 \\ 0 \\ \vdots \\ \vdots \\ 0 \\ C \end{bmatrix}$$

so that

$$[T]_B = \begin{bmatrix} C & 0 & 0 & \cdots & 0 & 0 \\ E & C & 0 & \cdots & 0 & 0 \\ 0 & E & C & \cdots & 0 & 0 \\ \cdot\cdot & \cdot\cdot & \cdot\cdot & \cdots & \cdot\cdot & \cdot\cdot \\ \cdot\cdot & \cdot\cdot & \cdot\cdot & \cdots & \cdot\cdot & \cdot\cdot \\ 0 & 0 & 0 & \cdots & C & 0 \\ 0 & 0 & 0 & \cdots & E & C \end{bmatrix}$$

Conversely, let $[T]_B$ be given as above.

By reversing the arguments, we see that if

$$B = \{\beta_0, \beta_1, \ldots, \beta_{nk-1}\}$$

then, given $1 \le s \le n,\ 0 \le j \le k-1$

$$T\beta_{(s-1)k+j} = \beta_{(s-1)k+j+1} \qquad \ldots (2)$$

If $s < n$,

$$T\beta_{sk-1} = \sum_{r=0}^{k-1} c_r \beta_{(s-1)k+r} + \beta_{sk} \qquad \ldots (3)$$

$$T\beta_{nk-1} = \sum_{r=0}^{k-1} c_r \beta_{(n-1)k+r} \qquad \ldots (4)$$

Let $\alpha = \beta_0.$

Consider $0 \le i \le nk-1.$

By the division algorithm, $i = qk + j$

for some uniquely determined non-negative integers q and j with $0 \le j \le k-1$.

As $i \le nk-1,\ 0 \le q \le n-1.$

By writing $s-1 = q$, we get

$$i = (s-1)k + j,\ 1 \le s \le n,\ 0 \le j \le k-1 \qquad \ldots (5)$$

We prove $\beta_{(s-1)k+j} = m(T)^{s-1} T^j \alpha$, by induction on s.

For $s = 1$, (2) gives $\beta_1 = T\beta_0$,

$$\beta_2 = T\beta_1 = T^2\beta_0, \ldots, \beta_{k-1} = T^{k-1}\beta_0.$$

So, the result is true for $s = 1$. Suppose that for some s, $1 \le s < n$, the result is true, for all $0 \le j \le k-1$.

In particular, $\beta_{sk-1} = \beta_{(s-1)k+k-1} = m(T)^{s-1} T^{k-1}\beta_0$.

This gives $T\beta_{sk-1} = m(T)^{s-1} T^k \beta_0$.

By using (3) and that $T^k = m(T) + \sum_{r=0}^{k-1} c_r T^r$,

we get
$$\sum_{r=0}^{k-1} c_r \beta_{(s-1)k+r} + \beta_{sk} = m(T)^s \beta_0 + \sum_{r=0}^{k-1} c_r m(T)^{s-1} T^r \beta_0.$$

By the induction hypothesis $\beta_{(s-1)k+r} = m(T)^{s-1} T^r \beta_0$.

Consequently, $\beta_{sk} = m(T)^s \beta_0 = m(T)^s \alpha$.

By repeated application of (2), we get

$$\beta_{sk+j} = m(T)^s T^j \alpha, \qquad 0 \le j \le k-1$$

so that
$$\beta_{(s-1)k+j} = m(T)^{s-1} T^j \alpha,\ 1 \le s \le n,\ 0 \le j \le k-1. \qquad \ldots (6)$$

This gives $V = F[T]\alpha$.

Further, (6) gives $\beta_{nk-1} = m(T)^{n-1} T^{k-1}\alpha$.

Hence, $T\beta_{nk-1} = m(T)^{n-1} T^k \alpha$

i.e.,
$$\sum_{r=0}^{k-1} c_r \beta_{(n-1)k+r} = m(T)^n \alpha + \sum_{r=0}^{k-1} c_r \beta_{(n-1)k+r}$$

so that $m(T)^n \alpha = 0$.

By using (6), we get $m(T)^n \beta_{(s-1)k+j} = 0$.

Hence, $m(T)^n = 0$.

Consider any non-zero polynomial $f(X) \in F[X]$ of degree $\le nk-1$.

By Theorem 15,
$$f(X) = \sum_{i=1}^{n} m(X)^{(i-1)} q_{(i-1)}(X) \qquad \ldots (8)$$

with $\deg q_{(s-1)}(X) \le k-1$.

For some s, $q_{(s-1)}(X) \ne 0$.

Write $$q_i(X) = \sum_{j=0}^{k-1} a_{ij} X^j,$$

so that $$f(T)\alpha = \sum_{i=1}^{n} m(T)^{i-1}\left[\sum_{j=0}^{k-1} a_{(i-1)j} T^j \alpha\right]$$

$$= \sum_{j=0}^{k-1}\sum_{i=1}^{n} a_{(i-1)j}\beta_{(i-1)k+j} \neq 0$$

as some $a_{(s-1)j} \neq 0$.

Thus, T does not satisfy any non-zero polynomial of degree less than nk. Hence, $m(X)^n$ is the minimal polynomial of T.

Keeping in view this result and the definition of an elementary Jordan matrix we have the following definition :

Definition 20. Let $m(X)$ be any monic polynomial in $F[X]$ of degree $k \geq 1$, C be the companion matrix of $m(X)$ and E_k be the $k \times k$-matrix whose $(1, k)^{\text{th}}$ entry is 1 and all the other entries are zero. For any positive integer n, the $nk \times nk$ -matrix

$$\begin{bmatrix} C & 0 & 0 & \cdots & 0 & 0 \\ E_k & C & 0 & \cdots & 0 & 0 \\ 0 & E_k & C & \cdots & & 0 \\ .. & .. & .. & \cdots & .. & .. \\ .. & .. & .. & \cdots & .. & .. \\ 0 & 0 & 0 & \cdots & E_k & C \end{bmatrix}$$

is called an *elementary Jordan matrix* of $m(X)$, and it is denoted by $J_n[m(X)]$; it is also known as elementary Jordan matrix relative to $m(X)$.

Clearly, $$J_n(c) = J_n(X - c).$$

Theorem 19 gives :

Theorem 21. Let $m(X)$ be any monic polynomial in $F[X]$ of degree $k \geq 1$ and n a positive integer. Then, an $nk \times nk$ -matrix A over F has minimal polynomial $m(X)^n$ iff A is similar to $J_n[m(X)]$. In this case, $m(X)^n$ is also the characteristic polynomial of A.

Example 21.a. Let $$m(X) = X^2 + 1$$ over **R**

$$C = \text{Com}[m(X)] = \begin{bmatrix} 0 & -1 \\ 1 & 0 \end{bmatrix},$$

$$E_2 = \begin{bmatrix} 0 & 1 \\ 0 & 0 \end{bmatrix},$$

so that $$J_2(X^2+1)=\begin{bmatrix} C & 0 \\ E_2 & C \end{bmatrix}=\begin{bmatrix} 0 & -1 & 0 & 0 \\ 1 & 0 & 0 & 0 \\ 0 & 1 & 0 & -1 \\ 0 & 0 & 1 & 0 \end{bmatrix}.$$

Its minimal polynomial is $g(X)=(X^2+1)^2=X^2+2X^2+1$.

The companion matrix of $g(X)$ is

$$A=\begin{bmatrix} 0 & 0 & 0 & -1 \\ 1 & 0 & 0 & 0 \\ 0 & 1 & 0 & -2 \\ 0 & 0 & 1 & 0 \end{bmatrix}.$$

A also has minimal polynomial $g(X)$. By Theorem 21, A and $J_2(X^2+1)$ are similar.

Remark 22. Let T be a linear operator on V_F, with minimal polynomial

$$m(X)=\prod_{i=1}^{k} p_i(X)^{r_i} \qquad \text{... (1)}$$

where $p_1(X), p_2(X), \ldots, p_k(X)$ are the distinct (monic) prime factors of $m(X)$.

By the primary decomposition theorem,

$$V=V_1 \oplus V_2 \oplus \ldots \oplus V_k \qquad \text{... (2)}$$

where each V_i is T-invariant and the minimal polynomial $m_i(X)$ of the restriction T_i of T to V_i is $p_i(X)^{r_i}$.

By the cyclic decomposition theorem,

$$V_i=\oplus\sum_{j=1}^{k_i} F[T_i]\alpha_{ij}=\oplus\sum_{j=1}^{k_i} F[T]\alpha_{ij} \qquad \text{... (3)}$$

such that the T_i-annihilator polynomial (which is same as the T-annihilator polynomial) of α_{ij} is

$$m_{ij}(X),\ m_{i1}(X)=m_i(X)=p_i(X)^{r_i}$$

and $$m_{i,j+1}(X)\,\big|\,m_{ij}(X).$$

Any factor of $p_i(X)^{r_i}$ is of the form

$$p_i(X)^r,\ 0 \le r \le r_i;$$

consequently, $$m_{ij}(X)=p_i(X)^{r_{ij}}$$

and $$r_{ij} \ge r_{i,j+1}.$$

By Theorem 19, we get a basis B_{ij} of $F[T]\alpha_{ij}$, such that if T_{ij} is the restriction of T to $F[T]\alpha_{ij}$, then $[T_{ij}]_{B_{ij}} = J_{r_{ij}}[p_i(X)]$. Consequently, for $B = \bigcup_{i,j} B_{ij}$, $[T]_B$ is a direct sum of elementary matrices of the type $J_{r_{ij}}[p_i(X)]$; for each i, atleast one $r_{ij} = r_i$, and all the $r_{ij} \le r_i$. In view of the primary decomposition theorem and the cyclic decomposition theorem, one can see that the collection of these $J_{r_{ij}}[p_i(X)]$ is uniquely determined by T. In terms of matrices, this gives that any $n \times n$-matrix A over a field is a direct sum of elementary Jordan matrices of prime factors of the minimal polynomial of A.

Exercise 5.4

1. Let T be the linear operator on $\mathbf{R}^2$, given by $A = \begin{bmatrix} 1 & 3 \\ 0 & 2 \end{bmatrix}$

   (*i*) Find the minimal polynomial of T

   (*ii*) Find the T-annihilator polynomials of ϵ_1, ϵ_2, $\epsilon_1 + \epsilon_2$

   (*iii*) Is $\mathbf{R}^2$ T-cyclic?

2. Each of the following matrices is a companion matrix

$$\begin{bmatrix} 0 & -1 \\ 1 & 2 \end{bmatrix}, \begin{bmatrix} 0 & 1 \\ 1 & 2 \end{bmatrix}, \begin{bmatrix} 0 & 0 & 1 \\ 1 & 0 & -3 \\ 0 & 1 & 3 \end{bmatrix}, \begin{bmatrix} 0 & 0 & 1 \\ 1 & 0 & 1 \\ 0 & 1 & -1 \end{bmatrix}$$

   (*i*) Write down corresponding polynomials of which they are companion matrices.

   (*ii*) Which of these matrices are similar to elementary Jordan matrices over **R**.

3. Let T be the linear operator on $\mathbf{R}^3$ given by

$$A = \begin{bmatrix} 0 & 1 & 1 \\ 0 & 0 & 0 \\ 0 & 0 & 0 \end{bmatrix}.$$

   (*i*) Prove that X^2 is the minimal polynomial of T.

   (*ii*) Find bases and dimensions of $\mathbf{R}[T]\epsilon_2$ and $\mathbf{R}[T](\epsilon_2 - \epsilon_3)$.

   (*iii*) Verify that $\mathbf{R}^3 = \mathbf{R}[T](\epsilon_2) \oplus \mathbf{R}[T](\epsilon_2 - \epsilon_3)$.

   (*iv*) Find a matrix C such that $C^{-1}AC$ is in Jordan form.

4. By using Exercise 3, find a matrix D such that

$$D^{-1}\begin{bmatrix} 2 & 1 & 1 \\ 0 & 2 & 0 \\ 0 & 0 & 2 \end{bmatrix}D$$

   is in Jordan form.

5. Write a 4 × 4-matrix in Jordan form, whose minimal polynomial is $(X+3)^2(X-4)$ but characteristic polynomials is $(X+3)^2(X-4)^2$.

6. Are the following two matrices similar?

$$A = \begin{bmatrix} -1 & 0 & 0 \\ 1 & -1 & 0 \\ 0 & 0 & 1 \end{bmatrix}$$

and
$$B = \begin{bmatrix} 1 & 0 & 0 \\ 0 & -1 & 0 \\ 0 & 1 & -1 \end{bmatrix}$$

If yes, find a matrix C over $\mathbf{R}$ such that $B = C^{-1}AC$.

7. Are the following two matrices similar?

$$\begin{bmatrix} 2 & 0 & 0 & 0 \\ 1 & 2 & 0 & 0 \\ 0 & 0 & 3 & 3 \\ 0 & 0 & 0 & 3 \end{bmatrix} \text{ and } \begin{bmatrix} 2 & 0 & 0 & 0 \\ 1 & 2 & 0 & 0 \\ 0 & 0 & 3 & 0 \\ 0 & 0 & 1 & 3 \end{bmatrix}.$$

8. By using the cyclic decomposition theorem, prove that the characteristic polynomial and the minimal polynomial of an $n \times n$-matrix over a field F have the same sets of prime factors.

9. If two 3×3-matrices A and B over a field F have the same characteristic polynomials and the same minimal polynomials, prove that A and B are similar matrices.

10 .Write down all possible 5 × 5-matrices in Jordan form with minimal polynomial X^2.

11. Let A be the complex matrix

$$\begin{bmatrix} 2 & 0 & 0 & 0 & 0 \\ 1 & 2 & 0 & 0 & 0 \\ -1 & 0 & 2 & 0 & 0 \\ 0 & 1 & 0 & 2 & 0 \\ 0 & 0 & 0 & 1 & -1 \end{bmatrix}$$

Find a matrix in Jordan matrices similar to A.

12. Write down the elementary Jordan matrices of

(*i*) $(X^2+1)^2$ relative to X^2+1,

(*ii*) $(X^2+3X+1)^3$ relative to X^2+3X+1,

(*iii*) $(X+1)^6$ relative to $(X+1)^2$.

13. Let T_1 and T_2 be two nilpotent operators on V_F such that they have same nullity and same minimal polynomials.

(*i*) If $\dim V \le 6$, prove that T_1 and T_2 are similar

(*ii*) Give examples of T_1 and T_2 on V_F with $\dim V_F = 7$, such that T_1 and T_2 are not similar.

14. Let A be any $n \times n$-matrix over F such that $A^n = 0$, but $A^{n-1} \neq 0$. Prove that A and A^T are similar.

[**Hint :** Let T and T' be the linear operators on F^n given by A and A^T respectively, then, F^n is T-cyclic, as well as T'-cyclic].

15. For any $n \times n$-matrix over A over **C**, prove that A and A^T are similar over **C**.

16. Find Jordan canonical form for

$$\begin{bmatrix} 4 & 2 & 0 & 0 \\ -2 & 0 & 0 & 0 \\ 0 & 0 & 0 & 1 \\ 0 & 0 & 1 & 0 \end{bmatrix}.$$

17. By direct computation, show that

$$\begin{bmatrix} 1 & 1 & 0 \\ 0 & 1 & 1 \\ 0 & 0 & 1 \end{bmatrix} \text{ and } \begin{bmatrix} 1 & 1 & 0 \\ 0 & 1 & 0 \\ 0 & 0 & 1 \end{bmatrix}$$

are not similar.

5.5 Semi-simple Linear Operators

Throughout T denotes a non-zero linear operator on a finite dimensional vector space V_F, with minimal polynomial $m(X)$. For $\alpha \in V_F$, whose T-annihilator polynomials is $m(X)$, $W = F[T]\alpha$ is a T-invariant subspace and by Theorem (5.4.9) there exists a T-invariant subspace W' of V such that $V = W \oplus W'$. This motivates the following definition :

Definition 1. Given any T-invariant subspace W of V, a T-invariant subspace W' of V is called a *T-complement* of W if $V = W \oplus W'$. Further, a T-invariant subspace of W of V is said to be *T-complemented* if there exists a T-complement W' of W.

Theorem (5.4.9) shows that if V is not T-cyclic, then, there exists T-invariant subspace W such that W is T-complemented and W is non-trivial in the sense that $W \neq \{0\}$ and $W \neq V$. The primary decomposition theorem shows that any T-primary component of V is T-complemented.

Example 1.a. Let T be a linear operator on $\mathbf{R}^2$ given by $A = \begin{bmatrix} 0 & -1 \\ 1 & 2 \end{bmatrix}$. Its minimal polynomial is $m(X) = (X-1)^2$. Let W be a non-trivial T-invariant subspace of $\mathbf{R}^2$. Then $\dim W = 1$. So, for some $\alpha \neq 0$, $W = F[T]\alpha$. The T-annihilator polynomial of α is of degree one; so, it, being a factor of $(X-1)^2$, must be $X-1$. Let W be T-complemented. There exists a T-invariant subspace W' of $V = \mathbf{R}^2$ such that $V = W \oplus W'$. Clearly, $\dim W' = 1$ and, hence, $W' = F[T]\beta$, with T-annihilator polynomial of β being $(X-1)$. Thus, if T_1 and T_2 are the restrictions of T to W and W', respectively, then, each of them has $X-1$ as its minimal polynomial. By Theorem (5.2.15) the minimal polynomial of T is $X-1$. This is a contradiction. Hence, V has no non-trivial T-invariant subspace that is T-complemented.

Lemma 2. Let W be a T-invariant subspace of V_F such that for any $f(X) \in F[X]$ and $\alpha \in V$, $f(T)\alpha \in W$ implies $f(T)\alpha = f(T)\alpha_1$ for some $a_1 \in W$. Let T be the linear operator on $\overline{V} = V/W$ induced by T. Then given any $\overline{\alpha} \in \overline{V}$, $\overline{\alpha} = \alpha + W$, there exists a $\beta \in V$ such that $\dim F[\overline{T}]\overline{\alpha} = \dim F[T]\beta$ and $\overline{\alpha} = \overline{\beta}$.

Proof. Let $m_1(X)$ be the $\overline{T}$-annihilator polynomial of $\overline{\alpha}$.

For any $f(X) \in F[X]$, $\quad f(T)\alpha \in W$

iff $\quad f(\overline{T})\overline{\alpha} = \overline{0}$

iff $\quad m_1(X) | f(X)$.

By Proposition (5.3.3), $\quad m_1(X) | m(X)$.

Now $\quad m_1(T)\alpha \in W$.

By the hypothesis, there exists an $\alpha_1 \in W$ such that $m_1(T)\alpha = m_1(T)\alpha_1$.

Consider $\quad \beta = \alpha - \alpha_1$.

Then $\quad \overline{\beta} = \alpha - \alpha_1 + W = \overline{\alpha}, \quad$ as $\alpha_1 \in W$

Now $\quad m_1(T)\beta = 0$,

gives that the T-annihilator polynomial $m_2(X)$ of β is a factor of $m_1(X)$.

As $\quad \overline{\alpha} = \overline{\beta}$,

by Proposition (5.3.3), $\quad m_1(X) | m_2(X)$.

Consequently, $\quad m_1(X) = m_2(X)$.

However, $\quad \dim F[\overline{T}]\overline{\alpha} = \deg m_1(X)$,

$$\dim F[T]\beta = \deg m_2(X).$$

Hence, $\quad \dim F[\overline{T}]\overline{\alpha} = \dim F[T]\beta$.

We start with a criterion for a subspace of V to be T-complemented.

Theorem 3. Let T be a linear operator on a finite dimensional vector space V_F and W be a T-invariant subspace of V. Then, W is T-complemented iff for any $f(X) \in F[X]$ and $\alpha \in V$, $f(T)\alpha \in W$ implies $f(T)\alpha = f(T)\alpha_1$ for some $\alpha_1 \in W$.

Proof. Let W be T-complemented, then $V = W \oplus W'$ for some T-invariant subspace W'.

Consider $\alpha \in V$.

Thus, $\alpha = \alpha_1 + \alpha_2,\ \alpha_1 \in W,\ \alpha_2 \in W'$

$$f(T)\alpha = f(T)\alpha_1 + f(T)\alpha_2,$$

$$f(T)\alpha_1 \in W,$$

$$f(T)\alpha_2 \in W',$$

so that if $f(T)\alpha \in W$, we get $f(T)\alpha = f(T)\alpha_1$

and $f(T)\alpha_2 = 0$.

Conversely, let the hypothesis be satisfied. Consider the linear operator $\overline{T}$ on $\overline{V} = V/W$ induced by T. By the cyclic decomposition theorem [Theorem (5.4.10)],

$$\overline{V} = \oplus\sum_{i=1}^{t} F[T]\overline{\alpha}_i$$

for some $\alpha_i \in V$ with $\overline{\alpha}_i \neq \overline{0}$.

By Lemma 1, we get $\beta_i \in V$

with $\overline{\alpha}_i = \overline{\beta}_i$

and $\dim F[T]\overline{\alpha}_i = \dim F[T]\beta_i$.

Consider
$$W' = \sum_{i=1}^{t} F[T]\beta_i \qquad \text{... (1)}$$

Then,
$$\frac{(W'+W)}{W} = \sum_{i=1}^{t} F[T]\,\overline{\beta}_i = \frac{V}{W},$$

yields $V = W + W'$.

Further,
$$\dim\overline{V} = \sum_{i=1}^{t} \dim F[T]\,\overline{\alpha}_i = \sum_{i=1}^{t} F[T]\,\beta_i = \dim W'.$$

Hence,
$$\dim V = \dim\overline{V} + \dim W = \dim W' + \dim W.$$

As $V = W + W'$,

we get $W \cap W' = \{0\}$.

Hence, $V = W \oplus W'$.

This proves that W is T-complemented.

Definition 4. A linear operator T on a finite dimensional vector space V_F is said to be *semi-simple* if every T-invariant subspace of V is T-complemented. An $n \times n$-matrix A over F is said to be a *semi-simple* matrix if the linear operator on F^n given by A is semi-simple.

Theorem 5. Let T be a linear operator on a finite dimensional vector space V_F. Then T is semi-simple iff the minimal polynomial of T is a product of finitely many distinct primes in $F[X]$.

Proof. Let $m(X)$ be the minimal polynomial of T. Suppose that T is semi-simple and $p(X)$ be prime factor of $m(X)$ of multiplicity $k > 1$.

Then, $$m(X) = p(X)^k q(X),\ q(X) \in F[X]$$

and $$\gcd[p(X),\ q(X)] = 1.$$

Choose an $\alpha \in V$ whose T-annihilator polynomial is $m(X)$. Then, the T-annihilator polynomial of $\beta = q(T)\alpha$ is $p(X)^k$. Now, $W = F[T]p(T)\beta$ is a T-invariant subspace. By the hypothesis, W is T-complemented.

By Theorem 4, we get $$\beta_1 \in W$$

such that $$p(T)\beta = p(T)\beta_1.$$

However, $$\beta_1 \in F[T]p(T)\beta$$

yields $$\beta_1 = f(T)p(T)\beta$$

for some $$f(X) \in F[X],$$

so that $$p(T)\beta = f(T)p(T)^2\beta.$$

This yields $$p(T)^{k-1}\beta = f(T)p(T)^k\beta = 0.$$

This contradicts the fact that $p(X)^k$ is the T-annihilator polynomial of β. Hence, $k = 1$. Consequently, $m(X)$ is a product of distinct primes.

Conversely, let $$m(X) = p_1(X)p_2(X)\ldots p_k(X)$$

for some distinct prime factors $p_i(X)$.

It can be easily seen that if $f(X)$ and $g(X)$ are any two factors of $m(X)$ such that $f(X)|g(X)$, then $g(X) = f(X)h(X)$ with $\gcd[f(X),\ h(X)] = 1$. Let W be any T-invariant subspace of V. Let $f(X)$ be any non-zero polynomial in $F[X]$ such that for some $\alpha \neq 0$ in V, $f(T)\alpha \in W$. Let $m_1(X)$ be the $\overline{T}$-annihilator polynomial of $\overline{\alpha} = \alpha + W$ and $m'(X)$ be the T-annihilator polynomial of α.

Then, $$m_1(X)|m'(X),$$

so that $$m'(X) = m_1(X)m_2(X),$$
$$\gcd[m_1(X),\ m_2(X)] = 1.$$

Consequently, $$1 = m_1(X)u(X) + m_2(X)v(X) \quad \ldots (1)$$

for some $$u(X),\ v(X) \in F[X].$$

This gives $$\alpha = m_1(T)u(T)\alpha + m_2(T)v(T)\alpha \quad \ldots (2)$$

Now, $\beta = m_1(T)u(T)\alpha \in W$, and (2) gives

$$m_1(T)\alpha = m_1(T)\beta,$$

as $$m_1(T)m_2(T)v(T)\alpha = 0.$$

Now, $$f(T)\alpha \in W,$$

gives $$f(X) = m_1(X)g(X),$$

so that $$f(T)\alpha = g(T)m_1(T)\alpha = g(T)m_1(T)\beta = f(T)\beta.$$

So, by the Theorem 3, W is T-complemented. Hence, T is semi-simple.

This theorem gives the following theorem :

Theorem 6. An $n \times n$ matrix A over a field F is *semi-simple* iff the minimal polynomial of A is a product of finitely many distinct primes in $F[X]$.

Theorem 7. A linear operator T on V_F is semi-simple iff $V = W_1 \oplus W_2 \oplus \ldots \oplus W_k$ for some non-zero T-invariant subspaces W_i, such that each W_i is T-cyclic, and the restriction T_i of T to W_i has minimal polynomial a prime in $F[X]$.

Proof. Let $$V = V_1 \oplus V_2 \oplus \ldots \oplus V_t,$$

where V_i are different primary components relative to T.

Let V_i be $p_i(X)$-primary component relative to T. Then, the minimal polynomial of the restriction T_i' of T to V_i is $p_i(X)^{r_i}$, $m(X) = \prod_{i=1}^{t} p_i(X)^{r_i}$ is the minimal polynomial of T.

By the cyclic decomposition theorem,

$$V_i = \oplus \sum_{j=1}^{k_i} F[T]\alpha_{ij}, \ \alpha_{ij} \neq 0 \qquad \ldots (1)$$

with T-annihilator polynomial of α_{ij}, $m_{ij}(X)$, such that

$$m_{i1}(X) = p_i(X)^{r_i},$$

$$m_{i,j+1}(X) \big| m_{ij}(X).$$

$$V = \oplus \sum_{j=1}^{k_i} \sum_{j=1}^{k_i} W_{ij}, \ W_{ij} = F[T]\alpha_{ij}. \qquad \ldots (2)$$

Now, $m_{ij}(X) = p_i(X)^{r_{ij}}$ with $0 \le r_{ij} \le r_i$. Let T be semi-simple.

Then, by Theorem 6, $$r_i = 1,$$

and, hence, $$m_{ij}(X) = p_i(X) \text{ or } 1.$$

As $\alpha_{ij} \neq 0, m_{ij}(X) \neq 1.$

So $m_{ij}(X) = p_i(X).$

By (2), V is a direct sum of T-invariant subspaces W_{ij} such that the restriction of T to W_{ij} has minimal polynomial $p_i(X)$, a prime in $F[X]$.

Conversely, let $V = W_1 \oplus W_2 \oplus ... \oplus W_s$,

with $W_i = F[T]\alpha_i \neq \{0\}$

such that the minimal polynomial of the restriction T_i of T to W_i is a prime $p_i'(X) \in F[X]$

If $q_1(X), q_2(X), ..., q_l(X)$

are all the different polynomials among

$$p_1'(X), p_2'(X), ..., p_s'(X),$$

then, by Theorem (5.2.15), $m(X) = l\text{cm}[p_1'(X), p_2'(X), ..., p_s'(X)]$

$$= l\text{cm}\{q_1(X), q_2(X), ..., q_l(X)\}$$

$$= q_1(X)q_2(X).....q_l(X).$$

By Theorem 6, T is semi-simple.

The matrix version of this theorem is below.

Theorem 8. Any $n \times n$-matrix A over a field F is semi-simple iff A is similar to a direct sum of some matrices $A_1, A_2, ..., A_k$ such that for each i, A_i is either 0 or a companion matrix of some prime $p_i(X) \neq X$ in $F[X]$.

Theorem 9. Let T be a linear operator on a finite dimensional vector space V_F with minimal polynomial

$$m(X) = (X - c_1)^{r_1}(X - c_2)^{r_2}...(X - c_k)^{r_k}$$

for some distinct elements $c_1, c_2, ..., c_k$ of F. Then T is semi-simple iff T is diagonalizable.

Proof. T is diagonalizable iff

$$m(X) = (X - c_1)(X - c_2)...(X - c_k) \qquad \text{[Theorem (5.3.23)]}$$

This fact along with Theorem 6 completes the proof.

As any monic polynomial $m(X)$ over **C** factors into linear factors over **C**, this theorem gives the following corollary :

Corollary 10. Any $n \times n$-matrix over **C**, is semi-simple iff it is similar to a diagonal matrix.

Exercise 5.5

1. Which of the following matrices over **R**, are semi-simple :

$$\begin{bmatrix}1 & 1\\ 1 & 1\end{bmatrix}, \begin{bmatrix}0 & 1\\ -1 & 0\end{bmatrix}, \begin{bmatrix}1 & 1\\ 0 & 1\end{bmatrix}.$$

2. Prove that any 2×2-matrix over **R** is either similar to an elementary Jordan matrix or is semi-simple.

3. Prove that a nilpotent linear operator is semi-simple iff it is zero.

4. Let T be a linear operator on a finite dimensional vector space V_F, such that $V = F[T]\alpha$ and the minimal polynomial of T is $p(X)^r$ for some prime $p(X)$ in $F[X]$. Prove the following :

   (*i*) Any T-invariant subspace W equals $F[T]p(T)^s\alpha$ for a uniquely determined integer s, $1 \le s \le r$.

   (*ii*) For any two T-invariant subspaces W_1 and W_2 of V, either $W_1 \subseteq W_2$ or $W_2 \subseteq W_1$.

   (*iii*) T is semi-simple iff $r = 1$.

5. Let T be a linear operator on a finite dimensional vector space V_F. Prove that T is semi-simple iff the restrictions of T to various primary components of V relative to T are semi-simple.

6. Prove Theorem 5, by using the results in Exercises 4 and 5.

7. Give an example of two-semi-simple real matrices, whose sum is not semi-simple.

8. Give an example of two 2×2-real matrices A and B such that both are semi-simple, but AB is not-semi-simple.

Chapter 6

Inner Product Space

6.1 Inner Product

In this chapter we intend to study the connection of linear algebra to geometry and analysis. It is customary that the underlying fields used in studying geometry and analysis are **R** or **C**. So, in this chapter we study vector spaces over **R** or **C** ; we call such vector spaces real or complex vector spaces, respectively. Consider $V_{\mathbf{R}} = \mathbf{R}^3$. For any $\alpha = (a_1, a_2, a_3) \in V$, the length $\|\alpha\|$ of α is given by

$$\|\alpha\| = \sqrt{a_1^2 + a_2^2 + a_3^2}.$$

Further, given $\qquad \beta = (b_1, b_2, b_3) \in V,$

the dot product (or inner product) $\alpha.\beta = a_1b_1 + a_2b_2 + a_3b_3$ is well known.

The angle θ between α and β is derived from the equation

$$\cos\theta = \frac{\alpha.\beta}{\|\alpha\|\,\|\beta\|}.$$

Here, we start with an arbitrary vector space V_F with $F = \mathbf{R}$ or $\mathbf{C}$. We are going to define a function $V \times V \to F$ which for each $(\alpha, \beta) \in V \times V$, determines $(\alpha|\beta) \in F$ such that $(\alpha|\beta)$ has some basic properties similar to those of the dot product in $\mathbf{R}^3$. This concept is, then, used to define various geometric concepts in V and to develop geometry in V. For any complex number $z = a + bi$, $a, b \in R$, $i = \sqrt{-1}$, $\bar{z}$ denotes the complex conjugate $a - bi$ of z.

Definition 1. Let V_F be a real or complex vector space. An *inner product* in V is a function from $V \times V$ to F which for each $(\alpha, \beta) \in V \times V$ determines $(\alpha|\beta)$ in F satisfying the following conditions. For any $\alpha, \beta, \gamma \in V$ and $a, b \in F$

(*i*) $(\beta|\alpha) = \overline{(\alpha|\beta)}$

(*ii*) $(\alpha|\alpha) > 0$ whenever $\alpha \neq 0$

(*iii*) $(a\alpha + b\beta|\gamma) = a(\alpha|\gamma) + b(\beta|\gamma)$

In case $F = \mathbf{R}$, (*i*) simply becomes

(*i*)′ $(\beta|\alpha) = (\alpha|\beta)$.

Definition 2

(*i*) A real or complex vector space V_F together with an inner product $(\ |\ )$ is called an *inner product space*.

(*ii*) For any vector α in an inner product space V_F, the *norm* of α, denoted by $\|\alpha\|$, is defined to be the positive square root of $(\alpha|\alpha)$.

Example 2.a. Let $V = \mathbf{R}^n$. This V is real vector space.

For any $\alpha = (a_1, a_2, \ldots, a_n)$,

$$\beta = (b_1, b_2, \ldots, b_n),$$

define $(\alpha|\beta) = \alpha\beta^T = a_1b_1 + a_2b_2 + \ldots + a_nb_n$.

This is an inner product in V. This makes V an inner product space. Further,

$$\|\alpha\| = (\alpha|\alpha) = \sqrt{a_1^2 + a_2^2 + \ldots.. + a_n^2}.$$

The inner product given above is called *standard inner product* in $\mathbf{R}^n$. The resulting inner product space is called a *Euclidean space*.

Example 2.b. Consider the complex space $V = \mathbf{C}^n$.

For any $\alpha = (a_1, a_2, \ldots, a_n)$,

$$\beta = (b_1, b_2, \ldots, \beta_n),$$

define $\bar{\beta} = (\bar{b}_1, \bar{b}_2, \ldots., \bar{b}_n)$,

and $(\alpha|\beta) = \alpha\bar{\beta}^T = a_1\bar{b}_1 + a_2\bar{b}_2 + \ldots + a_n\bar{b}_n$.

Let $\gamma = (c_1, c_2, \ldots, c_n) \in V$

and $a, b \in \mathbf{C}$.

(*i*) $$(\beta|\alpha) = \sum_{j=1}^{n} b_j\, \bar{a}_j = \overline{\sum_{j=1}^{n} a_j\, \bar{b}_j}, \qquad \text{as } \bar{\bar{z}} = z \in \mathbf{C}$$

Consequently, $(\beta|\alpha) = \overline{(\alpha|\beta)}$.

(*ii*) For $\alpha \neq 0$, $a_j \neq 0$ for some j, $a_j\, \bar{a}_j > 0$.

Consequently, $(\alpha|\alpha) = a_1\bar{a}_1 + a_2\bar{a}_2, \ldots, + a_n\bar{a}_n > 0$

(*iii*) $a\alpha + b\beta = (aa_1 + bb_1, aa_2 + bb_2, \ldots, aa_n + bb_n)$,

gives $$(a\alpha + b\beta|\gamma) = \sum^{n} (aa_j + bb_j)\bar{c}_j$$

$$= a\sum_{j=1}^{n} a_j\bar{c}_j + b\sum_{j=1}^{n} b_j\bar{c}_j$$

$$= a(\alpha|\gamma) + b(\beta|\gamma).$$

Thus, the function $(\ |\ )$ defined above is an inner product ; it is called the *standard inner product* in $\mathbf{C}^n$; the resulting inner product space $\mathbf{C}^n$ is called a *unitary space*.

Example 2.c. Let $V = C[a, b]$ be the complex vector space of all complex valued continous functions defined on the closed interval $[a, b]$, $a < b$. For any $f, g \in V$, define

$$(f|g) = \int_a^b f(t)\overline{g(t)}\,dt.$$

By using the following elementary properties of an integral, one can see that the, above expression is an inner product,

$$\int_a^b (f+g)(t)\,dt = \int_a^b f(t)\,dt + \int_a^b g(t)\,dt$$

$$\int_a^b cf(t)\,dt = c\int_a^b f(t)\,dt$$

$$\overline{\int_a^b f(t)\,dt} = \int_a^b \overline{f(t)}\,dt$$

and if $f(t) \geq 0$ for all t in $[a, b]$, and $f \neq 0$, then, $\int_a^b f(t)\,dt > 0$.

Throughout, $F = \mathbf{R}$ or $F = \mathbf{C}$. Further, V_F is an inner product space.

Lemma 3. For any $\alpha, \beta, \gamma \in V_F$, $a, b \in F$, the following relations hold :

(*i*) $(\alpha|\beta+\gamma) = (\alpha|\beta) + (\alpha|\gamma)$

(*ii*) $(\alpha|a\beta) = \bar{a}(\alpha|\beta)$

(*iii*) If for a fixed $\alpha \in V$, $(\alpha|\beta) = 0$ for all $\beta \in V$, then $\alpha = 0$.

Proof.

(*i*) $(\alpha|\beta+\gamma) = \overline{(\beta+\gamma|\alpha)}$ by the definition of an inner product. But also,

$$(\beta+\gamma|\alpha) = (\beta|\alpha) + (\gamma|\alpha)$$

Consequently, $(\alpha|\beta+\gamma) = \overline{(\beta|\alpha)} + \overline{(\gamma|\alpha)} = (\alpha|\beta) + (\alpha|\gamma)$

(*ii*) $(\alpha|a\beta) = \overline{(a\beta|\alpha)} = \overline{a(\beta|\alpha)} = \bar{a}\overline{(\beta|\alpha)} = \bar{a}(\alpha|\beta)$

(*iii*) Let $(\alpha|\beta) = 0$ for all $\beta \in V$. In particular, for $\beta = \alpha$, $(\alpha|\alpha) = 0$. This gives $\alpha = 0$, as by definition $(\alpha|\alpha) > 0$ if $\alpha \neq 0$.

Proposition 4.

(*i*) For any $\beta_1, \beta_2 \in V_F$ if $(\alpha|\beta_1) = (\alpha|\beta_2)$ for all $\alpha \in V$, then, $\beta_1 = \beta_2$.

(*ii*) For any $\gamma \in V$, the mapping

$$\sigma_\gamma : V \to F,$$

such that $$\sigma_\gamma(\alpha) = (\alpha|\gamma),$$

$$\alpha \in V,$$

is a linear functional on V.

(*iii*) $$\lambda : V \to V_F^*,$$

the dual space of V, given by $\lambda(\gamma) = \sigma_\gamma$

is a one-to-one mapping such that

$$\lambda(\beta+\gamma) = \lambda(\beta) + \lambda(\gamma)$$
$$\lambda(a\beta) = \bar{a}\lambda(\beta)$$

for all $\beta, \gamma \in V$ and $a \in F$. Further, if V_F is finite dimensional, then λ is also onto; if in addition $F = \mathbf{R}$, λ is a vector space isomorphism.

Proof.

(*i*) As $$\overline{(\beta_1|\alpha)} = (\alpha|\beta_1) = (\alpha|\beta_2) = \overline{(\beta_2|\alpha)},$$

we get $$(\beta_1|\alpha) = (\beta_2|\alpha).$$

So that $$(\beta_1 - \beta_2|\alpha) = (\beta_1|\alpha) - (\beta_2|\alpha) = 0$$

for all $$\alpha \in V.$$

So, by Lemma 3, $$\beta_1 - \beta_2 = 0,$$

i.e. $$\beta_1 = \beta_2.$$

(*ii*) For any $a \in F$, $\alpha, \beta \in V$,

$$(a\alpha + \beta|\gamma) = a(\alpha|\gamma) + (\beta|\gamma), \qquad \text{by definition}$$

yields $$\sigma_\gamma(a\alpha + \beta) = a\sigma_\gamma(\alpha) + \sigma_\gamma(\beta).$$

This proves that σ_γ is a linear functional on V.

(*iii*) For any $\gamma \in V$, σ_γ being a linear functional on V, is in V_F^*.

$$\sigma_{\beta+\gamma}(\alpha) = (\alpha|\beta+\gamma) = (\alpha|\beta) + (\alpha|\gamma)$$

yield $$\lambda(\beta+\gamma) = \lambda(\beta) + \lambda(\gamma).$$

Again, $$\sigma_{a\beta}(\alpha) = (\alpha|a\beta) = \bar{a}(\alpha|\beta) = \bar{a}\sigma_\alpha(\beta)$$

yields $$\lambda(a\beta) = \bar{a}\lambda(\beta).$$

Thus, at least, λ is a group homomorphisms. Let $K = \ker \lambda$.

If $\beta \in K$,

then $\lambda(\beta) = 0$,

gives $\sigma_\beta = 0$,

i.e., $\sigma_\beta(\alpha) = 0$

for all $\alpha \in V$,

i.e., $(\alpha|\beta) = 0$

for all $\alpha \in V$.

So, by Lemma 3, $\beta = 0$.

Hence, $K = \{0\}$.

This proves that λ is one-to-one.

Let $W = \text{range}\ (\lambda)$.

Then, W is a subgroup of V^* under addition.

For $\sigma_\gamma \in W$, and $a \in F$, $\lambda(\bar{a}\gamma) = a\,\sigma_\gamma \in W$.

This gives, W is a subspace of V. Let V be finite dimensional with dim $V = n$.

We have a basis $B = \{\alpha_1, \alpha_2, \ldots, \alpha_n\}$ of V.

By using the fact that $\lambda(\alpha+\beta) = \lambda(\alpha) + \lambda(\beta)$

and $\lambda(\alpha\beta) = \bar{a}\lambda(\beta)$,

It is immediately clear that $B' = \{\lambda(\alpha_1), \lambda(\alpha_2), \ldots, \lambda(\alpha_n)\}$ is a basis of W, so that $\dim W = n = \dim V^*$. Consequently, λ is onto.

For $a \in \mathbf{R}$, $\bar{a} = a$, and the last part follows.

As λ defined in Proposition 4 is a group homomorphism, $\lambda(0) = \bar{0}$, yields $(\alpha|0) = 0$ for all $\alpha \in V$.

Then, $(0|\alpha) = \overline{(\alpha|0)} = 0$.

For any complex number $z = a + ib$, a, $b \in \mathbf{R}$, a and b are called the *real* and *imaginary* components of z respectively.

We shall write $a = Re(z)$, $b = Im(z)$.

Clearly, $z + \bar{z} = 2\,Re(z)$

and $z - \bar{z} = 2i\,Im(z)$.

From the definition of inner product, it is immediate that

$$(-\alpha|\beta) = -(\alpha|\beta) = (\alpha|-\beta).$$

Theorem 5. For any α, β in an inner product space V_F, and $c \in F$, the following relations hold :

(*i*) $\|\alpha\| \geq 0$

(*ii*) $\|\alpha\| = 0$ iff $\alpha = 0$

(*iii*) $\|c\alpha\| = |c|\ \|\alpha\|$

(*iv*) $\|\alpha+\beta\|^2 = \|\alpha\|^2 + 2\,\mathrm{Re}(\alpha|\beta) + \|\beta\|^2$

(*v*) If $F = R$, then,

$$(\alpha|\beta) = \frac{1}{2}\|\alpha+\beta\|^2 - \frac{1}{2}\|\alpha\|^2 - \frac{1}{2}\|\beta\|^2$$

$$= \frac{1}{4}\|\alpha+\beta\|^2 - \frac{1}{4}\|\alpha-\beta\|^2. \qquad \text{... (1)}$$

(*vi*) If $F = \mathbf{C}$, then,

$$(\alpha|\beta) = \frac{1}{4}\|\alpha+\beta\|^2 - \frac{1}{4}\|\alpha-\beta\|^2 + \frac{1}{4}\|\alpha+i\beta\|^2 - \frac{1}{4}\|\alpha-i\beta\|^2. \qquad \text{... (2)}$$

(The identities (1) and (2) are called *polarization identities*).

Proof. (*i*) and (*ii*) are immediately clear from the definition of an inner product.

For (*iii*), $\quad (c\alpha|c\alpha) = c(\alpha|c\alpha) = c\bar{c}(\alpha|\alpha) = |c|^2\ \|\alpha\|^2$

by using Lemma 3 (*ii*).

Consequently, $\quad \|c\alpha\| = \sqrt{(c\alpha|c\alpha)} = |c|\ \|\alpha\|.$

Now, for (*iv*),

$$\|\alpha+\beta\|^2 = (\alpha+\beta|\alpha+\beta)$$

$$= (\alpha|\alpha) + (\alpha|\beta) + (\beta|\alpha) + (\beta|\beta)$$

$$= \|\alpha\|^2 + (\alpha|\beta) + \overline{(\alpha|\beta)} + \|\beta\|^2$$

$$= \|\alpha\|^2 + 2\,\mathrm{Re}(\alpha|\beta) + \|\beta\|^2. \qquad \text{... (3)}$$

By changing β to $-\beta$, and by observing that $\|-\beta\|^2 = \|\beta\|^2$, $(\alpha|-\beta) = -(\alpha|\beta)$, we get

$$\|\alpha-\beta\|^2 = \|\alpha\|^2 - 2\,\mathrm{Re}(\alpha|\beta) + \|\beta\|^2 \qquad \text{... (4)}$$

Subtraction of (4) from (3) gives

$$4\,\mathrm{Re}(\alpha|\beta) = \|\alpha+\beta\|^2 - \|\alpha-\beta\|^2. \qquad \text{... (5)}$$

If $F = \mathbf{R}$, $\quad Re(\alpha|\beta) = (\alpha|\beta).$

Consequently, $\quad (\alpha|\beta) = \frac{1}{4}\|\alpha+\beta\|^2 - \frac{1}{4}\|\alpha-\beta\|^2.$

Also, for $F = \mathbf{R}$, (3) gives $(\alpha|\beta) = \frac{1}{2}\|\alpha+\beta\|^2 - \frac{1}{2}\|\alpha\|^2 - \frac{1}{2}\|\beta\|^2$

This proves (*v*).

Let $F = \mathbf{C}$. In (5) replace β by $i\beta$.

Then, $\quad 4\,\mathrm{Re}(\alpha|i\beta) = \|\alpha+i\beta\|^2 - \|\alpha-i\beta\|^2.$

However, $$(\alpha|i\beta) = -i(\alpha|\beta),$$

gives $$\text{Re}(\alpha|i\beta) = \text{Im}(\alpha|\beta)$$

Consequently, $$4\,\text{Im}(\alpha|i\beta) = \|\alpha + i\beta\|^2 - \|\alpha - i\beta\|^2. \quad \ldots (6)$$

Thus, from (5) and (6), $$(\alpha|\beta) = \text{Re}(\alpha|\beta) + i\,\text{Im}(\alpha|\beta)$$

$$= \frac{1}{4}\|\alpha+\beta\|^2 - \frac{1}{4}\|\alpha-\beta\|^2 + \frac{i}{4}\|\alpha+i\beta\|^2 - \frac{i}{4}\|\alpha-i\beta\|^2.$$

Theorem 6. (Cauchy-Schwarz inequality). For any α, β in an inner product space V_F

$$|(\alpha|\beta)| \le \|\alpha\|\,\|\beta\|$$

Proof. If $(\alpha|\beta) = 0$, the result obviously holds. Let $(\alpha|\beta) \ne 0$. Then, $\alpha \ne 0 \ne \beta$. For any $c \in F$

$$0 \le \|\beta + c\alpha\|^2 = \|\beta\|^2 + |c|^2\|\alpha\|^2 + 2\,\text{Re}\,c(\alpha|\beta).$$

(see Theorem 5). This gives

$$0 \le \|\beta + c\alpha\|^2 = (\|\beta\| - |c|\,\|\alpha\|)^2 + 2|c|\,\|\alpha\|\,\|\beta\| + 2\,\text{Re}\,c(\alpha|\beta).$$

Choose a, $c \in F$ such that $$|c| = \frac{\|\beta\|}{\|\alpha\|}.$$

This gives $$0 \le 2\|\beta\|^2 + 2\,\text{Re}\,c(\alpha|\beta). \quad \ldots (1)$$

Now, $$c = -\frac{\|\beta\|\,|(\alpha|\beta)|}{\|\alpha\|\,(\alpha|\beta)}$$

satisfies $$|c| = \frac{\|\beta\|}{\|\alpha\|}.$$

Consider this value of c.

We get $$c(\alpha|\beta) = -\frac{\|\beta\|\,|(\alpha|\beta)|}{\|\alpha\|}$$

so that $$\text{Re}\,c(\alpha|\beta) = -(\|\beta\|/\|\alpha\|)|(\alpha|\beta)|.$$

Then (1) gives $$0 \le 2\|\beta\|^2 - 2(\|\beta\|/\|\alpha\|)|(\alpha|\beta)|.$$

Hence, $$|\alpha|\beta| \le \|\alpha\|\,\|\beta\|.$$

Remark 7. (*i*) By considering the inner product space $\mathbf{C}^n$ in Example 2.b, this theorem gives that for any

$$\alpha = (a_1, a_2, \ldots, a_n),$$

$$\beta = (b_1, b_2, \ldots, b_n) \qquad \text{in } \mathbf{C}^n.$$

$$|(\alpha|\beta)| = \sqrt{\sum_{i=1}^{n} a_i\bar{b}_i} \le \left(\sum_{i=1}^{n}|a_i|^2\right)^{\frac{1}{2}}\left(\sum_{i=1}^{n}|b_i|^2\right)^{\frac{1}{2}}$$

(*ii*) By using the inner product space in Example 2.c, we get that for any two continuous, complex valued functions *f* and *g*, defined on a closed interval $[a, b]$,

$$\left|\int_a^b f(t)\overline{g(t)}\,dt\right| \le \left[\int_a^b |f(t)|^2\,dt\right]^{\frac{1}{2}}\left[\int_a^b |g(t)|^2\,dt\right]^{\frac{1}{2}}.$$

Definition 8. For any $m\times n$ complex matrix $A = [a_{ij}]$, its *conjugate transpose A** is the $n\times m$-matrix $[a'_{ij}]$ such that $a'_{ij} = \bar{a}_{ji}$.

For example, if
$$A = \begin{bmatrix} i & 1 \\ 0 & i \end{bmatrix},$$
then,
$$A^* = \begin{bmatrix} -i & 0 \\ 1 & -i \end{bmatrix}.$$

By using the above definition, we give another example of an inner product space. Fix a positive integer *n*. Let *V* be the complex space of all $n\times n$-complex matrices.

For any $A = [a_{ij}]$, $B = [b_{ij}]$ in *V*, define the inner product

$$(A|B) = \text{trace}(AB^*) = \sum_{i=1}^{n}\sum_{k=1}^{n} a_{ik}\,\bar{b}_{ik}.$$

The definition is very much similar to that in Example 2.c.

For any set *X*, a function $d: X\times X \to \mathbf{R}$ is called a *metric* in *X*, if for any *a*, *b*, $c\in X$ it satisfies the following :

(*i*) $d(a, b) \ge 0$

(*ii*) $d(a, b) = 0$ iff $a = b$

(*iii*) $d(a, b) = d(b, a)$

(*iv*) $d(a, c) \le d(a, b) + d(b, c)$.

Theorem 9. Let V_F be an inner product space. The following relations hold :

(*i*) For any $\alpha, \beta \in V$ $\quad \|\alpha+\beta\| \le \|\alpha\| + \|\beta\|$.

(*ii*) The function $d: V\times V \to \mathbf{R}$ given by $d(\alpha, \beta) = \|\alpha - \beta\|$, $\alpha, \beta \in V$ is a *metric* in *V*.

Proof

(*i*) By Theorem 5, $\quad \|\alpha+\beta\|^2 = \|\alpha\|^2 + \|\beta\|^2 + 2\,\text{Re}(\alpha|\beta)$.

But $$\text{Re}(\alpha|\beta) \le |(\alpha|\beta)| \le \|\alpha\|\,\|\beta\|,$$

by the Cauchy-Schwarz inequality.

Consequently, $$\|\alpha+\beta\|^2 \le \|\alpha\|^2 + \|\beta\|^2 + 2\|\alpha\|\,\|\beta\| = (\|\alpha\|+\|\beta\|)^2.$$

By taking square roots, we get

$$\|\alpha+\beta\| \le \|\alpha\| + \|\beta\|.$$

(*ii*) That $$d(\alpha,\beta) \ge 0,\ d(\alpha,\beta) = 0$$

iff $\alpha = \beta$, and $d(\alpha,\beta) = d(\beta,\alpha)$ are immediate from the properties of norm listed in Theorem 5. Finally, for any $\alpha, \beta, \gamma \in V$

$$d(\alpha,\gamma) = \|\alpha-\gamma\| = \|(\alpha-\beta)+(\beta-\gamma)\|$$
$$\le \|\alpha-\beta\| + \|\beta-\gamma\| = d(\alpha,\beta) + d(\beta,\gamma).$$

Hence, d is a metric in V.

For any α, β in an inner product space V_F, $d(\alpha,\beta) = \|\alpha-\beta\|$ is called *distance between* α, β. We now define the angle between two vectors in a real inner product space $V_{\mathbf{R}}$. Let α, β be two non-zero vectors in $V_{\mathbf{R}}$. By the Cauchy-Schwarz inequality

$$\left|\frac{(\alpha|\beta)}{\|\alpha\|\,\|\beta\|}\right| \le 1.$$

Thus, $$-1 \le \frac{(\alpha|\beta)}{\|\alpha\|\,\|\beta\|} \le 1.$$

Cos θ strictly decreases from 1 to -1 as θ varies from 0 to π. The above inequality shows that there exists unique θ, $0 \le \theta \le \pi$ such that

$$\cos\theta = \frac{(\alpha|\beta)}{\|\alpha\|\,\|\beta\|}.$$

We call θ the angle between α and β. If $\cos\theta = 1$, α and β are said to be parallel having the same direction. If $\cos\theta = -1$, they are said to be anti-parallel, having opposite directions.

Theorem 10. Let α, β be two non-zero vectors in a real inner product space $V_{\mathbf{R}}$. Then, the following relations hold :

(*i*) α, β are in the same direction iff $\beta = c\alpha$ for some $c > 0$.

(*ii*) α, β are in opposite direction iff $\beta = c\alpha$ for some $c < 0$.

Proof

(*i*) For any $c \in \mathbf{R}$, $$0 \le \|\beta + c\alpha\|^2 = \|\beta\|^2 + c^2\|\alpha\|^2 + 2c(\alpha|\beta) \qquad \text{... (1)}$$

Let α, β be in the same direction.

Then, $$\cos\theta = 1,$$

gives $$(\alpha|\beta) = \|\alpha\|\,\|\beta\|$$

So that (1) becomes

$$0 \le \|\beta + c\alpha\|^2 = \|\beta\|^2 + c^2\|\alpha\|^2 + 2c\|\alpha\|\,\|\beta\| = (\|\beta\| + c\|\alpha\|)^2$$

for $$c = -\frac{\|\beta\|}{\|\alpha\|},$$

we get $$\|\beta\| + c\|\alpha\|^2 = 0,$$

so that $$0 \le \|\beta + c\alpha\|^2 \le 0.$$

Hence $$\beta + c\alpha = 0,$$

i.e., $$\beta = c'\|\alpha\|,$$

with $$c' = -c = \frac{\|\beta\|}{\|\alpha\|} > 0.$$

Conversely, let $\beta = c\alpha$ for some $c > 0$,

then, $$(\alpha|\beta) = (\alpha|c\alpha) = c(\alpha|\alpha) = c\|\alpha\|^2$$

At the same time, $$\|\beta\| = \|c\alpha\| = c\|\alpha\|$$

so that $$\|\alpha\|\|\beta\| = c\|\alpha\|^2 = (\alpha|\beta).$$

Consequently, $$\cos\theta = 1$$

and, hence, α and β are in the same direction. Similarly, one can prove (*ii*).

Any $\gamma \in V_F$ is called a *unit vector* if $\|\gamma\| = 1$. For any $\alpha \ne \acute{0}$ in V, $\overline{\alpha} = \frac{1}{\|\alpha\|}\alpha$ is a unit vector. In case $F = \mathbf{R}$, this theorem shows that $\overline{\alpha}$ is a unit vector in the direction of α. Further $-\overline{\alpha}$ is the unit vector in opposite direction of α.

Example 10.a. $\epsilon_1 = (1, 0, 0)$, $\epsilon_2 = (0, 1, 0)$, $\epsilon_3 = (0, 0, 1)$ are unit vectors in $\mathbf{R}^3$.

Now, for $$\alpha = (0, 2, 0), \ \|\alpha\| = 2.$$

So, α is not a unit vector.

But $$\overline{\alpha} = \frac{1}{\|\alpha\|}\alpha = (0, 1, 0) = \epsilon_2,$$

so that ϵ_2 is the unit vector in the direction of α.

Two vector α, β in V_F are said to be *orthogonal* to each other if $(\alpha|\beta) = 0$; equivalently, if $(\beta|\alpha) = 0$. If $\alpha \ne 0 \ne \beta$ and $F = \mathbf{R}$, $(\alpha|\beta) = 0$, is equivalent to

$$\cos\theta = \frac{(\alpha|\beta)}{\|\alpha\|\,\|\beta\|} = 0$$

i.e., $$\theta = \pi/2.$$

Thus, in geometric sense two, non-zero vectors α, β in $V_{\mathbf{R}}$ are orthogonal iff the angle between them is $\pi/2$. For example, in $\mathbf{R}^3$, ϵ_1, ϵ_2, ϵ_3 are pairwise orthogonal.

The vectors $\alpha = (1, 2, 0)$, $\beta = (2, -1, 3)$ are orthogonal.

Theorem 11. Let $\alpha_1, \alpha_2, \ldots, \alpha_n$ be any $n \geq 1$ non-zero vectors in an inner product space V_F, such that they are pairwise orthogonal. Then, $\alpha_1, \alpha_2, \ldots, \alpha_n$ are linearly independent.

Proof.

By the hypothesis $(\alpha_i|\alpha_j) = 0$ whenever $i \neq j$, but $(\alpha_i|\alpha_i) = \|\alpha_i\|^2 > 0$.

For some $c_1, c_2, \ldots, c_n \in F$,

let $$c_1\alpha_1 + c_2\alpha_2 + \ldots + c_n\alpha_n = 0.$$

Then, for $1 \leq j \leq n$

$$0 = \left(\sum_{i=1}^{n} c_i\alpha_i \middle| \alpha_j\right) = \sum_{i=1}^{n} c_i(\alpha_i|\alpha_j) = c_j(\alpha_j|\alpha_j) = c_j\|\alpha_j\|^2.$$

As $\|\alpha_j\| \neq 0$, we get $c_j = 0$. This proves that $\alpha_1, \alpha_2, \ldots, \alpha_n$ are linearly independent.

Theorem 12. Let $\alpha_1, \alpha_2, \ldots, \alpha_n$ be any $n \geq 1$, linearly independent vectors in an inner product space V_F. Define $\beta_1, \beta_2, \ldots, \beta_n \in V$ inductively as follows :

$$\beta_1 = \alpha_1,$$

and for $2 \leq i \leq n$ $$\beta_i = \alpha_i - \sum_{j=1}^{i-1} \frac{(\alpha_i|\beta_j)}{\|\beta_j\|^2}\beta_j.$$

Then, $\beta_1, \beta_2, \ldots, \beta_n$ are pairwise orthogonal vectors in V. Further, the subspaces spanned by $\{\alpha_1, \alpha_2, \ldots, \alpha_n\}$ and $\{\beta_1, \beta_2, \ldots, \beta_n\}$ are the same.

Proof. We apply induction on n. For $n = 1$, $\beta_1 = \alpha_1$ and the result obviously holds. Let $n > 1$ and the result hold for any $m < n$ linearly independent members of V. By applying the induction hypothesis on $\alpha_1, \alpha_2, \ldots, \alpha_{n-1}$, we get that $\beta_1, \beta_2, \ldots, \beta_{n-1}$ are linearly independent and

$$W = \text{span}\{\alpha_1, \alpha_2, \ldots, \alpha_{n-1}\} = \text{span}\{\beta_1, \beta_2, \ldots, \beta_{n-1}\}.$$

Now, $$\beta_n = \alpha_n - \sum_{j=1}^{n-1} \frac{(\alpha_n|\beta_j)}{\|\beta_j\|^2}\beta_j. \qquad \ldots (1)$$

If $\beta_n = 0$, we get $$\alpha_n = \sum_{j-1}^{n-1} \frac{(\alpha_n|\beta_j)}{\|\beta_j\|^2}\beta_j \in \text{span}\{\alpha_1, \alpha_2, \ldots, \alpha_{n-1}\}.$$

This gives $\alpha_1, \alpha_2, \ldots, \alpha_n$ are linearly dependent. This is a contradiction. Hence $\beta_n \neq 0$.

Further, for $1 \leq k \leq n-1$

$$(\beta_n|\beta_k) = (\alpha_n|\beta_k) - \sum_{j=1}^{n-1} \frac{(\alpha_n|\beta_j)}{||\beta_j||^2}(\beta_j|\beta_k).$$

For $1 \leq j \leq n-1$, $(\beta_j|\beta_k) = 0$ if $k \neq j$, by the induction hypothesis.

Thus, $$(\beta_n|\beta_k) = (\alpha_n|\beta_k) - \frac{(\alpha_n|\beta_k)}{\|\beta_k\|^2}(\beta_k|\beta_k) = 0, \qquad \text{as } (\beta_k|\beta_k) = \|\beta_k\|^2$$

Hence, $\beta_1, \beta_2, \ldots, \beta_n$ are pairwise orthogonal. By Theorem 11, they are linearly independent. So, the subspace W_1 spanned by these n vectors has dimension n. It is clear from (1) that $\alpha_n \in W_1$. Consequently, n linearly independent vectors $\alpha_1, \alpha_2, \ldots, \alpha_n$ are in W_1 so that $W_1 = \text{span}\{\alpha_1, \alpha_2, \ldots, \alpha_n\}$. This completes the proof.

In $\mathbf{R}^n$ (or $\mathbf{C}^n$), the standard basis $B = \{\epsilon_1, \epsilon_2, \ldots, \epsilon_n\}$ is such that its members are pairwise orthogonal, and each of them is a unit vector. For any $n \geq 2$, this vector space has infinitely many bases, whose members are pairwise orthogonal, and are unit vectors. We now discuss a method of obtaining such bases for any finite dimensional inner product space. A subset S of V_F is called an *orthogonal set* if for any $\alpha, \beta \in S$ with $\alpha \neq \beta, (\alpha|\beta) = 0$; if in addition S is a basis of V, it is called an *orthogonal basis*.

Definition 13. A subset S of an inner product space V_F is called an *orthonormal* subset if for any $\alpha, \beta \in S$.

(*i*) $\|\alpha\| = 1$;

(*ii*) for $\alpha \neq \beta$, $(\alpha|\beta) = 0$.

If in addition, S is a basis of V_F, it is called an *orthonormal basis* of V_F.

Thus, standard basis of $\mathbf{C}^n$ is an *orthonormal basis*.

14. *Gram-Schmidt Orthonormalization Process*. Consider any sequence (finite or infinite) of linearly independent vectors

$$\alpha_1, \alpha_2, \ldots, \alpha_k, \ldots \qquad \ldots (1)$$

in V. Let W_i be the subspace spanned by $\alpha_1, \alpha_2, \ldots, \alpha_i$, and W be the subspace spanned by all these given vectors. Then, W is union of the W_i's. By Theorem 12, if we define

$$\beta_1 = \alpha_1,$$

$$\beta_n = \alpha_n - \sum_{j-1}^{n-1} \frac{(\alpha_n|\beta_j)}{\|\beta_j\|^2}\beta_j,$$

$n \geq 2$, then, $\beta_1, \beta_2, ..., \beta_i$ form a basis of W_i.

This, in turn, gives that $\beta_1, \beta_2, ..., \beta_k, ...$

form a basis of W. Further, these vectors are pairwise orthogonal.

For any $n \geq 1$, define $$\gamma_n = \frac{\beta_n}{\|\beta_n\|}.$$

Then, γ_n is a unit vector.

Further, $$\gamma_1, \gamma_2, ..., \gamma_k, ... \qquad ...(2)$$

form an *orthonormal* basis of W. This process of constructing an *orthonormal* basis, starting from a basis of a countably generated subspace of V_F is called *Gram-Schmidt Orthonormalization Process.*

Example 14.a. Let $V = \mathbf{C}^4$. Now $\alpha_1 = (1, 1, 1, 0)$, $\alpha_2 = (0, 1, 1, 1)$, $\alpha_3 = (0, 0, 1, 1)$, $\alpha_4 = (0, 0, 3, 0)$ form a basis of V.

Put $$\beta_1 = \alpha_1 = (1, 1, 1, 0).$$

Then, $$\|\beta_1\|^2 = 3, \ (\alpha_2|\beta_1) = 2$$

so that $$\beta_2 = \alpha_2 - \frac{(\alpha_2|\beta_1)}{\|\beta_1\|^2}\beta_1$$

$$= (0, 1, 1, 1) - \frac{2}{3}(1, 1, 1, 0)$$

$$= \left(-\frac{2}{3}, \frac{1}{3}, \frac{1}{3}, 1\right)$$

check that β_1, β_2 are orthogonal.

$$\|\beta_2\|^2 = \frac{15}{9},$$

$$(\alpha_3|\beta_1) = 1,$$

$$(\alpha_3|\beta_2) = \frac{4}{3}$$

so that $$\beta_3 = \alpha_3 - \frac{(\alpha_3|\beta_1)}{\|\beta_1\|^2}\beta_1 - \frac{(\alpha_3|\beta_2)}{\|\beta_2\|^2}\beta_2$$

$$= (0, 0, 1, 1) - \frac{1}{3}(1, 1, 1, 0) - \frac{12}{15}\left(-\frac{2}{3}, \frac{1}{3}, \frac{1}{3}, 1\right)$$

$$= \left(\frac{1}{5}, -\frac{3}{5}, \frac{2}{5}, \frac{1}{5}\right).$$

Clearly, $$\|\beta_3\|^2 = \frac{3}{5}, \ (\beta_3|\beta_1) = 0, \ (\beta_3|\beta_2) = 0.$$

Now, $(\alpha_4|\beta_1) = 3,\ (\alpha_4|\beta_2) = 1,\ (\alpha_4|\beta_3) = \frac{6}{5}$,

so that
$$\beta_4 = \alpha_4 - \frac{(\alpha_4|\beta_1)}{\|\beta_1\|^2}\beta_1 - \frac{(\alpha_4|\beta_2)}{\|\beta_2\|^2}\beta_2 - \frac{(\alpha_4|\beta_3)}{\|\beta_3\|^2}\beta_3$$
$$= (0,\ 0,\ 3,\ 0) - (1,\ 1,\ 1,\ 0) - \frac{3}{5}\left(-\frac{2}{3},\ \frac{1}{3},\ \frac{1}{3},\ 1\right) - 2\left(\frac{1}{5},\ -\frac{3}{5},\ \frac{2}{5},\ \frac{1}{5}\right)$$
$$= (-1,\ 0,\ 1,\ -1).$$

Here $\|\beta_4\|^2 = 3$

Further, $\beta_1, \beta_2, \beta_3, \beta_4$ are pairwise orthogonal.

Finally
$$\gamma_1 = \frac{\beta_1}{\|\beta_1\|} = \left(\frac{1}{\sqrt{3}},\ \frac{1}{\sqrt{3}},\ \frac{1}{\sqrt{3}},\ 0\right)$$
$$\gamma_2 = \frac{\beta_2}{\|\beta_2\|} = \left(-\frac{2}{\sqrt{15}},\ \frac{1}{\sqrt{15}},\ \frac{1}{\sqrt{15}},\ \frac{\sqrt{3}}{\sqrt{5}}\right)$$
$$\gamma_3 = \frac{\beta_3}{\|\beta_3\|} = \left(\frac{1}{\sqrt{15}},\ -\frac{3}{\sqrt{15}},\ \frac{2}{\sqrt{15}},\ \frac{1}{\sqrt{5}}\right)$$
$$\gamma_4 = \frac{\beta_4}{\|\beta_4\|} = \left(-\frac{1}{\sqrt{3}},\ 0,\ \frac{1}{\sqrt{3}},\ -\frac{1}{\sqrt{3}}\right)$$

$\{\gamma_1, \gamma_2, \gamma_3, \gamma_4\}$ is an orthonormal basis of $\mathbf{C}^4$. If W is the subspace of $\mathbf{C}^4$ spanned by α_1, α_2 then $\{\gamma_1, \gamma_2\}$ is an orthonormal basis of W.

Lemma 15. Let $\alpha_1, \alpha_2, \ldots, \alpha_n$ be non zero pairwise orthogonal vectors in V_F. Then, the vectors $\beta_1, \beta_2, \ldots, \beta_n$ such that $\beta_1 = \alpha_1$, and
$$\beta_k = \alpha_k - \sum_{j=1}^{k-1} \frac{(\alpha_i|\beta_j)}{\|\beta_j\|^2}\beta_j,$$
$$2 \le k \le n$$
are such that $\alpha_i = \beta_i, 1 \le i \le n$.

Proof. Obviously, $\beta_1 = \alpha_1$. Let for some k, $1 \le k < n$, we have already proved that $\beta_i = \alpha_i$, $1 \le i \le k$.

Then,
$$(\alpha_{k+1}|\beta_j) = (\alpha_{k+1}|\alpha_j) = 0, \qquad \text{for } 1 \le j \le k.$$

Thus,
$$\beta_{k+1} = \alpha_{k+1} - \sum_{j=1}^{k} \frac{(\alpha_{k+1}|\beta_j)}{\|\beta_j\|^2}\beta_j = \alpha_{k+1}.$$

Hence, the result follows by induction.

We know that every linearly independent subset of a vector space can be extended to a basis of that vector space. Here, an analogous result for extension of an orthogonal subset to an orthogonal basis is proved.

Theorem 16. Let V_F be a finite dimensional inner product space with dim $V_F = n \geq 1$. If $S = \{\alpha_1, \alpha_2, ..., \alpha_k\}$ is an *orthonormal* subset of V_F, then, there exists an *orthonormal* basis B of V such that $S \subseteq B$.

Proof. As S is orthogonal, it is linearly independent (Theorem 11). So, there exists a basis $\overline{B} = \{\alpha'_1, \alpha'_2, ..., \alpha'_n\}$ of V_F such that $\alpha'_i = \alpha_i$, $1 \leq i \leq k$.

We define
$$\beta_1 = \alpha_1,$$
and
$$\beta_i = \alpha'_i - \sum_{j=1}^{i-1} \frac{(\alpha'_i | \beta_j)}{\|\beta_j\|^2} \beta_j$$
for
$$2 \leq i \leq n.$$

By Lemma 15, $\beta_i = \alpha_i$ for $1 \leq i \leq k$. Let $\gamma_j = \dfrac{\beta_j}{\|\beta_j\|}$. Then $\gamma_j = \alpha_j$ for $1 \leq j \leq k$, as $\|\alpha_j\| = 1$. By the Gram-Schmidt Process, $B = \{\gamma_1, \gamma_2, ..., \gamma_n\}$ is a *orthonormal* basis of V.

Theorem 17. Let $\alpha_1, \alpha_2, ..., \alpha_n$ be *orthonormal* in V_F.

If
$$\alpha = \sum_{i=1}^{n} a_i \alpha_i,$$
then
$$a_i = (\alpha | \alpha_i)$$
and
$$\|\alpha\|^2 = \sum_{i=1}^{n} |\alpha_i|^2 = \sum_{i=1}^{n} |(\alpha | \alpha_i)|^2.$$

Proof.
$$(\alpha | \alpha_i) = \left(\sum_{j=1}^{n} a_j \alpha_j \middle| \alpha_i \right) = \sum_{j=1}^{n} a_j (\alpha_j | \alpha_i) = a_i,$$
as $(\alpha_j | \alpha_i) = 0$ for $j \neq i$ and $(\alpha_i | \alpha_i) = 1$.

Then
$$a_i \bar{a}_i = |a_i|^2 = |(\alpha | a_i)|^2.$$

Therefore,
$$\|\alpha\|^2 = (\alpha | \alpha)$$
$$= \left(\sum_{i=1}^{n} a_i \alpha_i \middle| \sum_{j=1}^{n} a_j \alpha_j \right)$$

$$= \sum_{j=1}^{n} \sum_{i=1}^{n} a_i \bar{a}_j (\alpha_i|\alpha_j)$$

$$= \sum_{i=1}^{n} a_i \bar{a}_i$$

$$= \sum_{i=1}^{n} |(\alpha|\alpha_i)|^2$$

Definition 18. Let S be any subset of an inner product space V_F. Then, the *orthogonal complement* of S is

$$S^\perp = \{\alpha \in V : (\alpha|\beta) = 0 \text{ for all } \beta \in S\}.$$

For any $\alpha \in S$, $\alpha^\perp$ is defined to be $\{\alpha\}^\perp$.

Remark 19

(*i*) As $(\beta|\alpha) = \overline{(\alpha|\beta)}$, it is obvious that for any $S \subseteq V_F$,

$$S^\perp = \{\alpha \in V : (\beta|\alpha) = 0 \text{ for all } \beta \in S\}.$$

Further, $$S^\perp = \bigcap_{\alpha \in S} \alpha^\perp.$$

(*ii*) Since for $0 \neq \alpha \in V$, $(\alpha|\alpha) \neq 0$, we get $\alpha \notin V^\perp$. So that $V^\perp = \{0\}$. Obviously, $\phi^\perp = 0^\perp = V$.

(*iii*) For any subset S of V, $S^{\perp\perp}$ will denote $(S^\perp)^\perp$.

Theorem 20. Let S_1 and S_2 be any two subsets of an inner product space V_F. Then,

(*i*) $S_1^\perp$ is a subspace of V and $S_1 \cap S_1^\perp \subseteq \{0\}$.

(*ii*) If $S_1 \subseteq S_2$, then $S_2^\perp \subseteq S_1^\perp$

(*iii*) $S_1 \subseteq S_1^{\perp\perp}$.

Proof. Obviously, $0 \in S_1^\perp$. Let $\alpha, \beta \in S_1^\perp$ and $a \in F$. Consider $\gamma \in S$.

Then, $$(\alpha|\gamma) = 0 = (\beta|\gamma).$$

Then, $$(a\alpha + \beta|\gamma) = a(\alpha|\gamma) + (\beta|\gamma) = a0 + 0 = 0,$$

gives $$a\alpha + \beta \in S_1^\perp.$$

Hence, $S_1^\perp$ is a subspace of V. If $\alpha \in S_1 \cap S_1^\perp$, then $(\alpha|\alpha) = 0$ yields $\alpha = 0$.

Hence $$S_1 \cap S_1^\perp \subseteq \{0\}.$$

This proves (*i*).

Let $S_1 \subseteq S_2$. Consider $\alpha \in S_2^{\perp}$. By definition, $(\alpha|\gamma) = 0$ for $\gamma \in S_2$. As $S_1 \subseteq S_2$, we get $(\alpha|\gamma) = 0$ for $\gamma \in S_1$.

Consequently, $$S_2^{\perp} \subseteq S_1^{\perp}.$$

This proves (*ii*).

Finally, given $\alpha \in S_1$, $(\gamma|\alpha) = 0$ for all $\gamma \in S_1^{\perp}$ yield $\alpha \in S_1^{\perp\perp}$.

This proves (*iii*).

Theorem 21. Let S be a subset of an inner product space V_F and W be the subspace spanned by S. Then, $S^{\perp} = W^{\perp}$.

Proof. If $S = \phi$, then $W = \{0\}$ and, obviously, $S^{\perp} = V = W^{\perp}$. Let $S \neq \phi$. As $S \subseteq W$, by Theorem 20, $W^{\perp} \subseteq S^{\perp}$. Consider $\beta \in S^{\perp}$. Let $0 \neq \alpha \in W$. Then, $\alpha = \sum_{i=1}^{k} a_i \alpha_i$ for some $\alpha_i \in S$, $a_i \in F$.

This gives $$(\alpha|\beta) = \sum_{i=1}^{k} a_i (\alpha_i|\beta) = 0$$

so that $$\beta \in W^{\perp}.$$

Hence, $$S^{\perp} \subseteq W^{\perp} \subseteq S^{\perp},$$

i.e., $$S^{\perp} = W^{\perp}.$$

Example 21.a. Consider $V = \mathbf{C}^4$, $S = \{\epsilon_1, \epsilon_2\}$. Then, the subspace $W = \langle S \rangle$ spanned by S equals $\{(a, b, 0, 0) : a, b \in \mathbf{C}\}$. Now, $\alpha = (a, b, c, d) \in S^{\perp}$ iff $(\alpha|\epsilon_1) = 0 = (\alpha|\epsilon_2)$ But $(\alpha|\epsilon_1) = a$ and $(\alpha|\epsilon_2) = b$ (see Theorem 17). Thus, $\alpha \in S^{\perp}$ iff $a = 0 = b$.

Hence, $$S^{\perp} = W^{\perp} = \{(0, 0, c, d) : c, d \in \mathbf{C}\}.$$

Example 21.b. Let $V_{\mathbf{C}}$ be the inner product space of all polynomials over $\mathbf{C}$ of degree ≤ 4, such that for any f, $g \in V_{\mathbf{C}}$

$$(f|g) = \int_0^1 f(t)\overline{g(t)}dt.$$

Consider $$S = \{X, X^2 - 1\} \subset V.$$

Given $$f = \sum_{j=1}^{4} a_j X^j \in V, f \in S^{\perp}$$

iff $$(f|X) = 0 = (f|X^2 - 1).$$

Now, $$(f|X)=\int_0^1 \sum_{j=0}^{4} a_j\, t^{j+1}dt=\frac{a_0}{2}+\frac{a_1}{3}+\frac{a_2}{4}+\frac{a_3}{4}+\frac{a_4}{6} \qquad \text{... (1)}$$

$$(f|x^2-1)=\int_0^1 \sum_{j=0}^{4} a_j\, t^j(t^2-1)dt$$

$$=\int_0^1 \sum_{j=0}^{4} a_j\, t^{j+2}dt-\int_0^1 \sum_{j=0}^{4} a_j\, t^j dt$$

$$=\sum_{j=0}^{4}\frac{a_j}{j+3}-\sum_{j=0}^{4}\frac{a_j}{j+1}$$

$$=-\frac{2a_0}{3}-\frac{a_1}{4}-\frac{2a_2}{15}-\frac{a_3}{12}-\frac{2a_4}{35}. \qquad \text{... (2)}$$

Thus, $f\in S^{\perp}$ iff $$\frac{a_0}{2}+\frac{a_1}{3}+\frac{a_2}{4}+\frac{a_3}{4}+\frac{a_4}{6}=0, \qquad \text{... (3)}$$

$$\frac{-2a_0}{3}-\frac{a_1}{4}-\frac{2a_2}{15}-\frac{a_3}{12}-\frac{2a_4}{35}=0. \qquad \text{... (4)}$$

These are two homogeneous linear equations in five unknowns a_0, a_1, a_2, a_3, a_4. The coefficient matrix of this system is of rank 2. So, the solution space has dimension $5-2=3$. Hence, $S^{\perp}$ is a subspace of V of dimension 3.

Proposition 22. Let $B=\{\alpha_1,\alpha_2,\ldots,\alpha_n\}$ be any finite, non-empty orthonormal subset of V_F. For any $\alpha\in V$, let $\beta=\sum_{i=0}^{n}(\alpha|\alpha_i)\alpha_i$. Then,

(*i*) $(\alpha-\beta|\beta)=0$, and

(*ii*) $\|\beta\|\le\|\alpha\|$ [i.e. $\sum_{i=0}^{n}|(\alpha|\alpha_i)|^2\le\|\alpha\|^2$. *Bessell's inequality.*

Proof. Now $$(\alpha-\beta|\alpha_j)=(\alpha|\alpha_j)-\sum_{i=1}^{n}(\alpha|\alpha_i)(\alpha_i|\alpha_j)=(\alpha|\alpha_j)-(\alpha|\alpha_j)=0$$

by using the orthonormality of B.

So that $$\alpha_j\in(\alpha-\beta)^{\perp}.$$

As $(\alpha-\beta)^{\perp}$ is a subspace, we get

$$\beta=\sum_{j=1}^{n}(\alpha|\alpha_j)\alpha_j\in(\alpha-\beta)^{\perp}$$

i.e. $$(\alpha-\beta|\beta)=0.$$

Then, $$\alpha = (\alpha - \beta) + \beta,$$

yields $$\|\alpha\|^2 = \|\alpha - \beta\|^2 + \|\beta\|^2,$$

so that $$\|\beta\|^2 \leq \|\alpha\|^2,$$

i.e. $$\|\beta\| \leq \|\alpha\|.$$

Theorem 23. Let W be a finite dimensional subspace of an inner product space V_F. Then, $V = W \oplus W^{\perp}$ and $W^{\perp\perp} = W$.

Proof. If $W = \{0\}$, then, $W^{\perp} = V$ and $W^{\perp\perp} = V^{\perp} = \{0\}$ (see Remark 19).

Obviously $$V = \{0\} \oplus V = W \oplus W^{\perp}.$$

Thus, the result hold for $W = \{0\}$. Let $W \neq \{0\}$. Then, $\dim W = n \geq 1$.

Let $$B = \{\alpha_1, \alpha_2, \ldots, \alpha_n\}$$

be an orthonormal basis of W.

Consider $$\alpha \in V.$$

Then, $$\beta = \sum_{i=0}^{n} (\alpha|\alpha_i)\alpha_i \in W.$$

By Proposition 22, $$(\alpha - \beta|\alpha_i) = 0, 1 \leq j \leq n.$$

As α_j's span W, we get $$\alpha - \beta \in W^{\perp}.$$

Hence, $$\alpha = \beta + (\alpha - \beta) \in W + W^{\perp}$$

Thus, $$V = W + W^{\perp}.$$

By Theorem 20, $$W \cap W^{\perp} = \{0\}.$$

Consequently, $$V = W \oplus W^{\perp}.$$

Now $W \subseteq W^{\perp\perp}$,

yields $$W^{\perp\perp} = W^{\perp\perp} \cap V = W + (W^{\perp\perp} \cap W^{\perp}) = W$$

as $$W^{\perp\perp} \cap W^{\perp} = \{0\}.$$

This proves the result.

Theorem 24. Let V_F be a finite dimensional inner product space, and $L(V)$ be the lattice of all the subspaces of V.

Then, the mapping $$\sigma : L(V) \to L(V)$$

such that $\sigma(W) = W^{\perp}$, $W \in L(V)$ is a one-to-one inclusion reversing mapping of $L(V)$ onto $L(V)$. Further, σ^2 is identity on $L(V)$.

Proof. As every subspace W of V is finite dimensional, $W^{\perp\perp} = W$. Consider W_1, $W_2 \in L(V)$. If $W_1 \subseteq W_2$, then, by Theorem 20, $W_2^{\perp} \subseteq W_1^{\perp}$ i.e. $\sigma(W_2) \subseteq \sigma(W_1)$, so that σ is inclusion reversing. If $\sigma(W_1) = \sigma(W_2)$, then, $W_1^{\perp} = W_2^{\perp}$, $W_1^{\perp\perp} = W_2^{\perp\perp}$, i.e. $W_1 = W_2$. Thus, σ is one-to-one. For any $W \in L(V)$, $\sigma(W^{\perp}) = W^{\perp\perp} = W$, gives σ is also onto. As $\sigma^2(W) = W^{\perp\perp} = W$, the last part follows.

For any subspace W of V, the projection of p_W of V onto W via $W^{\perp}$ is called the *orthogonal projection* of V onto W.

For $\qquad \alpha \in V,\ \alpha = \beta + \gamma,\ \beta \in W,\ \gamma \in W^{\perp},\ p_W(\alpha) = \beta.$

Remark 25. This theorem is analogous to the Duality [Theorem (3.4.16)] given in the section on dual spaces.

To compare two inner product spaces, we introduce the following definition :

Definition 26. Let V and V' be two inner product spaces over the same field F. Then, V is said to be isomorphic to V', if there exists a vector space isomorphism T of V onto V', such that for any α, $\beta \in V$

$$(\alpha|\beta) = (T\alpha|T\beta).$$

Such an isomorphism T is called an *inner product space isomorphism.*

We know that any finite dimensional vector space over a field F is isomorphic to an F^n. We prove an analogous result for inner product spaces.

Theorem 27. Any inner product space V_F of dimension n, $1 \le n \le \infty$, is isomorphic to $\mathbf{R}^n$ or $\mathbf{C}^n$ according as $F = \mathbf{R}$ or $F = \mathbf{C}$.

Proof. V_F has an orthonormal basis $B = \{\alpha_1, \alpha_2, \dots, \alpha_n\}$.

Define
$$T : F^n \to V$$

such that
$$T(a_1, a_2, \dots, a_n) = \sum_{i=1}^{n} a_i \alpha_i$$

for every
$$\alpha = (a_1, a_2, \dots, a_n) \in V.$$

Clearly T is a vector space isomorphism.

For any $\alpha = (a_1, a_2, \dots, a_n)$, $\quad \beta = (b_1, b_2, \dots, b_n)$

$$(\alpha|\beta) = \sum_{i=1}^{n} a_i \bar{b}_i.$$

Also
$$(T\alpha|T\beta) = \left(\sum_{i=1}^{n} a_i \alpha_i \,\middle|\, \sum_{j=1}^{n} b_j \alpha_j \right) = \sum_{j=1}^{n} \sum_{i=1}^{n} a_i \bar{b}_j (\alpha_i|\alpha_j).$$

However, $(\alpha_i|\alpha_j) = 0$ for $i \ne j$, so that

$$(T\alpha|T\beta) = \sum_{i=1}^{n} a_i \bar{b}_i (\alpha_i|\alpha_i) = \sum_{i=1}^{n} a_i \bar{b}_i , \qquad \text{as } (\alpha_i|\alpha_i) = 1$$

Hence, $(\alpha|\beta) = (T\alpha|T\beta)$.

Hence, V_F is isomorphic to F^n.

The following is an immediate consequence of the above theorem :

Corollary 28. Any two finite dimensional inner product spaces V and V' over the same field are isomorphic iff $\dim V = \dim V'$.

Exercise 6.1

1. Consider the inner product space $\mathbf{R}^3$. Let $X_1 = (1, 2, 3)$, $X_2 = (-2, 1, 0)$, $X_3 = (3, 6, 9)$, $X_4 = (2, 2, -2)$. Answer the following :

   (*i*) Find a unit vector in the direction of X_1.

   (*ii*) Are X_1, X_2 orthogonal?

   (*iii*) Are X_2, X_4 parallel?

   (*iv*) Are X_1 and X_3 in the same direction?

2. Given $\alpha_1 = (1, 3)$, $\alpha_2 = (2, 1)$ in $\mathbf{R}^2$, find $\alpha \in \mathbf{R}^2$ such that
$$(\alpha|\alpha_1) = 3(\alpha|\alpha_2) = -1.$$

3. Consider $X_1 = (0,3,4)$, $X_2 = (1,0,7)$ in $\mathbf{R}^3$. By using Gram-Schmidt Orthonormalization process, find an orthonormal basis of the subspace W of $\mathbf{R}^3$ spanned by X_1, X_2. Find a basis of $W^\perp$.

4 Find an orthonormal basis of the subspace W of the complex inner product space $\mathbf{C}^4$, spanned by $X_1 = (3, -i, 0, 0)$, $X_2 = (2, 2i, 1, 0)$. Find an orthonormal basis of $\mathbf{C}^4$ that contains this orthonormal basis of W.

5. Let B an orthonormal basis of an inner product space V_F. If $B = B_1 \cup B_2$ such that $B_1 \cap B_2 = \phi$, prove that for the subspace W of V generated by B_1, $W^\perp$ is the subspace generated by B_2.

6. For any two subspaces W_1 and W_2 of a finite dimensional inner product space V_F, prove that $(W_1 + W_2)^\perp = W_1^\perp \cap W_2^\perp$, $(W_1 \cap W_2)^\perp = W_1^\perp + W_2^\perp$.

   [Hint : Use Theorem 24]

7. Given any orthonormal set $B = \{\alpha_1, \alpha_2, ..., \alpha_n\}$ in an inner product space V_F, then, for any $\alpha \in V$ prove that $\sum_{i=1}^{\infty} |(\alpha|\alpha_i)|^2 \le \|\alpha\|^2$.

[Hint : This follows from Bessell's inequality]

8. **Let $\alpha = (1, 3)$, $\beta = (2, 5) \in \mathbf{R}^2$. If $\gamma \in \mathbf{R}^2$ satisfy $(\alpha|\gamma) = 0$, $(\beta|\gamma) = 3$, find γ.**

9. **Find an inner product $(\ |\ )'$ on $\mathbf{R}^2$ such that for $\alpha_1 = (1, 3)$, $\alpha_2 = (4, 1)$, $B = \{\alpha_1, \alpha_2\}$ becomes an orthonormal basis.**

10. **Given a finite dimensional real vector space V and any basis B of V, define an inner product in V such that B becomes an orthonormal basis of V.**

11. Let A be any 2×2-symmetric matrix over R. For any $\alpha, \beta \in R^{2\times 1}$ define $(\alpha|\beta) = \alpha^T A\beta$. Prove that this is an inner product iff $\det(A) > 0$, and the diagonal entries of A are positive.

12. **Consider $V = \mathbf{R}^2$. For any $\alpha = (x_1, y_1)$, $\beta = (x_2, y_2)$ in V, define**

$$(\alpha|\beta) = (x_1, y_1)\begin{bmatrix} 1 & 2 \\ 3 & 8 \end{bmatrix}\begin{bmatrix} x_2 \\ y_2 \end{bmatrix}$$

$$= x_1x_2 + 2x_1y_1 + 3y_1x_2 + 8y_1y_2$$

verify that it is an inner product. Consider the linear functional $f: V \to \mathbf{R}$ such that $f(\alpha) = x_1 + y_1$. Find a $\gamma \in V$ such that $f(\alpha) = (\alpha|\gamma)$ for all $\alpha \in V$.

[Hint : Use Proposition 4 *(iii)*]

13. **Consider any $\alpha, \beta \in \mathbf{R}^2$ prove that**

$$\begin{vmatrix} (\alpha|\alpha) & (\alpha|\beta) \\ (\alpha|\beta) & (\beta|\beta) \end{vmatrix} = 0$$

iff α and β are linearly dependent.

6.2 Adjoint Operator

We start with the following theorem :

Theorem 1. **Let V_F be a finite dimensional inner product space and T be a linear operator on V. Then, there exists a unique linear operator T^* on V such that**

$$(T\alpha|\beta) = (\alpha|T^*\beta)$$

for all $\alpha, \beta \in V$.

Proof. Consider any $\beta \in V$. Then the mapping $\lambda_\beta : V \to F$ such that $\lambda_\beta(\alpha) = (T\alpha|\beta)$ is a linear functional on V. By Proposition (6.1.4) *(iii)*, there exists unique β', depending on β, such that

$$\lambda_\beta(\alpha) = (\alpha|\beta'), \alpha \in V.$$

Put $T^*\beta = \beta'$, so that $(T\alpha|\beta) = (\alpha|T^*\beta)$. For any α, β_1, $\beta_2 \in V$, $c \in F$.

$$\begin{aligned}\left[(\alpha|T^*(\beta_1 + c\beta_2)\right] &= (T\alpha|\beta_1 + c\beta_2)\\ &= (T\alpha|\beta_1) + \bar{c}(T\alpha|\beta_2)\\ &= (\alpha|T^*\beta_1) + \bar{c}(\alpha|T^*\beta_2)\\ &= (\alpha|T^*\beta_1 + cT^*\beta_2).\end{aligned}$$

Thus, $$\left[\alpha|T^*(\beta_1 + c\beta_2)\right] = (\alpha|T^*\beta_1 + cT^*\beta_2).$$

By Proposition (6.1.4) $\quad T^*(\beta_1 + c\beta_2) = T^*\beta_1 + cT^*\beta_2.$

Hence, T^* is a linear operator on V. Let T' be another linear operator on V such that

$$(T\alpha|\beta) = (\alpha|T'\beta).$$

Then, $$(\alpha|T^*\beta) = (\alpha|T'\beta).$$

Once again, by using Proposition (6.1.4), we get

$$T^*\beta = T'\beta,\ \beta \in V$$

i.e. $$T^* = T'.$$

This establishes the uniqueness of T^*.

Definition 2. Let T be a linear operator on an inner product space V_F. Then, a linear operator T^* on V is called the *adjoint* of T if

$$(T\alpha|\beta) = (\alpha|T^*\beta)$$

for all α, $\beta \in V$.

It is immediately clear that $0^* = 0$ and $I^* = I$. We write T^{**} for $(T^*)^*$. Theorem 1 shows that any linear operator on a finite dimensional inner space admits a unique adjoint.

Lemma 3. Let V_F be a finite dimensional inner product space, S and T be two linear operators on V, and $a \in F$. Then,

(i) $(T^*)^* = T$

(ii) $(S + T)^* = S^* + T^*$

(iii) $(aT)^* = \bar{a}T^*$

(iv) $(ST)^* = T^*S^*$.

Proof

(i) Let α, $\beta \in V$.

Then, $(T^*\beta|\alpha) = (\beta|T^{**}\alpha)$,

gives $\overline{(T^*\beta|\alpha)} = \overline{(\beta|T^{**}\alpha)}$

i.e. $(\alpha|T^*\beta) = (T^{**}\alpha|\beta)$.

i.e. $(T\alpha|\beta) = (T^{**}\alpha|\beta)$.

Hence, $T\alpha = T^{**}\alpha$.

This gives $T = T^{**}$. This proves (*i*).

(*ii*) $$\big(\alpha|(S+T)^*\beta\big) = \big((S+T)\alpha|\beta\big)$$
$$= (S\alpha + T\alpha|\beta)$$
$$= (S\alpha|\beta) + (T\alpha|\beta)$$
$$= (\alpha|S^*\beta) + (\alpha|T^*\beta)$$
$$= (\alpha|S^*\beta + T^*\beta)$$
$$= \left[\alpha|(S^* + T^*)\beta\right],$$

gives $\left[(\alpha|(S+T)^*\beta\right] = \left[\alpha|(S^* + T^*)\beta\right]$.

Hence, $(S+T)^* = S^* + T^*$.

This proves (*ii*).

(*iii*) $$\left[(\alpha|(aT)^*\beta\right] = \left[(aT)\alpha|\beta\right]$$
$$= a(T\alpha|\beta)$$
$$= a(\alpha|T^*\beta)$$
$$= (\alpha|\bar{a}\,T^*\beta),$$

gives $(aT)^* = \bar{a}\,T^*$.

This proves (*iii*).

(*iv*) Finally, $$\left[(\alpha|(ST)^*\beta\right] = \left[(ST)\alpha|\beta\right]$$
$$= \left[S(T\alpha)|\beta\right]$$
$$= (T\alpha|S^*\beta)$$
$$= \left[\alpha|(T^*S^*)\beta\right]$$

gives, $(ST)^* = T^*S^*$.

This gives (*iv*).

It is immediately clear from (*i*) and (*ii*) in this lemma that the mapping $T \to T^*$, $T \in L_F(V)$, is an automorphism of $L_F(V)$ under addition. Before we construct some examples of adjoints, we prove the following theorem :

Theorem 4. Let V_F be a non-zero finite dimensional inner product space and $B = \{\alpha_1, \alpha_2, ..., \alpha_n\}$ be an ordered orthonormal basis of *V*. If $A = [a_{ij}]$ is the matrix of a linear operator *T* on *V* relative to *B*, then, the matrix of T^* relative to *B* is the conjugate transpose $A' = [b_{ij}]$, $b_{ij} = \bar{a}_{ji}$, of *A*.

Proof. Now
$$T\alpha_j = \sum_{i=1}^{n} a_{ij}\alpha_i .$$

Let
$$[T^*]_B = [b_{ij}].$$

Then,
$$T^*\alpha_j = \sum_{j=1}^{n} b_{ij}\alpha_i .$$

Thus,
$$(T\alpha_k|\alpha_j) = \left[\sum_{i=1}^{n} a_{ik}\alpha_i \middle| \alpha_j\right] = \sum_{i=1}^{n} a_{ik}(\alpha_i|\alpha_j) = a_{jk},$$

as $(\alpha_i|\alpha_j) = 0$ for $i \neq j$ and $(\alpha_j|\alpha_j) = 1$.

Again,
$$(T\alpha_k|\alpha_j) = (\alpha_k|T^*\alpha_j)$$
$$= \left(\alpha_k \middle| \sum_{i=1}^{n} b_{ij}\alpha_i\right)$$
$$= \sum_{i=1}^{n} \bar{b}_{ij}(\alpha_k|\alpha_i)$$
$$= \bar{b}_{kj}.$$

Hence, $a_{jk} = \bar{b}_{kj}$, so that $b_{kj} = \bar{a}_{jk}$. This completes the proof.

For any complex matrix *A*, recall that A^* denotes the conjugate transpose.

Definition 5. A linear operator *T* on an inner product space V_F is said to be *self-adjoint* if $(T\alpha|\beta) = (\alpha|T\beta)$ for all $\alpha, \beta \in V_F$ (In case V_F is finite dimensional, it means $T = T^*$). An $n \times n$-matrix *A* over **C**, is said to be *Hermitian* if $A = A^*$.

Let *B* be an ordered orthonormal basis of V_F. Consider a linear operator *T* on V_F. Let $A = [T]_B$. Then $A^* = [T^*]_B$. Now, $T = T^*$ iff $[T]_B = [T^*]_B$ iff $A = A^*$.

Example 5.a. Consider $A = \begin{bmatrix} 0 & i \\ -i & 0 \end{bmatrix} A^T = \begin{bmatrix} 0 & -i \\ i & 0 \end{bmatrix}$.

So $$A^* = (\overline{A^T}) = \begin{bmatrix} 0 & i \\ -i & 0 \end{bmatrix} = A.$$

Consequently, the linear operator T on $\mathbf{C}^{2\times 1}$ such that $T\begin{bmatrix} a \\ b \end{bmatrix} = A\begin{bmatrix} a \\ b \end{bmatrix} = \begin{bmatrix} ib \\ -ia \end{bmatrix}$ is a self-adjoint operator.

Example 5.b. Let V be the complex inner product space of all polynomials over **C**.

For any $$f(x) = \sum_{i=1}^{k} a_i x^i,$$

let $$\bar{f}(x) = \sum \bar{a}_i x^i.$$

Then, $$(f|g) = \int_0^1 f(x)\bar{g}(x)dx$$

is an inner product in V.

Consider any $h(x) \in V$. Then $T : V \to V$ such that $T\{f(x)\} = h(x)f(x)$ is a linear operator on T. Then, for any $f(x)$, $g(x) \in V$.

$$\begin{aligned}(Tf(x)|g(x)) &= (h(x)f(x), g(x)) \\ &= \int_0^1 h(x)f(x)\bar{g}(x)dx \\ &= \int_0^1 f(x)\bar{\bar{h}}(x)\bar{g}(x)dx \\ &= (f(x)|\bar{h}(x)g(x)).\end{aligned}$$

This gives $$T^*(g(x)) = \bar{h}(x)g(x).$$

Proposition 6. Let V_F be a finite dimensional inner product space, S and T be two linear operators on V_F.

(*i*) If T is invertible, then, T^* is invertible and $(T^*)^{-1} = (T^{-1})^*$.

(*ii*) If S and T are self-adjoint, then $S + T$ is self-adjoint.

(*iii*) If S and T are self-adjoint, then ST is self-adjoint iff $ST = TS$.

(*iv*) For any real number a and any self-adjoint operator T, aT is self-adjoint.

Proof. Let T be invertible. Then, $TT^{-1} = I$, $(TT^{-1})^* = I^* = I$, i.e. $$(T^{-1})^* T^* = I.$$

Hence, $$(T^*)^{-1} = (T^{-1})^*.$$

Let $$T^* = T,\ S^* = S.$$

Then, $$(S+T)^* = S^* + T^* = S + T,$$

yields $S + T$ is self-adjoint.

$$(ST)^* = T^* S^* = TS.$$

Thus, ST is self-adjoint iff $$(ST)^* = ST.$$

iff $$TS = ST.$$

Finally, $(aT)^* = \bar{a}T^* = aT$ yields aT is self-adjoint.

The following theorem on linear operators is analogous to finding real and imaginary parts of a complex numbers.

Theorem 7. Let T be a linear operator on a finite dimensional complex inner product space V. Then, $T = T_1 + iT_2$ for some uniquely determined self-adjoint linear operators T_1, T_2 on V.

Proof. Let $$T_1 = \frac{1}{2}(T + T^*).$$

Then, $$T_1^* = \frac{1}{2}(T + T^*)^* = \frac{1}{2}(T^* + T^{**}) = \frac{1}{2}(T + T^*) = T_1.$$

Thus, T_1 is self-adjoint. Similarly, $T_2 = \frac{i}{2}(T^* - T)$ is self-adjoint.

Obviously, $$T = T_1 + iT_2. \qquad \dots (1)$$

Let also $T = T_1' + iT_2'$ for some self-adjoint linear operations T_1' and T_2'.

Then, $$T_1 + iT_2 = T_1' + iT_2', \qquad \dots (2)$$

$$T_1 - T_1' = i(T_2' - T_2)$$

$$(T_1 - T_1')^* = \left[i(T_2' - T_2)\right]^*$$

i.e., $$T_1 - T_1' = -i(T_2' - T_2) = -(T_1 - T_1')$$

Thus, $$2(T_1 - T_1') = 0,\ T_1 - T_1' = 0,$$

so $$T_1 = T_1'.$$

Then, (2) gives $iT_2 = iT_2'$, so $T_2 = T_2'$. This completes the proof.

Remark 8. In this theorem T_1 and T_2 play the role similar to that of real and imaginary parts of a complex number. Observe that $T^* = T_1 - iT_2$, so that T^* is analogous to the conjugate of a complex number.

Theorem 9. Let T be a linear operator on a finite dimensional inner product space. Then, the following are equivalent :

(*i*) T is an inner product space automorphism

(*ii*) $TT^* = T^*T = I$

(*iii*) $\|T\alpha\| = \|\alpha\|$ for every $\alpha \in V$.

Proof. Let T be an inner product space automorphism of V.

By definition $(\alpha|\beta) = (T\alpha|T\beta)$ for all $\alpha, \beta \in V$.

Thus, $$(\alpha|\beta) = (\alpha|T^*T\beta).$$

Consequently, $$I = T^*T,\ T^* = T^{-1}.$$

Thus, $$TT^* = I.$$

Hence, (*i*) implies (*ii*).

Let $$TT^* = T^*T = I.$$

Then, for $\alpha \in V$, $\|\alpha\|^2 = (\alpha|\alpha) = (\alpha|T^*T\alpha) = (T\alpha|T\alpha) = \|T\alpha\|^2$.

So, $$\|\alpha\| = \|T\alpha\|.$$

Hence, (*ii*) implies (*iii*).

Finally, let $\|T\alpha\| = \|\alpha\|$ for all $\alpha \in V$. Obviously, for $\alpha \neq 0$, $T\alpha \neq 0$. Thus T is one-to-one. As V is finite dimensional, T is an automorphism of V.

Case I. $F = \mathbf{R}$. Recall polarization identity given in Theorem (6.1.5) (v)

$$(\alpha|\beta) = \frac{1}{4}\|\alpha+\beta\|^2 - \frac{1}{4}\|\alpha-\beta\|^2.$$

We get
$$\begin{aligned}(T\alpha|T\beta) &= \frac{1}{4}\|T\alpha+T\beta\|^2 - \frac{1}{4}\|T\alpha-T\beta)\|^2 \\ &= \frac{1}{4}\|T(\alpha+\beta)\|^2 - \frac{1}{4}\|T(\alpha-\beta)\|^2 \\ &= (\alpha|\beta)\end{aligned}$$

Case II. $F = \mathbf{C}$. The polarization identity is

$$(\alpha|\beta) = \frac{1}{4}\|\alpha+\beta\|^2 - \frac{1}{4}\|\alpha-\beta\|^2 + \frac{i}{4}\|\alpha+i\beta\|^2 - \frac{i}{4}\|\alpha-i\beta\|^2.$$

As in case I, once again we get $(\alpha|\beta) = (T\alpha|T\beta)$.

Hence, (*iii*) implies (*i*).

This theorem and the isomorphism between $L_F(V)$ and $F^{n\times n}$, $n = \dim V$, given by Theorem (3.3.5), give the following theorem :

Theorem 10. Let T be a linear operator on a n dimensional inner product space V_F, B be an orthonormal basis of V and $A = [T]_B$. Then, T is an inner product space automorphism of V iff $AA^* = I_n = A^*A$.

Example 10.a. For $A = \begin{bmatrix} i & 0 \\ 0 & i \end{bmatrix}$, $A^* = \begin{bmatrix} -i & 0 \\ 0 & -i \end{bmatrix}$.

Clearly $$AA^* = A^*A = I_2.$$

Thus, the linear operator T on $\mathbf{C}^2$ given by A is an inner product space automorphism of $\mathbf{C}^2$.

Definition 11. An $n \times n$-matrix A over $\mathbf{C}$ is called a *unitary* matrix if $AA^* = A^*A = I_n$. Any linear operator T on a finite dimensional inner product space V is called a *unitary operator* if $TT^* = I = T^*T$.

If A is a real $n \times n$-matrix, then, $A^* = A^T$, and A is unitary iff $AA^T = I_n = A^T A$. However, a unitary matrix over $\mathbf{R}$ is usually called an *orthogonal matrix*. It is immediately clear from Definition 11, that an $n \times n$-matrix A over $\mathbf{C}$ is unitary iff its n columns form an orthonormal basis of $\mathbf{C}^{n \times 1}$.

Theorem 12. Let V_F be a finite dimensional inner product space, B and B' be any two ordered orthonormal bases of V. Then, the matrix P of B' relative to B is a unitary matrix.

Proof. Let $$B = \{\alpha_1, \alpha_2, \ldots, \alpha_n\}$$
and $$B' = \{\beta_1, \beta_2, \ldots, \beta_n\}.$$

For any $\alpha = \sum_{i=1}^{n} a_i \alpha_i$, $\|\alpha\|^2 = \sum_{i=1}^{n} |a_i|^2$, by Theorem (6.1.17).

The mapping $$T : V \to V$$
such that $$T(\alpha) = \sum_{i=1}^{n} a_i \beta_i$$
is an isomorphism.

Again, by Theorem (6.1.17) $$\|T\alpha\|^2 = \sum_{i=1}^{n} |a_i|^2,$$
so that $$\|T\alpha\| = \|\alpha\|.$$

Hence, By Theorem 9, T is a unitary linear operator. Let $P = [P_{ij}]$ be the matrix of B' relative to B.

By definition, $$T(\alpha_j) = \beta_j = \sum_{i=1}^{n} P_{ij}\alpha_i$$

so that $$P = [T]_B.$$

By Theorem 10, $$PP^* = I_n = P^*P,$$

i.e. P **is unitary.**

Lemma 13. Let A be any complex $n \times n$-matrix

(*i*) If $f(X) = X^n + a_1X^{n-1} + \ldots + a_n$ is the characteristic polynomial of A,

then $$\bar{f}(X) = X^n + \bar{a}_1X^{n-1} + \ldots + \bar{a}_n$$

is the characteristic polynomial of A^*.

(*ii*) If $a \in \mathbf{C}$ is a characteristic value of A, then $\bar{a}$ is a characteristic value of A^*.

(*iii*) If A is Hermitian, then, all the coefficients of $f(X)$ are real.

Proof. $$f(X) = \det(XI_n - A).$$

Then, $$\overline{(XI_n - A)} = \overline{f(X)} = \bar{f}(X) = X^n + \bar{a}_1X^{n-1} + \ldots + \bar{a}_n.$$

However, $$\overline{(XI_n - A)} = XI_n - \bar{A}.$$

So, $\bar{f}(X)$ is the characteristic polynomial of $\bar{A}$.

As $$A^* = (\bar{A})^T,$$

$$\begin{aligned}\bar{f}(X) &= \det(XI_n - \bar{A}) \\ &= \det\{(XI_n - \bar{A})\}^T \\ &= \det\{XI_n - (\bar{A})^T\} \\ &= \det(XI_n - A^*).\end{aligned}$$

So, $\bar{f}(X)$ is the characteristic polynomial of A^*. This proves (*i*).

Let $a \in \mathbf{C}$ be a characteristic value of A.

Then, $$f(X) = (X - a)g(X)$$

for some polynomial $g(X)$ of degree $n-1$

Then, $$\bar{f}(X) = (X - \bar{a})\bar{g}(X),$$

gives $\bar{a}$ is a root of $\bar{f}(X)$ and, hence, it is a characteristic value of A^*.

Let $$A = A^*.$$

Then, by (*i*), $$f(X) = \bar{f}(X).$$

Thus, $$a_i = \bar{a}_i, 1 \le i \le n$$

i.e., all the coefficients of $f(X)$ are real.

Lemma 14. If an $n \times n$-matrix A is unitary, then, any characteristic value of A satisfies $|a| = 1$.

Proof.

Let $AX = aX$ for some $X \neq 0$ in $\mathbf{C}^{n\times 1}$.

This gives $$\overline{X}^T A^* = \overline{X}^T \overline{a}$$

so that $$\overline{X}^T A^* AX = \overline{a}\overline{X}^T Xa = |a|^2 \overline{X}^T X$$

i.e., $$\overline{X}^T X = |a|^2 \overline{X}^T X, \qquad \text{as } AA^* = I_n$$

Thus, $$|a| = 1, \qquad \text{as } \overline{X}^T X > 0$$

In terms of linear operators, these two lemmas give the following :

Lemma 15. Let T be a linear operator on a finite dimensional inner product space V_F.

(*i*) If $f(X)$ is the characteristic polynomial of T, then, $\bar{f}(X)$ is the characteristic polynomial of T^*.

(*ii*) If $a \in F$ is a characteristic value of T, then, $\overline{a}$ is a characteristic value of T^*.

(*iii*) If T is self-adjoint, then, the coefficients of its characteristic polynomial are real.

(*iv*) If T is unitary, then, for any characteristic value a of T, $|a| = 1$.

We shall see that every unitary operator T on V_F is diagonalizable in some what stronger sense. We shall prove this for more general type of operators, known as normal operators.

Definition 16. A linear operator T on a finite dimensional inner product space V_F is called a *normal operator* if $TT^* = T^*T$. If a matrix A over $\mathbf{C}$ satisfies $AA^* = A^*A$, it is called a *normal matrix*.

Example 16.a.

Let $$a = \begin{bmatrix} 2 & 3 \\ -3 & 2 \end{bmatrix}.$$

Then, $$A^* = A^T = \begin{bmatrix} 2 & -3 \\ 3 & 2 \end{bmatrix} \neq A.$$

So, A is not Hermitian.

But $$AA^* = A^*A = \begin{bmatrix} 13 & 0 \\ 0 & 13 \end{bmatrix} \neq I_2.$$

Thus, A is not unitary. However, A is normal.

Lemma 17. Let T be a linear operator on V_F and W be a subspace of V, invariant under T. Then, $W^{\perp}$ is invariant under T^*. If T is self-adjoint, then $W^{\perp}$ is invariant under T.

Proof. Let $\alpha \in W$ and $\beta \in W^{\perp}$. As $T\alpha \in W$, $(T\alpha|\beta)=0$. But $(T\alpha|\beta)=(\alpha|T^*\beta)$. Thus, $(\alpha|T^*\beta)=0$. This gives $T^*\beta \in W^{\perp}$. If T is self-adjoint, then, $T=T^*$ and, hence, $W^{\perp}$ is invariant under T.

Lemma 18. Let T be a normal operator on V_F. Then, the following relations hold :

(*i*) For any $\alpha \in V$, $\|T\alpha\| = \|T^*\alpha\|$.

(*ii*) If c is a characteristic value of T, then, the eigen-space $W_c = \{\alpha \in V : T\alpha = c\alpha\}$ $W_c^{\perp}$ both are invariant under T and T^*. For any $\alpha \in W_c$, $T^*\alpha = \bar{c}\alpha$.

Proof.

(*i*) Let $\alpha \in V$

Now,
$$T^*T = TT^*,$$
$$\|T\alpha\|^2 = (T\alpha|T\alpha) = (\alpha|T^*T\alpha) = (\alpha|TT^*\alpha) = (T^*\alpha|T^*\alpha) = \|T^*\alpha\|^2.$$
Hence,
$$\|T\alpha\| = \|T^*\alpha\|.$$

(*ii*) Let $\alpha \in W_c$. Then $T\alpha = c\alpha \in W_c$ gives W_c is T-invariant.
$$T(T^*\alpha) = (TT^*)\alpha = (T^*T)(|\alpha|) = T^*(c\alpha) = cT^*\alpha,$$
yields
$$T^*\alpha \in W_c.$$

Thus, W_c is T^*-invariant. By Lemma 17, $W_c^{\perp}$ is T^*-invariant and $T^{**}(=T)$-invariant. Consider $T' = (T - cI)$, then, $(T')^* = T^* - \bar{c}I$. Obviously, T' is normal.

Also,
$$T'\alpha = T\alpha - c\alpha = 0.$$

So, $\|T'\alpha\| = 0$ and by (*i*) $\|(T')^*\alpha\| = 0$, i.e., $\|T^*\alpha - \bar{c}\alpha\| = 0$, i.e., $T^*\alpha - \bar{c}\alpha = 0$. Hence $T^*\alpha = \bar{c}\alpha$.

Theorem 19 (Spectral Theorem). Let T be a linear operator on a finite dimensional complex inner product space V. Then, T is normal iff there exists an orthonormal basis B of V, each of whose member is a characteristic vector of T.

Proof. Let V an orthonormal basis B, each of whose member is a characteristic vector of T. By regarding B an ordered basis, we get, $A = [T]_B$ is a diagonal matrix. So A^* is also a diagonal matrix. Consequently, $AA^* = A^*A$. As $A^* = [T^*]_B$, we get $TT^* = T^*T$ i.e. T is normal. Conversely, let T be normal. We apply induction on $\dim V = n$. Let $n = 1$. Choose any $\|\alpha_1\| = 1$. Then $T\alpha_1 = c\alpha_1$ for some $c \in \mathbf{C}$. Then, $B = \{\alpha_1\}$ is a desired orthonormal basis of V. So, the result holds if $n = 1$. Let $n > 1$ and to apply induction, let the result hold for any complex inner product space of dimension less than n. Now, T has a characteristic value a_1.

Then, $$V = W_{a_1} \oplus W_{a_1}^{\perp} \quad \cdots (1)$$

where W_{a_1} is the eigen-space of T relative to a_1.

Now, $$\dim W_{a_1} = m \geq 1.$$

So $$\dim W_{a_1}^{\perp} = n - m < n.$$

Then, W_{a_1} has an orthonormal basis

$$B_1 = \{\alpha_1, \alpha_2, \ldots, \alpha_m\}.$$

By definition, $$T\alpha_j = a_1\alpha_j, \ 1 \leq j \leq m.$$

Now $W_{a_1}^{\perp}$ is invariant under T and T^*. If T_1 is the restriction of T to $W_{a_1}^{\perp}$, then, T_1 is a linear operator on $W_{a_1}^{\perp}$ and T_1^* is the restriction of T^* to $W_{a_1}^{\perp}$. Obviously, $T_1T_1^* = T_1^*T_1$ as $TT^* = T^*T$. Thus T_1 is a normal operator on $W_{a_1}^{\perp}$. By the induction hypothesis, $W_{a_1}^{\perp}$ has an orthonormal basis,

$$B_2 = \{\alpha_{m+1}, \alpha_{m+2}, \ldots, \alpha_n\}$$

each of whose member is a characteristic vector of T_1 and, hence, of T.

Then, $B = B_1 \cup B_2$ is a desired orthonormal basis of V.

Theorem 20. Let A be any complex $n \times n$-matrix. Then, A is normal iff for some unitary matrix P, P^*AP is a diagonal matrix.

Proof.

Let T be the linear operator on $\mathbf{C}^{n\times 1}$ given by A. Let $A = [T]_B$ where B is the standard ordered basis of $\mathbf{C}^{n\times 1}$. Let A be normal, then, T is normal. So, by Theorem 19, there exists an ordered orthonormal basis B' of $\mathbf{C}^{n\times 1}$ such that C is a diagonal matrix. Let P be the matrix of B' relative to B. By Corollary (3.3.7), $[T]_{B'} = P^{-1}AP$. By Theorem 12, P is unitary, so, $P^* = P^{-1}$. Hence, P^*AP is a diagonal matrix.

Conversely, let for some unitary matrix P, $D = P^*AP$ be a diagonal matrix. Then, D^* is also a diagonal matrix. So, $DD^* = D^*D$.

But $$D^* = (P^*AP)^* = P^*A^*P^{**} = P^*A^*P.$$

Thus, $$DD^* = D^*D$$

yield $$P^*APP^*A^*P = P^*A^*PP^*AP,$$

i.e., $$P^*AA^*P = P^*A^*AP$$

By cancelling P and P^*, we get $AA^* = A^*A$.

Difference between diagonalizability of a linear operator, in general, and of a normal operator needs to be emphasized. Thus, in terms of a square matrix A over C, there may exist a non-singular matrix P such that $P^{-1}AP$ is a diagonal matrix, but if A is normal, P can be chosen to be a unitary matrix.

Example 20.a. Let $A = \begin{bmatrix} 1 & 1 \\ 0 & 2 \end{bmatrix}$. It has two distinct characteristic values. So, A is diagonalizable. However, A is not normal. Hence, there exists no unitary matrix P such that $P^{-1}AP$ is a diagonal matrix.

Example 20.b. Let $A = \begin{bmatrix} 1 & 2 \\ 2 & 1 \end{bmatrix}$. Its characteristic values are 3 and −1. The corresponding eigen spaces in $\mathbf{C}^{2+1}$ are

$$W_3 = \left\{ \begin{bmatrix} a \\ a \end{bmatrix} : a \in \mathbf{C} \right\}$$

$$W_{-1} = \left\{ \begin{bmatrix} a \\ -a \end{bmatrix} : a \in \mathbf{C} \right\}.$$

Their respective orthonormal basis are $\left\{ \begin{bmatrix} \frac{1}{\sqrt{2}} \\ \frac{1}{\sqrt{2}} \end{bmatrix} \right\}$ and $\left\{ \begin{bmatrix} \frac{1}{\sqrt{2}} \\ -\frac{1}{\sqrt{2}} \end{bmatrix} \right\}$.

Their union $$B = \left\{ \begin{bmatrix} \frac{1}{\sqrt{2}} \\ \frac{1}{\sqrt{2}} \end{bmatrix}, \begin{bmatrix} \frac{1}{\sqrt{2}} \\ -\frac{1}{\sqrt{2}} \end{bmatrix} \right\}$$

is an ordered orthonormal basis of $\mathbf{C}^{2\times1}$.

$$P = \begin{bmatrix} \frac{1}{\sqrt{2}} & \frac{1}{\sqrt{2}} \\ \frac{1}{\sqrt{2}} & -\frac{1}{\sqrt{2}} \end{bmatrix}$$

is the matrix of B relative to the standard ordered basis of $\mathbf{C}^{2\times1}$

Then, $$P^* = P^T = P$$

$$P^*AP = \begin{bmatrix} 3 & 0 \\ 0 & -1 \end{bmatrix}.$$

Theorem 21. An $n \times n$-matrix A over **C** is Hermitian iff P^*AP is a real diagonal matrix for some unitary matrix P. All the characteristic values of a Hermitian matrix are real.

Proof. Let A be Hermitian. As A is normal, there exists a unitary matrix P such that $D = P^*AP$ is a diagonal matrix. Then, $D^* = D$, as $A^* = A$. As D is diagonal, $D^* = \overline{D}$, so that $\overline{D} = D$. Hence, D is a real diagonal matrix. Conversely, let for some unitary matrix B, $D_1 = B^*AB$ be a real diagonal matrix.

Then, $$D_1 = D_1^* = B^* A^* B,$$

i.e., $$B^* AB = B^* A^* B$$

This gives $A = A^*$.

Finally, as an Hermittian matrix A is similar to a real diagonal matrix $P^* AP$, so its characteristic values must be real.

If A is a symmetric real matrix, obviously, A is Hermittian. Is it possible to find a real unitary matrix P (i.e., an orthogonal matrix P) such that $P^T AP$ is a diagonal matrix. We answer this in the following theorem :

Theorem 22. Let T be a self-adjoint linear operator on a finite dimensional real inner product space $V_{\mathbf{R}}$. Then, there exists an ordered orthonormal basis B of V such that $[T]_B$ is a real diagonal matrix ; each member of B is a characteristic vector of T.

Proof. Consider an ordered orthonormal basis B' of V. Then, $A = [T]_{B'}$ is a real self-adjoint matrix, i.e., A is a real symmetric matrix. By Theorem 21, all the characterisuc values of A in $\mathbf{C}$ are real. We now apply induction on $\dim V = n$. We follow the arguments in Theorem 19. As in there the result obviously holds for $n = 1$. Let $n > 1$, and the result be true for all real inner product spaces of dimension less than n. Let a_1 be a characteristic value of A. As a_1 is real, it is a characteristic value of T.

Thus, $$W_{a_1} = \{\alpha \in V : T\alpha = a_1\alpha\}$$

is a non-zero subspace of V.

Then, $$V = W_{a_1} \oplus W_{a_1}^{\perp}.$$

By Lemma 18, W_{a_1} and $W_{a_1}^{\perp}$ both are invariant under T. After this, follow the arguments in Theorem 19 and complete the proof.

The results in Theorems 21 and 22 are known as *Principal Axis Theorems*. The following is immediately clear.

Corollary 23. Let A be any real symmetric $n \times n$-matrix. Then, there exists an orthogonal $n \times n$-matrix P such that $P^T AP$ is a diagonal matrix.

Theorem 24. Let A be a real symmetric, $n \times n$-matrix and $f(X)$ be its characteristic polynomial. Let $f(X) = X^k g(X)$, where $g(X)$ is a polynomial such that $g(0) \neq 0$. Then,

(*i*) rank$A$ = deg $g(X)$;

(*ii*) all the roots of $g(X)$ are real and non-zero

(*iii*) If deg$(X) = r$, and s roots of $g(X)$ are positive, then, there exists an orthogonal matrix P such that

$$P^T AP = \text{diag}[c_1, c_2, ..., c_r, 0, 0, ..., 0]$$

where $c_1, c_2, \ldots, c_s$ are positive characteristic values and $c_{s+1}, c_{s+2}, \ldots, c_r$ are the negative characteristic values.

Proof.

Let T be the linear operator on $\mathbf{R}^{n\times 1}$ given by A. By Theorem 22, there exists an orthonormal basis B of $\mathbf{R}^{n\times 1}$ whose members are characteristic vectors of T. Each $\alpha \in B$ corresponds to some characteristic value of T. Let r members, say $\alpha_1, \alpha_2, \ldots, \alpha_r$ of B correspond to non-zero characteristic values, and the last $n-r$ members $\alpha_{r+1}, \alpha_{r+2}, \ldots, \alpha_n$ correspond to the characteristic value 0. For $1 \le i \le r, T\alpha_i = c_i\alpha_i$ for some $c_i \ne 0$. By re-indexing, we can suppose that $c_1, c_2, \ldots, c_s$ are positive, and c_j, $s+1 \le j \le r$ are negative. Then, for the ordered basis.

$$B = \{\alpha_1, \alpha_2, \ldots, \alpha_r, \alpha_{r+1}, \ldots, \alpha_n\}$$

$$[T]_B = \text{diag}[c_1, c_2, \ldots, c_r, 0, 0, \ldots, 0] = P^T AP,$$

where P is the matrix of B relative to the standard ordered basis.

Then,
$$\begin{aligned} f(X) &= \text{characteristic polynomial of } A \\ &= \text{characteristic polynomial of } P^T AP \\ &= \prod_{i=1}^{r} (X - c_i) X^{n-r} \\ &= X^{n-r} g(X), \end{aligned}$$

$$\text{rank}(A) = \text{rank}(P^T AP) = r = \deg g(X).$$

We end this section by describing some special types of normal matrices.

Theorem 25. If a triangular matrix is normal, it is a diagonal matrix.

Proof. As transpose of a normal matrix is normal, it is enough to prove the result for an upper triangular matrix. Let $A = [a_{ij}]$ be a complex, upper triangular, $n\times n$-matrix. The result is obvious for $n = 1$. Let $n > 1$ and to apply induction, let the result hold for all $m\times m$-matrices with $m < n$. Now, $a_{ij} = 0$ for $i > j$.

$$A^* = [a'_{ij}],\ a'_{ij} = \bar{a}_{ji}.\ AA^* = [c_{ij}],$$

$$c_{ij} = \sum_{k=1}^{n} a_{ik}\bar{a}_{jk} = \sum a_{ik}\bar{a}_{jk},\ k \ge \max(i, j) \quad \ldots (1)$$

$$A^*A = [d_{ij}],\ d_{ij} = \sum \bar{a}_{ki}\, a_{kj},\ k \le \min(i, j) \quad \ldots (2)$$

As $$AA^* = A^*A,\ c_{ij} = d_{ij}.$$

In particular, $$c_{11} = d_{11}.$$

Thus, $$a_{11}\bar{a}_{11} + a_{12}\bar{a}_{12} + a_{1n}\bar{a}_{1n} = \bar{a}_{11}a_{11}.$$

This immediately gives $a_{1j} = 0$ for $j \geq 2$. To prove that A is diagonal, we need to prove is that $a_{ij} = 0$ for $j > i$. We have already proved it for $i = 1$. suppose for some $t < n$, we have already proved that $a_{ij} = 0$ for $j > i$, whenever $i \leq t$.

By (1) $$c_{t+1,\,t+1} = a_{t+1,\,t+1}\,\bar{a}_{t+1,\,t+1} + \sum_{j=t+2}^{n} a_{t+1,\,j}\,\bar{a}_{t+1,\,j} = \left|a_{t+1,\,t+1}\right|^2 + \sum_{j=t+2}^{n}\left|a_{t+1,\,j}\right|^2 \quad \text{... (3)}$$

By the induction hypothesis, $a_{j,\,t+1} = 0$ for $j \leq t$.

So, by (2), $$d_{t+1,\,t+1} = \sum_{j<t+1}\left|a_{j,\,t+1}\right|^2 + \left|a_{t+1,\,t+1}\right|^2 = \left|a_{t+1,\,t+1}\right|^2 \quad \text{... (4)}$$

As $$c_{t+1,\,t+1} = d_{t+1,\,t+1},$$

(3) and (4) yield $$\sum_{j=t+2}^{n}\left|a_{t+1,\,j}\right|^2 = 0.$$

Thus, $a_{t+1,\,j} = 0$ for $j > t+1$. Hence, by induction $a_{ij} = 0$, whenever $j > i$. This proves that A is a diagonal matrix.

Proposition 26. A real normal 2 × 2-matrices is either a symmetric matrix or is of the form $\begin{bmatrix} a & b \\ -b & a \end{bmatrix}$.

Proof. Let $A = \begin{bmatrix} a & b \\ c & d \end{bmatrix}$ be a real, normal matrix.

Then, $$A^* = A^T = \begin{bmatrix} a & c \\ b & d \end{bmatrix},$$

$$AA^* = A^*A,$$

gives $$\begin{bmatrix} a^2+b^2 & ac+bd \\ ca+bd & c^2+d^2 \end{bmatrix} = \begin{bmatrix} a^2+c^2 & ab+cd \\ ba+cd & b^2+d^2 \end{bmatrix},$$

so that $$a^2+b^2 = a^2+c^2,$$

$$c^2+d^2 = b^2+d^2,$$

so $$b^2 = c^2.$$

Thus, $$b = \pm c.$$

If $b = c$, A is symmetric. Let A be not symmetric.

Then, $$b = -c \neq 0.$$

Then, $$ac+bd = ab+cd$$

gives $$-ab+bd = ab-bd,$$

$$2ab = 2bd$$

and $$a = d.$$

Hence, $$A = \begin{bmatrix} a & b \\ -b & a \end{bmatrix}.$$

Proposition 27. An orthogonal 2×2-matrix A is of one of the following forms

$$\begin{bmatrix} \cos\theta & -\sin\theta \\ \sin\theta & \cos\theta \end{bmatrix}, \begin{bmatrix} \cos\theta & -\sin\theta \\ -\sin\theta & -\cos\theta \end{bmatrix}.$$

Proof. Let $$A = \begin{bmatrix} a & b \\ c & d \end{bmatrix}.$$

As $$AA^T = I_2,$$

$$a^2 + b^2 = 1,$$
$$c^2 + d^2 = 1,$$
$$ac + bd = 0.$$

We get θ and ψ, such that $a = \cos\theta$, $b = -\sin\theta$, $c = \cos\psi$, $d = -\sin\psi$.

Further, $$ac + bd = 0,$$

yields $$\cos(\psi - \theta),\ \psi = \theta \pm \frac{\pi}{2}.$$

If $\psi = \theta + \frac{\pi}{2}$ $$A = \begin{bmatrix} \cos\theta & -\sin\theta \\ -\sin\theta & -\cos\theta \end{bmatrix}$$

If $\psi = \theta - \frac{\pi}{2}$ $$A = \begin{bmatrix} \cos\theta & -\sin\theta \\ \sin\theta & \cos\theta \end{bmatrix}.$$

Corollary 28. The linear transformation on $\mathbf{R}^2$, given by an orthogonal matrix A is a rotation or a reflection according as $\det(A) = 1$ or $\det(A) = -1$.

Proof. If $\det(A) = 1$, it is immediate from Proposition 27, that $A = \begin{bmatrix} \cos\theta & -\sin\theta \\ \sin\theta & \cos\theta \end{bmatrix}$.

Consider $$\begin{bmatrix} x \\ y \end{bmatrix} \in \mathbf{R}^{2\times 1}.$$

Then, $$x = r\cos\psi,\ y = r\sin\psi,\ r = \sqrt{x^2 + y^2}.$$

$$A\begin{bmatrix} x \\ y \end{bmatrix} = \begin{bmatrix} \cos\theta & -\sin\theta \\ \sin\theta & \cos\theta \end{bmatrix}\begin{bmatrix} r\cos\psi \\ r\sin\psi \end{bmatrix} = \begin{bmatrix} r\cos(\psi + \theta) \\ r\sin(\psi + \theta) \end{bmatrix}.$$

Thus, if O is the origin, P is the point (x, y), its image $P'(x', y')$ under A is given by

$$x' = r\cos(\psi + \theta),\ y' = \sin(\psi + \theta)$$

which is obtained by rotating OP around O through the angle θ.

If $\det(A) = -1$, then, by Proposition 27,

$$A = \begin{bmatrix} \cos\theta & -\sin\theta \\ -\sin\theta & -\cos\theta \end{bmatrix}.$$

In this case, the image of $P(x, y)$ is $P''(x'', y'')$ with

$$x'' = r\cos(\theta + \psi) y'' = -r\sin(\theta + \psi),$$

i.e.,
$$x'' = r\cos[-(\psi + \theta)] y'' = r\sin[-(\psi + \theta)].$$

Clearly, P'' is obtained by reflecting OP in the line OB such that $\angle XOB = -\theta/2$.

This corollary motivates the following :

Definition 29. Let A be any $n \times n$-orthogonal matrix. Then, the linear operator on $\mathbf{R}^n$ given by A is called a *rotation* or a *reflection* according as $\det(A) = 1$ or $\det(A) = -1$.

Remark 30.

(*i*) Let T be a normal operator on a finite dimensional complex inner product space V. By the spectral theorem (Theorem 19), T is diagonalizable.

If
$$c_1, c_2, \ldots, c_k$$

are distinct characteristic values of T, and $W_1, W_2, \ldots, W_k$ are corresponding eigen-spaces, then,

$$V = W_1 \oplus W_2 \oplus \ldots, \oplus W_k.$$

[Theorem (4.4.12)]. Further, by Theorem (3.5.2) we have idempotent linear operators $E_1, E_2, \ldots, E_k$ on V such that range $(E_i) = W_i$, $E_iE_j = 0$ for $i \neq j$, $I = E_1 + E_2 + \ldots + E_k$.

Consider
$$T' = c_1E_1 + c_2E_2 + \ldots + c_kE_k.$$

Let $\alpha = \sum_{j=1}^{k} \alpha_j$, $\alpha_j \in W_j$ be any member of V.

Then,
$$T\alpha = \sum_{j=1}^{k} T\alpha_j = \sum_{j=1}^{k} c_j\alpha_j.$$

As $E_j(\alpha_i) = 0$ for $j \neq i$,

$$T'(\alpha_i) = \sum_{i=1}^{k} c_j E_j(\alpha_i) = c_i E_i(\alpha_i) = c_i\alpha_i$$

so that
$$T'(\alpha) = \sum_{i=1}^{k} c_i\alpha_i = T(\alpha).$$

Hence,
$$T = c_1E_1 + c_2E_2 + \ldots + c_kE_k.$$

This expression of T is known as *spectral resolution* of T.

For $\alpha_i \in W_i$, $\alpha_j \in W_j$,

$$i \neq j \ \ (T\alpha_i|\alpha_j) = (c_i\alpha_i|\alpha_j) = c_i(\alpha_i|\alpha_j).$$

Also, by Lemma 18,

$$(T\alpha_i|\alpha_j) = (\alpha_i|T^*\alpha_j) = (\alpha_i|\bar{c}_j\alpha_j) = c_j(\alpha_i|\alpha_j)$$

so that $(c_i - c_j)(\alpha_i|\alpha_j) = 0.$

However, $c_i - c_j \neq 0.$

Hence, $(\alpha_i|\alpha_j) = 0.$

(*ii*) Let T be a self-adjoint linear operator on a finite dimensional inner product space V_F. Because of Theorem 22, we do not need F to be **C**. In this case, we always have spectral resolution $T = c_1E_1 + c_2E_2 + ... + c_kE_k$, where $c_1, c_2, ..., c_k$ are the distinct characteristic values of T, $E_1, E_2, ..., E_k$ are orthogonal idempotent operators of V_F, such that each $W_i = \text{range}(E_i)$ is the eigen-space of T relative to c_i.

(*iii*) Consider a normal operator T on $V_{\mathbf{C}}$.

Let $T = c_1E_1 + c_2E_2 + ... + c_kE_k$

be its spectral resolution.

Then,
$$T^2 = \sum_{j=1}^{k} \sum_{i=1}^{k} c_iE_ic_jE_j$$
$$= \sum_{j=1}^{k} \sum_{i=1}^{k} c_ic_jE_iE_j$$
$$= \sum_{i=1}^{k} c_i^2E_i^2$$
$$= \sum_{i=1}^{k} c_i^2E_i,$$

as $E_iE_j = 0$

for $i \neq j$.

By continuing this process, we see that

$$T^s = \sum_{i=1}^{k} c_i^sE_i.$$

Then, for any polynomial $f(X) = \sum_{s=0}^{m} a_sX^s$

$$f(T) = \sum_{s=0}^{m} a_s T^s = \sum_{s=0}^{m} a_s \left[\sum_{i=1}^{m} c_i^s E_i \right] = \left(\sum_{s=0}^{m} a_s c_i^s \right) E_i$$

i.e., $$f(T) = \sum_{i=1}^{k} f(c_i) E_i .$$

Suppose T is not a scalar operator, then, $k \geq 2$.

Consider $$g_1(X) = \frac{\left[\prod_{j \neq i} (X - c_j) \right]}{\prod_{j \neq i} (c_i - c_j)}$$

Then, $$g_i(c_i) = 1$$

and $$g_j(c_i) = 0 \qquad \text{for } i \neq j$$

We get $$g_j(T) = \sum_{i=1}^{k} g_j(c_i) E_i = E_j .$$

Thus, E_j is expressible as a polynomial in T.

(*iv*) For any polynomial, $f(X) = \sum_{s=0}^{m} a_s X^s$ over C and any linear operator T on $V_{\mathbf{C}}$,

$$f(T) = \sum_{s=0}^{m} a_s T^s$$

and $$[f(T)]^* = \sum_{s=0}^{m} \bar{a}_s (T^*)^s = \bar{f}(T^*).$$

If T is normal, then $$TT^* = T^* T$$

gives $$f(T)\bar{f}(T^*) = \bar{f}(T^*) f(T),$$

i.e., $f(T)$ is normal.

Exercise 6.2

1. Let T be the linear operator on $\mathbf{C}^2$ such that $T\epsilon_1 = (2, -2)$, $T\epsilon_2 = (1, 1)$.

(*i*) For any $\alpha = (a_1, a_2) \in \mathbf{C}^2$, find $T^* \alpha$.

(*ii*) Is T unitary?

(*iii*) Is T normal?

2. For each of the following matrices find its conjugate transpose

$$\begin{bmatrix} 3 & 2 \\ -2 & 3 \end{bmatrix}, \begin{bmatrix} 1 & 5 \\ 5 & 2 \end{bmatrix}, \begin{bmatrix} 1+i & i \\ ie^{-i\theta} & 1+i \end{bmatrix}$$

$$\begin{bmatrix} 0 & 1 & 2 \\ -1 & 0 & 3 \\ -2 & -3 & 1 \end{bmatrix}, \begin{bmatrix} -1 & 1 & 2 \\ -1 & -1 & 3 \\ -2 & -3 & -1 \end{bmatrix}, \begin{bmatrix} \frac{4}{5} & \frac{3}{5} \\ \frac{-3}{5} & \frac{4}{5} \end{bmatrix}.$$

(*i*) Which of these are unitary?

(*ii*) Which of these are self-adjoint?

(*iii*) Which of these are normal?

3. Give an example of a complex 2 × 2-matrix A, which is not unitary, not Hermittian, but is normal.

4. Prove that a complex 2 × 2-matrix $\begin{bmatrix} a & b \\ c & d \end{bmatrix}$ is normal

iff $$c = be^{i\theta},\ \bar{d}b - d\bar{b}e^{-i\theta} = \bar{a}b - a\bar{b}e^{-i\theta}.$$

5. Prove that each of the following types of linear operators on $\mathbf{R}^n$ is normal :

(*i*) Idempotent linear operators.

(*ii*) Any linear operator on $\mathbf{R}^n$ given by a skew-symmetric matrix.

6. Let A be a real skew-symmetric $n \times n$-matrix and D be a diagonal, real $n \times n$-matrix. Prove that $D + A$ is normal iff $DA = AD$.

7. For the matrix $A = \begin{bmatrix} 2 & \sqrt{2} \\ \sqrt{2} & 3 \end{bmatrix}$, find an orthogonal matrix P such that $P^T AP$ is a diagonal matrix.

8. Give example of two real 2 × 2-matrices, which are normal, but their sum is not normal.

9. Give example of two real 2 × 2-matrices, which are normal, but their product is not normal.

10. For $A = \begin{bmatrix} 0 & 1 & 0 \\ 0 & 0 & 2 \\ -1 & 0 & 0 \end{bmatrix}$, find, if possible, an orthogonal matrix P such that $P^T AP$ is a diagonal matrix.

11. Let T be a normal operator on a finite dimensional inner product space V_F. Prove that T is self-adjoint or unitary according as every characteristic value of T is real or of absolute value one.

12. Let T be a linear operator on a finite dimensional complex inner product space V. Prove that T is normal iff $T = T_1 + iT_2$ for some self-adjoint linear operators T_1, T_2 such that $T_1T_2 = T_2T_1$.

13. If a self-adjoint linear operator is nilpotent, prove that it is zero.

14. Let A be any $n \times n$-matrix over **C** such that $A(AA^*) = (AA^*)A$. Prove that A is normal.

    [Hint : show first $A^*(AA^*) = (AA^*)A^*$, and, then, $(AA^* - A^*A)^2 = 0$].

15. Let T be a self-adjoint operator on a finite dimensional inner product space V_F. Let c_1, c_2 be two distinct characteristic values of T. For any α_1, $\alpha_2 \in V$ satisfying $T\alpha_1 = c_1\alpha_1$, $T\alpha_2 = c_2\alpha_2$, prove that $(\alpha_1|\alpha_2) = 0$.

16. For any real, symmetric $n \times n$-matrix A, prove that there exists a real, symmetric $n \times n$-matrix B such that $A = B^3$.

17. Let V be the complex inner product space of all polynomials of degree ≤ 2, such that

$$(f|g) = \int_0^1 f(t)\,\overline{g(t)}\,dt$$

    for any f, $g \in V$. For any fixed a_0, a_1, $a_2 \in \mathbf{C}$, consider the linear functional L on V such that

$$L(d_0 + d_1x + d_2x^2) = a_0d_0 + a_1d_1 + a_2d_2.$$

    Find $\qquad h(X) = c_0 + c_1X + c_2X^2 \in V$

    such that $\qquad L(f) = (f|h)$.

18. Let V be the complex inner product space of all polynomials over **C**, such that for any f, $g \in V$, $(f|g) = \int_0^1 f(t)\,\overline{g(t)}\,dt$.

    Prove the following :

    (*i*) Let z be any fixed complex number and $L \in V^*$ be such that $L(f) = f(z)$. Then, there exist no $h \in V$ such that $L(f) = (f|h)$ for all $f \in V$.

    (*ii*) The differentiation operator D on V has no adjoint.

19. Let A be a 2 × 2-matrix over **C**. Prove the following :

    (*i*) A is unitary iff $A = \begin{bmatrix} a & b \\ -\lambda\bar{a} & \lambda\bar{b} \end{bmatrix}$ such that $|a|^2 + |b|^2 = 1 = |\lambda|^2$.

    (*ii*) If A is real, then, A is orthogonal iff

$$A = \begin{bmatrix} \cos\theta & -\sin\theta \\ \sin\theta & -\cos\theta \end{bmatrix} \quad \text{for some real } \theta.$$

    [Hint : $(a, b)(\bar{c}, \bar{d}) = (0, 0)$ with $(a, b) \neq (0, 0)$ holds iff $(c, d) = (-\lambda\bar{a}, \lambda\bar{b})$. For any two real numbers r, s, $r^2 + s^2 = 1$ iff $r = \cos\theta$, $s = \sin\theta$ for some real θ].

6.3 Hermittian Form

Throughout $\qquad F = \mathbf{R}$

or $$F = \mathbf{C}.$$

In this section, se generalize the inner product to the following definition :

Definition 1. A *sesquilinear* $\left(1\frac{1}{2}\right)$ *form* on a real or complex vector space V_F is a function f on $V \times V$ to F such that

$$f(c\alpha + \beta, \gamma) = cf(\alpha, \gamma) + f(\beta, \gamma)$$
$$f(\alpha,\ c\beta + \gamma) = \bar{c}f(\alpha, \beta) + f(\alpha, \gamma)$$

for all $\alpha,\ \beta,\ \gamma \in V,\ c \in F$.

Consider a linear operator T on an inner product space V.

Then, $$f(\alpha,\ \beta) = (T\alpha|\beta)$$

is a sesquilinear form ; if T is not one-to-one, then, for some $\alpha \neq 0$, $T\alpha = 0$, gives that $f(\alpha,\ \beta) = 0$ for every $\beta \in V$. It is unlike the inner product. An inner product is a sesquilinear form f such that $f(\alpha,\ \alpha) > 0$ for every $\alpha \neq 0$ in V and $f(\beta,\ \alpha) = \overline{f(\alpha,\ \beta)}$.

The following result gives constructions of sesquilinear forms on a finite dimensional vector space V_F.

Proposition 2. Let V_F be any finite dimensional vector space, $F = \mathbf{R}$ or $\mathbf{C}$, and

$$B = \{\alpha_1,\ \alpha_2, \ldots,\ \alpha_n\}$$

be an ordered basis of V.

Then, any sesquilinear form f determines an $n \times n$-matrix A over F such that for any $\alpha,\ \beta \in V$, if $X = [\alpha]_B$ and $Y = [\beta]_B$, then,

$$f(\alpha,\ \beta) = Y^* AX.$$

Conversely, any $n \times n$-matrix A' over F, defines a sesquilinear form

$$f'(\alpha,\ \beta) = Y^* A'X.$$

Proof. Let f be a sesquilinear form. Let $\alpha,\ \beta \in V$.

Then, $$\alpha = \sum_{i=1}^{n} x_i \alpha_i,$$

$$\beta = \sum_{i=1}^{n} y_i \alpha_i,$$

$$X = [\alpha]_B = \begin{bmatrix} x_1 \\ x_1 \\ \vdots \\ x_n \end{bmatrix},$$

$$Y = [\beta]_B$$

$$f(\alpha, \beta) = f\left(\sum_{i=1}^{n} x_i \alpha_i, \beta\right)$$

$$= \sum_{i=1}^{n} x_i f(\alpha_i, \beta)$$

$$= \sum_{i=1}^{n} x_i f\left(\alpha_i, \sum y_j \alpha_j\right)$$

$$= \sum_{j=1}^{n} \sum_{i=1}^{n} x_i \overline{y}_j f(\alpha_i, \alpha_j)$$

Put $A = \left[a_{ij}\right]$,

$$a_{ij} = f(\alpha_j, \alpha_i).$$

Obviously, $f(\alpha, \beta) = \sum_{j=1}^{n} \sum_{i=1}^{n} \overline{y}_j a_{ji} x_i$

$$= \left[\overline{y}_1, \overline{y}_2, \ldots, \overline{y}_n\right] A \begin{bmatrix} x_1 \\ x_2 \\ \vdots \\ x_n \end{bmatrix}$$

$$= Y^* AX.$$

Conversely, if $A' = \left[a'_{ij}\right]$,

it is immediate that $f'(\alpha, \beta) = Y^* AX$

defines a sesqui-linear form on V.

Definition 3. Let f be a sesqui-linear form on a finite -dimensional vector space V_F.

If $B = \{\alpha_1, \alpha_2, \ldots, \alpha_n\}$

is, an ordered basis of V_F.

Then, the $n \times n$-matrix $A = \left[a_{ij}\right]$, with $a_{ij} = f\left(\alpha_j, \alpha_i\right)$ is called the *matrix of A relative to f*; we write $A = [f]_B$.

Given two sesqui-linear forms f **and** g **on** V **and any scalar** $c \in F$,

$$f + g : V \times V \to F$$

$$cf : V \times V \to F$$

such that
$$(f+g)(\alpha, \beta) = f(\alpha, \beta) + g(\alpha, \beta)$$
$$(cf)(\alpha, \beta) = c[f(\alpha, \beta)]$$

are sesqui-linear form.

This makes the set S of all sesqui-linear forms on V_F a vector space over F. For a fixed ordered basis B of V, $[f]_B = [g]_B$ iff $f = g$, so that the vector space S is isomorphic to $F^{n\times n}$ and, hence, $\dim S = n^2 = (\dim V)^2$. Further, $f^*(\alpha, \beta) = \overline{f(\beta,\alpha)}$ defines a sesqui-linear form such that $\left[f^*\right]_B = [f]_B^*$. Thus, $f = f^*$ iff $A = [f]_B$ is a *Hermittian matrix* ; such an f is called a *Hermittian form*. Thus, an inner product is a Hermittian form.

Example 3.a. Consider $V = \mathbf{C}^{2\times 1}$, $A = \begin{bmatrix} 1 & 2 \\ 0 & 1 \end{bmatrix}$.

We get sesqui-linear form

$$f\left[\begin{pmatrix} x_1 \\ x_2 \end{pmatrix}, \begin{pmatrix} y_1 \\ y_2 \end{pmatrix}\right] = [\bar{y}_1, \bar{y}_2]A\begin{bmatrix} x_1 \\ x_2 \end{bmatrix}$$
$$= x_1\bar{y}_1 + 2x_2\bar{y}_1 + x_2\bar{y}_2.$$

As $A \neq A^*$, f is not a Hermittian form. $B = \begin{bmatrix} 2 & i \\ -i & 3 \end{bmatrix}$ is a Hermittian matrix.

So,
$$g\left[\begin{pmatrix} x_1 \\ x_2 \end{pmatrix}, \begin{pmatrix} y_1 \\ y_2 \end{pmatrix}\right] = [\bar{y}_1, \bar{y}_2]B\begin{bmatrix} x_1 \\ x_2 \end{bmatrix}$$
$$= 2x_1\bar{y}_1 - ix_1\bar{y}_2 + ix_2\bar{y}_1 + 3x_2\bar{y}_2$$

is a Hermittian form.

Proposition 4. (change of basis). Let f be a sesqui-linear form on a finite dimensional vector space V_F. If B and B' be any two ordered bases of V.

Then,
$$[f]_{B'} = P^*[f]_B P$$
where P is the matrix of B' relative to B.

Proof. Let
$$B = \{\alpha_1, \alpha_2, \ldots, \alpha_n\}$$
$$B' = \{\beta_1, \beta_2, \ldots, \beta_n\},$$
$$[f]_B = \left[a_{ij}\right] = A,$$
$$[f]_{B'} = \left[b_{ij}\right]$$

and
$$P = \left[p_{ij}\right]$$

the matrix of B' relative to B.

Then, $$a_{ij} = f(\alpha_j, \alpha_i),$$

$$b_{ij} = f(\beta_j, \beta_i),$$

$$\beta_j = \sum_{i=1}^{n} p_{ij}\alpha_i$$

so that $$b_{ij} = f(\beta_j, \beta_i)$$

$$= f\left[\sum_{k=1}^{n} p_{kj}\,\alpha_k,\ \sum_{l=1}^{n} p_{li}\,\alpha_l\right]$$

$$= \sum_{l=1}^{n}\sum_{k=1}^{n} p_{kj}\overline{p}_{li}\, f(\alpha_k, \alpha_l)$$

$$= \sum_{l=1}^{n}\sum_{k=1}^{n} \overline{p}_{li}\, a_{lk}\, p_{kj},$$

the $(i, j)^{\text{th}}$ element of $P^* AP$.

Hence, $$[f]_{B'} = P^*[f]_B\, P.$$

In this result, $$\operatorname{rank}[f]_B = \operatorname{rank}[f]_{B'},$$

so that given a sesqui-linear form on a finite dimensional vector space V_F, the rank of the matrix $[f]_B$, for any ordered basis B of V is called the *rank* of *f*. In the case of an inner product space, the connection between sesqui-linear forms and linear operators is given by the following result :

Theorem 5. Let f be a sesqui-linear form on a finite dimensional inner product space V_F. Then, there exists a unique linear operator T on V such that $f(\alpha, \beta) = (T\alpha|\beta)$ for all $\alpha, \beta \in V$.

Proof. For any $\beta \in V$, the mapping $\sigma_\beta : V \to F$ such that $\sigma_\beta(\alpha) = f(\alpha, \beta)$ is a linear functional on V_F. By Proposition (6.1.4), there exists unique β' depending upon β such that

$$\sigma_\beta(\alpha) = f(\alpha|\beta') \qquad \text{for all } \alpha \in V.$$

Define $T' : V \to V$ such that $T'(\beta) = \beta'$. By using the definition of a sesqui-linear form, it can be easily seen that T' is a linear operator on V.

Let $$T = (T')^*.$$

Then, $$T^* = T',$$

$$f(\alpha, \beta) = (\alpha|T'\beta) = (T\alpha|\beta) \qquad \text{... (1)}$$

Let T_1 be another linear operator on V such that

$$f(\alpha, \beta) = (T_1\alpha|\beta). \quad \text{... (2)}$$

Then, (1) and (2), give $$(T\alpha|\beta) = (T_1\alpha|\beta) \quad \text{... (3)}$$

for all $\alpha, \beta \in V$.

Proposition (6.1.4) gives $$T\alpha = T_1\alpha.$$

Hence, $$T = T_1.$$

The following is immediately clear.

Corollary 6. Let V_F be any finite dimensional inner product space, and Ses(V) be the vector space of all sesqui-linear forms on V.

Then, the mapping $$\lambda : \text{Ses}(V) \to L(V)$$

such that for any $$f \in \text{Ses}(f),\ \lambda(f) = T_f$$

given by $$f(\alpha|\beta) = (T_f\alpha|\beta),$$

is a vector space isomorphism.

Theorem 7. Let f be a sesqui-linear form on a finite dimensional inner product space V_F, and B be an ordered orthonormal basis of V. Then,

(*i*) $[f]_B = [T_f]_B$

(*iii*) rank (f) = rank (T_f)

(*iii*) f is Hermittian iff T_f is self adjoint

(*iv*) For any $\alpha \in V$, $f(\alpha, \beta) = 0$ for all $\beta \in V$ implies $\alpha = 0$ iff T_f is invertible.

Proof. (*i*) & (*ii*).

Let $$B = \{\alpha_1, \alpha_2, ..., \alpha_n\},$$

$$[f]_B = [a_{ij}],$$

$$[T_j]_B = [b_{ij}].$$

Then, $$a_{ij} = f(\alpha_j, \alpha_i)$$

$$T_f\,\alpha_j = \sum_{i=1}^{n} b_{ij}\alpha_i$$

By definition, $$a_{ij} = f(\alpha_j, \alpha_i)$$
$$= (T_f\alpha_j|\alpha_i)$$

$$= \left(\sum_{k=1}^{n} b_{kj} \alpha_k | \alpha_i \right)$$

$$= \sum_{k=1}^{n} b_{kj} (\alpha_k | \alpha_i)$$

$$= b_{ij},$$

by the orthonormality of B.

Hence, $[f]_B = [T_f]_B$.

Hence, by the definition,

$$\text{rank}(f) = \text{rank}[f]_B = \text{rank}\left[(T_f)_B\right] = \text{rank}(T_f)$$

This proves (*i*) and (*ii*).

Let f be such that for any $\alpha \in V$, $f(\alpha, \beta) = 0$ for all $\beta \in V$ implies $\alpha = 0$. If $T_f\alpha = 0$ for some α, then $f(\alpha, \beta) = (T_f\alpha|\beta) = 0$.

So, by the hypothesis $\alpha = 0$.

Consequently, $\ker T_f = 0$; hence, T_f is invertible. Conversely, let T_f be invertible. Consider $\alpha \in V$ such that $f(\alpha, \beta) = 0$ for all $\beta \in V$.

Then, $(T_f\alpha|\beta) = 0$.

This yields $T_f\alpha = 0$, so $\alpha = 0$ as T_f is invertible.

This proves (*iv*). Finally, let f be Hermittian.

Thus, $f^*(\alpha, \beta) = f(\alpha, \beta)$.

Consequently, $T_{f^*} = T_f$.

Thus, $f^*(\alpha, \beta) = \left(T_{f^*}\alpha|\beta\right) = \left(T_f\, \alpha|\beta\right) = f(\alpha, \beta)$.

Also $f^*(\alpha,\beta) = \overline{f(\beta,\alpha)} = \overline{\left(T_f\beta|\alpha\right)} = \left(\alpha| T_f\beta\right) = \left(T_f^*\alpha|\beta\right)$.

Consequently, $\left(T_f\alpha|\beta\right) = \left(T_f^*\, \alpha|\beta\right)$.

Hence, $T_f = T_f^*$.

Conversely, let $T_f = T_f^*$.

Then,

$$f^*(\alpha, \beta) = \overline{f(\beta, \alpha)}$$

$$= \overline{\left(T_f\, \beta|\alpha\right)}$$

$$= \left(\alpha \middle| T_f \beta\right)$$

$$= \left(T_f^* \alpha \middle| \beta\right)$$

$$= \left(T_f \alpha \middle| \beta\right)$$

$$= f(\alpha, \beta)$$

So, $$f^* = f.$$

Theorem 8. Let f be a sesqui-linear form on a finite dimensional complex inner product space V_F. Then, f is Hermittian iff $f(\alpha, \alpha)$ is real every $\alpha \in V$.

Proof.

Let f be Hermittian.

Then, given $\alpha, \beta \in V$, $\overline{f(\alpha, \beta)} = f(\beta, \alpha)$.

In particular, $\overline{f(\alpha, \alpha)} = f(\alpha, \alpha)$

so that $f(\alpha, \alpha)$ is real.

Conversely, let $f(\alpha, \alpha)$ be real.

Now, $f(\alpha+\beta, \alpha+\beta) = f(\alpha, \alpha) + f(\beta, \beta) + f(\alpha, \beta) + f(\beta, \alpha)$.

As $f(\alpha+\beta, \alpha+\beta), f(\alpha, \alpha), f(\beta, \beta)$ are real,

$$f(\alpha, \beta) + f(\beta, \alpha)$$

is real so that

$$f(\alpha, \beta) = u + iv,$$

$$f(\beta, \alpha) = w - iv$$

for some real numbers u, w, v.

Then also, $f(\alpha, i\beta) + f(i\beta, \alpha)$ is real.

However,

$$f(\alpha, i\beta) + f(i\beta, \alpha) = -if(\alpha, \beta) + if(\beta, \alpha)$$

$$= -i(u+iv) + i(w-iv)$$

$$= i(w-u) + 2v,$$

so that $$w - u = 0$$

i.e. $$u = w.$$

Consequently,

$$\overline{f(\beta, \alpha)} = u + iv$$

$$= f(\alpha, \beta).$$

Hence, f is Hermittian.

Analogous to the above theorem, we have the following useful criterion for certain sesqui-linear forms to be zero.

Theorem 9. A sesqui-linear form f on a finite dimensional complex inner product space is zero iff $f(\alpha, \alpha) = 0$ for every $\alpha \in V$.

Proof. If $f = 0$, obviously, $f(\alpha, \alpha) = 0$ for every $\alpha \in V$. Conversely, let $f(\alpha, \alpha) = 0$ for any $\alpha \in V$. By Theorem 8, f is Hermittian. Consider $\alpha, \beta \in V$.

Then,
$$\begin{aligned} 0 &= f(\alpha+\beta, \alpha+\beta) \\ &= f(\alpha, \alpha) + f(\beta, \beta) + f(\alpha, \beta) + f(\beta, \alpha) \\ &= f(\alpha, \beta) + f(\beta, \alpha). \end{aligned}$$

As f is Hermittian, $f(\beta, \alpha) = \overline{f(\alpha, \beta)}$.

Consequently, $f(\alpha, \beta) + \overline{f(\alpha, \beta)} = 0$

So that $f(\alpha, \beta) = iu$ for some real u.

Then, $f(i\alpha, \beta) = iv$ for some real v.

On the other hand $f(i\alpha, \beta) = if(\alpha, \beta) = -u$.

So $-u = iv$.

This gives $u = 0$, so, $f(\alpha, \beta) = 0$.

Lemma 10. Let f be a sesqui-linear form on a finite dimensional vector space V_F.

Then, $$W_f = \{\alpha \in V : f(\alpha, \beta) = 0 \text{ for every } \beta \in V\}$$

is a subspace of V.

If f is Hermittian, then, $W_f = \{\alpha \in V : f(\beta, \alpha) = 0 \text{ for all } \beta \in V\}$.

Proof. That W_f is a subspace of V is obvious.

If f is Hermittian, then, $f(\beta, \alpha) = \overline{f(\alpha, \beta)}$

gives the second part.

If for a sesqui-linear form f, $W_f = 0$, it is called a *non-singular form*. In general, W_f is called the *degenerator* of f; we write $Dn(f)$ for W_f. We now describe relationship between degenerators and non-singular forms in case of a Hermittian form. Observe that any finite dimensional real or complex vector space can be turned into an inner product space. This is why we confine ourself to inner product spaces.

Proposition 11. Let f be a Hermittian form on a finite dimensional inner product space V_F. Then,

(*i*) $V = Dn(f)^{\perp} \oplus Dn(f)$.

(*ii*) T_f is zero on $Dn(f)$ and restriction of f to $Dn(f)$ is zero.

(*iii*) The restriction of f to $Dn(f)^{\perp}$ is non-singular.

(*iv*) $Dn(f)^{\perp}$ is invariant under T_f

(*v*) rank(f) = $\dim Dn(f)^{\perp}$

Proof. (*i*) is a special case of Theorem (6.1.23). That restriction of f to $Dn(f)$ is zero, is obvious from the definition of $Dn(f)$.

For any
$$\alpha \in Dn(f),$$
$$\beta \in V,$$
$$0 = f(\alpha, \beta) = \left(T_f\alpha \middle| \beta\right) = 0$$

gives
$$T_f \alpha = 0$$

In particular, it gives that $Dn(f)$ is invariant under T_f. As $T_f = T_f^*$, by Lemma (6.2.17), $Dn(f)^{\perp}$ is invariant under T_f. Consider $\alpha \neq 0$ in $Dn(f)^{\perp}$.

As $\alpha \notin Dn(f)$, we get $\beta \in V$ such that $f(\alpha,\beta) \neq 0$. Now $\beta = \beta_1 + \beta_2$, with $\beta_1 \in Dn(f), \beta_2 \in Dn(f)^{\perp}$, so $f(\alpha,\beta) = f(\alpha,\beta_2) \neq 0$.

Hence, the restriction of T_f to $Dn(f)^{\perp}$ is non-singular.

Further,
$$\ker(T_f) = Dn(f).$$

Hence,
$$\text{rank}(T_f) = \dim Dn(f)^{\perp},$$
$$\text{rank}(f) = \text{rank}(T_f) = \dim Dn(f)^{\perp}$$

As T_f is non singular on $Dn(f)^{\perp}$, f is non-singular on $Dn(f)^{\perp}$. This proves the result.

Remark 12. Let A be any Hermittian $n \times n$-matrix. Then, the Hermittian form f on $\mathbf{C}^{n\times 1}$ given by $f(X, Y) = Y^* AX$ is said to be given by A.

Clearly,
$$f(X, Y) = (AX|Y) = (TX|Y)$$

where T is the linear operator on $\mathbf{C}^{n\times 1}$ given by A.

Thus,
$$T = T_f.\ Dn(f) = \ker T = \left\{X \in \mathbf{C}^{n\times 1} : AX = 0\right\}$$

Find an orthonormal basis B_1 of ker T and extend it to an orthonormal basis B of $\mathbf{C}^{n\times 1}$. The subspace spanned by the vectors in $B - B_1$ is $Dn(f)^{\perp}$.

Example 12. Let $A = \begin{bmatrix} 1 & -2 & 1 \\ -2 & 0 & 0 \\ 1 & 0 & 0 \end{bmatrix}$ and f be the Hermittian form on $\mathbf{C}^{3\times 1}$ given by A.

Then,
$$f(X,Y) = Y * AX = x_1\bar{y}_1 - 2x_1\bar{y}_2 + x_1\bar{y}_3 - 2x_2\bar{y}_1 + x_3\bar{y}_1.$$
$$Dn(f) = \left[X \in \mathbf{C}^{3\times 1} : AX = 0\right]$$

Its orthonormal basis is $$B_1 = \left\{ \begin{bmatrix} 0 \\ \frac{1}{\sqrt{5}} \\ \frac{2}{\sqrt{5}} \end{bmatrix} \right\}.$$

By applying Gram-Schmidt orthonormalization to the basis

$$B' = \left\{ \begin{bmatrix} 0 \\ \frac{1}{\sqrt{5}} \\ \frac{2}{\sqrt{5}} \end{bmatrix}, \begin{bmatrix} 1 \\ 0 \\ 0 \end{bmatrix}, \begin{bmatrix} 0 \\ 1 \\ 0 \end{bmatrix} \right\},$$

we get orthonormal basis $$B = \left\{ \begin{bmatrix} 0 \\ \frac{1}{\sqrt{5}} \\ \frac{2}{\sqrt{5}} \end{bmatrix}, \begin{bmatrix} 1 \\ 0 \\ 0 \end{bmatrix}, \begin{bmatrix} 0 \\ \frac{2}{\sqrt{5}} \\ \frac{-1}{\sqrt{5}} \end{bmatrix} \right\}$$ of $\mathbf{C}^{3\times 1}$.

Then, $Dn(f)^{\perp}$ has basis $$B_2 = \left\{ \begin{bmatrix} 1 \\ 0 \\ 0 \end{bmatrix}, \begin{bmatrix} 0 \\ \frac{2}{\sqrt{5}} \\ \frac{-1}{\sqrt{5}} \end{bmatrix} \right\}.$$

For $$\alpha_1 = \begin{bmatrix} 0 \\ \frac{1}{\sqrt{5}} \\ \frac{2}{\sqrt{5}} \end{bmatrix}, \alpha_2 = \begin{bmatrix} 1 \\ 0 \\ 0 \end{bmatrix}, \alpha_3 = \begin{bmatrix} 0 \\ \frac{2}{\sqrt{5}} \\ \frac{-1}{\sqrt{5}} \end{bmatrix}$$

observe that

$$f(\alpha_1, \alpha_1) = 0$$

$$f(\alpha_2, \alpha_2) = 1$$

$$f(\alpha_3, \alpha_3) = 0$$

$$f\left(\begin{bmatrix} 1 \\ 2 \\ 0 \end{bmatrix}, \begin{bmatrix} 1 \\ 2 \\ 0 \end{bmatrix} \right) = -7 < 0.$$

We now go a step further in the decomposition of a Hermittian form into some canonical Hermittian forms. For this, we introduce the following :

Definition 13. Let f be a Hermittian form on a finite dimensional vector space V_F. Then,

(*i*) f is said to be *positive* on a subspace W of V if $f(\alpha, \alpha) > 0$ for any $\alpha \neq 0$ in W ; if f is positive on V, it is called a *positive form*.

(*ii*) f is said to be *non-negative* if $f(\alpha, \alpha) \geq 0$ for every $\alpha \in V$.

Similarly, one defines a *negative form* and a *non-positive form*.

In Example 12, $f(\alpha_2, \alpha_2) = 1 > 0,$

$$f(\alpha, \alpha) - 7 < 0$$

for $\alpha = \begin{bmatrix} 1 \\ 2 \\ 0 \end{bmatrix}$

so that f is neither a positive nor a negative form.

The form g on $\mathbf{C}^{3\times 1}$ given by

$$g(X, Y) = x_1\bar{y}_1 + 2\bar{x}_2 y_2 + x_3\bar{y}_3$$

is positive, and $-g$ is negative.

Any inner product on V is a positive form.

Theorem 14. Let f be a Hermittian form on a finite dimensional inner product space V_F. Then, $V = V_1 \oplus V_2 \oplus V_3$ such that

(*i*) for any $\alpha \in V_3$, $\beta \in V$, $f(\alpha, \beta) = 0$;

(*ii*) f is positive on V_1;

(*iii*) f is negative on V_2;

(*iv*) given any decomposition $V = W_1 \oplus W_2 \oplus W_3$ such that W_3 satisfies (*i*), f is positive on W_1 and negative on W_2, then,

$$W_3 = V_3 = Dn(f).$$

$$\dim W_1 = \dim V_1$$

and $\dim W_2 = \dim V_2.$

Proof. By Theorem 7, T_f is self-adjoint. By Theorem (6.2.22), all the characteristic values of T_f are real and for some orthonormal basis B of V, $A = [T_f]_B$ is a diagonal matrix. Let $\text{rank}(T_f) = r$. Then, r entries on the diagonal of A are non-zero real numbers and the others are zeros. If s of these non-zero entries are positive, we can choose B such that

$$A = \text{diag}[c_1, c_2, \ldots, c_r, 0, 0, \ldots, 0]$$

with $c_i > 0$

for $1 \le i \le s$, $c_j < 0$

for $s + 1 \le j \le r$.

Let $B = \{\alpha_1, \alpha_2, \ldots, \alpha_n\}.$

Then, $T_f\,\alpha_i = c_i\alpha_i,$

$$1 \le i \le r,\ T_f\,\alpha_j = 0$$

for $j \ge r + 1.$

By Theorem 7, $$[f]_B = A.$$

So, $$f(\alpha_i, \alpha_j) = 0 \quad \text{for } i \neq j$$

$$f(\alpha_i, \alpha_i) = c_i > 0 \quad \text{for } 1 \leq i \leq s$$

$$f(\alpha_i, \alpha_i) = c_i < 0 \quad \text{for } s+1 \leq i \leq r$$

$$f(\alpha_i, \alpha_j) = a_{ji} = 0 \quad \text{for } j \geq r+1, \text{ and any } i.$$

As *f* is Hermittian, $$f(\alpha_j, \alpha_i) = \overline{f(\alpha_i, \alpha_j)} = 0 \quad \text{for } j \geq r+1 \text{ and any } i.$$

Let V_3 be the subspace spanned by all the α_j, $j \geq r+1$.

Then, $$\dim V_3 = n - r.$$

Given $$\alpha \in V_3,$$

$$\beta \in V,$$

$$\alpha = \sum_{j=r+1}^{n} a_j \alpha_j,$$

$$\beta = \sum_{j=1}^{n} b_j \alpha_j,$$

gives $$f(\alpha, \beta) = \sum_{k=1}^{n} \sum_{j=r+1}^{n} a_j \bar{b}_k f(\alpha_j, \alpha_k) = 0.$$

Thus, V_3 satisfy (*i*).

Let V_1 be the subspace spanned by $B = \{\alpha_i : 1 \leq i \leq s\}$

Then, $$\dim V_1 = s.$$

For $$0 \neq \alpha \in V_1, \ \alpha = \sum_{i=1}^{n} a_i \alpha_j$$

with some $$a_i \neq 0, \ f(\alpha, \alpha) = \sum_{i=1}^{s} a_i \bar{a}_i \, c_i > 0,$$

as every $$c_i > 0.$$

Thus, *f* is positive on V_1.

Let V_2 be the subspace spanned by

$$B_2 = \{\alpha_i : s+1 \leq i \leq r\}.$$

As for V_1, *f* is negative on V_2.

$$\dim V_2 = r - s.$$

Obviously, $$V = V_1 \oplus V_2 \oplus V_3,$$

$$V_3 \subseteq Dn(f).$$

However, $\operatorname{rank}(T_f) = \dim Dn(f)^{\perp} = r,$

gives $\dim Dn(f) = n - r = \dim V_3.$

So $V_3 = D_n(f).$

Consider any decomposition $V = W_1 \oplus W_2 \oplus W_3$ given in (*iv*).

Clearly, $W_3 \subseteq D_n(f) = V_3.$

We claim that $W_1 + V_2 + V_3$ is a direct sum.

Let $\alpha + \beta + \gamma = 0$

for some $\alpha \in W_1,\ \beta \in V_2,\ \gamma \in V_3.$

Then.
$$\begin{aligned} 0 &= f(\alpha, \alpha + \beta + \gamma) \\ &= f(\alpha, \alpha) + f(\alpha, \beta) + f(\alpha, \gamma) \\ &= f(\alpha, \alpha) + f(\alpha, \beta). \end{aligned}$$

Also,
$$\begin{aligned} 0 &= \overline{f(\beta, \alpha + \beta + \gamma)} \\ &= \overline{f(\beta, \alpha)} + \overline{f(\beta, \beta)} \\ &= f(\beta, \beta) + f(\alpha, \beta). \end{aligned}$$

Thus,
$$\begin{aligned} f(\alpha, \alpha) + f(\alpha, \beta) &= 0 \\ &= f(\beta, \beta) + f(\alpha, \beta), \end{aligned}$$

$$f(\alpha, \alpha) = f(\beta, \beta).$$

However, $f(\alpha, \alpha) \geq 0$

and $f(\beta, \beta) \leq 0.$

As f is positive on W_1, we get $\alpha = 0$; as f is negative on V_2, we get $\beta = 0$.

Consequently, $\gamma = 0.$

Hence, $W_1 + V_2 + V_3 = W_1 \oplus V_2 \oplus V_3 \subseteq V_1 \oplus V_2 \oplus V_3 = V.$

This yields $\dim W_1 \leq \dim V_1.$

Similarly, $\dim W_2 \leq \dim V_2.$

Then, $\dim W_1 + \dim W_2 + \dim W_3 = \dim V_1 + \dim V_2 + \dim V_3$

yields $\dim W_3 \geq \dim V_3.$

However, $W_3 \subseteq V_3.$

Hence, $\dim W_3 = \dim V_3$

and $W_3 = V_3.$

This, in turn, yields $\dim W_1 = \dim V_1$

and $\dim W_2 = \dim V_2$.

Example 14. Consider $A = \begin{bmatrix} 1 & -2 & 1 \\ -2 & 0 & 0 \\ 1 & 0 & 0 \end{bmatrix}$ as in Example 12.a. Its characteristic values are 0, $\frac{1}{2}+\frac{\sqrt{21}}{2}$, $\frac{1}{2}-\frac{\sqrt{21}}{2}$; the corresponding characteristic vectors are

$$X_1 = \begin{bmatrix} 0 \\ \frac{1}{2} \\ 1 \end{bmatrix}, X_2 = \begin{bmatrix} \frac{1}{2}+\frac{\sqrt{21}}{2} \\ -2 \\ 1 \end{bmatrix}, X_3 = \begin{bmatrix} \frac{1}{2}-\frac{\sqrt{21}}{2} \\ -2 \\ 1 \end{bmatrix}.$$

One can see that these three vectors are pairwise orthogonal. For the Hermittian form f on $\mathbf{C}^3$ given by A, $Dn(f)$ is the subspace spanned by X_1.

So, $V_3 = Dn(f)$ has dimension 1.

$$f(X_2, X_2) > 0,\ f(X_3, X_3) < 0.$$

In the notation of the above theorem, we get

$$V_1 = \langle X_2 \rangle,\ V_2 = \langle X_3 \rangle.$$

Once again, consider $Y_1 = \begin{bmatrix} 1 \\ 0 \\ 0 \end{bmatrix}, Y_2 = \begin{bmatrix} 2 \\ 1 \\ 0 \end{bmatrix}, Y_3 = \begin{bmatrix} 0 \\ \frac{1}{2} \\ 1 \end{bmatrix} = X_1.$

Then, $f(Y_1, Y_1) = 1,$

$$f(Y_2, Y_2) = -4,$$

$$f(Y_3, Y_3) = 0.$$

$$f(Y_i, Y_j) = 0 \qquad \text{for } i \neq j.$$

Further, $B = \{Y_1, Y_2, Y_3\}$ is a basis of $\mathbf{C}^3$; this is not an orthogonal basis. It is clear that f is positive on $W_1 = \langle Y_1 \rangle$, negative on $W_2 = \langle Y_2 \rangle$, and $W_3 = D_n(f) = \langle Y_3 \rangle$.

Observe that $[f]_B = \begin{bmatrix} 1 & 0 & 0 \\ 0 & -4 & 0 \\ 0 & 0 & 0 \end{bmatrix}$ a diagonal matrix. If B' is the standard ordered basis of $\mathbf{C}^3$, $[f]_{B'} = A$. Clearly, $[f]_B$ and $[f]_{B'}$ are not similar matrices, the reason being that B is not an ordered orthonormal basis of $\mathbf{C}^3$.

We now give some criteria for a Hermittian form to be positive.

Theorem 15. A Hermittian form f on a finite dimensional inner product space V_F is positive iff for any ordered basis B of V_F

$$[f]_B = P^* P$$

for some non-singular matrix P over F.

Proof.

Let $$A = [f]_B.$$

Then, $$A^* = A.$$

By using Theorem (6.2.21) and Corollary (6.2.23), one gets a unitary matrix B over f such that

$$B^* AB = \text{diag}[c_1, c_2, \ldots, c_n].$$

Let f be positive. Then, $c_i > 0$ for every i.

Let $$d_i = \sqrt{c_i} > 0,$$

then, $$D = \text{diag}[d_1, d_2, \ldots, d_n] = D^*$$

and $$B^* AB = DD^*.$$

Consequently, $$A = P^* P,\ P = D^* B^*.$$

Conversely, let $A = P^* P$ for some non-singular matrix P over F. Let $\alpha \neq 0$ in V, $X = [\alpha]_B$.

Then, $$f(\alpha, \alpha) = X^* AX = X^* P^* PX = Y^* Y,$$

with $$Y = PX \neq 0.$$

Obviously, $$Y^* Y > 0.$$

Hence, f is positive.

We now give a criterion in terms of determinants, for a Hermittian form to be positive. This criterion is very easy to apply.

Definition 16. Let $A = [a_{ij}]$ be any $n \times n$-matrix over a field F. For any positive integer $k \leq n$, the *principal submatrix* $A(k)$ of A is defined to be

$$A(k) = \begin{bmatrix} a_{11} & a_{12} & \ldots\ldots & a_{1k} \\ a_{21} & a_{22} & \ldots\ldots & a_{2k} \\ \ldots & \ldots & \ldots\ldots & \ldots \\ \ldots & \ldots & \ldots\ldots & \ldots \\ a_{k1} & a_{k2} & \ldots\ldots & a_{kk} \end{bmatrix}.$$

The determinant of $A(k)$ is called the kth *principal minor* of A, and is denoted by $\Delta_k(A)$.

Thus, for
$$A = \begin{bmatrix} 0 & 1 & 2 \\ 1 & 0 & 1 \\ 3 & 0 & 0 \end{bmatrix}, \ A(1) = [0],$$

$$\Delta_1(A) = 0,$$

$$A(2) = \begin{bmatrix} 0 & 1 \\ 1 & 0 \end{bmatrix},$$

$$\Delta_2(A) = -1,$$

$$\Delta_3(A) = \begin{vmatrix} 0 & 1 & 2 \\ 1 & 0 & 1 \\ 3 & 0 & 0 \end{vmatrix} = |A| = 3.$$

Lemma 17. Let A be any $n \times n$-matrix over a field F such that each of its principal minor is non-zero. Then, there exists a unique upper triangular matrix $P = \left[p_{ij}\right]$ such that $p_{ii} = 1$ for $1 \le i \le n$ and $B = AP$ is a lower triangular matrix ; in this situation, $\Delta_k(B) = \Delta_k(A)$ for every $k \le n$.

Proof. Let $P = \left[p_{ij}\right]$ be any upper triangular matrix over F with $p_{ii} = 1$. Also $p_{ij} = 0$ for $i > j$.

Consider
$$B = AP = \left[b_{ij}\right],$$

$$b_{ij} = \sum_{k=1}^{n} a_{ik} p_{kj} = \sum_{k=1}^{j} a_{ik} p_{kj} = \sum_{k=1}^{j-1} a_{ik} p_{kj} + a_{ij} \qquad \text{... (1)}$$

so that
$$\sum_{k=1}^{j-1} a_{ik} p_{kj} = b_{ij} - a_{ij}. \qquad \text{... (2)}$$

We want to solve for p_{kj} and b_{ij} in (1), so that B becomes a lower triangular matrix.

For
$$j = 1,\ b_{i1} = a_{i1}.$$

We require $b_{ij} = 0$ for $j > i$,

i.e., we need,
$$\sum_{k=1}^{j-1} a_{ik} p_{kj} = -a_{ij}$$

i.e.,
$$[A(j-1)] \begin{bmatrix} p_{1j} \\ p_{2j} \\ \vdots \\ \vdots \\ p_{j-1j} \end{bmatrix} = -\begin{bmatrix} a_{1j} \\ a_{2j} \\ \vdots \\ \vdots \\ a_{j-1j} \end{bmatrix}. \qquad \text{... (3)}$$

By the assumption $\Delta_{j-1}(A) = |A(j-1)| \neq 0,$

so $A(j-1)$ is non-singular, hence, (3) has a unique solutions for $(p_{1j}, p_{2j}, \ldots, p_{j-1,j})$.

Also, $p_{jj} = 1.$

Thus, for $j > 1$, j^{th} column of P is uniquely determined. As $p_{11} = 1$, the first column of P is also determined. This proves the first part. We now prove that $\Delta_k(B) = \Delta_k(A)$.

As $B(1) = [b_{11}] = [a_{11}] = A(1),$

obviously, $\Delta_1(B) = \Delta_1(A).$

For $r \geq 2$, the jth column of $B(r)$, because of (1), is

$$\begin{bmatrix} b_{1j} \\ b_{2j} \\ \vdots \\ \vdots \\ b_{rj} \end{bmatrix} = \sum_{k=1}^{j-1} \begin{bmatrix} a_{1k} \\ a_{2k} \\ \vdots \\ \vdots \\ a_{rj} \end{bmatrix} p_{kj} + \begin{bmatrix} a_{1j} \\ a_{2j} \\ \vdots \\ \vdots \\ a_{rj} \end{bmatrix}. \qquad \ldots (4)$$

Thus, in $B(r)$, its jth column is obtained by adding a linear combination of columns in $A(r)$ preceding the jth column of $A(r)$ so that $|B(r)| = |A(r)|$, i.e., $\Delta_r(B) = \Delta_r(A)$. This completes the proof.

We now prove the main theorem.

Theorem 18. **Let f be a Hermittian form on a finite dimensional vector space V_F and $A = [f]_B$ for some ordered basis B of V_F. Then, f is positive iff all the principal minors of A are positive.**

Proof. **Let $B = \{\alpha_1, \alpha_2, \ldots, \alpha_n\}$, $A = [a_{ij}]$ and V_k be the subspace of V spanned by the ordered set $B_k = \{\alpha_1, \alpha_2, \ldots, \alpha_k\}$. Let f_k be the restriction of f to V_k. Let f be positive.**

$$A_k = [f_k]_{B_k} = [a'_{ij}]_{k \times k}$$

satisfies $\quad a'_{ij} = f_k(\alpha_j, \alpha_i) = f(\alpha_j, \alpha_i) = a_{ij}$

so that A_k is the principal $k \times k$-submatrix of A.

Thus, $|A_k| = \Delta_k(A).$

As f_k is positive, $A_k = P_k^* P_k$ for some non-singular matrix P_k, by Theorem 15, $\Delta_k(A) = |P_k|^2 > 0$. This proves that every principal minor of A is positive.

Conversely, let every principal minor of A be positive. By Lemma 17 there exists a upper triangular matrix $P = [p_{ij}]$ over F such that $p_{ii} = 1$ and $A' = AP$ is a lower triangular matrix. Further, $\Delta_k(A') = \Delta_k(A)$ for $1 \leq k \leq n$.

Then, $D = P^* A'$ being a product of two lower triangular matrices is lower triangular. As $D = P^* AP$ and $A^* = A$, we get $D^* = D$. Consequently, D is a diagonal matrix.

Further, $$D = D^* = (A')^* P.$$

By using Lemma 17, we get $\Delta_k(D) = \Delta_k(A') > 0$

so that if $$D = \text{diag}[d_1, d_2, ..., d_n],$$

then, $$\Delta_k(D) = d_1 d_2 ... d_k > 0.$$

In particular, $$\Delta_1(D) = d_1 > 0.$$

Then, $$\Delta_2(D) = d_1 d_2 > 0,$$

gives, $$d_2 > 0.$$

Continue this process. We get $d_i > 0$ for $1 \le i \le n$.

Now, $$D = [f]_{B'}$$

where $$B' = \{\alpha'_1, \alpha'_2, ..., \alpha'_n\}$$

with $$\alpha'_j = \sum_{i=1}^{n} p_{ij} \alpha_i.$$

For any $\alpha \ne 0$ in F, if $$X = \begin{bmatrix} x_1 \\ x_2 \\ : \\ : \\ x_n \end{bmatrix} = [\alpha]_{B'}$$

then, $$f(\alpha, \alpha) = X^* DX = \sum_{i=1}^{n} d_i x_i \bar{x}_i > 0.$$

Hence, f is a positive form.

A $n \times n$-matrix A over **C** is called a *positive matrix* if A is Hermittian and all its principal minors are positive. Thus, a Hermittian form f on V_F is positive iff its matrix relative to an ordered basis of V is a positive matrix.

Exercise 6.3

1. Which of the following functions defined on $\mathbf{C}^{2\times 1}$ are sesqui-linear forms. For $X = \begin{bmatrix} x_1 \\ x_2 \end{bmatrix}, Y = \begin{bmatrix} y_1 \\ y_2 \end{bmatrix}$

(*i*) $f_1(X, Y) = 0$

(*ii*) $f_2(X, Y) = x_1 + x_2 + \bar{y}_1 + \bar{y}_2$

(*iii*) $f_3(X, Y) = (x_1 - x_2)^2 + (\overline{y}_1 - \overline{y}_2)^2$

(*iv*) $f_4(X, Y) = x_1\overline{y}_1 - x_2\overline{y}_2$

(*v*) $f_5(X, Y) = (x_1\overline{y}_1 + x_2\overline{y}_2)\cos\theta + (x_1\overline{y}_2 - x_2\overline{y}_1)\sin\theta$, θ fixed.

2. Consider the sesqui-linear form *f* on $\mathbf{C}^{2\times1}$ given by $f(X, Y) = x_1\overline{y}_1 + ix_2\overline{y}_2$.

   (*i*) Find the matrix of *f* relative to the standard ordered basis.

   (*ii*) Find an $X \in \mathbf{C}^{2\times1}$ and a $Y \in \mathbf{C}^{2\times1}$ such that $f(X, X) > 0$ and $f(Y, Y) > 0$.

   (*iii*) Find the matrix of *f* relative to the ordered basis $B = \left\{\begin{bmatrix}1\\1\end{bmatrix}, \begin{bmatrix}1\\-1\end{bmatrix}\right\}$.

   (*iv*) Is *f* Hermittian?

3. If a sesqui-linear form over a complex vector space V is such that $f(\alpha, \beta) = f(\beta, \alpha)$ for all $\alpha, \beta \in V$, prove that $f = 0$.

4. A sesqui-linear form *f* on $\mathbf{R}^{3\times1}$ is given by $f(X, Y) = Y^T AX$,

   where
$$A = \begin{bmatrix}0 & 1 & -1\\1 & 0 & 1\\-1 & 1 & 0\end{bmatrix}.$$

   (*i*) Is *f* non-degenerate?

   (*ii*) find an ordered basis B' of $\mathbf{R}^{3\times1}$ such that $[f]_{B'}$ is a diagonal matrix.

   (*iii*) Find $Dn(f)$.

   (*iv*) Find V_1, V_2 such that $\mathbf{R}^{3\times1} = V_1 \oplus V_2 \oplus Dn(f)$, *f* is positive on V_1 and *f* is negative on V_2.

5. Let *T* be the linear operator on $\mathbf{C}^{2\times1}$ whose matrix relative to the standard ordered basis is $\begin{bmatrix}3 & 0\\0 & -3\end{bmatrix}$. Find the sesqui-linear form *f* on $\mathbf{C}^{2\times1}$ such that $f(X, Y) = (TX|Y)$.

6. Let *T* be the linear operator on $\mathbf{C}^{2\times1}$ whose matrix relative to the ordered basis
$$B = \left\{\begin{bmatrix}i\\1\end{bmatrix}, \begin{bmatrix}0\\-i\end{bmatrix}\right\} \text{ is } \begin{bmatrix}3 & 0\\0 & -3\end{bmatrix}$$
   Find the sesqui-linear form *f* on $\mathbf{C}^{2\times1}$ such that $f(X, Y) = (TX|Y)$.

7. A sesqui-linear form *f* on a finite dimensional vector space V_F is said to be *skew-Hermittian* if $f(\beta, \alpha) = -\overline{f(\alpha, \beta)}$.

   Prove the following :

   (*i*) If *f* is skew-Hermittian, then its matrix *A* relative to any ordered basis is skew-Hermittian in the sense that $A^* = -A$.

(*ii*) The set W_s of all skew-Hermittian forms on V_F is an **R**-subspace of $\text{Ses}(V_F)$

8. Let f be a Hermittian form on $\mathbf{C}^3$, such that its matrix relative to the standard order basis is

$$A = \begin{bmatrix} 0 & i & 0 \\ -i & 0 & 0 \\ 0 & 0 & 1 \end{bmatrix}.$$

(*i*) Find a unitary matrix P such that P^*AP is a diagonal matrix.

(*ii*) Find $\alpha, \beta \in \mathbf{C}^3$ such that $f(\alpha, \alpha) > 0$, $f(\beta, \beta) < 0$.

(*iii*) Is A a positive matrix?

9. Which of the following are positive matrices?

$$\begin{bmatrix} 0 & 1 \\ 1 & 0 \end{bmatrix}, \begin{bmatrix} 1 & 2 \\ -1 & 1 \end{bmatrix}, \begin{bmatrix} 2i & 1 \\ 1 & -2i \end{bmatrix}, \begin{bmatrix} 2 & i \\ -i & 2 \end{bmatrix}$$

10. If A and B are positive matrices, prove that $\begin{bmatrix} A & 0 \\ 0 & B \end{bmatrix}$ is a positive matrix.

11. If A and B are two positive $n \times n$-matrices prove that $A + B$ is positive.

12. Give an example of two 2 × 2-matrices A and B which are positive but AB is not a positive matrix.

[**Hint:** $A = \begin{bmatrix} 1 & i \\ -i & 2 \end{bmatrix}, B = \begin{bmatrix} 2 & 1 \\ 1 & 1 \end{bmatrix}$]

13. Prove that $A = \begin{bmatrix} 1 & i & -i \\ -i & 5 & 2 \\ i & 2 & 4 \end{bmatrix}$ is positive.

Find a 3 × 3-matrix P such that $A = P^*P$.

14. Prove that every entry on the main diagonal of a positive matrix A is positive.

[**Hint :** $A = P^*P$]

15. Let A be any $n \times n$-Hermittian matrix. Prove that there exists $c > 0$ such that $cI + A$ is a positive matrix.

16. Find a non-singular real 2 × 2-matrix P such that

$$\begin{bmatrix} 1 & \frac{1}{3} \\ \frac{1}{3} & 1 \end{bmatrix} = P^T P.$$

17. For any positive integer n prove that the $n \times n$-matrix

$$A_n = \begin{bmatrix} 1 & \frac{1}{2} & \frac{1}{3} & \cdots & \frac{1}{n} \\ \frac{1}{2} & \frac{1}{3} & \frac{1}{4} & \cdots & \frac{1}{n+1} \\ \vdots & \vdots & \vdots & \vdots & \vdots \\ \frac{1}{n} & \frac{1}{n+1} & \frac{1}{n+2} & \cdots & \frac{1}{2n-1} \end{bmatrix}$$

is positive.

Chapter 7

Bilinear Form

7.1 Bilinear Form

Throughout, F is a field and V is a finite dimensional non-zero vector space over F. We have studied linear functionals on V_F. Here, we study functions over $V \times V$, which, in each variable, are linear functionals. One can notice a certain amount of similarity between bilinear forms and sesqui-linear forms, discussed in chapter VI.

Definition 1. Any function $f: V_F \times V_F \to F$ is called a ***bilinear form*** on V if for any $\alpha, \beta, \gamma \in V$ and $c \in F$.

$$f(c\alpha+\beta, \gamma) = cf(\alpha, \gamma) + f(\beta, \gamma)$$

$$f(\alpha, c\beta+\gamma) = cf(\alpha, \beta) + f(\alpha, \gamma).$$

(Observe that the definition does not depend on the finite dimensionality of V_F).

For any bilinear form f on V and any fixed $\gamma \in V$, the mapping $\sigma_\gamma : V \to F$ such that $\sigma_\gamma(\alpha) = f(\alpha, \gamma)$ is a linear functional on V, as

$$\sigma_\gamma(c\alpha+\beta) = f(c\alpha+\beta, \gamma) = cf(\alpha, \gamma) + f(\beta, \gamma) = c\sigma_\gamma(\alpha) + \sigma_\gamma(\beta);$$

similarly, $$\eta_\gamma : V \to F$$

such that $$\eta_\gamma(\alpha) = f(\gamma, \alpha)$$

is a linear functional on V.

Consider any two linear functionals L_1 and L_2 on V. Then, the function $g : V \to F$ such that $g(\alpha, \beta) = L_1(\alpha) L_2(\beta)$ is a bilinear form on V. If $F = \mathbf{R}$. Then, any sesqui-linear form on V is a bilinear form.

Let $BL_F(V)$ [or simply $BL(V)$] denote the set of all bilinear forms on V. For any f, $g \in BL(V)$ and $c \in F$, define $f+g$, cf on $V \times V$ such that

$$(f+g)(\alpha, \beta) = f(\alpha, \beta) + g(\alpha, \beta)$$

$$(cf)(\alpha, \beta) = cf(\alpha, \beta)$$

for $(\alpha, \beta) \in V \times V$. These operations turn $BL(V)$ into a vector space over F. We now introduce certain special kinds of bilinear forms.

Definition 2. Let f be a bilinear form on V_F. Then,

(*i*) f is said to be *symmetric* if $f(\beta, \alpha) = f(\alpha, \beta)$

(*ii*) f is said to be *skew-symmetric* if $f(\beta, \alpha) = -f(\alpha, \beta)$

(*iii*) f is said to be *alternate* if $f(\alpha, \alpha) = 0$.

For any $n \times n$-matrix A over F, $f: F^{n\times 1} \times F^{n\times 1} \to F$ such that $f(x, y) = Y^T AX$ is a bilinear form on $F^{n\times 1}$. This f is said to be given by A. This f is symmetric or skew-symmetric according as A symmetric or skew-symmetric.

Definition 3. Let f be a bilinear form on V_F and $B = \{\alpha_1, \alpha_2, ..., \alpha_n\}$ be an ordered basis of V_F. Then $A = \left[a_{ij}\right]_{n\times n}$ such that $a_{ij} = f(\alpha_j, \alpha_i)$ is called the matrix of f relative to B, we write $A = [f]_B$.

The results, given below, are analogous to those given in Propositions (6.3.2), (6.3.4) and those given just after Definition (6.3.3).

Proposition 4. Let B and B' be two ordered bases of a finite dimensional vector space V_F and f be a bilinear form on V. Then,

(*i*) For any $\alpha, \beta \in V$, $f(\alpha, \beta) = Y^T AX$, where

$$A = [f]_B, X = [\alpha]_B, Y = [\beta]_B.$$

(*ii*) $[f]_{B'} = P^T [f]_B P$, where P is the matrix of B' relative to B.

(*iii*) $\lambda : BL(V) \to F^{n\times n}$, $n = \dim V$, given by $\lambda(f) = [f]_B$ is a vector space isomorphism.

Proposition 5. Let f be a bilinear form on a finite dimensional vector space V_F, and B be an ordered basis of V. Then,

(*i*) f is symmetric iff $A = [f]_B$ is symmetric

(*ii*) f is skew-symmetric iff $A = [f]_B$ is skew-symmetric.

Proof.

Let $\alpha, \beta \in V$, $X = [\alpha]_B$, $Y = [\beta]_B$. Then, $f(\alpha, \beta) = Y^T AX$.

For $g(\alpha, \beta) = f(\beta, \alpha)$, we have

$$g(\alpha, \beta) = X^T AY + (X^T AY)^T = Y^T A^T X.$$

Thus,
$$[g]_B = A^T.$$

Now, *f* is symmetric iff $g = f$ iff $[g]_B = [f]_B$ by Proposition 4 iff $A = A^T$ iff $[f]_B$ is symmetric. Again, *f* is skew-symmetric iff $g = -f$ iff $[g]_B = -[f]_B$ iff $A^T = -A$ iff A is skew-symmetric.

Example 5a. Consider *f* defined on $\mathbf{C}^2 \times \mathbf{C}^2$ such that for $\alpha = (x_1, x_2)$, $\beta \in (y_1, y_2)$ in $\mathbf{C}^2$, $f(\alpha, \beta) = x_1y_2 + x_2y_1 - x_1y_1$.

Then, $$f(\alpha, \beta) = [y_1, y_2]\begin{bmatrix} -1 & 1 \\ 1 & 0 \end{bmatrix}\begin{bmatrix} x_1 \\ x_2 \end{bmatrix} = Y^TAX,$$

where X and Y are the respective co-ordinate vectors of α, β relative to the standard ordered basis B of $\mathbf{C}^2$. $[f]_B = \begin{bmatrix} -1 & 1 \\ 1 & 0 \end{bmatrix}$ is a symmetric matrix, so that *f* is a symmetric bilinear form on $\mathbf{C}^2$.

The function g on $\mathbf{C}^2 \times \mathbf{C}^2$ such that

$$g(\alpha, \beta) = x_1y_2 + 2x_2y_1 - x_1y_1 = [y_1, y_2]\begin{bmatrix} -1 & 2 \\ 1 & 0 \end{bmatrix}\begin{bmatrix} x_1 \\ x_2 \end{bmatrix}$$

is a bilinear form with $[g]_B = \begin{bmatrix} -1 & 2 \\ 1 & 0 \end{bmatrix}$.

As $[g]_B$ is not symmetric, g is not symmetric. Indeed, for $\alpha = (1, 0)$, $\beta = (0, 1)$, $g(\alpha, \beta) = 1$, $g(\beta, \alpha) = 2 \neq g(\alpha, \beta)$.

Recall that a field F is of characteristic a positive integer p, if the order under addition of its identity element is p; such a p is always a prime number. If the order of the identity element of F, under addition, is infinite, then, F is said to be of characteristic zero. For any space V_F, if the characteristic of F is different from 2, then, $\alpha \neq -\alpha$ for any $\alpha \neq 0$ in V.

Proposition 6. Let *f* be a bilinear form on a vector space V_F and characteristic of F be different from 2. Then, *f* is skew-symmetric iff *f* is alternate.

Proof. Let *f* be skew-symmetric. For $\alpha, \beta \in V$, $f(\beta, \alpha) = -f(\alpha, \beta)$. In particular, for $\beta = \alpha$, $f(\alpha, \alpha) = -f(\alpha, \alpha)$. As char $(F) \neq 2$, we get $f(\alpha, \alpha) = 0$. Hence, *f* is alternate. Conversely, let *f* be alternate.

Then,
$$\begin{aligned} 0 &= f(\alpha+\beta, \alpha+\beta) \\ &= f(\alpha, \alpha) + f(\alpha, \beta) + f(\beta, \alpha) + f(\beta, \beta) \\ &= f(\alpha, \beta) + f(\beta, \alpha), \end{aligned}$$

gives $$f(\beta, \alpha) = -f(\alpha, \beta).$$

Hence, *f* is skew-symmetric.

Example 6.a. Let $V_{\mathbf{R}}$ be the real vector space of all continuous real valued functions defined on **R**. $V_{\mathbf{R}}$ is not finite dimensional. For any $\alpha, \beta \in V_R$, define

$$f(\alpha, \beta) = \int_0^1 \alpha(t-1)\beta(t)dt$$

f is a bilinear form on V. It is neither symmetric nor skew-symmetric.

Define $$g(\alpha, \beta) = \int_0^1 \{\alpha(t-1)\beta(t) + \beta(t-1)\alpha(t)\}dt.$$

Obviously, g is a symmetric bilinear form.

Further, $$h(\alpha, \beta) = \int_0^1 \{\alpha(t-1)\beta(t) - \beta(t-1)\alpha(t)\}dt$$

defines a skew-symmetric bilinear form on $V_{\mathbf{R}}$.

Observe that $$f(\alpha, \beta) = \frac{1}{2}g(\alpha, \beta) + \frac{1}{2}h(\alpha, \beta).$$

so that $$f = \frac{1}{2}g + \frac{1}{2}h,$$

with $\frac{1}{2}g$ symmetric, and $\frac{1}{2}h$ skew-symmetric.

Definition 7. Let A and B be any two $n \times n$-matrices over a field. Then, A is said to be *congruent* to B if $B = P^T AP$ for some non-singular matrix P over F.

The relation of congruence in $F^{n\times n}$ is an equivalence relation. As any two congruent matrices A and B are equivalent, by Theorem (2.3.1), rank (A) = rank (B). Given a bilinear form f on V_F, matrices $[f]_B$ and $[f]_{B'}$ relative to any two ordered bases B and B' of V are congruent by Proposition 3, so that rank $[f]_B$ = rank $[f]_{B'}$. This leads to the following definition :

Definition 8. Let f be a bilinear form on a finite dimensional vector space V_F. Then, the *rank of f* is the rank of $[f]_B$, where B is any ordered basis of V_F.

Theorem 9. Let f be a bilinear form on a finite dimensional vector space V_F. Then,

(*i*) $L_f : V \to V^*$ and $R_f : V \to V^*$ such that $[L_f(\alpha)](\beta) = f(\alpha, \beta)$, $[R_f(\alpha)](\beta) = f(\beta, \alpha)$ for $\alpha, \beta \in V$, are linear transformations

(*ii*) rank (L_f) = rank (R_f) = rank (f)

Proof. That L_f and R_f are linear transformations, is immediate from the definition of a bilinear form. For any $\alpha \in V$, $\alpha \in \ker L_f$ iff $f(\alpha, \beta) = 0$ for every $\beta \in V$. Let $A = [f]_B$, where B is some fixed ordered basis of V_F. Let $X = [\alpha]_B$. Then, $f(\alpha, \beta) = Y^T AX$, $Y = [\beta]_B$, so that $\alpha \in \ker L_f$ iff $Y^T AX = 0$ for every $Y \in F^{n\times 1}$ iff

$AX = 0$ iff $X \in$ null-space (A). Consequently, nullity (L_f) = nullity (A), so that rank $(L_f) = n -$ nullity (A) = rank (A), where $n = \dim V_F$. Similarly, $\beta \in \ker R_f$ iff for $Y = [\beta]_B$, $Y^T A = 0$ iff $A^T Y = 0$, so that rank (R_f) = rank (A^T). However, rank (A) = rank (A^T). Hence, rank (L_f) = rank (R_f) = rank (A) = rank (f).

Corollary 10. For any bilinear form f on a finite dimensional vector space V_F, the following are equivalent :

(*i*) Rank $(f) = \dim V_F$

(*ii*) $\{\alpha \in V : f(\alpha, \beta) = 0 \text{ for every } \beta \in V\} = 0$

(*iii*) $\{\beta \in V : f(\alpha, \beta) = 0 \text{ for every } \alpha \in V\} = 0$.

Proof. Let $A = [f]_B$, where B is an ordered basis of V_F, $n = \dim V_F$. Let $L_f : V \to V^*$ and $R_f : V \to V^*$ be as given in the above theorem. As rank (A) = rank (L_f) = rank (R_f), rank $(A) = \dim V_F$ iff nullity (L_f) = nullity $(R_f) = 0$ iff null space $(L_f) = 0 =$ null – space (R_f) iff

$$\{\alpha \in V : f(\alpha, \beta) = 0 \text{ for every } \beta \in V\} = 0 = \{\beta \in V : f(\alpha, \beta) = 0 \text{ for every } \alpha \in V\}.$$

This proves the result.

Definition 11. Let f be a bilinear form on V_F.

The subspace $\quad W = \{\alpha \in V : f(\alpha, \beta) = 0 \text{ for every } \beta \in V\}$

is called the left degenerator of f ; we write $Dn(f)$ for this subspace. If $Dn(f) = \{0\}$, then, f is called a *non-singular* (or *non-degenerate*) bilinear form. The dimension of $Dn(f)$ is called the *nullity* of f.

The following inference is obvious :

Corollary 12. For any bilinear form f on a finite-dimensional vector space V_F, the following hold :

(*i*) Dim $Dn(f) = n -$ rank (f), $n = \dim V_F$

(*ii*) If the restriction of f to a subspace V' of V is non-singular, then, $V' \cap Dn(f) = \{0\}$.

Theorem 13. Let f be a non-singular bilinear form on a finite dimensional vector space V_F. Then, for any linear functional λ, on V, there exists unique $\alpha_0 \in V$ such that

$$\lambda(\alpha) = f(\alpha, \alpha_0)$$

for all $\alpha \in V$.

Proof. As f is non-degenerate, rank $(R_f) = n = \dim V$. So that $R_f : V \to V^*$ is an isomorphism. Consequently, given $\lambda \in V^*$, there exists unique $\alpha_0 \in V$ such that $R_f(\alpha_0)$

$= \lambda$, i.e. $\lambda(\alpha) = \left[R_f(\alpha_0)\right](\alpha) = f(\alpha, \alpha_0)$. (This result is known as *Reisz representation theorem*).

Let f be a bilinear form on V_F, B and B' be two ordered basis of V, $A = [f]_B$, $A' = [f]_{B'}$ and P be the matrix of B' relative to B. As in Proposition (6.3.4), $A' = P^T AP$. In case A is symmetric, we shall prove in the next section that for some non-singular matrix P, A' is a diagonal matrix, whenever the characteristic of F is different from 2. Observe that the set $SBL(V)$ of symmetric bilinear forms on V is a subspace of $BL(V)$, and so is the set $SKBL(V)$ of skew-symmetric bilinear forms on V.

Exercise 7.1

1. Which of the following functions f on $\mathbf{C}^2 \times \mathbf{C}^2$ are bilinear forms. For $\alpha = (x_1, x_2)$, $\beta = (y_1, y_2)$ in $\mathbf{C}^2$.

   (*i*) $f(\alpha, \beta) = 1$

   (*ii*) $f(\alpha, \beta) = x_1 + x_2 + y_1 + y_2$

   (*iii*) $f(\alpha, \beta) = x_1y_1 + x_2y_2$

   (*iv*) $f(\alpha, \beta) = x_1\overline{y}_1 + x_2\overline{y}_2$

   (*v*) $f(\alpha, \beta) = x_1y_1 + x_2y_2 + 2x_2y_1 + 3x_2\overline{y}_2$

   (*vi*) $f(\alpha, \beta) = x_1^2 + y_1^2 + x_1y_1$.

   If any one is a bilinear form, find its matrix relative to the standard ordered basis.

2. Let f be a bilinear form on V_F. For any subset W of V, prove that

$$W^0 = \{\alpha \in V : f(\alpha, \beta) = 0 \text{ for every } \beta \in W\}$$

   is a subspace of V. If f is symmetric, prove that $W \subseteq (W^0)^0 = W^{00}$

3. Let f be the bilinear form on $\mathbf{R}^{2\times 1}$ such that

$$f(X, Y) = X^T \begin{bmatrix} 1 & 2 \\ 0 & 3 \end{bmatrix} Y$$

   Find matrix of f relative to the standard ordered basis of $\mathbf{R}^{2\times 1}$. Is f symmetric?

4. Let f be a bilinear form on $\mathbf{C}^{2\times 1}$ such that

$$f(X, Y) = x_1y_1 + x_1y_2 + x_2y_1 + 3x_2y_2$$

   Find, if possible, an ordered basis B of $\mathbf{C}^{2\times 1}$ such that $[f]_B$ is a diagonal matrix.

5. Let $V_{\mathbf{R}}$ be the real vector space of all real valued continuous functions defined on the closed interval $[0, \pi]$. For any $f, g \in V$, define $\lambda(f, g) = \int_0^\pi f(t)g(t)dt$

   (*i*) Prove that λ is a bilinear form.

(*ii*) For $\sigma_n = \sin nt$, $\eta_m = \cos mt$, where *n*, *m* are any integers, find

$$\lambda(\sigma_n, \sigma_m),\ \lambda(\sigma_n, \eta_n) \text{ and } \lambda(\eta_n, \eta_m).$$

(*iii*) Let *V'* be the subspace of *V* consisting of all polynomial functions of degree ≤ 2 and let λ′ be the restriction of λ to *V'*. Find the matrix of λ′ belative to the ordered basis $\{1, t, t^2\}$. Find an ordered basis *B* of *V'* such that $[\lambda']_B$ is a diagonal matrix.

6. Let *f* be the bilinear form on $\mathbf{R}^4$ such that for any $X, Y \in \mathbf{R}^4$

$$f(X, Y) = x_1y_1 + x_2y_2 + x_3y_3 - x_4y_4$$

($\mathbf{R}^4$ along with *f* is called a *Minkowski space*)

(*i*) Find $X \neq 0$ in $\mathbf{R}^4$ such that $f(X, X) = 0$

(*ii*) Given linear functional *g* on $\mathbf{R}^4$ such that $g(X) = x_1 + x_2$, find a vector $Y_0 \in \mathbf{R}^4$ such that $g(X) = f(X, Y_0)$ for all $X \in \mathbf{R}^4$.

7. Let *f* be a bilinear form on a finite dimensional vector space V_F. Prove that there exist linear functionals L_1, L_2 on *V* such that $f(\alpha, \beta) = L_1(\alpha)L_2(\beta)$ for all α, $\beta \in V$ iff rank (*f*) ≤ 1.

8. Let $B = \{\alpha_1, \alpha_2, ..., \alpha_n\}$ be an ordered basis of V_F. For any $\alpha, \beta \in V$ with

$$X = [\alpha]_B = \{x_1, x_2, ..., x_n\}^T,$$

$$Y = [\beta]_B = \{y_1, y_2, ..., y_n\}^T,$$

$$1 \leq i, j \leq n,$$

define $f_{ij}(\alpha, \beta) = x_iy_j$. Prove the following :

(*i*) Each f_{ij} is a bilinear form.

(*ii*) $\{f_{ij} : 1 \leq i, j \leq n\}$ is a basis of *BL*(*V*).

9. Let *f* be a non-degenerate bilinear form on V_F and *T be a* linear operator *on* V_F. Prove that there exists a unique linear operator *T'* on V_F such that $f(T\alpha, \beta) = f(\alpha, T'\beta)$ for all $\alpha, \beta \in V$. (*T'* is called the *adjoint of T relative to f*).

7.2 Symmetric Bilinear Form

Throughout, *F* is a field of characteristic different form 2 and V_F is a non-zero vector space of dimension *n*. We start with the following :

Definition ·. Any function $q : F^{n\times 1} \to F$ such that for some $n \times n$-matrix $A = [a_{ij}]$ over *F*,

$$q(X) = \sum_{i=1}^{n} \sum_{j=1}^{n} a_{ij} x_i x_j = X^T AX$$

for all $$X = [x_1, x_2, ..., x_n]^T \in F^{n\times 1},$$

is called a *quadratic form* on $F^{n\times 1}$. The matrix A is said to be associated with q and we write q_A for q.

Let f be a bilinear form on $F^{n\times 1}$ and A be the matrix of f relative to the standard ordered basis of $F^{n\times 1}$. Then $q_A(X) = X^T AX = f(X, X)$ is a quadratic form associated with the bilinear form f.

For example, $$q(x_1, x_2, x_3) = x_1^2 + x_2^2 - x_3^2$$

defines a quadratic form on $\mathbf{R}^{3\times 1}$ (identified with $\mathbf{R}^{1\times 3}$); here $A = \text{diag}[1, 1, -1]$.

Proposition 2. For a bilinear form f on a vector space V_F, the following statements are equivalent :

(*i*) f is symmetric

(*ii*) $f(\alpha, \beta) = \frac{1}{2}[f(\alpha+\beta, \alpha+\beta) - f(\alpha, \alpha) - f(\beta, \beta)]$

(*iii*) $f(\alpha, \beta) = \frac{1}{4}[f(\alpha+\beta, \alpha+\beta) - f(\alpha-\beta, \alpha-\beta)]$

[Here, $\text{char}(F) \neq 2$].

Proof. Let f be symmetric.

Then, $$f(\alpha, \beta) = f(\beta, \alpha).$$

So $$\begin{aligned} f(\alpha+\beta, \alpha+\beta) &= f(\alpha, \alpha) + f(\alpha, \beta) + f(\beta, \alpha) + f(\beta, \beta) \\ &= 2f(\alpha, \beta) + f(\alpha, \alpha) + f(\beta, \beta). \end{aligned}$$

This gives $$f(\alpha, \beta) = \frac{1}{2}[f(\alpha+\beta, \alpha+\beta) - f(\alpha, \alpha) - f(\beta, \beta)].$$

This proves, (*i*) implies (*ii*). Let (*ii*) hold. As $\alpha + \beta = \beta + \alpha$, it is immediately clear from (*ii*) that $f(\alpha, \beta) = f(\beta, \alpha)$. Hence, (*i*) and (*ii*) are equivalent. Similarity, (*i*) and (*iii*) are equivalent.

The identity $$f(\alpha, \beta) = \frac{1}{4}[f(\alpha+\beta, \alpha+\beta) - f(\alpha-\beta, \alpha-\beta)]$$

for a symmetric bilinear form f is called a *polarization identity* for f.

Thus, polarization identity gives that if $f(\alpha, \alpha) = 0$ for every $\alpha \in V$, then, the symmetric bilinear form $f = 0$.

The theorem, given below, provides a connection between symmetric provides bilinear forms and quadratic forms.

Theorem 3. Let q be a quadratic form on $F^{n\times 1}$. Then, (*i*) $q(cX) = c^2 q(X)$ for any $c \in F$ and $X \in F^{n\times 1}$

(*ii*) $f : F^{n\times 1} \times F^{n\times 1} \to F$ such that

$$f(X, Y) = \frac{1}{2}[q(X+Y) - q(X) - q(Y)]$$

is a symmetric bilinear form on $F^{n\times 1}$ such that $q(X) = f(X, X)$.

Proof. (*i*) By definition there exists an $n \times n$-matrix $A = [a_{ij}]$ over F such that

$$q(X) = \sum_{j=1}^{n} \sum_{i=1}^{n} a_{ij} x_i x_j ,$$

$$X = [x_1, x_2, \ldots, x_n]^T .$$

Obviously,
$$q(cX) = \sum_{j=1}^{n} \sum_{i=1}^{n} a_{ij} (cx_i)(cx_j)$$

$$= c^2 \sum_{j=1}^{n} \sum_{i=1}^{n} a_{ij} x_i x_j$$

$$= c^2 q(X).$$

This proves (*i*). Now, define $B = [b_{ij}]$ such that $b_{ij} = \frac{1}{2}(a_{ij} + a_{ji})$.

Then, B is a symmetric matrix, and

$$a_{ij}x_i x_j + a_{ji}x_j x_i = (a_{ij} + a_{ji})x_i x_j = b_{ij}x_i x_j + b_{ji}x_j x_i,$$

$$a_{ii}x_i x_i = b_{ii}x_i x_i .$$

This gives
$$q(X) = \sum_{j=1}^{n} \sum_{i=1}^{n} b_{ij} x_i x_j = X^T B X .$$

As B is a symmetric matrix $\quad f(X, Y) = Y^T B X$

is a symmetric bilinear form on $F^{n\times 1}$.

Finally,
$$q(X+Y) - q(X) - q(Y) = (X+Y)^T B(X+Y) - X^T BX - Y^T BY$$

$$= (X^T + Y^T)B(X+Y) - X^T BX - Y^T BY$$

$$= X^T BY + Y^T BX .$$

But
$$X^T BY = (X^T BY)^T = Y^T BY .$$

So,
$$q(X+Y) - q(X) - q(Y) = 2Y^T BX ,$$

and, hence,
$$f(X, Y) = \frac{1}{2}[q(X+Y) - q(X) - q(Y)]$$

$$f(X, X) = \frac{1}{2}[q(2X) - 2q(X)] = \frac{1}{2}[4q(X) - 2q(X)] = q(X).$$

The result, given above, motivates the following definition of a quadratic form :

Definition 4. Let V_F be any vector space. Then, a function $q : V_F \times V_F \to F$ is called a *quadratic form* on V, if for any $c \in F$, $\alpha, \beta \in V$

(*i*) $q(c\alpha) = c^2 q(\alpha)$

(*ii*) $g(\alpha, \beta) = q(\alpha+\beta) - q(\alpha) - q(\beta)$, is a symmetric bilinear form on V_F.

Theorem 5. Let V_F be a finite dimensional vector space and B be an ordered basis of V.

(*i*) If f is a symmetric bilinear form on V with $A = [f]_B$, then, for $\alpha \in V$, $q(\alpha) = f(\alpha, \alpha)$ gives a quadratic form on V, such that for

$$X = [\alpha]_B,\ q(\alpha) = X^T AX$$

(*ii*) If q is a quadratic form on V_F, then,

$$f(\alpha, \beta) = \frac{1}{2}[q(\alpha+\beta) - q(\alpha) - q(\beta)]$$

is a symmetric bilinear form on V_F such that $q(\alpha) = f(\alpha, \alpha)$.

Further, if g is a symmetric bilinear form on V, satisfying

$$q(\alpha) = g(\alpha, \alpha),$$

then, $$g = f$$

Proof.

(*i*) For $$c \in F,$$

$$\alpha \in V,\ q(c\alpha) = f(c\alpha, c\alpha) = c^2 f(\alpha, \alpha) = c^2 q(\alpha).$$

By Proposition 2, $f(\alpha, \beta) = \frac{1}{2}[q(\alpha+\beta) - q(\alpha) - q(\beta)]$,

and this is a bilinear form, so that q is a quadratic form.

(*ii*) Let q be a quadratic form. By the definition of q,

$$f(\alpha, \beta) = \frac{1}{2}[q(\alpha+\beta) - q(\alpha) - q(\beta)]$$

is a symmetric bilinear form.

Obviously, $f(\alpha, \alpha) = \frac{1}{2}[q(2\alpha) - q(\alpha) - q(\alpha)] = q(\alpha)$.

Let g be another symmetric bilinear form on V_F such that $q(\alpha) = g(\alpha,\alpha)$. Then, by Proposition 2,

$$\begin{aligned} g(\alpha, \beta) &= \frac{1}{2}[g(\alpha+\beta, \alpha+\beta) - g(\alpha, \alpha) - g(\beta, \beta)] \\ &= \frac{1}{2}[q(\alpha+\beta) - q(\alpha) - q(\beta)] \\ &= f(\alpha, \beta). \end{aligned}$$

Hence, $$g = f$$

We now prove a diagonalization theorem for a symmetric bilinear form analogous to the spectral theorem for a Hermittian form.

Lemma 6. Let f be a symmetric non-zero bilinear form on V_F. Then,

$$V = W_1 \oplus V_1$$

such that $\dim W_1 = 1, f(\alpha, \alpha) \neq 0$ for any non-zero $\alpha \in W_1$, and $f(\alpha, \beta) = 0$ for $\alpha \in W_1$, $\beta \in V_1$. Further, the rank of f', the restriction of f to V_1, is $r - 1$, where $r = \text{rank}\ (f)$.

Proof. As $f \neq 0$, the polarization identity for f gives $f(\alpha, \alpha) \neq 0$ for some $\alpha \neq 0$. Choose α_1 such an α. Let $W_1 = \langle \alpha_1 \rangle$. Then, for $0 \neq \alpha \in W_1$, $\alpha = c\alpha_1$, $c \neq 0$, $f(\alpha, \alpha) = c^2 f(\alpha_1, \alpha_1) \neq 0$. We have an ordered basis $B = \{\alpha_1, \alpha_2, ..., \alpha_n\}$. For $i \geq 2$, define

$$\beta_i = \alpha_i - \frac{f(\alpha_1, \alpha_i)}{f(\alpha_1, \alpha_1)} \alpha_1,$$

$$\beta_1 = \alpha_1.$$

Then, $$B = \{\beta_1, \beta_2, ..., \beta_n\}$$

is on ordered basis of V. Clearly, for $i \geq 2$.

$$\begin{aligned} f(\beta_1, \beta_i) &= f\left(\alpha_1, \alpha_i - \frac{f(\alpha_1, \alpha_i)}{f(\alpha_1, \alpha_1)} \alpha_1\right) \\ &= f(\alpha_1, \alpha_i) - f(\alpha_1, \alpha_i) \\ &= 0. \end{aligned}$$

Thus, for $$V_1 = \langle \beta_2, \beta_3, ..., \beta_n \rangle,$$

we have $f(\alpha_1, \beta) = 0$, for $\beta \in V_1$.

Then, for $$\alpha \in W_1, \beta \in V_1, \alpha = c\alpha_1,$$

yields $$f(\alpha, \beta) = cf(\alpha_1, \beta) = 0.$$

Clearly, $$V = W_1 \oplus V_1.$$

Consider $$A = [a_{ij}] = [f]_{B'}.$$

Then, $$a_{11} = f(\alpha_1, \alpha_1) \neq 0,$$

for $$j \geq 2,$$

$$a_{1j} = f(\beta_j, \beta_1) = f(\beta_1, \beta_j) = 0 = a_{j1}.$$

This gives $$A = \begin{bmatrix} a_{11} & 0 \\ 0 & D \end{bmatrix}$$

for some $(n-1) \times (n-1)$-matrix D.

Obviously, D is the matrix of f' the restriction of f to V_1 relative to the ordered basis $B' = \{\beta_2, \beta_3, ..., \beta_n\}$ of V_1, so thatrank (f') = rank (D).

However, $\text{rank}(D) = \text{rank}[a_{11}] + \text{rank}(D) = 1 + \text{rank}(D)$;

Hence, $\text{rank}(f') = \text{rank}(f) - 1$

Theorem 7. Let f be a non-zero symmetric bilinear form on a finite dimensional vector space V_F, with char $F \neq 2$. If r = rank (f), then, there exists an ordered basis B of V such that

$$[f]_B = \text{diag}[c_1, c_2, ..., c_r, 0, 0, ..., 0]$$

and $c_i \neq 0$ **for** $1 \le i \le r$

Proof.

We apply induction on the dimension of the vector space. Let $\dim V_F = n > 1$ and the result hold for all vector spaces of dimensions less than n. By Lemma 6, $V = W_1 \oplus V_1$ such that $\dim W_1 = 1$, $f(\alpha, \alpha) \neq 0$ and $f(\alpha, \beta) = 0$ for $0 \neq \alpha \in W_1$ and $\beta \in V_1$, and the rank of f' the restriction of f to V_1 is r-1. By the induction hypothesis, V_1 has an ordered basis

$$B' = \{\alpha_2, \alpha_3, ..., \alpha_n\}$$

such that $[f']_{B'} = \text{diag}[c_2, c_3, ..., c_r, 0, 0, ..., 0]$

with $c_i \neq 0$ for $2 \le i \le r$.

This gives $f(\alpha_i, \alpha_i) = c_i$

for $2 \le i \le r$, $f(\alpha_i, \alpha_j) = 0$

for $i \neq j$

and $f(\alpha_j, \alpha_j) = 0$

for $j \ge r + 1$.

Choose any $\alpha_1 \neq 0$ in W_1.

Then, $f(\alpha_1, \alpha_1) = c_1 \neq 0$

$$f(\alpha_1, \alpha_j) = 0$$

for $j \ge 2$.

This gives that for the ordered basis $B = \{\alpha_1, \alpha_2, ..., \alpha_n\}$ of V,

$$[f]_B = \text{diag}[c_1, c_2, ..., c_r, 0, 0, ..., 0].$$

Theorem 8. Let A be any symmetric $n \times n$-matrix over a field F of characteristic different from 2. If rank $(A) = r$, then, there exists a non-singular $n \times n$-matrix P over F such that

$$P^T AP = \text{diag}[c_1,\ c_2, \ldots, c_r,\ 0,\ 0, \ldots,\ 0],$$

and $\qquad c_i \neq 0 \qquad$ for $1 \leq i \leq r$

Proof. The matrix A defines the symmetric bilinear form f on $F^{n\times 1}$ such that $f(X, Y) = Y^T AX$. Then, $A = [f]_B$, where B is the standard ordered basis of $F^{n\times 1}$. If $f = 0$, $A = 0$, and we finish. Otherwise, by Theorem 7, there exists an ordered basis B' of $F^{n\times 1}$ such that

$$[f]_{B'} = \text{diag}[c_1,\ c_2, \ldots, c_r,\ 0,\ 0, \ldots,\ 0],$$

$$c_i \neq 0 \qquad \text{for } 1 \leq i \leq r$$

Let P be the matrix of B' relative to B. Then, by Proposition (7.1.4),

$$P^T AP = [f]_{B'} = \text{diag}[c_1,\ c_2, \ldots, c_r,\ 0,\ 0, \ldots,\ 0,\ 0].$$

This theorem does not hold for fields of characteristic 2, as is shown by the following :

Example 8.a. Let $F = Z_2$.

Now, $$A = \begin{bmatrix} 0 & 1 \\ 1 & 0 \end{bmatrix}$$

is a symmetric matrix over F.

For any non-singular $$P = \begin{bmatrix} a & b \\ c & d \end{bmatrix}$$

over F $$P^T AP = \begin{bmatrix} 2ac & ad+bc \\ ad+bc & 2bd \end{bmatrix} = \begin{bmatrix} 0 & x \\ y & 0 \end{bmatrix},$$

never a diagonal matrix.

Example 8.b. Consider $A = \begin{bmatrix} 0 & 1 \\ 1 & 0 \end{bmatrix}$ over **R**.

Let $P = \begin{bmatrix} a & b \\ c & d \end{bmatrix}$ be a non-singular matrix over **R** such that

$$P^T AP = \begin{bmatrix} x & 0 \\ 0 & y \end{bmatrix}.$$

Then $x \neq 0 \neq y$, as rank $(A) = 2$.

This gives $$\begin{bmatrix} 2ac & ad+bc \\ ad+bc & 2bd \end{bmatrix}\begin{bmatrix} x & 0 \\ 0 & y \end{bmatrix}.$$

Consequently, $$2ac = x,\ 2bd = y,\ ad+bc = 0.$$

Thus $$a \neq 0,\ b \neq 0,\ c \neq 0,\ d = -\frac{bc}{a}.$$

Hence $$P = \begin{bmatrix} a & b \\ c & -\frac{bc}{a} \end{bmatrix}$$

and $$P^T AP = \begin{bmatrix} 2ac & 0 \\ 0 & -\frac{2b^2c}{a} \end{bmatrix}$$

As a and c are arbitrary non-zero real number. so, $2ac$ need not be a characteristic value of A. Similarly, $\dfrac{-2b^2c}{a}$ need not be a characteristic value of A. As ac and $\dfrac{c}{a}$ have the same signs; obviously, $x = 2ac$, $y = \dfrac{-2b^2c}{a}$ have opposite signs.

For any bilinear form f on V_F, let us recall that the left degenerator of f denoted $Dn(f)$ is the subspace $\{\alpha \in V : f(\alpha, \beta) = 0 \text{ for every } \beta \in V\}$. This f is non-singular in case $Dn(f) = \{0\}$.

Theorem 9. Let f be a symmetric bilinear form on a finite dimensional vector space V_F, with char $F \neq 2$.

Then, $$V = Dn(f) \oplus V_1$$

such that the restriction of f to V_1 is non-degenerate and $\dim V_1 = \text{rank}(f)$.

Proof. By Theorem 8, V admits an ordered basis $B = \{\alpha_1, \alpha_2, ..., \alpha_n\}$ such that

$$[f]_B = \text{diag}[c_1, c_2, ..., c_r, 0, 0, ..., 0]$$

$$r = \text{rank}(f),\ c_i \neq 0,\ 1 \leq i \leq r.$$

Let V_1 be the subspace $\langle \alpha_1, \alpha_2, ..., \alpha_r \rangle$,

and $$W = \langle \alpha_{r+1}, \alpha_{r+2}, ..., \alpha_n \rangle.$$

Clearly, $$V = V_1 \oplus W \text{ and } W \subseteq Dn(f)$$

Let f' be the restriction of f to V_1. For the ordered basis $B' = \{\alpha_1, \alpha_2, ..., \alpha_r\}$ of V_1,

$$[f']_{B'} = \text{diag}[c_1, c_2, ..., c_r]$$

is a non-singular matrix. So, by Corollary (7.1.10), f' is non-degenerate, for any $0 \neq \alpha \in V_1$, there exists $\beta \in V_1$ such that $f(\alpha, \beta) \neq 0$.

Hence, $$V_1 \cap Dn(f) = \{0\}.$$

Further, $$\dim\{V_1 \oplus Dn(f)\} = \dim V_1 + \dim D_n(f) = n, \quad \text{by (7.1.12)}$$

As $$n = \dim V,$$

we get $$V = V_1 \oplus Dn(f) \text{ and } W = Dn(f).$$

As remarked in Example 8.b, given a symmetric bilinear form on V_F, if for some ordered basis B of V, $A = [f]_B$ and for some ordered basis B' of V,

$$A' = [f]_{B'} = \text{diag}[c_1, c_2, ..., c_r, 0, 0, ..., 0],$$

the c_i need not be the characteristic values of A.

We now examine the case of $F = \mathbf{R}$, to find some invariances in A'.

Theorem 10. Let f be a symmetric bilinear form on a finite dimensional real vector space $V_{\mathbf{R}}$.

Then, $$V = V^+ \oplus V^- \oplus V_0$$

for some subspaces V^+, V^- and V_0 such that

(i) $f(\alpha, \alpha) > 0$ for any $\alpha \neq 0$ in V^+

(ii) $f(\beta, \beta) < 0$ for any $\beta \neq 0$ in V^-

(iii) For any $\alpha \in V_0$, $\beta \in V$, $f(\alpha, \beta) = 0$.

Further, for any other similar decomposition $V = W^+ \oplus W^- \oplus W_0$ the following relations hold :

(iv) $V_0 = W_0 = Dn(f)$

(v) $\dim V^+ = \dim W^+$ and $\dim V^- = \dim W^-$.

Proof. By Theorem 9, we have an ordered basis $B = \{\alpha_1, \alpha_2, ..., \alpha_n\}$ such that

$$V = V_1 \oplus Dn(f) \qquad ... (1)$$

$$V_1 = \langle \alpha_1, \alpha_2, ..., \alpha_r \rangle,$$

$$r = \text{rank}(f),$$

$$Dn(f) = \langle \alpha_{r+1}, ..., \alpha_n \rangle$$

$$[f]_B = \text{diag}[c_1, c_2, ..., c_r, 0, 0, ..., 0].$$

Let V^+ be the subspace spanned by those α_i's such that $f(\alpha_i, \alpha_i) = c_i > 0$, and V^- be the subspace spanned by those α_j's such that $f(\alpha_j, \alpha_j) = c_j < 0$. Clearly, $V_1 = V^+ \oplus V^-$

Then, $$V = V^+ \oplus V^- \oplus V_0,\ Dn(f) = V_0 \qquad ... (2)$$

Consider $$0 \neq \alpha \in V^+$$

Then, $$\alpha = \sum_{i=1}^{n} a_i \alpha_i,$$

such that if an $a_i \neq 0$, then $f(\alpha_i, \alpha_i) = c_i > 0$.

This gives $$f(\alpha, \alpha) = \sum_{i=1}^{n} a_i^2 f(\alpha_i, \alpha_i) = \sum_{i=1}^{n} a_i^2 c_i > 0,$$

as $$a_i^2 \geq 0,\ c_j > 0,$$

wherever $$a_j \neq 0,$$

and some $$a_j \neq 0.$$

Thus V^+ satisfies (*i*). Similarly, V^- satisfies (*ii*). As $V_0 = Dn(f)$, by definition of $Dn(f)$, V_0 satisfies (*iii*).

Let $$V = W^+ \oplus W^- \oplus W_0 \qquad \text{... (3)}$$

be another decomposition similar to (2).

By condition (*iii*), $$W_0 \subseteq Dn(f) = V_0.$$

We prove that $$W^+ \cap (V^- \oplus V_0) = \{0\}.$$

To see this, let $$\alpha \in W^+ \cap (V^- \oplus V_0).$$

Then, $$\alpha = \beta + \gamma,\ \beta \in V^-,\ \gamma \in V_0.$$

By using (*iii*), we get $f(\alpha, \alpha) = f(\beta+\gamma, \beta+\gamma) = f(\beta, \beta)$ ... (4)

As $$\alpha \in W^+, f(\alpha, \alpha) \geq 0,$$

and as $$\beta \in V^-, f(\beta, \beta) \leq 0.$$

Consequently, (4) gives $f(\alpha, \alpha) = 0$, so $\alpha = 0$.

This proves that $$W^+ \cap (V^- \oplus V_0) = \{0\}.$$

As $$V = V^+ \oplus (V^- \oplus V_0),$$

we get $$\dim W^+ \leq \dim V^+.$$

By interchanging the role of (2) and (3), we also get $\dim V^+ = \dim W^+$

Similarly, $$\dim V^- = \dim W^-$$

This proves (*v*).

By (2) and (3),

$$\dim V^+ + \dim V^- + \dim V_0 = \dim W^+ + \dim W^- + \dim W_0.$$

Consequently, $$\dim V_0 = \dim W_0.$$

However, $$W_0 \subseteq V_0.$$

Hence, $$W_0 = V_0 = Dn(f)$$

This theorem motivates the following definition :

Definition 11. Let f be a symmetric bilinear form on a finite dimensional real vector space V.

Let $$V = V^+ \oplus V^- \oplus Dn(f),$$

such that $f(\alpha, \alpha) > 0$ for every $\alpha \neq 0$ in V^+ and $f(\beta, \beta) < 0$ for every $\beta \neq 0$ in V^-. Then,

(*i*) dim V^+ is called the *index* of *f*.

(*ii*) dim V^+ – dim V^- is called the *signature* of *f*

(*iii*) *f* is said to be *positive definite* if $V = V^+$

(*iv*) *f* is said to be *negative definite* if $V = V^-$

(*v*) *f* is said to be *indefinite* if $V^+ \neq \{0\}$ and $V^- \neq \{0\}$

(*vi*) *f* is said to be positive semi-definite if $V^- = \{0\}$ but $Dn(f) \neq \{0\}$

(*vii*) *f* is said to be negative semi definite if $V^+ = \{0\}$ but $Dn(f) \neq \{0\}$.

The construction of V^+, V^- and V_0 in the proof of the theorem, given above, shows the existence of ordered bases

$B_1 = \{\alpha'_1, \alpha'_2, ..., \alpha'_s\}$ of V^+,

$B_2 = \{\alpha'_{s+1}, \alpha'_{s+2}, ..., \alpha'_r\}$ of V^-, and

$B_3 = \{\alpha'_{r+1}, \alpha'_{r+2}, ..., \alpha'_n\}$ of V_0,

such that for $$B = B_1 \cup B_2 \cup B_3,$$

$$[f]_B = \text{diag}[c'_1, c'_2, ..., c'_r, 0, 0, ..., 0]$$

with $$f(\alpha'_i, \alpha'_i) = c_i$$

satisfying $$c'_i > 0$$

for $$1 \leq i \leq s,\ c'_i < 0$$

for $$s+1 \leq i \leq r,$$

so that for $$\alpha \in V^+,\ \beta \in V^- + V_0,\ f(\alpha, \beta) = 0,$$

for $$\alpha_1 \in V^-,\ \beta_1 \in V^+ + V_0,\ f(\alpha_1, \beta_1) = 0.$$

Hence, we get the following corollary :

Corollary 12. Let *f* be a symmetric bilinear form on a finite dimensional real vector space V_F.

Then, $$V = V^+ \oplus V^- \oplus V_0$$

such that

(*i*) $V_0 = D_n(f)$,

(*ii*) f is positive on V^+,

(*iii*) *f* is negative on V^-,

(*iv*) for $\alpha \in V^+$, $\beta \in V^- + V_0$, $f(\alpha, \beta) = 0$, and (*v*) for $\alpha \in V^-$, $\beta \in V^+ + V_0, f(\alpha, \beta) = 0$.

Theorem 13. Let A be any non-zero symmetric real $n \times n$-matrix. Then, there exists a non-singular real $n \times n$-matrix P such that

$$P^T AP = \text{diag}[1, 1, \ldots, 1, -1, -1, \ldots, -1, 0, 0, \ldots, 0]$$

with number s of entries equal to 1 uniquely determined by A, for r = rank (A), $r - s$ number of entries equal -1.

Proof. Let f be the symmetric bilinear form on $V = \mathbf{R}^{n \times 1}$ given by A, i.e. $f(X, Y) = Y^T AX$ for all $X, Y \in V$.

By using Corollary 12, we have

$$V = V^+ \oplus V^- \oplus Dn(f),$$

and an ordered basis $$B' = \{\alpha'_1, \alpha'_2, \ldots, \alpha'_n\}$$

such that $$f(\alpha'_i, \alpha'_j) = 0$$

for $$i \neq j,$$

$$B_1 = \{\alpha'_1, \alpha'_2, \ldots, \alpha'_s\}$$

is a basis of V^+,

$$B_2 = \{\alpha'_{s+1}, \ldots, \alpha'_r\}$$

is a basis of V^-,

and $$B_3 = \{\alpha'_{r+1}, \ldots, \alpha'_n\}$$

is a basis of $Dn(f)$.

Thus, if $$f(\alpha'_i, \alpha'_i) = d_i$$

then, $$d_i > 0$$

for $$1 \leq i \leq s, \ d_i < 0,$$

for $$s+1 \leq i \leq r$$

and $$d_i = 0$$

for $$i > r.$$

Define β_i, $1 \leq i \leq n$ as follows :

$$\beta_i = \frac{1}{\sqrt{d_i}}\alpha' \qquad \text{for } 1 \leq i \leq s$$

$$= \frac{1}{\sqrt{-d_i}}\alpha' \qquad \text{for } s+1 \leq i \leq r$$

$$= \alpha_i \qquad \text{for } i \geq r+1.$$

For $1 \le i \le s$,
$$f(\beta_i, \beta_i) = f\left(\frac{1}{\sqrt{d_i}}\alpha_i, \frac{1}{\sqrt{d_i}}\alpha_i\right)$$
$$= \frac{1}{\left(\sqrt{d_i}\right)^2} f(\alpha_i, \alpha_i)$$
$$= \frac{1}{d_i} d_i$$
$$= 1, \qquad \text{for } s+1 \le i \le r,$$
$$f(\beta_i, \beta_i) = \frac{1}{\left(\sqrt{-d_i}\right)^2} f(\alpha_i, \alpha_i)$$
$$= \frac{1}{-d_i} d_i$$
$$= -1, \qquad \text{for } i \ge r+1,$$
$$f(\beta_i, \beta_i) = f(\alpha_i, \alpha_i)$$
$$= 0.$$

For $i \ne j$, $$f(\beta_i, \beta_j) = 0.$$

Hence for $$B'' = \{\beta_1, \beta_2, \ldots, \beta_n\},$$
$$[f]_{B''} = \text{diag}\left[\overbrace{1, 1, \ldots, 1}^{s}, \overbrace{-1, -1, \ldots, -1}^{r-s}, 0, 0, \ldots, 0\right].$$

Let P be the matrix of B' relative to the standard ordered basis of $\mathbf{R}^{n\times 1}$.

Then, by Proposition (7.1.4) $$[f]_{B''} = P^T AP.$$
$$[f]_{B''} = \text{diag}[1, 1, \ldots, 1, -1, -1, \ldots, -1, 0, 0, \ldots, 0]$$

Let also $P'^T AP' = \text{diag}[1,1,\ldots, 1, -1, -1,\ldots,-1, 0, 0,\ldots, 0]$ for some non-singular real matrix P'.

As $$\text{rank}(A) = \text{rank}(P^T AP) = \text{rank}(P'^T AP'),$$
$$s' + t' = r,$$
so $$t' = r - s'.$$

Let $$B''' = \{\gamma_1, \gamma_2, \ldots, \gamma_n\}$$
be such that its matrix relative to the standard ordered basis is P'.

Then, $$[f]_{B'''} = P'^T AP'. \qquad \ldots (4)$$

Let W^+ be the subspace $\langle \gamma_1, \gamma_2, \ldots, \gamma_{s'} \rangle$, W^- the subspace $\langle \gamma_{s'+1}, \ldots, \gamma_r \rangle$ and W_0 the subspace $\langle \gamma_{r+1}, \ldots, \gamma_n \rangle$. It can be easily seen that f is positive on W^+, negative on W^-, and $W_0 = D_n(f)$. By Theorem 10, $\dim W^+ = \dim V^+$. So $s = s'$. This proves the theorem.

Definition 14. Let A be any non-zero, symmetric real $n \times n$-matrix. If for some non-singular, real $n \times n$-matrix P.

$$P^T AP = \text{diag}[1, 1, \ldots, 1, -1, -1, \ldots, -1, 0, 0, \ldots, 0],$$

then, the number s of entries equal to 1 in $P^T AP$ is called the *index* of A, and for rank $(A) = r$, $2s - r$ is called the *signature* of A. For a zero $n \times n$-matrix, the index and signature are both defined to be equal to zero.

Theorem 13 shows that the index and signature of a real symmetric matrix are uniquely determined. This result is known as *Sylvester's law of inertia*. The diagonal matrix, in Definition 14, is denoted by $Z_{r,s,p}$, where $p = n - r - s$.

Corollary 15. Any two real, symmetric $n \times n$-matrices A and B are congruent over **R** iff they have the same ranks and the same signatures.

Proof. Let A and B be congruent, so rank (A) = rank $(B) = r$. [see the comment after Definition (7.1.7)]. Let s = index (A). Then, by Theorem 13, A is congruent to

$$C = \text{diag}\left[\overbrace{1, 1, \ldots, 1}^{s}, \overbrace{-1, -1, \ldots, -1}^{r-s}, 0, 0, \ldots, 0\right].$$

Then, B is also congruent to C. So by Theorem 13, index $(B) = s$. Consequently, signature$(A) = 2s - r$ = signature(B). Conversely, let A and B have the same rank r and the same signature t. If s and s' are indices of A and B, respectively, we get $2s - r = t = 2s' - r$, so that $s = s'$. By Theorem 12, A and B are both congruent to C. Hence, A and B are congruent.

Practical Method 16. For a symmetric $n \times n$-matrix A over a field F of characteristic different from 2, we give a practical method of finding a non-singular matrix P over F such that $P^T AP$ is a diagonal matrix. We explain this method by working with a specific real matrix. Consider

$$B = \begin{bmatrix} 0 & 1 & 2 \\ 1 & 1 & 0 \\ 2 & 0 & 2 \end{bmatrix},$$

$$C = \begin{bmatrix} 0 & -1 & 1 \\ -1 & 0 & 1 \\ 1 & 1 & 0 \end{bmatrix}.$$

Look at the matrix B. Its (1, 1) entry is 0. In such a situation, the first step is to construct a matrix B', that is congruent to B and has non-zero (1, 1) entry.

Case I. The matrix B has a non-zero entry on the diagonal, say at $(i, i)^{\text{th}}$ place. For B, we take $i = 2$. First interchange the Ist and the ith columns, follow it by the interchange of the Ist and the ith rows of the resulting matrix. Let E be the elementary $n \times n$-matrix corresponding to the elementary column operation $c_1 \leftrightarrow c_i$. Then, E^T is the ele-

mentary row operation $r_1 \leftrightarrow r_i$. These two operations give $A' = E^T AE$ whose (1, 1)th entry being the $(i, i)^{\text{th}}$ entry of A, is non-zero. Here

$$B' = E^T BE = \begin{bmatrix} 1 & 1 & 0 \\ 1 & 0 & 2 \\ 0 & 2 & 2 \end{bmatrix}.$$

Case II. The diagonal of the matrix A is zero, as is the case with **C**. Find any, $(i, j)^{\text{th}}$ entry that is non-zero. For C, we consider $(2, 1)^{\text{th}}$ entry which equals -1. Add j^{th} column to the i^{th}column, we get A_1 whose i^{th} column has k^{th} entry $a_{ki} + a_{kj}$. In particular, the $(i, i)^{\text{th}}$ entry of A_1 is $a_{ii} + a_{ij} = a_{ij} \neq 0$. The i^{th}column of A_1 is

$$X_i = \begin{bmatrix} a_{1i} + a_{1j} \\ a_{2i} + a_{2j} \\ \vdots \\ a_{ni} + a_{nj} \end{bmatrix}$$

In A_1, add j^{th} row to the ith row, this gives a matrix A' which is symmetric and has $(i, i)^{\text{th}}$ entry $(a_{ii} + a_{ij}) + (a_{ji} + a_{jj}) = 2a_{ij}$, as $a_{ii} = 0 = a_{jj}$, $a_{ij} = a_{ji}$. If E is the elementary $n \times n$-matrix corresponding to the elementary column operation $c_i \to c_i + c_j$, then, $A' = E^T AE$ is congruent to A, and A' has a non-zero diagonal. By starting with A' and applying case I, we get a symmetric matrix A'' congruent to A such that (1, 1)-entry of A'' is non-zero. Specifically, in C, add the Ist column to the 2nd column.

This gives $$C_1 = \begin{bmatrix} 0 & -1 & 1 \\ -1 & -1 & 1 \\ 1 & 2 & 0 \end{bmatrix}.$$

In C_1, add Ist row to the 2nd row.

We get $$C' = \begin{bmatrix} 0 & -1 & 1 \\ -1 & -2 & 2 \\ 1 & 2 & 0 \end{bmatrix},$$

$$E = \begin{bmatrix} 1 & 1 & 0 \\ 0 & 1 & 0 \\ 0 & 0 & 1 \end{bmatrix}.$$

In C', interchange the Ist and 2nd columns.

This gives $$C_2 = \begin{bmatrix} -1 & 0 & 1 \\ -2 & -1 & 2 \\ 2 & 1 & 0 \end{bmatrix}$$

In C_2, interchange the Ist row and the 2nd row.

This gives $$C'' = \begin{bmatrix} -2 & -1 & 2 \\ -1 & 0 & 1 \\ 2 & 1 & 0 \end{bmatrix}.$$

Thus, by starting with A, we get $A' = \left[b_{ij}\right]_{n\times n}$ congruent to A with $b_{11} \neq 0$. In A' apply the column operations $c_i \to c_i + (-b_{11}^{-1})b_{1i}c_1,\ 1 \le i \le n$, and let E_i be the corresponding elementary $n\times n$-matrices. We get matrix $D = \left[d_{ij}\right]$ whose Ist row is

$$[b_{11},\ 0,\ 0,\ldots,\ 0],$$
$$D = AE_2E_3\ldots E_n.$$

The first column of D is the same as that of A. Apply the elementary row operations

$$r_i \to r_i + (-b_{11}^{-1})b_{i1}\,r_1 = \left\{r_i + (-b_{11}^{-1})b_{1i}\,r_1\right\} \text{ to } D.$$

This gives $$D' = E_n^T E_{n-i}^T \ldots E_2^T D = E^T A' E,$$
$$E = E_2E_3\ldots E_n,$$

which is congruent to A and has the form

$$D' = \begin{bmatrix} a_{11} & 0 & 0 & \ldots & 0 \\ 0 & d_{22} & d_{23} & \ldots & d_{2n} \\ 0 & d_{32} & d_{33} & \ldots & d_{3n} \\ \vdots & \vdots & \vdots & & \vdots \\ 0 & d_{n2} & d_{n3} & \ldots & d_{nn} \end{bmatrix} = \begin{bmatrix} a_{11} & 0 & 0 & \ldots & 0 \\ 0 & & & & \\ 0 & & D'' & & \\ \vdots & & & & \\ 0 & & & & \end{bmatrix}$$

where D'' is a symmetric $(n-1)\times(n-1)$-matrix.

To be specific, in B' apply elementary column operations $c_2 \to c_2 - c_1,\ c_3 \to c_3 - 0c_1$.

We get $$B_1 = \begin{bmatrix} 1 & 1 & 0 \\ 1 & -1 & 2 \\ 0 & 2 & 2 \end{bmatrix}.$$

In B_1, apply elementary row operations $r_2 \to r_2 - r_1,\ r_3 \to r_3 - or_1$.

This gives $$B'' = \begin{bmatrix} 1 & 0 & 0 \\ 0 & -1 & 2 \\ 0 & 2 & 2 \end{bmatrix}.$$

If $D'' = 0$, D' is a diagonal matrix and we finish. Let $D'' \neq 0$. Follow the methods in Case I and Case II, by using rows and columns other than the first ones, to get a matrix $T = \left[t_{ij}\right]$, which is congruent to D' and, hence, so to A, such that in T, $t_{11} \neq 0,\ t_{22} \neq 0$, and all the other entries in the first and second rows as well as columns are zeros.

If $$T = \text{diag}\left[t_{11},\ t_{22},\ 0,\ 0,\ldots,\ 0\right]$$

we finish, otherwise, continue the process.

After a finite number of steps we get a diagonal matrix A_0 congruent to A. To be specific, consider B''. In B'', already $(2, 2)^{\text{th}}$ entry is non-zero.

Apply $c_3 \to c_3 + 2c_2$ to get $\begin{bmatrix} 1 & 0 & 0 \\ 0 & -1 & 0 \\ 0 & 2 & 6 \end{bmatrix}$.

To this, apply $r_3 \to r_3 + 2r_2$.

We get
$$\begin{bmatrix} 1 & 0 & 0 \\ 0 & -1 & 0 \\ 0 & 0 & 6 \end{bmatrix} = B_0,$$
a diagonal matrix.

To find P such that
$$B_0 = P^T BP,$$
start with I_3 and apply the elementary column operations on I_3, those applied on B, to get
$$\begin{bmatrix} 1 & 0 & 0 \\ 0 & 1 & 0 \\ 0 & 0 & 1 \end{bmatrix} \xrightarrow{c_1 \leftrightarrow c_2} \begin{bmatrix} 0 & 1 & 0 \\ 1 & 0 & 0 \\ 0 & 0 & 1 \end{bmatrix} \xrightarrow{c_2 \to c_2 - c_1} \begin{bmatrix} 0 & 1 & 0 \\ 1 & -1 & 0 \\ 0 & 0 & 1 \end{bmatrix}$$
$$\xrightarrow{c_3 \to c_3 + 2c_2} \begin{bmatrix} 0 & 1 & 2 \\ 1 & -1 & -2 \\ 0 & 0 & 1 \end{bmatrix} = P.$$

Clearly,
$$B_0 = P^T BP.$$

There are three non-zero entries on the main diagonal of B_0.

So, $r = \text{rank}\,(B) = 3$.

Two entries are positive, so, the index of B is 2, signature of B is $2s - r = 2 \times 2 - 3 = 1$.

Coming back to the general case of the symmetric matrix A, specifically over $\mathbf{R}$, $A_0 = P^T AP$, the columns $C_1, C_2, \ldots, C_n$ of P form a basis of $\mathbf{R}^n$. Let rank $(A) = r$, index $(A) = s$.

We get s of the columns of P which we rename as
$$\beta_1, \beta_2, \ldots, \beta_s,$$
satisfy
$$\beta_i^T A \beta_i = c_i > 0.$$
Put
$$\alpha_i = \frac{1}{\sqrt{c_i}} \beta_i.$$
Then
$$\alpha_i^T A \alpha_i = 1$$
for
$$1 \le i \le s.$$

Further, $r - s$ columns of P, named as

$$\beta_{s+1},\ \beta_{s+2},\ldots,\ \beta_r,$$

satisfy $$\beta_j^T A\beta_j = -c_j,\ c_j > 0.$$

Put $$\alpha_j = \frac{1}{\sqrt{c_j}}\beta_j,\ s+1 \le j \le r,$$

then $$\alpha_j^T A\alpha_j = -1$$

for $$s+1 \le j \le r.$$

Rename the remaining $n-r$ columns of P as $\alpha_{r+1},\ \alpha_{r+2},\ldots,\ \alpha_n$ in any order. we get a non-singular matrix P' whose i^{th} column is α_i, $1 \le i \le n$.

Then, $$(P')^T AP' = \text{diag}[1,\ 1,\ldots,\ 1,\ -1,\ -1,\ldots,\ -1,\ 0,\ 0,\ldots,\ 0].$$

Specifically, for matrix B, we have $P = \begin{bmatrix} 0 & 1 & 2 \\ 1 & -1 & -2 \\ 0 & 0 & 1 \end{bmatrix}$.

In B_0, the first and the third entries, on the main diagonal, are positive.

Take $$\beta_1 = C_1 = \begin{bmatrix} 0 \\ 1 \\ 0 \end{bmatrix},$$

$$\beta_2 = C_3 = \begin{bmatrix} 2 \\ -2 \\ 1 \end{bmatrix}.$$

As the entries on B_0 corresponding to C_1 and C_3 are 1 and 6, respectively, we put

$$\alpha_i = \frac{1}{\sqrt{1}}\beta_1 = \beta_1,$$

$$\alpha_2 = \frac{1}{\sqrt{6}}\beta_2.$$

In B_0, the second entry on the main diagonal is -1.

Put $$\beta_3 = C_2,\ c_3 = 1,\ \alpha_3 = \frac{1}{\sqrt{1}}\beta_3.$$

Then, $$P' = [\alpha_1\ \alpha_2\ \alpha_3] = \begin{bmatrix} 0 & \frac{2}{\sqrt{6}} & 1 \\ 1 & \frac{-2}{\sqrt{6}} & -1 \\ 0 & \frac{1}{\sqrt{6}} & 0 \end{bmatrix},$$

gives
$$(P')^T BP' = \begin{bmatrix} 1 & 0 & 0 \\ 0 & 1 & 0 \\ 0 & 0 & -1 \end{bmatrix}.$$

This process to get a diagonal matrix A_0 from a symmetric matrix A can be further streamlined. For this, form the $n \times 2n$-matrix $D = [A : I_n]$. Perform elementary column operations on D as follows. Choose those elementary column or row operations that lead us from A to A_0. Apply any such elementary column operation on A as well as I_n. If a row operation is selected, apply it only on A. After a finite number of step, we shall get $D' = [A_0 : P]$, where A_0 is a diagonal matrix, and P is a non-singular matrix satisfying $A_0 = P^T AP$.

Example 16a. Let
$$A = \begin{bmatrix} 1 & 0 & 2 \\ 0 & -1 & 3 \\ 2 & 3 & 0 \end{bmatrix}.$$

We have
$$[A : I_3] = \begin{bmatrix} 1 & 0 & 2 & : & 1 & 0 & 0 \\ 0 & -1 & 3 & : & 0 & 1 & 0 \\ 2 & 3 & 0 & : & 0 & 0 & 1 \end{bmatrix}$$

apply $c_3 \to c_3 - 2c_1$ on both sides.
$$\begin{bmatrix} 1 & 0 & 0 & : & 1 & 0 & -2 \\ 0 & -1 & 3 & : & 0 & 1 & 0 \\ 2 & 3 & -4 & : & 0 & 0 & 1 \end{bmatrix}$$

apply $r_3 \to r_3 - 2r_1$ on the left.
$$\begin{bmatrix} 1 & 0 & 0 & : & 1 & 0 & -2 \\ 0 & -1 & 3 & : & 0 & 1 & 0 \\ 0 & 3 & -4 & : & 0 & 0 & 1 \end{bmatrix}$$

apply $c_3 \to c_3 + 3c_2$ on both sides.
$$\begin{bmatrix} 1 & 0 & 0 & : & 1 & 0 & -2 \\ 0 & -1 & 0 & : & 0 & 1 & 3 \\ 0 & 3 & 5 & : & 0 & 0 & 1 \end{bmatrix}$$

apply $r_3 \to r_3 + 3r_2$ on the left.
$$\begin{bmatrix} 1 & 0 & 0 & : & 1 & 0 & -2 \\ 0 & -1 & 0 & : & 0 & 1 & 3 \\ 0 & 0 & 5 & : & 0 & 0 & 1 \end{bmatrix}.$$

$$A_0 = \begin{bmatrix} 1 & 0 & 0 \\ 0 & -1 & 0 \\ 0 & 0 & 5 \end{bmatrix} = P^T AP,$$

$$P = \begin{bmatrix} 1 & 0 & -2 \\ 0 & 1 & 3 \\ 0 & 0 & 1 \end{bmatrix}.$$

As A is a real matrix, we continue the process to get

$$\begin{bmatrix} 1 & 0 & 0 \\ 0 & 1 & 0 \\ 0 & 0 & -1 \end{bmatrix}.$$

For this, consider $[A_0 : P] = \begin{bmatrix} 1 & 0 & 0 & : & 1 & 0 & -2 \\ 0 & -1 & 0 & : & 0 & 1 & 3 \\ 0 & 0 & 5 & : & 0 & 0 & 1 \end{bmatrix}$

apply $c_2 \leftrightarrow c_3$ on both sides.

$$\begin{bmatrix} 1 & 0 & 0 & : & 1 & -2 & 0 \\ 0 & 0 & -1 & : & 0 & 3 & 1 \\ 0 & 5 & 0 & : & 0 & 1 & 0 \end{bmatrix}$$

apply $r_2 \leftrightarrow r_3$ on the left.

$$\begin{bmatrix} 1 & 0 & 0 & : & 1 & -2 & 0 \\ 0 & 5 & 0 & : & 0 & 3 & 1 \\ 0 & 0 & -1 & : & 0 & 1 & 0 \end{bmatrix}$$

apply $c_2 \to \frac{1}{\sqrt{5}} c_2$ on both sides.

$$\begin{bmatrix} 1 & 0 & 0 & : & 1 & \frac{-2}{\sqrt{5}} & 0 \\ 0 & \sqrt{5} & 0 & : & 0 & \frac{3}{\sqrt{5}} & 1 \\ 0 & 0 & -1 & : & 0 & \frac{1}{\sqrt{5}} & 0 \end{bmatrix}$$

apply $r_2 \to \frac{1}{\sqrt{5}} r_2$ on the left.

$$\begin{bmatrix} 1 & 0 & 0 & : & 1 & \frac{-2}{\sqrt{5}} & 0 \\ 0 & 1 & 0 & : & 0 & \frac{3}{\sqrt{5}} & 1 \\ 0 & 0 & -1 & : & 0 & \frac{1}{\sqrt{5}} & 0 \end{bmatrix}.$$

For
$$P' = \begin{bmatrix} 1 & \frac{-2}{\sqrt{5}} & 0 \\ 0 & \frac{3}{\sqrt{5}} & 1 \\ 0 & \frac{1}{\sqrt{5}} & 0 \end{bmatrix},$$

$$(P')^T AP = \begin{bmatrix} 1 & 0 & 0 \\ 0 & 1 & 0 \\ 0 & 0 & -1 \end{bmatrix}.$$

Observe that rank $(A) = 3$, index $(A) = 2$ and signature $(A) = 1$.

Theorem 17. Let A be any complex, symmetric $n \times n$-matrix. Then, there exists a complex non-singular matrix P such that

$$P^T AP = \text{diag}\left[\overbrace{1, 1, \ldots, 1}^{r}, 0, 0, \ldots, 0\right]$$

where $r = \text{rank}\,(A)$.

Proof. By Theorem 8, there exists a complex $n \times n$-non-singular matrix B such that

$$B^T AB = \text{diag}[c_1, c_2, \ldots, c_r, 0, 0, \ldots, 0]$$

with $$c_i \neq 0 \qquad \text{for } 1 \leq i \leq r$$

Consider $$d_i = \frac{1}{\sqrt{c_i}} \in \mathbf{C}.$$

Then, $$U = \text{diag}_{n \times n}[d_1, d_2, \ldots, d_r, 0, 0, \ldots, 0]$$

satisfy $$U = U^T$$

and $$U^T B^T ABU = \text{diag}\left[\overbrace{1, 1, \ldots, 1}^{r}, 0, 0, \ldots, 0\right].$$

Then, $P = BU$ is a desired matrix.

The following is immediately clear from the above result :

Theorem 18. Any two complex, symmetric $n \times n$-matrices are congruent over C, iff they have the same ranks.

Remark 19. The above theorem is valid for symmetric matrices over any field F of characteristic different from 2, such that every $\alpha \in F$ has a square root $\beta \in F(\beta^2 = \alpha)$; this is true in case F is algebraically closed.

Definition 20. A finite dimensional vector space V_F together with a bilinear form f on V, is called a *metric vector space*, in case f is symmetric, skew-symmetric or alternate.

In a metric vector space V_F, the associated bilinear form is denoted by $< ; >$. If $< ; >$ is symmetric, then, V is called an *orthogonal geometry*. If $< ; >$ is alternate, then, V is called a *sympletic geometry*. Theorem 7 gives that, in case char $F \neq 2$, and V is an orthogonal geometry, there exists an ordered basis $B = \{\alpha_1, \alpha_2, ..., \alpha_n\}$ of V such that

$$\langle \alpha_i ; \alpha_j \rangle = 0 \qquad \text{for } i \neq j$$

$$\langle \alpha_i , \alpha_i \rangle = c_i \neq 0 \qquad \text{for } 1 \leq i \leq r$$

$$= 0 \qquad \text{for } i > r$$

where r is the rank of the bilinear form.

The space $\mathbf{R}^n$ along with the bilinear form

$$\langle X ; Y \rangle = \sum_{j=1}^{n} \sum_{j=1}^{n} x_i y_j$$

is the *Euclidean n-space* $\mathbf{R}^n$.

In a metric vector space V_F, vector α is said to be *orthogonal* to a vector β if $\langle \alpha ; \beta \rangle = 0$; a non-zero vector $\alpha \in V$ is called an *isotropic vector* or a *null vector* if $\langle \alpha ; \alpha \rangle = 0$. A Euclidean n-space has no null vector, but the Minkowski space has null vectors.

Exercise 7.2

1. For each of the following symmetric bilinear forms over R, find the associated quadratic form :

   (*i*) $f(X, Y) = x_1 y_1 + x_1 y_2 + x_2 y_1$ on $\mathbf{R}^2$

   (*ii*) $f(X, Y) = -2x_1 y_1 + x_2 y_2 - x_3 y_3 + 4x_1 y_2 + 4x_2 y_1$ on $\mathbf{R}^3$

   (*iii*) $f(X, Y) = x_1 y_2 + x_2 y_1$ on $\mathbf{R}^2$.

2. For each of the following quadratic forms, find the corresponding symmetric bilinear form f:

   (*i*) $q(X) = x_1^2 + x_2^2 - x_3^2$ on $\mathbf{R}^2$

   (*ii*) $q(X) = ax_1 x_2 + bx_1 x_3 + cx_3^2$, $a, b, c \in \mathbf{R}$, on $\mathbf{R}^3$

   (*iii*) $q(X) = x_2^2$ on $\mathbf{R}^4$

   (*iv*) $q(X) = 2x_1 x_2 - x_2^2$ on $\mathbf{R}^3$

   (*v*) $q(X) = x_1 x_2 + x_2 x_3 + x_3 x_4 + x_4 x_1$ on $\mathbf{R}^4$.

3. Find matrix, the rank, of each of the bilinear form in Exercise 2.

4. If dim $V_F = n$, prove that the space *SBL*(V) of all symmetric bilinear forms on V has dimension $\dfrac{n(n+1)}{2}$.

5. Let T be a linear operator on V_F and B be an ordered basis of V_F. For any bilinear form f on V_F, prove the following :

   (*i*) $g(\alpha, \beta) = f(T\alpha, T\beta)$ defines a bilinear form on V_F such that $[g]_B = P^T[f]_B P$, where $P = [T]_B$.

   (*ii*) If f is symmetric, then, g is symmetric

   (*iii*) If f is symmetric and q is the bilinear form associated with f, then, $q'(\alpha) = q(T\alpha)$, gives the quadratic form associated with g.

6. If in the matrix $A = \begin{bmatrix} a & b \\ b & c \end{bmatrix}$ over F, $a \neq 0$, prove that A is congruent to $A' = \begin{bmatrix} a & 0 \\ 0 & c - \frac{b^2}{a} \end{bmatrix}$. Further, find a non-singular matrix P such that $A' = P^T AP$.

7. If in a matrix $A = \begin{bmatrix} a & b \\ b & c \end{bmatrix}$ over F, $c \neq 0$, prove that A is congruent to $A' = \begin{bmatrix} c & 0 \\ 0 & a - \frac{b^2}{c} \end{bmatrix}$. Further, find a non-singular matrix P over F such $A' = P^T AP$.

8. Prove that matrix $A = \begin{bmatrix} 0 & b \\ b & 0 \end{bmatrix}$ over a field F with char $F \neq 2$, is congruent to $A' = \begin{bmatrix} 2b & 0 \\ 0 & -\frac{b}{2} \end{bmatrix}$. Further, find a non-singular matrix P over F such that $A' = P^T AP$.

7.3 Skew-Symmetric Bilinear Form

As in the last section, F is a field of characteristic different from 2 and V_F is a (non-zero) finite dimensional vector of dimension n. Further, $B = \{\alpha_1, \alpha_2, ..., \alpha_n\}$ is an ordered basis of V. Recall that a bilinear form f on V is said to be skew-symmetric if $f(\alpha, \beta) = -f(\beta, \alpha)$. In such a situation, $f(\alpha, \alpha) = 0$ and $[f]_B$ is a skew-symmetric matrix. (See section 7.1). We are interested in finding canonical forms of $[f]_B$.

Lemma 1. Let f be a skew-symmetric bilinear on V_F and for some $\alpha_1, \beta_1 \in V$, $f(\alpha_1, \beta_1) \neq 0$. Then, the subspace W of V spanned by α_1, β_1 is of dimension 2.

Proof. To prove that dim $W = 2$, we prove that α_1, β_1 are linearly independent. Consider any $\alpha = c\alpha_1 + d\beta_1 \in W$, $c, d \in F$.

Then, $$f(\alpha, \alpha_1) = cf(\alpha_1, \alpha_1) + df(\beta_1, \alpha_1) = -df(\alpha_1, \beta_1) \qquad ... (1)$$

and $$f(\alpha, \beta_1) = cf(\alpha_1, \beta_1). \qquad ... (2)$$

(1) and (2) show that if $c \neq 0$ or $d \neq 0$, then $f(\alpha, \alpha_1) \neq 0$ or $f(\alpha, \beta_1) \neq 0$, so that $\alpha \neq 0$ whenever $c \neq 0$ or $d \neq 0$. Hence, α_1, β_1 are linearly independent.

Lemma 2. Let f be a non-zero skew-symmetric bilinear form on V_F. Then, there exist $\alpha_1, \beta_1 \in V$ such that

(*i*) $f(\alpha_1, \beta_1) = 1$

(*ii*) W the subspace spanned by α_1, β_1 is of dimension 2.

Further, $$W^{\perp} = \{\alpha \in V : f(\alpha, \beta) = 0 \text{ for } \beta \in W\}$$

is a subspace of V such that $V = W \oplus W^{\perp}$.

Proof.

(*i*) As $f \neq 0$, for some α, β in V, $f(\alpha, \beta) \neq 0$. Let $c = f(\alpha, \beta)$, then $\alpha_1 = \frac{1}{c}\alpha$, $\beta_1 = \beta$, satisfy $f(\alpha_1, \beta_1) = 1$. That $W = \langle \alpha_1, \beta_1 \rangle$ has dimension 2, has been proved in Lemma 1. This proves (*i*). As W is spanned by α_1, β_1, it is clear that $W^{\perp}$ is the set of those $\alpha \in V$ such that $f(\alpha, \alpha_1) = 0$ and $f(\alpha, \beta_1) = 0$. For any $\alpha \neq 0$ in W, we have seen in Lemma 1 that $f(\alpha, \alpha_1) \neq 0$ or $f(\alpha, \beta_1) \neq 0$, so that $\alpha \notin W^{\perp}$, hence, $W \cap W^{\perp} = \{0\}$.

Consider any $\alpha \in V$, $\beta = f(\alpha, \beta_1)\alpha_1 - f(\alpha, \alpha_1)\beta_1 \in W$.

$$\begin{aligned} f(\alpha - \beta, \alpha_1) &= f(\alpha, \alpha_1) - f(\beta, \alpha_1) \\ &= f(\alpha, \alpha_1) - f(\alpha, \beta_1) f(\alpha_1, \alpha_1) + f(\alpha, \alpha_1) f(\beta_1, \alpha_1) \\ &= f(\alpha, \alpha_1) - f(\alpha, \alpha_1) f(\alpha_1, \beta_1) \\ &= 0, \qquad \text{as } f(\alpha_1, \beta_1) = 1 \end{aligned}$$

Similarly, $$f(\alpha - \beta, \beta_1) = 0.$$

Consequently, $$\alpha - \beta \in W^{\perp},$$

$$\alpha = \beta + (\alpha - \beta) \in W + W^{\perp}.$$

Hence, $$V = W \oplus W^{\perp}.$$

Theorem 3. Let f be a non-zero skew-symmetric bilinear form on a finite dimensional vector space V_F, char $F \neq 2$. Then,

$$V = W_1 \oplus W_2 \oplus \ldots \oplus W_k \oplus V_0$$

for some subspaces V_0, W_i, $1 \le i \le k$, such that

(*i*) $V_0 = V^{\perp} = Dn(f)$

(*ii*) Each W_i has a basis $\{\alpha_i, \beta_i\}$ satisfying $f(\alpha_i, \beta_i) = 1$.

(*iii*) $f(\alpha_i, \alpha_j) = 0$, $f(\beta_i, \beta_j) = 0$ for $1 \le i, j \le k$

(*iv*) $f(\alpha_i, \beta_j) = 0$ for $i \neq j$.

Further, $\quad \text{rank}(f) = 2k, \dim V_0 = \dim V\text{-rank}(f)$

Proof. Let dim $V = n$ and to apply induction, let the result hold for all vector spaces of dimension less that n. As $f \neq 0$, there exist $\alpha_1, \beta_1 \in V$ such that $f(\alpha_1, \beta_1) = 1$, $W = \langle \alpha_1, \beta_1 \rangle$ is a subspace of V of dimension 2 and $V = W_1 \oplus W_1^{\perp}$ (see Lemma 2). In particular, we get $V^{\perp} \subseteq W_1^{\perp}$. As $\dim W_1^{\perp} = n - 2 < n$, by the induction hypothesis

$$W_1^{\perp} = W_2 \oplus W_3 \oplus \ldots \oplus W_k \oplus V_0 \quad \ldots (1)$$

where $\quad V_0 = \{\alpha \in W_1^{\perp} : f(\alpha, \beta) = 0 \text{ for every } \beta \in W_1^{\perp}\},$

each $\quad W_i = \langle \alpha_i, \beta_i \rangle,$

with $\quad f(\alpha_i, \beta_i) = 1,$

$$f(\alpha_i, \alpha_j) = 0 = f(\beta_i, \beta_j)$$

for $2 \leq i, \ j \leq k$;

further, $\quad f(\alpha_i, \beta_j) = 0$

whenever $\quad i \neq j.$

Then, $\quad V = W_1 \oplus W_2 \oplus \ldots \oplus W_k \oplus V_0.$

It is immediately clear that (*ii*), (*iii*) and (*iv*) hold. As $V^{\perp} \subseteq W_1^{\perp}$, the definition of V_0 gives $V^{\perp} \subseteq V_0$. Consider $\alpha \in V_0$ and $\beta \in V$.

Then, $\quad \beta = \beta' + \beta'', \beta' \in W_1, \beta'' \in W_1^{\perp}.$

As $\quad \alpha \in W_1^{\perp}, f(\alpha, \beta') = 0,$

as $\quad \beta'' \in W_1^{\perp}, f(\alpha, \beta'') = 0,$

so $\quad f(\alpha, \beta) = f(\alpha, \beta') + f(\alpha, \beta'') = 0,$

and, hence, $\quad \alpha \in V^{\perp}.$

This proves, $\quad V_0 = V^{\perp} = D_n(f).$

$$\begin{aligned} \text{rank}(f) &= \dim V - \dim Dn(f) \\ &= \dim V - \dim V_0 \\ &= \sum_{i=1}^{k} \dim W_i \\ &= 2k, \qquad \text{as } \dim W_i = 2 \end{aligned}$$

In this theorem, $B_i = \{\alpha_i, \beta_i\}$ is an ordered basis of W_i, for the restriction f_i of f to W_i, $[f_i]_{B_i} = \begin{bmatrix} 0 & -1 \\ 1 & 0 \end{bmatrix}$. Let f_0 be the restriction of f to V_0, as f_0 is zero, for any ordered basis B_0 of V_0, $[f_0]_{B_0} = 0$.

Then, $$B = B_1 \cup B_2 \cup ... \cup B_k \cup B_0$$

is an ordered basis of V, and $$[f]_B = J_1 \oplus J_2 \oplus ... \oplus J_k \oplus [f_0]_{B_0},$$

with $$J_i = [f_i]_{B_i} = \begin{bmatrix} 0 & -1 \\ 1 & 0 \end{bmatrix},$$

for $$1 \le i \le k.$$

This observation gives the following theorem :

Theorem 4. Let A be any non-zero, skew-symmetric $n \times n$-matrix over a field F of characteristic different from two. Then, A is congruent to

$$\begin{bmatrix} J_1 & 0 & ... & 0 & 0 \\ 0 & J_2 & ... & 0 & 0 \\ \vdots & \vdots & ... & \vdots & \vdots \\ 0 & 0 & ... & J_k & 0 \\ 0 & 0 & ... & 0 & 0 \end{bmatrix}$$

where $$J_i = \begin{bmatrix} 0 & -1 \\ 1 & 0 \end{bmatrix}.$$

Further, rank (A) is even.

Practical Method 5. The method to get a matrix A' congruent to a skew-symmetric matrix A over a field F, such that A' is in the canonical form given in Theorem 4, is similar to the one described in (7.2.16) for symmetric matrices. Let $A = [a_{ij}]$ be a non-zero $n \times n$-skew-symmetric matrix , with $n \ge 2$. The first step is to get A_1 congruent to A with (1, 2)th entry -1 ; then, the (2, 1)th entry of A_1 is 1. Appropriate multiples of the second column of A_1 are added to the each jth column for $j \ge 3$, to get an A_1' in which every (1, j)th entry for $j \ge 3$ is 0. There-after similar operations on the rows of A_1' are performed. This gives a matrix B_2 congruent to A. In B_2, appropriate multiples of first column are added to the other columns, to get a matrix whose second row is [1, 0, ... 0]. This gives a matrix B_3. Similar operations on the rows of B_3 are performed. This gives a matrix A_2 congruent to A, such that

$$A_2 = \begin{bmatrix} J & 0 \\ 0 & A' \end{bmatrix},$$

$$J = \begin{bmatrix} 0 & -1 \\ 1 & 0 \end{bmatrix}.$$

This process is repeated on A'. We illustrate it by example.

Consider $$A = \begin{bmatrix} 0 & 2 & 4 \\ -2 & 0 & 6 \\ -4 & -6 & 0 \end{bmatrix}$$

$$[A:I_3]=\left[\begin{array}{ccc:ccc}0&2&4&1&0&0\\-2&0&6&0&1&0\\-4&-6&0&0&0&1\end{array}\right]\quad \text{apply } c_1\to\frac{-1}{2}c_1 \text{ on both sides}$$

$$\left[\begin{array}{ccc:ccc}0&2&4&-\frac{1}{2}&0&0\\1&0&6&0&1&0\\2&-6&0&0&0&1\end{array}\right]\quad \text{apply } r_1\to\frac{-1}{2}r_1 \text{ on the left.}$$

$$\left[\begin{array}{ccc:ccc}0&-1&-2&-\frac{1}{2}&0&0\\1&0&6&0&1&0\\2&-6&0&0&0&1\end{array}\right]\quad \text{apply } c_3\to c_3-2c_2 \text{ on both sides}$$

$$\left[\begin{array}{ccc:ccc}0&-1&0&-\frac{1}{2}&0&0\\1&0&6&0&1&-2\\2&-6&12&0&0&1\end{array}\right]\quad \text{apply } r_3\to r_3-2r_2 \text{ on the left}$$

$$\left[\begin{array}{ccc:ccc}0&-1&0&-\frac{1}{2}&0&0\\1&0&6&0&1&-2\\0&-6&0&0&0&1\end{array}\right]\quad \text{apply } c_3\to c_3-6c_1 \text{ on both sides}$$

$$\left[\begin{array}{ccc:ccc}0&-1&0&-\frac{1}{2}&0&3\\1&0&0&0&1&-2\\0&-6&0&0&0&1\end{array}\right]\quad \text{apply } r_3\to r_3-6r_1 \text{ on the left}$$

$$\left[\begin{array}{ccc:ccc}0&-1&0&-\frac{1}{2}&0&3\\1&0&0&0&1&-2\\0&0&0&0&0&1\end{array}\right].$$

Then, $$A'=\begin{bmatrix}0&-1&0\\1&0&0\\0&0&0\end{bmatrix}=P^TAP,$$

$$P=\begin{bmatrix}-\frac{1}{2}&0&3\\0&1&-2\\0&0&1\end{bmatrix}.$$

Consider the skew-symmetric bilinear form f on $\mathbf{R}^3$, given by A.

Then, $$f(X,Y)=2X_2Y_1-2X_1Y_2+4X_3Y_1-4X_1Y_3+6X_3Y_2-6X_2Y_3$$

$$\text{rank}(f)=\text{rank}(A')=2$$

so that $Dn(f)=V^{\perp}$ has dimension one.

As the third column and the third row of A' are zeros, indeed, $V^{\perp}$ is generated by the third column of P. To see this, take $Y=(3,-2,1)$.

Then, $f(X, Y) = 6x_2 + 4x_1 + 12x_3 - 4x_1 - 12x_3 - 6x_2 = 0.$

Hence, $V^{\perp} = \langle(3, -2, 1)\rangle.$

For a skew-symmetric bilinear form f on V_F, char $F \neq 2$, we have used the fact that $f(\alpha,\alpha) = 0$ for every $\alpha = V$. However, if char$(F) = 2$, then, a skew-symmetric bilinear form on V_F is symmetric, as $a = -a$ for every $a \in F$. If we assume that $f(\alpha,\alpha) = 0$ for every $\alpha \in F$, we can, on the similar lines, prove an analogue of Theorem 3. Indeed, we have the following.

Theorem 6. Let char(F) = 2, and f be a non-zero skew-symmetric (i.e. symmetric) bilinear form on a finite dimensional vector space V_F, such that f is alternate (i.e., $f(\alpha, \alpha) = 0$ for every $\alpha \in V$).

Then, $$V = W_1 \oplus W_2 \oplus \ldots \oplus W_k \oplus V_0$$
for some subspaces V_0, W_i, $1 \leq i \leq k$ such that

(*i*) $V_0 = V^{\perp} = Dn(f)$

(*ii*) Each W_i has a basis $\{\alpha_i, \beta_i\}$ satisfying $f(\alpha_i, \beta_i) = 1$

(*iii*) $f(\alpha_i, \alpha_j) = 0 = f(\beta_i, \beta_j)$ for $1 \leq 1, j \leq k$

(*iv*) $f(\alpha_i, \beta_j) = 0$ for $i \neq j$.

Exercise 7.3

1. Show that the set W of all skew-symmetric bilinear forms on a vector space V_F is a subspace of $BL(V)$, the vector space of all bilinear forms on V. If $V = \mathbf{R}^n$, find a basis and dimension of W.

2. Prove that any non-zero skew-symmetric, real 3 × 3-matrix has rank 2.

3. Does there exist a skew-symmetric real 4 × 4-matrix of rank 3?

4. For each of the following skew-symmetric matrices A, find skew-symmetric A' in canonical form, congruent to A. Further, find, non-singular matrix P such that $A' = P^T AP$.

$$\begin{bmatrix} 0 & -5 \\ 5 & 0 \end{bmatrix}, \begin{bmatrix} 0 & 1 & 4 \\ -1 & 0 & 0 \\ -4 & 0 & 0 \end{bmatrix}, \begin{bmatrix} 0 & 0 & 3 \\ 0 & 0 & 0 \\ -3 & 0 & 0 \end{bmatrix}.$$

5. For $A = \begin{bmatrix} 0 & i & -i & 2i \\ -i & 0 & 1 & 0 \\ i & -1 & 0 & 4 \\ -2i & 0 & -4 & 0 \end{bmatrix}$ find a non-singular complex 4 × 4-matric P, such that $P^T AP$ is a skew-symmetric matrix in canonical form. Find rank (A).

6. Consider the skew-symmetric bilinear form on $V = \mathbf{R}^3$ given by

$$f(X, Y) = x_1 y_2 - x_2 y_1 + 4x_1 y_3 - 4x_3 y_1.$$

Find *Dn(f)* and its dimension.

7. Let $A = \begin{bmatrix} 0 & -1 & 0 & 0 \\ 1 & 0 & 3 & 5 \\ 0 & -3 & 0 & -1 \\ 0 & -5 & 1 & 0 \end{bmatrix}$

(*i*) Write down the skew-symmetric bilinear form f on $\mathbf{R}^4$ given by A.

(*ii*) Find the canonical form of f.

(*iii*) Is f non-singular?

8. Let F be any field of characteristic different from 2. Define $P : BL(V_F) \to BL(V_F)$ such that for $f \in BL(V_F)$.

$$(Pf)(\alpha, \beta) = \frac{1}{2}[f(\alpha, \beta) - f(\beta, \alpha)].$$

Prove the following :

(*i*) $P^2 = P$

(*ii*) $\text{rank}(P) = \dfrac{n(n-1)}{2}$, $n = \dim V$

(*iii*) Range $(P) = BSK(V)$, the space of all skew-symmetric bilinear forms on V.

9. Let f be a symmetric or skew-symmetric bilinear form on V_F, and W be a subspace of *Dn(f)*. Prove that $\bar{f}$ defined on V/W by $\bar{f}(\alpha + W, \beta + W) = f(\alpha, \beta)$ is a bilinear form. Further, $\bar{f}$ is skew-symmetric or symmetric. Prove that $\bar{f}$ is non-singular iff $W = D_n(f)$.

10. Let L_1 and L_2 be two linear functionals on V_F, prove that

$$f(\alpha, \beta) = L_1(\alpha) L_2(\beta) - L_2(\alpha) L_1(\beta)$$

defines a skew-symmetric bilinear form on V_F. Further, $f = 0$ iff L_1, L_2 are linearly dependent.

[**Hint :** If $\qquad L_1(\alpha) L_2(\beta) = L_2(\alpha) L_1(\beta)$

for all $\alpha, \beta \in V$ and $L_1 \neq 0$, $L_2 \neq 0$, show that $\ker L_1 = \ker L_2$]

11. Let f be the bilinear form on $\mathbf{R}^2$ given by the matrix $J = \begin{bmatrix} 0 & -1 \\ 1 & 0 \end{bmatrix}$, L_1 and L_2 be the linear functionals on $\mathbf{R}^2$ such that

$$L_1(x_1, x_2) = -x_1, \; L_2(x_1, x_2) = x_2.$$

For any $X, Y \in \mathbf{R}^2$, prove that

$$f(X, Y) = L_1(X)L_2(Y) - L_2(X)L_1(Y).$$

12. By using the above exercise and Theorem 4, prove that for any non-zero skew-symmetric bilinear form f on a real vector space V of dimension 2 or 3, there exist linear functionals L_1, L_2 on V such $f(\alpha, \beta) = L_1(\alpha)L_2(\beta) - L_1(\beta)L_2(\alpha)$.

INDEX